WileyPLUS Learning Space

An easy way to help your students learn, collaborate, and grow.

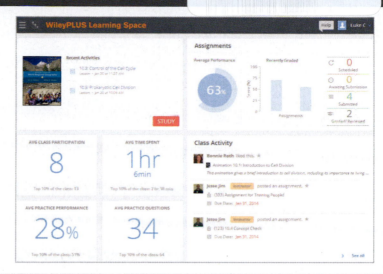

 **Esri ArcGIS Online Webmaps:** *Seamlessly integrated into the e-textbook, these dynamic webmaps enhance the beautiful cartography in our map program by providing students with the ability to explore thematic content at different levels of scale. As students interact with the maps, they learn how to "think like a geographer" by asking geographic questions and using webmaps to find answers.*

Personalized Experience

Students create their own study guide while they interact with course content and work on learning activities.

Flexible Course Design

Educators can quickly organize learning activities, manage student collaboration, and customize their course—giving them full control over content as well as the amount of interactivity among students.

Clear Path to Action

With visual reports, it's easy for both students and educators to gauge problem areas and act on what's most important.

Instructor Benefits

- Assign activities and add your own materials
- Guide students through what's important in the interactive e-textbook by easily assigning specific content
- Set up and monitor collaborative learning groups
- Assess learner engagement
- Gain immediate insights to help inform teaching

Student Benefits

- Instantly know what you need to work on
- Create a personal study plan
- Assess progress along the way
- Participate in class discussions
- Remember what you have learned because you have made deeper connections to the content

'e are dedicated to supporting you from idea to outcome.

 WILEY

UNDERSTANDING

World Regional Geography

Erin H. Fouberg | William G. Moseley

Vice President & Director	Petra Recter
Executive Editor	Jessica Fiorillo, Ryan Flahive
Assistant Development Editor	Julia Nollen
Market Solutions Assistant	Kathryn Hancox
Assoc. Director, Content Production	Kevin Holm
Production Editor	Sandra Dumas
Aptara Project Manager	Dennis Free
Senior Photo Editor	Billy Ray
Product Design Manager	Geraldine Osnato
Marketing Manager	Christine Kushner
Cover & Interior Design	Wendy Lai
Front Cover Photo	© David Sanger Photography/Alamy
Back Cover Photo	Courtesy of Boka Ilic

This book was set in Jansen by Aptara, Inc., and printed and bound by Quad Graphics. The cover was printed by Quad Graphics.

This book is printed on acid-free paper. ∞

Founded in 1807, John Wiley & Sons, Inc. has been a valued source of knowledge and understanding for more than 200 years, helping people around the world meet their needs and fulfill their aspirations. Our company is built on a foundation of principles that include responsibility to the communities we serve and where we live and work. In 2008, we launched a Corporate Citizenship Initiative, a global effort to address the environmental, social, economic, and ethical challenges we face in our business. Among the issues we are addressing are carbon impact, paper specifications and procurement, ethical conduct within our business and among our vendors, and community and charitable support. For more information, please visit our website: www.wiley.com/go/citizenship.

Main Book ISBN   978-0-471-73517-5

Binder Version ISBN   978-1-118-9664-5

Printed in the United States of America

10 9 8 7 6 5 4 3 2 1

UNDERSTANDING WORLD REGIONAL GEOGRAPHY

Five hundred years ago, the greatest library in Europe was at Queens' College in Cambridge and had only 199 books in the entire collection. In 2013, publishers released more than 300,000 new titles or editions of books, and authors self-published more than 390,000 books in the United States alone. Even more remarkably, more than 1.1 million print on-demand or Internet-access only books were published, with marketing done chiefly on the Internet. Five hundred years ago, a scholar could claim to have read every book in Europe's greatest library. Today, no one can claim to have read each of the more than 36 million books in the Library of Congress in the United States. At the turn of the millennium, Librarian of Congress James Billington quipped that we are no longer in the Age of the Renaissance or the Information Age; rather, we currently live in a "Too Much Information Age."[1]

Too much information. Your generation, which cannot remember a time before the Internet and may have never used a card catalog in a library, is adept at quickly finding information. The vast majority of the information you find on the Internet is less than 15 years old. All of this recent information can give society a short-term memory that lacks the depth of geographic context and historical knowledge. Does all of the information with which you are inundated each day seep into your brain and improve your understanding of the world? Not likely.

One goal of this book is to help you sort through the "too much information" and gain understanding by thinking geographically about the world. Many general education curriculums at colleges and universities throughout North America require students to take "globalization" courses or to become better "global citizens." We are biased, as geographers, but we contend that the best possible way to gain a global perspective and to organize the vast amount of information that floods your lives each day is through world regional geography.

Understanding World Regional Geography helps students begin to grasp the complexities of our world. If you have ever watched the news or read a newspaper and thought, "I cannot possibly understand what is going on in Syria;" "I cannot understand how China became so economically powerful in the last 30 years;" or "Why are some people and places incredibly poor while others are ridiculously

wealthy?" you know the feeling of thinking the world is simply too complex to understand.

How can *Understanding World Regional Geography* help you on your journey of better understanding our complex world? We designed this book and the corresponding online content to introduce you to the integrative way geographers gather and process information. To do this, we build from the geography education literature, which has established that thinking geographically requires two things: applying geographic concepts to real-life situations and going into the field and seeing the world as geographers.

Understanding World Regional Geography introduces you to dozens of geographic concepts that you can integrate and apply to real-world situations. Across 14 chapters we highlight 25 **Threshold Concepts** that will help you learn to think geographically. Once you learn one of these concepts and apply it yourself, you will begin to integrate the concept in your thinking and can draw from it

Threshold Concepts in Geography

CHAPTER	THRESHOLD CONCEPTS
Introduction to World Regional Geography	Context Region Cultural Landscape Scale
Global Connections	Anthropocene Globalization Networks
Geography of Development	Development Unequal Exchange Mental Map
Subsaharan Africa	Site Situation
Southwest Asia and North Africa	Diffusion Hearth
South Asia	Green Revolution
Southeast Asia	Tourism Authenticity
East Asia	Commodity Chain
Europe	Population Pyramid
North and Central Eurasia	Identity
North America	Migration
Latin America and the Caribbean	Race Gender
Pacific	Time–Space Compression
World Regions and World Cities	World Cities

[1]Achenbach, Joel. "The Too-Much-Information Age: Today's Data Glut Jams Libraries and Lives. But Is Anyone Getting Any Wiser?" *Washington Post*, March 12, 1999, A1.

to learn new material and think geographically. When a Threshold Concept is introduced in depth, an icon appears in the margin and a definition is given at the bottom of the page. At the end of each chapter, the **Creative and Critical Thinking Questions** each integrate Threshold Concepts, which affords you an opportunity to apply Threshold Concepts to your surroundings and case studies from each world region.

WileyPLUS Learning Space includes to a matrix for the 25 Threshold Concepts that links you through to a variety of case studies from *Understanding World Regional Geography* that use each threshold concept.

We established the mapping program in *Understanding World Regional Geography* through partnerships with Esri and Maps.com. Maps.com used ArcGIS to create six thematic maps for print and digital in each chapter of *Understanding World Regional Geography*. Clicking on a webmap in the *Understanding World Regional Geography* online environment opens the map in ArcGIS Online, making the map dynamic and interactive.

Dynamic webmaps give students ownership of their learning. In the ArcGIS Online environment, students can interact with the maps, turning on and off layers, zooming in and out, panning across maps, adding their own layers of data, and using spatial analysis tools in ArcGIS Online to ask and answer questions. An expert in geography education and spatial thinking designed map exercises in each chapter that take students through steps in Bloom's taxonomy, first asking students to look for patterns on a map, then asking students to compare and contrast different data or different areas of the map, and later asking students to infer and predict based on the map.

ArcGIS Online is simple to use while also being robust. Students can work through map questions using really basic commands in ArcGIS Online. Students who grow curious and explore ArcGIS Online or whose professors or discussion leaders create their own assignments using *Understanding World Regional Geography's* dynamic webmaps can ask and answer new questions by adding layers from outside, creating class (virtual) field trips, designing Esri storymaps to track a history, pattern, or phenomenon, and using ArcGIS Online's spatial analysis tools to ask and answer complex questions.

As students interact with the maps, they learn how to think geographically by asking geographic questions and using webmaps to find answers. Instead of telling students geography is something you memorize, we show students that geography is something you do.

UNDERSTANDING WORLD REGIONAL GEOGRAPHY FEATURES:

- **Chapter Opener** Featuring a single photograph and a short vignette, the authors draw students into reading the chapter. The goal of each chapter opener is to get students to think twice about their existing perception of a region.

- **Guest Field Notes:** The authors invited colleagues to discuss their fieldwork-based research. Each guest field note includes a photograph and a vignette describing the fieldwork.

- **Using Geographic Tools** This feature demonstrates, through real-world examples, how geographers use tools, including regions, mental maps, GIS, crowdsourced maps, statistics, surveys, landscape analysis, and planning, to do geography. Each Using Geographical Tools entry includes a photo or map and two Thinking Geographically questions.

- **Reading the Physical Landscape** Students are walked through the process of examining a physical landscape, identifying physical features, and understanding the processes that created the features in the landscape.

- **Reading the Cultural Landscape** Students are walked through the process of examining a cultural landscape, identifying the visible imprint of humans and cultures on the landscape, and understanding the processes that shaped the landscape.

- **Your Turn: Geography in the Field** Students are given the opportunity to analyze a photograph, make observations, ask questions, and think through answers. Each Your Turn: Geography in the Field feature includes at least two Thinking Geographically questions.

Learning to think geographically requires practice. The Guest Field Notes, Reading the Cultural Landscape, Reading the Physical Landscape, and the Your Turn: Geography in the Field features may inspire you to look at your campus or neighborhood in a new light. Start making observations, asking why something is where, and seeking answers.

A global perspective is not easily achieved. It may take you a lifetime. Our hope is that this class and this book will set you on the path to seeing how you fit into the world, to becoming curious about the varied people and places in our world, and to gaining a deeper understanding of this vast place we call home.

ACKNOWLEDGMENTS

We began work on *Understanding World Regional Geography* about eight years ago when Erin envisioned designing a book around how students learn. Inspired by the National Research Council's *How Students Learn: History, Mathematics, and Science in the Classroom*, by articles and presentations in geography education emanating from the National Council for Geography Education and the Association of American Geographers, and by her own research in geography education based on metacognition assignments she has used for approximately 2,000 students in her introductory geography classes over the last 20 years, Erin wrote the outline for the book and the introductory chapter around the best practices she had found to help students understand world regional geography.

At the time, Erin was already working with Ryan Flahive, Wiley Geosciences editor, on *Human Geography: People, Place, and Culture* (with Alexander B. Murphy and H. J. de Blij), and Ryan called Erin and asked that she consider writing a world regional geography book. The project started in earnest in 2006.

Dozens of people at Wiley have played a hand in bringing this book to fruition. Ryan Flahive was a partner in developing *Understanding World Regional Geography* from the beginning. Ryan recognized and helped refine Erin's vision for a book designed around how students learn. He and Erin engaged in countless discussions about the limitations of a PDF-based, traditional e-textbook. Ryan played a central role in developing Wiley's dynamic learning platform, WileyPLUS Learning Space, and we are proud that *Understanding World Regional Geography* is the pioneer first edition book to be designed and published using WileyPLUS Learning Space. Erin met with Jay O'Callaghan at the Association of American Geographers meeting in 2012. Erin and Jay were walking through the convention center and mulling over how to make maps with which students could take ownership of their learning. Erin was lamenting the fact that ArcGIS would be the best possible platform but that it would have too high of a learning curve for an introductory class, when they were drawn into the Esri booth to watch a demonstration of ArcGIS Online. As Erin soon discovered, Esri's ArcGIS Online is perfectly suited for the novice, armchair user with no background in GIS. This generation of college students, the Millennials, can navigate well-designed software intuitively, and ArcGIS Online is a well-designed, dynamic geographic information systems (GIS) environment. Jay gave his support to a revolutionary mapping program designed to fully function in ArcGIS Online, and Ryan and Erin met with Esri in Redlands, California.

David DiBiase, Esri Director of Education, and the education team at Esri, including Charlie Fitzpatrick, Angela Lee, and Joseph Kerski, have given exceptional support to the *Understanding World Regional Geography* mapping program. Sean Breyer, Esri Program Manager for Online Content, and his team searched for hard to find data and vetted layers for the best sources to integrate into our maps. We thank Jack Dangermond, founder and president of Esri, for his continually evolving and progressive vision of maps as spatial thinking, analysis, and decision-making tools and for his unwavering support for our project. David DiBiase served as lead vocal and Ryan Flahive was lead guitar for the mapping program band. Erin periodically beat the drum to keep moving us forward, but it was the trust established between David and Ryan that allowed Esri, the leading GIS company, and Wiley, the leading geography publishing company, to take a leap of faith and provide the resources and effort necessary to make our robust mapping program. We thank them for creating a partnership that made our groundbreaking mapping program possible. In a conversation in 2014, David quipped, "Maps are not trivial things." Thanks to David's guidance and the support of the education team at Esri, students will recognize how maps can transform their learning of facts into an understanding of connections, patterns, and context, while using the dynamic webmaps and accompanying exercises in WileyPLUS Learning Space.

Wiley has been incredibly supportive of *Understanding World Regional Geography* since its inception. We were assigned several of the best development editors in the business. Mary O'Sullivan, Ellen Ford, and Nancy Perry helped envision the flow of the chapters, adeptly coordinated reviews, thoughtfully guided our progress, and learned to think geographically along the way. Once we moved from development to production, Sandra Dumas and Janet Foxman served as production editors, organizing the manuscript schedule and map and art programs to bring the book together. Jackie Henry stepped in, and we benefitted from her organization, support, and ability to juggle. While moving *Understanding World Regional Geography* through production, Jackie also served as production editor for *Human Geography: People, Place, and Culture*. Dennis Free was the final production editor. His pragmatism, communication skills, and organizational abilities brought the book to print, and we are thankful for his work.

Wiley Vice President and Director Petra Recter championed this first edition. Petra recognized potential bandwidth issues and supported us through periods of overload. Wiley Geosciences editor Jess Fiorillo hit the ground running in fall 2014. We appreciate her candor, intelligence, kindness, and optimism. Veronica Armour has a keen sense of curiosity, which helped improve this book. Darnell Sessom's kindness and support are truly appreciated. Julia Nollen adeptly handled the guest field notes and stepped up in the last few weeks to help us reach our goal. Kathryn Hancox was invaluable to Erin as coordinator for permissions, art, and photos. The amount of work that goes into permissions has increased significantly since the beginning of this project, and Kathryn and permissions editor Craig Leonard were thorough. Photo editor Billy Ray was a delight to work with, as he took the time to really think about how a photo needed to be framed to convey a certain geographic concept. He has a great eye for photography and displayed a willingness to dig when we asked him to do so. After working on a book so long, the authors hope that the designer will present the material in an engaging manner. Senior designer Wendy Lai created a layout that engages the reader, and she drew from her impeccable taste and aesthetic to design a beautiful book. Behind the scenes, Geraldine Osnato, Kevin Holm, and Harry Nolan moved our book forward, and we are so thankful for their work. Suzanne Bochet and Christine Kushner developed a marketing plan that conveyed the message of our goal—to help students *understand* world regional geography. We tell our students that they can only improve their writing by being open to review, and we appreciate Karen Fein took the time to read our every word and give us constructive feedback. Both of us improved our writing thanks to Karen's guidance. During the production process very little can make an author laugh as the stress level builds, but Betty Pessagno, who served as copyeditor and proofreader, made Erin laugh out loud, alone in her basement at 4 in the morning many times. Erin actually looked forward to getting Betty's edits because they were always spot on and were often tinged with a sarcasm that made the process so much more enjoyable.

Erin's dear friend and past president of the National Council for Geography Education, Jan Smith, authored a white paper on best practices for textbook maps that informed our mapping program. Jan and Erin engaged in research in threshold concepts to test the pedagogy of *Understanding World Regional Geography* among faculty, instructors, and students. We thank Jan for stepping up every time she was asked, participating in conference calls and coffee meetings and responding to emails and phone calls.

We are indebted to all our colleagues who served as reviewers or in focus groups. At the very beginning, Jason Dittmer, Cary Komoto, and Eric Fournier gave us valuable feedback that helped establish a sound footing for our writing. Thank you to Alan Arbogast who wrote the drafts of "Reading the Physical Landscape" throughout the 14 chapters. Thank you to every reviewer who took the time to read carefully, critically analyze, and offer suggestions for our chapters:

Iddrisu Adam	*University of Wisconsin– Marshfield*
John Agnew	*UCLA*
Ola Ahlqvist	*Ohio State University Columbus*
Victoria Alapo	*Metropolitan Community College*
Heike Alberts	*University of Wisconsin Oshkosh*
Michele Barnaby	*Pittsburg State University*
Richard Benfield	*Central Connecticut State University*
Kate Berry	*University of Nevada– Reno*
Paul Bimal	*Kansas State University*
Sarah Blue	*Texas St University–San Marcos*
Xuwei Chen	*Northern Illinois University*
Jill Coleman	*Ball State University*
Debbie Corcoran	*Missouri State University*
William Courter	*Santa Ana College*
Adam Dastrup	*Salt Lake Community College*
James Doerner	*University Of Northern Colorado*
Elizabeth Dudley-Murphy	*University of Utah*
Eric Fournier	*Samford University*
Chad Garick	*Jones County Junior College*
Jerry Gerlach	*Winona State University*
Sarah Goggin	*Cypress College*
Emily Good	*Northeastern Illinois University*
Ellen Hansen	*Emporia State University*
Shireen Hyrapiet	*Oregon State University*
Juana Ibanez	*University of New Orleans*
Injeong Jo	*Texas A & M University College Station*
Karen Johnson	*North Hennepin Community College*
Cary Komoto	*University of Wisconsin Barron County*
Richard Lisichenko	*Fort Hays State University*
Alan Marcus	*Towson University*
Kent Mathewson	*Louisiana State University*
Deb Matthews	*Boise State University*
Molly McGraw	*Southeastern Louisiana University*
Neusa McWilliams	*University of Toledo*
Douglas Munski	*University of North Dakota*
Katherine Nashleanas	*Southeast Community College Lincoln*
Joe Naumann	*University of Missouri St Louis*
Diane O'Connell	*Schoolcraft College*
Kathleen O'Reilly	*Texas A & M University College Station*
Thomas Orf	*Las Positas College*
Judith Otto	*Framingham State University*
Thomas Owusu	*William Paterson University–New Jersey*
Adam Parrillo	*University of Wisconsin Green Bay*
Sonja Porter	*Central Oregon Community College*
William Price	*North Country Community College*
Sumanth Reddy	*Coppin State University*
William C. Rowe, Jr.	*Louisiana State University*
James Saku	*Frostburg State University*
Dmitrii Sidorov	*California State University–Long Beach*
Janet Smith	*Shippensburg University of Pennsylvania*
Tim Strauss	*University of Northern Iowa*
Pat Wurth	*Roane State Community College*
Gang Xu	*Grand Valley State University*
Leon Yacher	*Southern Connecticut State University*
Laura Zeeman	*Red Rocks Community College*
Jackson Zimmerman	*Divine Word College*

After establishing the Esri–Wiley partnership, we needed to find the best possible cartographers to create our maps. We were aided by two well-respected cartography teams: Maps.com and Mapping Specialists. Bennett Moe at Maps.com worked with Esri and Wiley and other data vendors to create systems that made our maps work in ArcGIS Online and took the lead on creating the dynamic webmaps. Cartographer Martha Bostwick designed the map palette and design specifications based on Wendy Lai's book design. Martha designed the six print thematic maps and Jesse Wickizer created the dynamic webmaps for each chapter, and we appreciate the thought they put into designing them. Erin has worked with Don Larson at Mapping Specialists since 2004, and we appreciate his ability to convey data and concepts through beautifully designed maps. We trusted Don's team, including Beth

Robertson, Paula Robbins, Terry Bush, and Glen Pawelski to use their design aesthetic and skills in cartography and GIS to update or create more than 100 maps and figures in this first edition. An additional thank you to PhD student Kim Johnson, who researched the data used in many of the maps and figures in this first edition.

Erin thanks her professor, mentor, and colleague, H.J. de Blij, from whose unwavering support she drew strength. She also wishes to acknowledge the influence of her colleagues and mentors, Alec Murphy, Clark Archer, Marshall Bowen, David Wishart, Dawn Bowen, Stephen Hanna, Fred Shelley, Jason Dittmer, David Grettler, and Jonathan Leib. Erin is grateful for her husband, Robert, who would listen to a particular concern of hers and say "How can I help?" Erin recognizes that many of the formative years of her children's (Maggie and Henry) lives were punctuated by replies of "As soon as I finish this chapter." They likely think the book is 500 chapters long, not recognizing that each chapter was "finished" several times before it went to print. In numerous ways, Maggie and Henry are in this book, as they described what they saw in photos to help with photo selections, gave feedback on drafts of maps, and even listened to passages and suggested revisions. A special thanks to Amanda Jacobs, who along with Maggie, helped explain how a Millennial would interpret a particular passage. Every academic needs friends who can appreciate their nerdiness, and Erin relies on Barb Magera and Molly Richter to accept her for who she is—in panic and in calm. Erin's father, Ed, who is a geographer, trained her to think geographically while she sat in the backseat of the family station wagon on long drives around the country. The results of his encouragement to ask questions, make connections, and reason through answers and his tip to read certain geographers whom he considered "mad geniuses" are found throughout this book. Erin's mother, Joan, knew exactly when to ask about the project and when to step back. Her constant intercession to "Keep the faith" propelled her forward during some dodgy parts of the process. Thank you to Glenna and Rod, who lovingly provided for the "care and keeping of Maggie and Henry" many times over the course of this project. Erin thanks her sisters, Molly and Bridget, and her brothers, Tim and Eddie, for believing in her.

Bill thanks his spouse, Julia Earl, and children, Ben and Sophie, for their understanding, patience and support throughout this project. He also wishes to acknowledge his many influential mentors and colleagues in geography whose thinking is reflected throughout this book, especially Ikubolajeh Logan, Kavita Pandit, Dave Lanegran, Alec Murphy, Michael Watts, Diana Liverman, Tom Bassett, Piers Blaikie and Paul Richards.

Finally, we thank the professors who taught us, the colleagues who challenged us, and the students who inspired us, each helping to shape us into the geographers we are.

Erin H. Fouberg
Aberdeen, South Dakota

William G. Moseley
St. Paul, Minnesota

ERIN H. FOUBERG is professor of geography and director of the Honors Program at Northern State University where she teaches courses in world regional geography, human geography, physical geography, geographic information systems (GIS), and political geography. She graduated from the Georgetown University School of Foreign Service and then earned her master's and Ph.D. at the University of Nebraska-Lincoln (1997). Her research interests include the political geography of American Indian sovereignty, geography education, and sacred sites. She served as Vice President of Publications and Products for the National Council for Geography Education. She has co-authored four editions of *Human Geography: People, Place, and Culture* (with Alexander B. Murphy and H. J. de Blij, Wiley, 2006, 2009, 2012, 2015). She served as an editor on *The Atlas of the 2012 Elections* (with J. Clark Archer, Robert H. Watrel, Fiona Davidson, Kenneth C. Martis, Richard L. Morrill, Fred M. Shelley, and Gerald R. Webster, Rowan and Littlefield, 2014) and co-edited *The Tribes and the States* (with Brad A. Bays, Rowman and Littlefield, 2002). Dr. Fouberg excels in teaching and advising undergraduate students, earning teaching awards from the University of Nebraska-Lincoln as a graduate student, from the University of Mary Washington in Fredericksburg, Virginia, as an assistant professor, and from Northern State University in Aberdeen, South Dakota, as an associate professor. She is active in her community, serving leadership roles on the soccer board, PTA, and fundraising campaigns for children's charities. Erin lives with her husband and two children in Aberdeen, South Dakota, where she enjoys traveling, reading, golfing, and watching athletic and theater events at Northern State.

WILLIAM G. MOSELEY is a professor and chair of geography, and director of African studies, at Macalester College where he teaches courses on human geography, environment, development, and Africa. His research interests include political ecology, tropical agriculture, environment and development policy, and livelihood security. His research and work experiences have led to extended stays in Mali, Zimbabwe, South Africa, Botswana, Malawi, Niger, and Lesotho. He is the author of over 70 peer-reviewed articles and book chapters. His books include: *An Introduction to Human-Environment Geography: Local Dynamics and Global Processes* (with Eric Perramond, Holly Hapke, and Paul Laris) (Wiley-Blackwell, 2013); four editions of *Taking Sides: Clashing Views on African Issues* (McGraw-Hill/Dushkin, 2004, 2006, 2008, 2011); *Hanging by a Thread: Cotton, Globalization and Poverty in Africa* (with Leslie Gray) (Ohio University Press, 2008); *The Introductory Reader in Human Geography: Contemporary Debates and Classic Writings* (with David Lanegran and Kavita Pandit) (Wiley-Blackwell, 2007); and *African Environment and Development: Rhetoric, Programs, Realities* (with B. Ikubolajeh Logan) (Ashgate, 2004). His fieldwork has been funded by the National Science Foundation and the Fulbright-Hays program. He has served as editor of the *African Geographical Review*, as a national councilor to the Association of American Geographers, and as chair of the cultural and political ecology specialty group. In 2011, he won the Educator of the Year award from students at Macalester College, and in 2013 he won the Media award from the Association of American Geographers for his work communicating geography to the general public via essays that have appeared in outlets such as the *New York Times, Washington Post,* and *Al Jazeera English*. Bill lives with his wife and two children in Saint Paul, Minnesota, where he enjoys running, cross-country skiing, and camping in his spare time.

DEDICATION

For
Robert John Fouberg
my husband and best friend
and
James F. Moseley
my little brother and inspiration

CONTENTS

1

INTRODUCTION TO WORLD REGIONAL GEOGRAPHY

This may be your first college-level geography class, but your mind is not a blank slate. You come into this class with a set of perceptions about peoples, places, cultures, and regions. Your understanding of the world comes from your experiences, as well as from news stories, magazines, movies, novels, and books. Images you have seen in your life—an Afghani girl with piercing green eyes or a Maasi woman with a beaded collar on her neck—help create a structure in your brain, and this structure helps you make sense of the world.

These structures in our brains are like scaffolding. Educators across disciplines call these scaffolds schemata. Pieces of information create the base, then we build upon it with other perceptions and ideas we gain, and finally, we have schemata through which we build ideas and attach any new information we learn about places and peoples. If we read a novel set in southern Africa and the characters, stories, and descriptions contrast with our schemata of Africa, we may choose to restructure our conceptualization of Africa, reject the image of Africa presented in the novel, or pick and choose what fits and what does not fit in our perception.

The goal of *Understanding World Regional Geography* is to help you build your knowledge and appreciation of the world. As global citizens, each of us has opportunities every day to be aware of our schema, question it, wonder about it, challenge it, and ultimately, revise it so that we may have a deeper understanding of the world we call home.

Steve McCurry/Magnum Photos, Inc.

Nasir Bagh, Pakistan. Photographed in 1984, in the midst of the Soviet Union's occupation of Afghanistan, Sharbat Gula came to symbolize the pain of a refugee's life and the resilience of the Afghan people. Americans were drawn to Sharbat Gula's piercing green eyes on the cover of *National Geographic* magazine. Gula is a member of the Pashtun ethnic group in Afghanistan, and her parents died in the war with the Soviet Union. As a young child, she and her siblings and grandmother fled to the Nasir Bagh refugee camp across the border in Pakistan, where she lived until after she married as a teenager. The photograph of Gula is considered one of the most famous photographs in the United States. Something about her eyes spoke to Americans. Perhaps we imagined the life of the young woman in a refugee camp and thought of girls in America who were the same age. Perhaps we could not look away from those eyes, and we realized the war happening so far away was impacting people who seemed real in her face. Perhaps her eyes were engaging enough that we wanted to learn more about the war in Afghanistan. Or perhaps the glimmer of hope behind the strain in her bright eyes made us hopeful for an end to conflict in the region. Regardless of how we interpreted Gula's eyes, we paused and considered where Afghanistan and the war fit in our schemata of the world.

WHAT IS GEOGRAPHY?

LEARNING OBJECTIVES

1. **Describe** what geography is.
2. **Explain** the differences between physical geography and human geography.
3. **Understand** what it means to think geographically.

Geographers are interested in studying the world, specifically, the multitude of peoples and the places they create. Most people new to the field of geography imagine that geographers study maps in order to memorize places, and read statistics in order to gather facts about people and places, such as chief exports, capitals, and major languages. However, geographers do not study peoples and places simply to gather sets of data. Geographers study peoples and places in order to *understand* how people create places, how cultures are reflected in places, how people and environment interact, and how people's perceptions of places and of others influence the ways they interact.

STUDYING GEOGRAPHY

Geography is the study of people, place, environment, and space. **People** incorporate the 7 billion people on Earth. Geographers study the ways people identify themselves and define others, the cultural practices and norms they follow, and the actions they take whether on a mundane day or in the context of a major event or crisis. **Place** is the uniqueness of a location, and to understand place, geographers think about how places are shaped by people and their interaction with environment. A place has a character, a visual aesthetic that typically reflects the people, their cultural values, and their reciprocal interaction with the physical environment (Figure 1.1).

Reading the **CULTURAL** Landscape

Ifugao Province, Philippines: Rice Terraces

These rice fields in the Philippines look nothing like the vast, flat corn fields of Iowa and Illinois. How do people farm in a hilly or mountainous environment? Approximately 2,000 years ago, Ifugao people on the island of Luzon in the present-day Philippines carved terraces into the hillside to shape flat lands they could farm. Since then, farmers have used the terraces to cultivate rice. The terraces are irrigated by rainfall at higher elevations that is then pooled and released over time down the steps of the landscape. The engineering feat of rice terraces and the balance between nature and society exemplified by this form of agriculture were reasons the United Nations recognized the Ifugao rice terraces as a World Heritage Site in 1995.

© Skip Nall/Corbis.

FIGURE 1.1 **Banaue, Philippines.** These rice terraces have been farmed for 2,000 years.

Environment is the physical context of Earth, but environment is not a static stage on which people act. Environment incorporates Earth processes, such as erosion caused by rivers, and the human impact on environment. **Space** is an abstract, boundless set of connections and construction of relations where processes occur. Geographers think about what and who creates space—the structures of everyday society, the economic, political, and social constraints in which people operate.

Geography incorporates two major fields of study: **physical geography** and **human geography**. Physical geographers take a geographic approach (analyzing people, place, and space) to the study of Earth, environment, and human–environment interactions. Physical geographers study weather (meteorology), climate (climatology), natural hazards, landforms (geomorphology), human–environment interactions, and more. A physical geographer approaches an issue such as flooding in Pakistan differently from other physical scientists (Figure 1.2), because a physical geographer is concerned with why something is happening precisely where it is happening, why it matters to that place and the people, and how what is happening in this one place is connected to other peoples, places, and space.

Khalid Tanveer/AP Photo.

FIGURE 1.2 Multan, Pakistan. The Indus River is the lifeblood of Pakistan, providing fresh water to over 100 million people. In August 2010, the Indus experienced massive flooding, caused by incredibly heavy rains from the summer monsoon. Monsoon rains are intensifying as the temperatures of the Indian Ocean have warmed with recent global climate change. Approximately 20 million people were displaced by the flooding in Pakistan.

Human geographers takes a geographic approach (analyzing people, place, and space) to the study of peoples, cultures, and their shaping of place and environment. Human geographers study economics, politics, culture, health and medicine, history, and more. A human geographer approaches an issue such as guest workers migrating from Indonesia to Saudi Arabia (Figure 1.3) differently from other social scientists because a human geographer is concerned with where something is happening, why it is happening there, and why it matters (to the place migrants came from, the migrants themselves, and the place where the migrants went to).

Human geographer Rachel Silvey interviewed Indonesian women who worked as domestic help in Saudi Arabia and found that, although reports of cases of abuse of Indonesian women in Saudi Arabia were high enough for non-government organizations to ask the Indonesian government to stop women from becoming guest workers, thousands of women were still willing to migrate. The women believed earning **remittances**, money earned through work in Saudi Arabia that they could send home to Indonesia, was worth the risk of abuse. They also believed working in Saudi Arabia gave them opportunity to make the pilgrimage to Mecca (located in Saudi Arabia), which is expected of all Muslims if they can afford to go.

Geographers also study the reciprocal relationship between physical and human, in a field of study called **human–environment** relations. Catastrophes, such as the 2010 flooding in Pakistan, can have global environmental causes with very real local impacts on people. Drawing from the Indus River to irrigate lands makes

REUTERS/Newscom.

FIGURE 1.3 Tangerang, Indonesia. The largest Muslim country in the world, based on total followers of Islam, is Indonesia, with over 200 million Muslims. The government of Saudi Arabia claims approximately 100,000 Indonesians work in Saudi Arabia. Other sources estimate that more than 500,000 Indonesians work in Saudi Arabia as guest workers, invited by the government to work temporarily, mainly as domestic help. The women in this photograph are filing papers to leave Indonesia and become guest workers in Saudi Arabia.

agriculture possible in swaths of the otherwise semiarid to arid Pakistan. The Indus River also provides fresh water to over 100 million people. Beginning in the Himalayas, water in the Indus River comes from the melting snow and glaciers and from the annual summer monsoon. Monsoon rains bring more than 80 percent of precipitation that falls in the lower Indus River valley in just three months each summer: July, August, and September.

In August 2010, extremely heavy rains flooded the Indus River. Monsoon rains are intensifying as Indian Ocean temperatures rise with global climate change. Warmer air holds more moisture, bringing more rainfall during the summer, or wet monsoon season, throughout South Asia. The United Nations (2010) estimated 15 to 20 million people were affected by the Pakistan flooding, "more than the entire populations hit by the Indian Ocean tsunami (2004), the Kashmir earthquake (2005), Cyclone Nargis (2008), and the Haiti earthquake (2010) combined." While a relatively small number of people died in the Pakistan floods (1,700 people compared with over 250,000 deaths in the Indian Ocean tsunami), the Pakistan floods caused billions of dollars of damage to agricultural fields, infrastructure, buildings, and livestock.

In addition to heavier rainfall, caused in part by a warming Indian Ocean, human changes to environment in Pakistan intensified the flooding in 2010. Pakistanis have deforested much of northern Pakistan, where the heaviest rains fell, leaving little vegetation to absorb the rain and allowing much more runoff of rainfall into the Indus River. In southern Pakistan, people had drained wetlands along the Indus in order to farm and build settlements. The number of people displaced by the flood was higher than it would have been if the flood had occurred ten years earlier because more people live along the river now than in years past.

What humans do affects environment, and environmental change affects what humans do. In the case of Pakistan, human action helped warm the Indian Ocean and altered the summer monsoon, leading to heavier rainfall. The heavy rainfall led to floods, and the damage of those floods was intensified because of human changes to the landscape, in this case deforesting and draining wetlands. In response, people in Pakistan are now working to rebuild their economy, establish a stable food supply and access to clean water, and find a way to survive floods of this magnitude in the future.

WHAT DOES IT MEAN TO THINK GEOGRAPHICALLY?

Thinking geographically means considering and understanding the context of what is going on in the world. Historians focus on the historical context and geographers focus on the spatial context. Geographers use geographic concepts to find connections and make distinctions among different people and places. A geographic concept is a way of thinking about or understanding something that has a geography, a location, a spatial difference, or a place distinction.

Threshold Concepts

Understanding World Regional Geography emphasizes 25 **threshold concepts** as key to thinking geographically. Meyer and Land (2006) established that once a student understands threshold concepts in geography, it opens the door to thinking and learning like a geographer. We wrote and designed the book to highlight 25 threshold concepts and show how to apply them to specific examples. Our discipline has far more than 25 geographic concepts, but we chose these 25 as the pivotal concepts for thinking geographically in a world regional geography course. Each chapter highlights one or more threshold concepts and uses case studies to demonstrate the threshold concepts.

The creative and critical thinking questions at the end of each chapter challenge you to apply threshold concepts across regions. The concept checks within each chapter test whether you have learned the material and can apply threshold concepts and other geographic concepts. The online component also has graded questions regarding specific threshold concepts and case studies to help you practice employing them. In this chapter, we will discuss threshold concepts of context, scale, region, and cultural landscape.

CONTEXT

Reading statistics such as 4 out of 5 children with HIV/AIDS live in Subsaharan Africa may lead us to believe the disease is really an issue for Africa and less so for the rest of the world. However, HIV/AIDS diffuses in each region of the world in varying ways, and, only by understanding each **context** can we grasp the impact of the disease and design programs to help combat its spread. To geographers, the world is not a stage on which people act and diseases spread. Rather, in each of the world's regions, people's experiences, shared histories, cultural values and norms, and interactions with each other and places, as well as human–environment interactions, create and define a regional context in which a disease diffuses.

In regional contexts, we can better understand why diseases spread in the patterns they do. The diffusion of HIV/AIDS is a global phenomenon, interacting with many governments that work to create programs to slow the transmission of the disease in their countries. Regionally, HIV/AIDS is most widespread in Subsaharan Africa, with 31 million of the 38 million worldwide cases. New research gives two major reasons for its high concentration in Africa: high rates of malaria and lack of male circumcision.

Studies report that people infected with malaria who also have HIV/AIDS can become "supercontagious," thus increasing transmission rates of HIV/AIDS among people in the region. Scientists have also confirmed that uncircumcised men are more likely to contract HIV/AIDS than circumcised men (Figure 1.4).

 context   The physical and human geographies that give meaning to the place, environment, and space in which events occur and people act.

USING GEOGRAPHIC TOOLS

Using Regions to Analyze Differences: The Diffusion of HIV/AIDS by Region

The first cases of HIV/AIDS in the world were found in Central Africa, and the first major diffusion of HIV/AIDS occurred in eastern Africa along a well-traveled highway from Kenya to Uganda. This area of Africa became known as the first HIV/AIDS belt. A number of factors led to the first HIV/AIDS belt developing in Africa. A person who has contracted HIV does not know until they test positive or start showing symptoms, so they can spread the disease unknowingly. Before HIV/AIDS education was widespread, the disease spread relatively quickly in Subsaharan Africa.

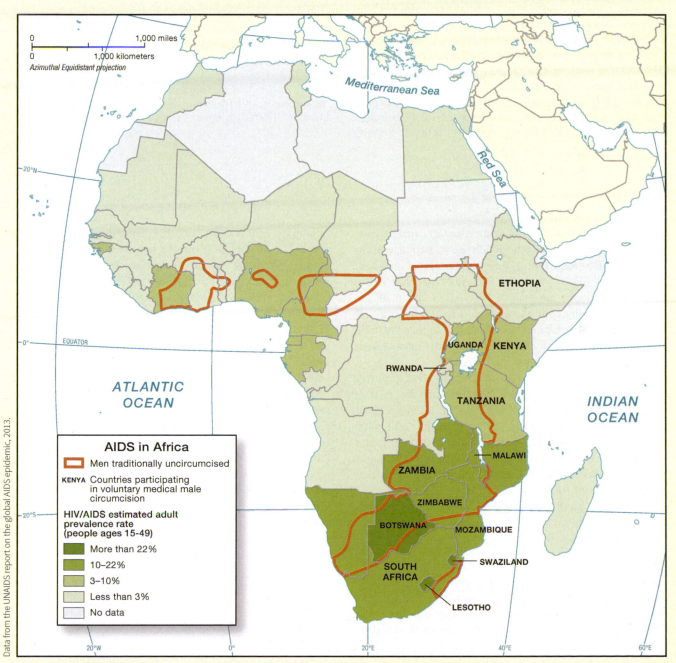

FIGURE 1.4 **HIV/AIDS in Africa.** Researchers mapped adult prevalence rates and areas where males are traditionally uncircumcised to see if the two were related.

(Continued)

In 1996, researchers mapped areas where males are not commonly circumcised and overlaid the map with the first HIV/AIDS belt to demonstrate a spatial correlation (Figure 1.4). Knowing this helped scientists make connections between disease transmission and male circumcision and led to further research.

Examining the differences in methods of transmission of HIV/AIDS in other world regions can help geographers gain a sense of how social, economic, cultural, gendered, and sexual norms vary by world region. For example, in Subsaharan Africa, HIV/AIDS is largely disseminated through heterosexual (male–female) sex. The diffusion of HIV/AIDS in eastern Europe is caused largely by intravenous drug use and then is secondarily diffused by male–female sex. In western Europe, the diffusion of HIV/AIDS reflects the same pattern found in North America, with male–male sex being the leading transmission method, followed by intravenous drug use. In East Asia, Southeast Asia, and South Asia, the sex industry generates the highest transmission rates. Designing programs to effectively combat the spread of HIV/AIDS depends on a regional understanding of how the disease most commonly diffuses.

Thinking Geographically

1. Knowing that the primary transmission methods of HIV/AIDS varies by world region, how could you design a program to combat the diffusion of HIV/AIDS that takes into account the primary transmission methods in each world region?

2. What is unique about HIV/AIDS relative to other diseases that would explain why many analysts claim that the economic impact of HIV/AIDS on a country is more severe than most diseases?

Sources: United Nations Programme on HIV/AIDS (UNAIDS). 2013. "Global Report: UNAIDS Report on the Global AIDS Epidemic 2013." BBC. 2005. "HIV Impact: Region-by-Region." BBC News. November 30.

The World Health Organization reports that 85 percent of all malaria cases worldwide are in Africa, and the Centers for Disease Control reports circumcised men have a 42 percent lower chance of HIV infection than uncircumcised men. An analysis of HIV/AIDS and male circumcision rates by country and region found that males in Subsaharan Africa, where the dominant religions are traditionalist and Christianity, have relatively low rates of male circumcision and high rates of HIV/AIDS.

In 2007, the World Health Organization reported that medical male circumcision "reduces the risk of female-to-male sexual transmission of HIV-AIDS by approximately 60 percent." Governments and non-governmental organizations have established programs and clinics to circumcise adult males in Subsaharan Africa. The World Health Organization is approving new medical devices to aid in adult male circumcision and seeking more qualified, surgically trained practitioners to deliver services.

(countries of the world versus counties in a country). In this book, we use the second approach to scale. The scale of analysis is the extent of a study, the scope of Earth (local, national, or global) we are considering when looking at circumstances, patterns, and relationships.

Understanding an issue as complex as the diffusion of HIV/AIDS depends on grasping the context of the disease and also on integrating our knowledge of circumstances at multiple scales: individual, family, local, national, regional, and global.

At the same time that HIV/AIDS diffuses globally, regional and local processes (the diffusion of malaria, the preparedness of a health care clinic, the access to antiretroviral medicine) are interacting with global processes (sometimes creating global processes of their own). All of these processes, occurring across all of these scales, affect one another and create a context for understanding what is happening in a given place at a given time and why.

SCALE

Geographers use **scale** in two ways: first, to explain the relationship between the distance on a map and distance on Earth (one inch on the map equals one mile on Earth, for example), and second, to describe the extent of a study

 scale The geographical scope (local, national, or global) in which we analyze and understand a phenomenon.

CONCEPT CHECK

1. **Why** does place matter in the study of geography?

2. **How** do human geographers and physical geographers study Earth differently?

3. **How** does understanding the context and scale of the diffusion of HIV/AIDS help you better understand the global AIDS crisis?

WHAT ARE REGIONS?

LEARNING OBJECTIVES

1. **Identify** what a world region is.
2. **Explain** the role humans play in creating regions.
3. **Understand** why scale matters when defining boundaries.

Are **regions** natural, built into Earth, or are regions created by humans? Historically, geographers believed regions were real or natural—that you were born into a region as you were born into a family. Geographers used regions as the basic units for studying the world. Studying world regional geography became a way of sorting the world into similar groupings whereby each region had its own distinct character or set of characteristics. People who thought of world regions as natural entities saw them as relatively static spaces, each standing alone, created through the collection of smaller cohesive spatial units.

Researchers in the field of new regional geography have come to see regions as things humans construct to categorize people and places into conveniently sized pieces of the world. New regional geographies question who draws the world's regions, how regions are drawn, what they represent, and how they are used. A new regional geographer would see a region like "the Middle East" in its historical context as something Europeans created and defined during the Middle Ages in order to separate themselves (Europeans) from the "other" people in the Orient, or what journalists now call the Middle East.

REGIONS

In *Understanding World Regional Geography*, we take from both the old and new regional geographies, defining regions as a balance of real and constructed. Regions exhibit actual differences in languages, religions, climates, and vegetation. Regions also have perceived differences because humans decide what is important in classifying them.

If we think of regions only as real and built into Earth, we miss how the movement of people and ideas has changed the world and how people in power can decide what is important and construct a place or region as they choose. If we think of regions only as constructed, we miss how much local circumstances and important places, including cultural hearths (regions of origin of cultural traits) and sacred sites, directly influence peoples and regions.

For example, envision the parts of North and South America where Spanish is the predominant language. The Spanish-speaking region has a real, similar cultural trait within it, but people perceived that the linguistic similarity

was an important enough difference to make it the defining criterion for the region.

The regions used in world regional geography books are constructs of the authors who determine the regions based on a combination of physical or cultural traits. We can consider physical geography separately from human geography in constructing world regions (Figure 1.5); however, if we do, we overlook the countless ways humans have changed earth over time.

To understand that regions are dynamic and not static, think about regions at a different scale. In a world regional geography class, all of Europe is treated as one region. But in a class on the geography of Europe, the professor will define distinct regions within Europe, such as Scandinavia or the Mediterranean. If you shifted scales to the Mediterranean region of Europe, you would see distinct regions within that space, for example, the coastal tourist region, the coastal nontourist region, and the interior. Shifting scales again to a single city in the Mediterranean such as Nice, France, you will find further distinct regions: the gambling region, the research and development park, the tourist hotels, and the locals' neighborhoods (Gade 1982).

In building your schematic understanding of world regional geography this semester, you will improve your understanding of the physical and human world and also determine how you construct world regions in your own mind.

To define world regions using physical geographic traits, a geographer might start with this satellite image of earth (Figure 1.5a). Patterns of climate, vegetation, and soil are reflected in this image, where green represents vegetation and tan indicates areas that are sparsely vegetated. Physical geographers can use the concept of **relative humidity**, the ratio of water vapor in a parcel of air compared to the total amount of water vapor air at that temperature can hold, to understand why some areas of Earth receive more precipitation than others.

How much water vapor air can hold (its capacity) depends on the air temperature, as warmer air can hold more water vapor than cold air. Condensation occurs when air cools, the capacity decreases, and air reaches 100 percent relative humidity. Once water vapor is condensing into liquid water, precipitation can follow. On the satellite image of Earth, the equatorial regions are bright green, dense with plant life because this part of the world receives the most direct sunlight over the course of the year. The warm air along the equator consistently rises, cools, decreases its capacity, reaches 100 percent relative humidity, and precipitation follows, creating rain forests (Figure 1.5b). Soils in this part of the world are typically old and deeply weathered because the abundant rainfall leaches nutrients over time but leaves the iron (Figure 1.5c).

In contrast, the northern parts of Africa, Central Asia, central Australia, and the southwestern United States are deserts (Figure 1.5d). The winds in these areas are high-pressure cells. Because air flows from high pressure to low pressure,

---

region an area of Earth with a degree of similarity that differentiates it from surrounding areas.

Reading the **PHYSICAL** Landscape

Defining World Regions Using Physical Geographic Traits

The satellite image of Earth is colored according to vegetative regions, with green representing lush vegetation and tan representing sparse vegetation (Figure 1.5a).

The green region of the map marks rainforest around the equator. The area around the equator receives more direct sunlight over the course of the year than other

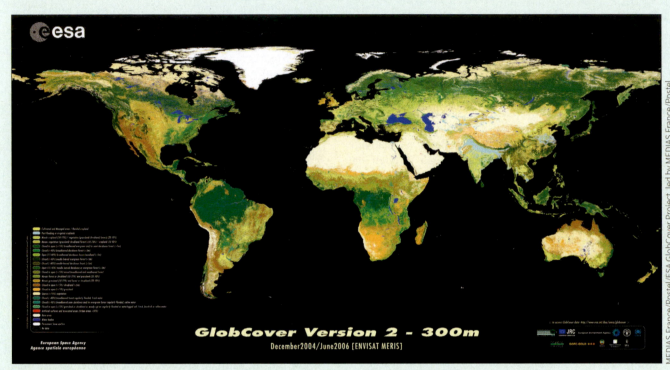

FIGURE 1.5a   This satellite image shows areas of dense vegetation in green and sparsely vegetated areas in tan.

MEDIAS France/Postel/ESA GlobCover Project, led by MEDIAS France/Postel.

Michael Boyny/LOOK/Getty Images.

FIGURE 1.5b   Costa Rica.   Monteverde Cloud Forest Reserve demonstrates the layers of vegetation found in a rainforest climate.

© Boaz Rottem/Alamy.

FIGURE 1.5c   Western Africa.   The red soil of this dirt road is typically found in tropical regions.

latitudes. Earth absorbs incoming solar radiation, then releases it as warm air, and when that warm air rises, it cools, condenses, and then precipitation can follow. Consistent precipitation creates rainforests (Figure 1.5b). The soils in the equatorial or tropical region are deeply weathered by abundant rainfall. The weathered, iron-rich oxisols of the tropics are orange or reddish in color (Figure 1.5c). The tan region of the map marks the desert and semiarid regions (Figure 1.5d). The poorly developed soils are light brown in color and have limited weathering because of limited rainfall. Only the narrow, top band of soil is dark with organic, weathered material in the semi-arid and arid regions (Figure 1.5e).

FIGURE 1.5e **Eastern Colorado.** The soil profile of a semiarid climate region shows a narrow band of dark soil on the top supporting vegetation.

FIGURE 1.5d **Libya, Africa.** Date palm trees dot the true desert Mandara Lakes region of the Sahara Desert.

Frank Krahmer/The Image Bank/Getty Images.

USDA.

air descends from the upper atmosphere into the high pressure cell and then air flows out away from these areas. Air warms as it descends from the upper atmosphere to the surface, increasing the air's capacity and decreasing its relative humidity. Precipitation is unlikely (the relative humidity of the air is decreasing instead of increasing), resulting in sparse vegetation and poorly developed soils that are lighter in color than their tropical counterparts (Figure 1.5e).

Using the example of precipitation, we can divide the world into two basic physical geography regions: areas that receive a lot of precipitation and areas that receive very little precipitation. These two regions, wet and dry, describe precipitation patterns, but they do not account for a multitude of other physical geographic traits (landforms, hazards, temperatures) nor do they explain human

geography, how people in different places interact with and make sense of the world's physical diversity.

INSTITUTIONALIZING REGIONS

Regions created in people's minds gain permanence as governments and organizations build institutions to support them (Paasi 2002). For example, the United Nations (UN) divides the world into regions (and subregions) to gather data, report statistics and also for operating purposes (Figure 1.6). The United Nations reports data by region and by country.

In the day-to-day activities of UN organizations, workers use regions. For instance, the UN World Food Program divides the world into five regions: Asia, Oceania, Africa, the Americas, and Europe. The most important decisions and policies in

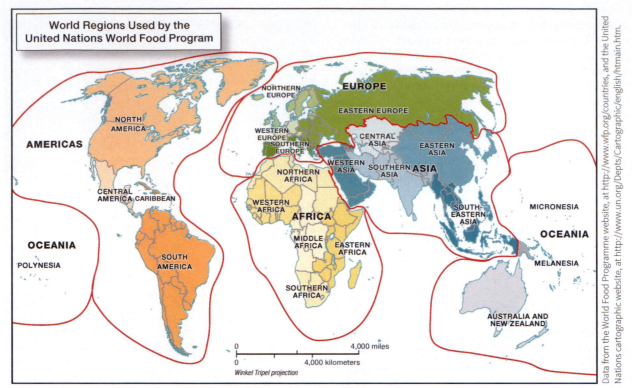

FIGURE 1.6   United Nations Regions and the United Nations World Food Program.   The United Nations collects data about the world by region, marked with the burgandy lines on this map. The United Nations World Food Program administers food aid to the world based on regional clusters of countries where the program operates (marked by color and name on the map). Using these distinct regions to operate food distribution helps institutionalize these regions in the minds of employees in the World Food Program.

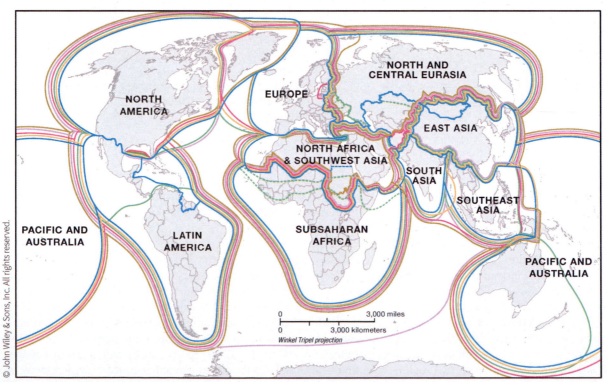

FIGURE 1.7   World Regions used in North American world regional geography books.   Maps of the world regions defined by authors of several world regional geography books demonstrate how the boundaries of regions are subjective. The overlap in regions across the books shows relative agreement among authors in North America.

the operations of the World Food Program (Figure 1.6), however, are made at the country level, by the country directors.

Data are collected within countries, policies are instituted by country, and the governments of countries make the laws within which policies are carried out and global organizations such as the UN World Food Program operate. So, countries usually end up defining boundary lines between world regions.

World regional geography books aid in **institutionalizing** world regions. Books written by North American geographers divide the world into similar regions (Figure 1.7). Authors tend to use continents and contiguous lands and keep countries intact as guidelines for world regions. Although some regions such as Asia may be divided into smaller regional units, the regions of the world that most geography books use are contiguous, nonoverlapping, country-border-following, rather neat agglomerations of places.

Scale matters when defining regions. For example, if we look at the Spanish-speaking population at the global scale (Figure 1.8), the entire United States is shaded as having a relatively large percent (10.7 percent) of Spanish speakers. However, if we map the Spanish-speaking population according to counties within a country, we find the Spanish-speaking population is concentrated in certain areas of the country, and in these areas, Spanish speakers compose up to 92 percent of a county's population. The patterns we see in maps are limited by the scale at which we collect data. If we collect data by country, the data we find for a country is spread across the territory of the country as its average.

Choosing a scale for analysis is key because answers to questions such as "where is the Spanish-speaking region of the Americas?" depend on the scale of our data. Analyzing the maps in Figure 1.8a and b may lead us to two different conclusions about the place of the United States in the Spanish-speaking region.

Because data about physical and human phenomena are gathered and reported according to specific territories, like countries or counties, called "areal units," geographers call this problem the **modifiable areal unit problem (MAUP)**. The MAUP happens when geographers change scales. Patterns seen at the global scale are not necessarily evident at the local scale (or vice versa), and the problem also happens when geographers aggregate data in different ways; for example, crime statistics reported by school district appear differently from the same statistics reported by police district. A continuous phenomenon may be measured over several areal units, or the data themselves appear differently based on the areal unit used for reporting and mapping.

WHY DO GEOGRAPHERS USE REGIONS?

Knowing what regions are and the limitations of statistically defining them, you might question why geographers use

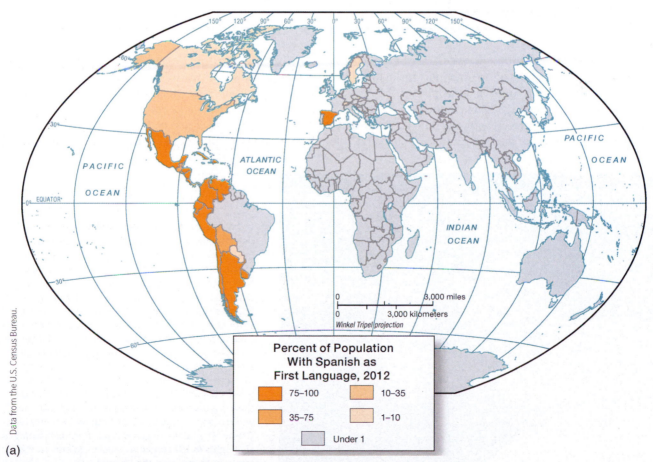

Data from the U.S. Census Bureau.

Percent of Population With Spanish as First Language, 2012

75–100   10–35

35–75   1–10

Under 1

(a)

FIGURE 1.8　Spanish-speaking population.　Shifting scales from countries on a world map (a) to counties within the United States (b) reveals a different pattern in the Spanish-speaking population of the United States.

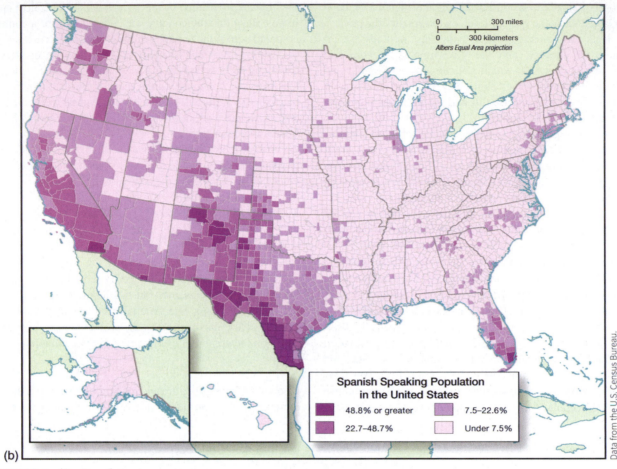

FIGURE 1.8 (*Continued*)

them. Each of us uses regions to order our world, to imagine the vastness of the world and our place in it. As the U.S. National Geography Standards contend, "The geographically informed person knows and understands that people create regions to interpret Earth's complexity."

Regions are powerful concepts: People use them to order and describe their world, and regions reflect the social, political, and economic processes weaving through places and across peoples. We have perceptions in our minds of places like "the South" in the United

FIGURE 1.9   Montgomery, Alabama.   The civil rights memorial across from the Southern Poverty Law Center demonstrates the complexities of the region Americans call the South. The memorial was built to commemorate those who lost their lives in the U.S. civil rights movement between 1954 (*Brown v. Board of Education*) and 1968 (assassination of Martin Luther King Jr.). Each inscription on the black granite circular table is the name of a person who died in the civil rights struggle. The civil rights movement took place largely in the South and has paved the way for a groundswell of civil rights attainment by social groups beyond African Americans. Today, some of the most progressive thinkers on civil rights in the world are in the South, at the Southern Poverty Law Center. The Center began as a law firm specializing in civil rights in 1971 and now teaches tolerance; it also tracks hate groups in the United States.

States. People who have not experienced or lived in the South may simplify the region in their minds, perhaps in a negative way. However, the South, like every region, is complex and is created and re-created by a combination of multifaceted social, political, and economic processes (Figure 1.9). People's **schemata** have multiple, overlapping regions based on positive and negative perceptions within them, and people use these regions and their schemata to make sense of the world and their place in it.

The greatest value in studying the world region by region is that we can better interpret and understand our world by comparing and contrasting the processes operating across global, regional, and local scales. Regions give us a way to talk about and understand our world.

◉

CONCEPT CHECK

1. **How** is the concept of region useful in the study of geography?

2. **Are** regions real or are they constructed?

3. **Why** does scale matter in defining regions?

WHAT REGIONS ARE USED IN *UNDERSTANDING WORLD REGIONAL GEOGRAPHY?*

LEARNING OBJECTIVES

1. **Explain** the differences between formal and functional regions.

2. **Describe** why perceptual regions are a powerful influence in constructing schemata.

3. **Identify** the regions used in this book and **explain** why they are used.

Despite the limitations of dividing the world into regions, geographers rely on them to organize and digest the complexities of the world. To organize ideas and places, we use three types of geographic regions: formal, functional, and perceptual.

FORMAL REGIONS

A **formal region** is an area of land distinguished by either cultural or physical criteria. For example, physical geographers take places with similar temperature, precipitation, and seasonality of precipitation over at least a 30-year period of time and divide the world into climate regions. Figure 1.10

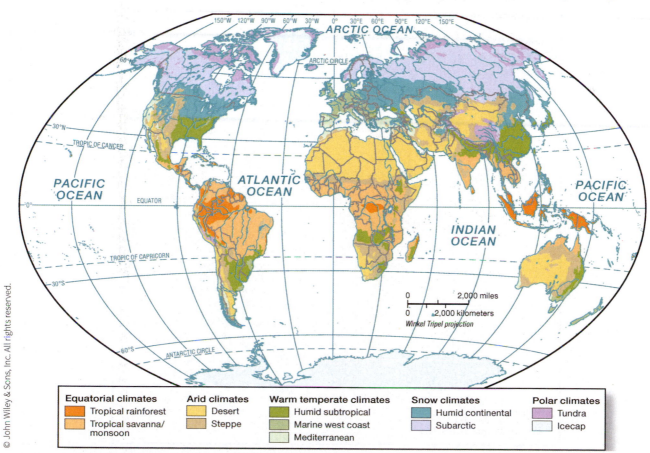

Equatorial climates
- Tropical rainforest
- Tropical savanna/ monsoon

Arid climates
- Desert
- Steppe

Warm temperate climates
- Humid subtropical
- Marine west coast
- Mediterranean

Snow climates
- Humid continental
- Subarctic

Polar climates
- Tundra
- Icecap

FIGURE 1.10   World climate regions.   Climates are defined based on averages over at least a 30-year period of time. Climate regions are classified by and seasonal precipitation and temperatures.

maps the climate regions originally drawn by Russian climatologist Wladimir Köppen and revised over time. Places in the same climate region, such as the humid subtropical climate zone, are found on the east coasts of continents because high-pressure systems in the oceans bring warm, moist air to these places. The subtropical high gets stronger in the summer months, creating a season of extremely humid summer days for the humid subtropical climate region.

Traditionally, regional geography has been based on formal regions. A geographer chooses criteria and then constructs a formal region based on them. Since the 1980s, geographers studying regions have realized how imprecise the boundaries of formal regions are and have dispelled the myth that formal regions are natural. The *Encyclopedia of the Great Plains* pulled together over 50 different studies of the Great Plains to create a map of the formal region of the Great Plains (Wishart 2004) (Figure 1.11). The map gives us an idea of what places 50 different researchers thought were definitely in the Great Plains, and it also shows us how fuzzy boundaries for formal regions can be.

FUNCTIONAL REGIONS

A **functional region** is an area that has a shared political, economic, or social purpose. For example, the jurisdiction of a circuit court or a county court is considered a functional region. A city is a functional region defined by city boundaries. Functional regions generally have nodes or points where the function is clustered. In the functional region of a city, headquarters of city government would be the node.

Many of the parts of everyday life that Westerners take for granted are provided through functional regions: school districts, water treatment districts, and waste management districts (Figure 1.12). Standing in any place in the United States, you could, if you took the time, easily name at least 100 functional regions you are in. Functional regions can cover larger spaces, such as an entire state, or even the entire country or broader North America. For example, the North American Free Trade Agreement (NAFTA) creates a functional region for trade in North America.

Functional regions can provide or limit opportunities for the people in the region. Take the example of school districts in the United States. Public schools in the United States

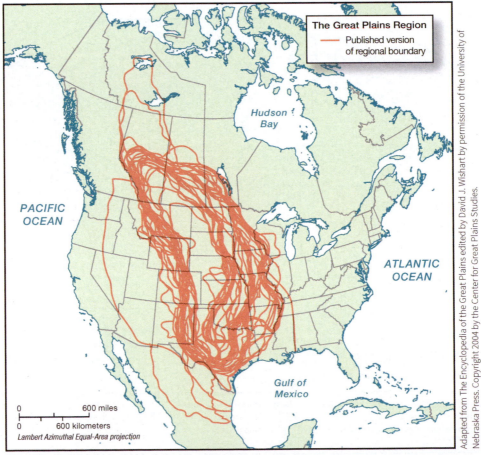

FIGURE 1.11   The Great Plains region.   Fifty different ways the formal region of the Great Plains has been defined over time.

© Ian Dagnall/Alamy.

© nik wheeler/Alamy.

(a)   (b)

FIGURE 1.15   **Toponym and Place.**   (a) **Winter Park, Colorado.** Winter Park Ski Resort offers accommodations on the slope as well as vacation homes for rental. (b) **Durango, Colorado.** Main Avenue in downtown Durango features architecture dating back to the 1800s, including the beautiful Strater Hotel in the background, which was built in 1887.

Ute Indians. The city of Durango was founded in 1880 by non-Indian Americans encouraged by gold and westward movement to settle in the area.

A single perception of a region rarely reflects its richness or complexity. In this case, both winter ski land and Old West town may create conflicting perceptions of Colorado, but both places are part of the American West and should factor into our perception of the region.

REGIONS IN *UNDERSTANDING WORLD REGIONAL GEOGRAPHY*

Institutions divide the world into regions in order to deliver services, and geographers offer maps of world regions to help make sense of the world (see Figure 1.6). No one way of regionalizing the world is the "right" way. The regions in *Understanding World Regional Geography* draw from historical spatial interaction and physical and cultural processes that have created areas of greater interaction, deeper connection, similarities in geography, and a perceived commonality.

Spatial Interaction Throughout History

The degree of contact among peoples and places, which geographers call **spatial interaction**, depends on distance, connectivity, and accessibility. The likelihood that a particular trait or innovation will spread depends on the amount of spatial interaction between two places. The probability of diffusion declines with distance from the hearth, which is demonstrated in the concept of **time–distance decay**.

Connectivity means people in two places are tied together through a cultural process, such as trade or religion, or a physical feature, such as a river or sea, that encourages spatial interaction. In Africa, Europe, and Asia from 500 BCE to 1500 CE, the most connected places, in terms of trade, were coastal locations in and around the Mediterranean Sea, the Arabian Sea, and the Indian Ocean. Connectivity through trade also eased the diffusion of other aspects of life, including contagious diseases. One way to see how closely the people in Africa, Europe, and Asia interacted with each other is to examine a map of the diffusion of contagious diseases, such as smallpox and the plague (Figure 1.16). The map shows areas of major spatial interaction at the time, as contagious diseases diffused between areas with high degrees of spatial interaction.

The map of disease diffusion also shows the places where little, if any, spatial interaction occurred during this time frame. The areas in gray on the map do not imply a lack of people in the place. Rather, the areas in gray reveal little accessibility. *Accessibility* points to what must be traversed between two places (mountains or deserts, for example), and it also points to whether people in one place are welcome or able to engage with people in another, which often depends on religion, language, or social status. For example, during the time frame of this map thousands of people lived in Subsaharan Africa, and West Africa had a series of empires that rose and fell. However, the Sahara Desert divided Subsaharan Africa from North Africa, making accessibility difficult and resulting

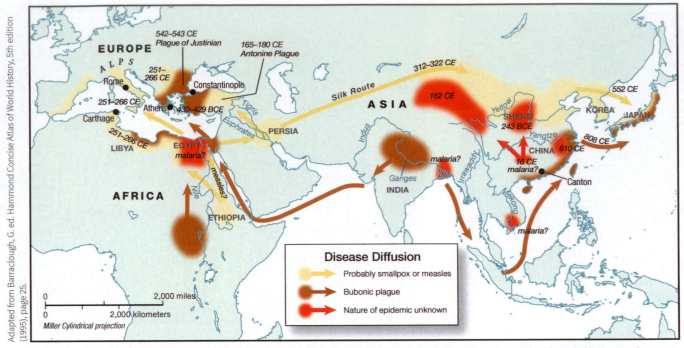

FIGURE 1.16 **Disease diffusion in Africa, Asia, and Europe circa 300 CE.** The diffusion of smallpox, measles, and bubonic plague followed trade routes on land and by sea. The arrows on the map reflect major trade routes at the time. Trade flowed overseas, as between eastern India and the east coast of China, and over land, along the Silk Road between Persia and China.

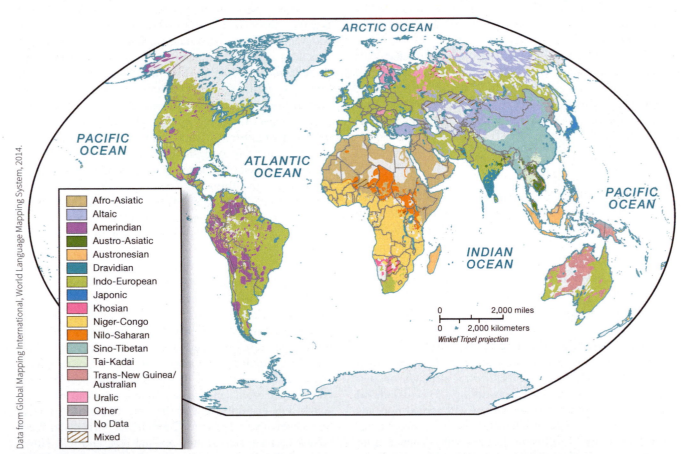

FIGURE 1.17 **World languages.** The distribution of the major language families reflects historical spatial interaction (or lack thereof) across the world.

in a lack of spatial interaction. Subsaharan Africa thus avoided the plague and smallpox in the timeframe of the map. North and South America are not on the map. When Europeans brought smallpox and plague to the Americas after 1492, the diseases ravaged the people of the Americas because having lacked prior exposure, they had not built up immunity.

By studying historical spatial interaction, we can see places that were tied together and places that were not. A map of historical civilizations from 500 BCE to 1500 CE points to the regional clusters of religions, languages, and other cultural attributes present in the world today. Analyzing modern spatial distribution of cultural attributes can illustrate how people and cultures interacted in the past.

The map of world languages (Figure 1.17) shows the extent of the Indo-European language family in Europe and Asia. As different as English is from Hindi, they are actually part of the same language family. Linguists believe the **hearth** of the Indo-European language family was near

the Caspian Sea, and from there, speakers of the language migrated both west and east, diffusing the language into Europe and South Asia.

The Sino-Tibetan language family is concentrated in East, southeast, and South Asia, near its hearth in the Himalayan Plateau to the north and east of the Himalayas. Linguists believed the speakers of the proto-Sino-Tibetan language migrated from its hearth along rivers into China, Southeast Asia, and South Asia.

The map of world religions (Figure 1.18) demonstrates the **diffusion** of ideas and innovations among people and places in two major ways: **expansion diffusion** and **relocation diffusion**. With relocation diffusion, people move and take ideas and innovations from their homeland with them. Expansion diffusion occurs in two ways: ideas and innovations spread from the hearth to nearby places and people first (*contagious diffusion*), or ideas and innovations spread from the most important places to the next most important places, skipping over people and places

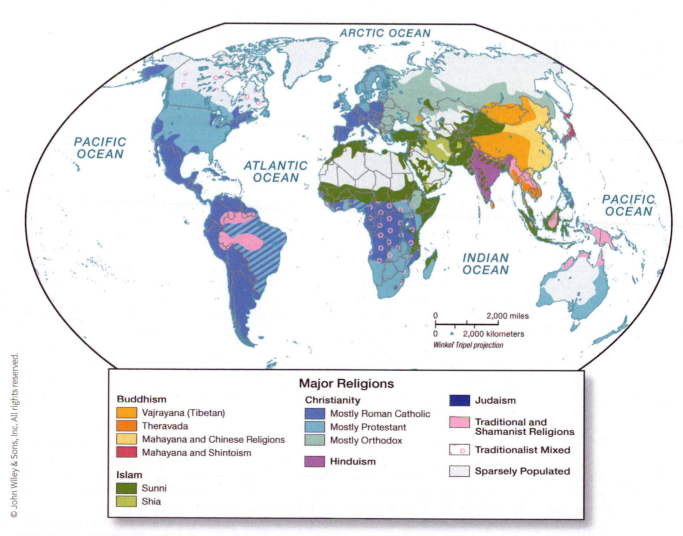

Major Religions

Buddhism
- Vajrayana (Tibetan)
- Theravada
- Mahayana and Chinese Religions
- Mahayana and Shintoism

Islam
- Sunni
- Shia

Christianity
- Mostly Roman Catholic
- Mostly Protestant
- Mostly Orthodox

- Hinduism

- Judaism

- Traditional and Shamanist Religions

- Traditionalist Mixed

- Sparsely Populated

FIGURE 1.18   **World religions.**   While Islam and Judaism persist in their hearths (the Arabian peninsula and the eastern Mediterranean, respectively), other major world religions have diffused broadly but are not the majority religion in their hearths, including Christianity (eastern Mediterranean) and Buddhism (northern India).

Courtesy Erin Fouberg.

FIGURE 1.19 **Seattle, Washington, United States.** The first Starbucks store opened just outside Pike Place Market in 1971. From its hearth in Seattle, Starbucks diffused around the world and is now located in 65 countries and has more than 21,000 stores.

in between (*hierarchical diffusion*). For example, Islam spread contagiously from its hearth in Mecca to encompass half of the Arabian Peninsula by the time of Muhammad's death in 632.

The spread of Christianity from its hearth in Jerusalem to people and lands along the northern Mediterranean Sea is a good example of hierarchical diffusion (see Chapter 9). When Christianity diffused into Europe near the end of the Roman Empire, it traveled from the most important cities to the next most important cities, skipping over the countryside between. Christianity followed the same pattern among people. The Emperor Constantine converted to Christianity, and from that point, conversions in the empire diffused hierarchically, from most important person to next most important.

Relocation diffusion also helped spread both Christianity and Islam. Christianity came to the Americas by relocation diffusion. Europeans who traded with the Americas, colonized the region, and migrated from Europe brought their religions (and languages and other cultural attributes) with them. Islam spread out from the Arabian Peninsula when Muslims conducted trade with people in South Asia and Southeast Asia. As the traders set up ports in South and Southeast Asia, they taught people about Muhammad's teachings, spreading the religion to new areas of the world.

The paths of diffusion reveal the underlying political, economic, and social structures that shape the spread of ideas, innovations, and traits. If a practice diffuses from one person or place to the next contagiously, it demonstrates that spatial interaction is occurring. When something diffuses hierarchically, it shows that an underlying, hierarchical political, economic, or social structure exists and influences the acceptance or rejection of new ideas (Figure 1.19).

REGIONAL MANIFESTATIONS OF GLOBAL PROCESSES

A map of major civilizations between 1000 CE and 1500 CE (Figure 1.20) shows the connections established before the onset of globalization and can be seen in the world regions used in this book, as they reflect long-term physical and cultural processes carried out in place.

Between 1500 and 1950, growing connections among people and places around the world established **globalization** (see Chapter 2). Civilizations reached beyond regional seas and coastlines to cross oceans and create global networks of trade, culture, and society. By 1950, as a result of global trade established through colonization and innovations in shipping, trade had come to encompass the entire world. With global trade, interactions among people and places continue to increase across the world, often ignoring state (country) borders.

Traders from Europe, Africa, North America, South America, and the Caribbean all crisscrossed the Atlantic Ocean. A triangular trade network brought minerals (gold) and enslaved people from Africa to the Americas, agricultural products (sugar and cotton) to Europe, and finished products (rum, cigars, and clothing) into European cities and colonial port cities in Africa and the Americas.

Along with trade, a multitude of cultural and physical processes have been exchanged across the Atlantic since 1500. African slaves brought music, language, and economic, social, political, and religious practices to the Americas. For example, enslaved Africans brought rice cultivation methods from West Africa to coastal North Carolina and South Carolina, helping establish rice plantations in the American South (Moseley 2005; Carney 2001).

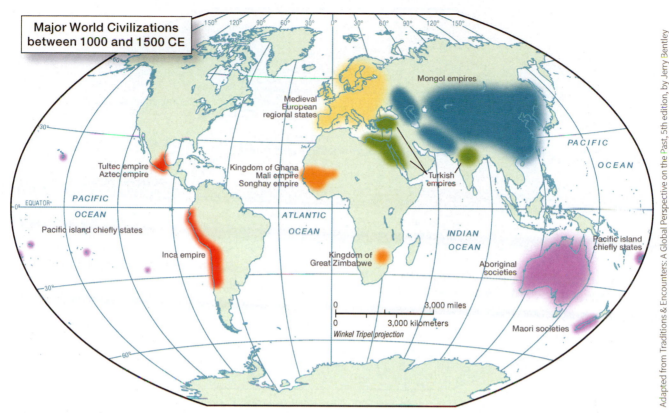

Major World Civilizations between 1000 and 1500 CE

Mongol empires

Medieval European regional states

Turkish empires

Tultec empire
Aztec empire

Kingdom of Ghana
Mali empire
Songhay empire

PACIFIC OCEAN

PACIFIC OCEAN

EQUATOR

Pacific island chiefly states

Inca empire

ATLANTIC OCEAN

INDIAN OCEAN

Kingdom of Great Zimbabwe

Aboriginal societies

PACIFIC OCEAN

Pacific island chiefly states

Maori societies

0   3,000 miles
0   3,000 kilometers
Winkel Tripel projection

Adapted from *Traditions & Encounters: A Global Perspective on the Past*, 5th edition, by Jerry Bentley and Herbert Ziegler, New York: McGraw-Hill, 2011.

FIGURE 1.20   Major civilizations between 1000 CE and 1500 CE.   Spatial interaction occurred within the major civilizations and helped establish cultural commonalities, including language, in East Asia, Southeast Asia, South Asia, Australia and the Pacific, Russia, Europe, the Arabian Peninsula, North Africa, West Africa, South Africa, South America, and Central America.

From the Americas, agricultural crops, including potatoes and corn, diffused to Europe where farmers found the crops well suited for European soils (potatoes in Ireland and corn in France). The introduction of American crops altered agricultural practices and economies in Europe. Agricultural land use in the Americas, which was based on efficient, large-scale plantations, even motivated governments to legislate consolidation of land in Europe. On the negative side, Europeans brought diseases, especially animal-borne diseases, to which people in the Americas had never been exposed. The exchange of goods, people, diseases, and ideas in a triangular pattern around the Atlantic is called the Columbian Exchange because it dates back to Columbus's travels to the "New World."

Between 1500 and 1950, several regional empires around the world peaked and then declined. The Ottoman Empire rose to colonize much of southwest Asia and southeastern Europe. Japan grew to colonize Korea, parts of China, and much of the Pacific. Europe colonized the Americas and then Africa and Asia. The experiences of the colonies versus the colonizer and the differences between preindustrial growth and industrial growth helped to distinguish the world regions defined in this book (see Chapters 2 and 3).

Trade around the Atlantic Ocean and European colonization had distinct regional outcomes as local cultural groups, civilizations, growing nations, and new countries interacted with global processes and shaped them in different ways to create unique regional effects. The processes of globalization operate across country borders, but the outcomes of globalization are not evenly distributed, as some places are more connected to global processes and others shape the way globalization affects them.

WORLD REGIONS

We delineated the major world regions around which we organize chapters (Figure 1.21) according to historical spatial interaction before 1500 and colonization and trade during the growth of globalization between 1500 and 1950. World regions reflect historical connectedness that helped create the fundamental cultural differences among each of the regions today.

The boundaries between world regions are transition zones. In some cases, such as the divide between Subsaharan Africa and North Africa, the physical boundary (the Sahara Desert) has been a barrier between peoples and

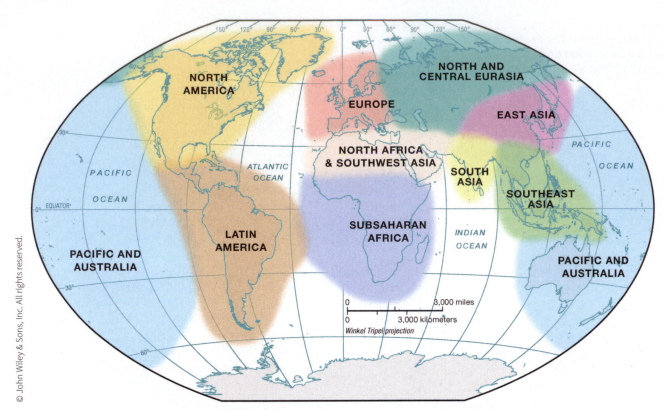

FIGURE 1.21   World regions.   Regions of the world used in *Understanding World Regional Geography*. The regions are based on a combination of historical civilizations, the growth of globalization between 1500 and 1950, the hearths and diffusion of major cultural traits, and regional organizations established since World War II.

places at various times in history. In other cases, a lack of spatial interaction between two regions occurred because of political differences, not physical boundaries. Between 1500 and 1900, tsarist Russia and western Europe had a weak relationship and experienced relatively little spatial interaction. From 1924 to 1991, when Russia was part of the Soviet Union, the Cold War divided Russia from western Europe, and again little spatial interaction occurred. These periods of division help explain why Europe and Russia are different world regions. In some cases, diverse areas encompassing many peoples and places—such as the countries throughout South America, Central America, and the Caribbean—are joined together because of an overwhelming set of shared historical, cultural, and physical experiences. In this case, Spanish and Portuguese colonization helped define the region of Latin America.

Because world regions are dynamic and boundaries are transition zones and not firm limits, some countries appear in more than one region of this book. For example, Mexico and parts of the Caribbean are discussed in both the North America chapter and the Latin America chapter. Countries in East Asia are discussed in the chapter on East Asia and in the chapter on the Pacific Rim.

The Pacific Rim is a new region that is a product of globalization. Geographer John Agnew contends that the "changing character of the world economy" is "redrawing the regional map of the world." The Pacific Rim offers an insightful snapshot of the changing character of the world economy (see Chapter 13).

Overall, the regions in *Understanding World Regional Geography* reflect historical spatial interaction, shared experiences of colonialism, distinct local cultural identities, the growth of world religions, the evolution of world languages, and significant interactions among people and places.

CONCEPT CHECK

1. **What** are the differences between formal and functional regions?

2. **How** do perceptual regions help shape your schemata?

3. **What** criteria are used to delineate the regions in *Understanding World Regional Geography* and why?

WHAT TOOLS DO GEOGRAPHERS USE TO STUDY WORLD REGIONAL GEOGRAPHY?

LEARNING OBJECTIVES

1. **Define** GIS and remote sensing.
2. **Explain** what a physical landscape is and why geographers study them.
3. **Explain** what a cultural landscape is and why geographers study them.

Photographs, maps, satellite images, and graphics can help you gain an understanding of world regional geography. To understand the images in this book, you need to learn to look at maps and photographs like a geographer. Read each caption carefully and try to see on the map or in the photograph what the caption discusses. Look at the images when they are called out in the text because the text gives you a greater context for understanding the images. We also recommend that you practice analyzing maps and photographs using the exercises available online.

The tools geographers use vary greatly because geographic studies cover the entire earth, including people, place, environment, and space. The range of tools in the discipline of geography includes fieldwork, map and air photo analysis, cultural and physical landscapes, geographic information systems (GIS), remote sensing, surveys, interviews, ethnography, computer modeling, and statistical analysis. We will discuss fieldwork, cartography, GIS and remote sensing, and cultural and physical landscapes in this section.

FIELDWORK

Geographers go outside, into the field, and observe how people, place, environment, and space vary across the world. As early as the 1800s, geographers argued that students learn more when they study their own community to observe and apply geographic concepts. In 1861, educator James Currie wrote that students should start understanding geography by experiencing the environment in their town.

Currie reasoned that after understanding local places, geography students can make connections to remote places they have read about but have not visited.

We cannot memorize world regional geography because it is constantly in flux. But we can keep up with the changing world by applying geographic concepts, and one of the best ways to learn concepts is to look at the world around you, your town or city, or even your college campus. Take your blinders off as you walk around outside.

Put your cell phone in your bag and start observing. Can you see examples of functional regions in your town? How do humans and environment interact in your region, and how do your observations change if you switch scales of analysis?

Through **fieldwork** we make observations and then we analyze these observations. Geographers do fieldwork systematically by mapping change over time, categorizing observations, applying geographic questions, or analyzing differences in phenomena spatially. At the end of each chapter in this book, you will have an opportunity to explore fieldwork and think through it on your own in the *Your Turn: Geography in the Field* features.

CARTOGRAPHY

Geographers use maps as tools to present data spatially. Maps are powerful visuals because cartographers (map-makers) design maps with a certain purpose in mind—such as showing data, swaying an opinion, or conveying an insight. **Cartography** is the subdiscipline of geography concerned with the construction of maps.

Cartographers divide maps into two basic types: *reference maps* and *thematic maps*. A reference map shows locations of places and geographic features. A map in a road atlas is a good example of a reference map. The goal of a reference map is to convey **absolute locations** of places. Absolute locations are usually described in latitude (degrees north or south of the equator) and longitude (degrees east or west of the Prime Meridian). With a global positioning system (GPS), finding the precise latitude and longitude of a place is simple and can be easily mapped. By using a reference map, a map reader can tell where places are, where they are in reference to each other, and whether and how the places are connected (Figure 1.22a).

A thematic map tells a story about the degree of an attribute such as the percent of retired people by county, the pattern of its distribution, or its movement. A thematic map shows **relative locations** of places. Instead of placing a dot for every city and town in the world, a thematic map tells the story of the distribution of the world's population by placing and coloring dots to represent the distribution of people (Figure 1.22b) Thematic maps are used to tell the story of data whether by location, distribution, or movement.

All maps, whether reference or thematic, are representations of the world, and no map is truly accurate. Earth is three dimensional, and maps are two dimensional. Because each map has a projection that transforms earth from three dimensions to two, earth loses shape or distance, or becomes distorted in area (size) or direction when we flatten it onto a map. Maps are also only as accurate as the data that go into them. They lose accuracy because a cartographer must generalize and standardize data and because the areal unit the data represents limits what can be included.

GIS AND REMOTE SENSING

Analysts in the field of geographic information sciences (GISc), which encompasses both **geographic information systems (GIS)** and remote sensing, create maps on computer programs that organize and integrate data spatially. The technologies developed in these fields have revolution-

ized the process of mapmaking. GIS allows geographers to store massive amounts of spatial data (Where is it?) and attribute data (What is it like? How is it?) and map it easily. Through GIS analysis, geographers can query both spatial and attribute data to answer questions. GIS analysts create models, using tools in GIS, which help them predict human

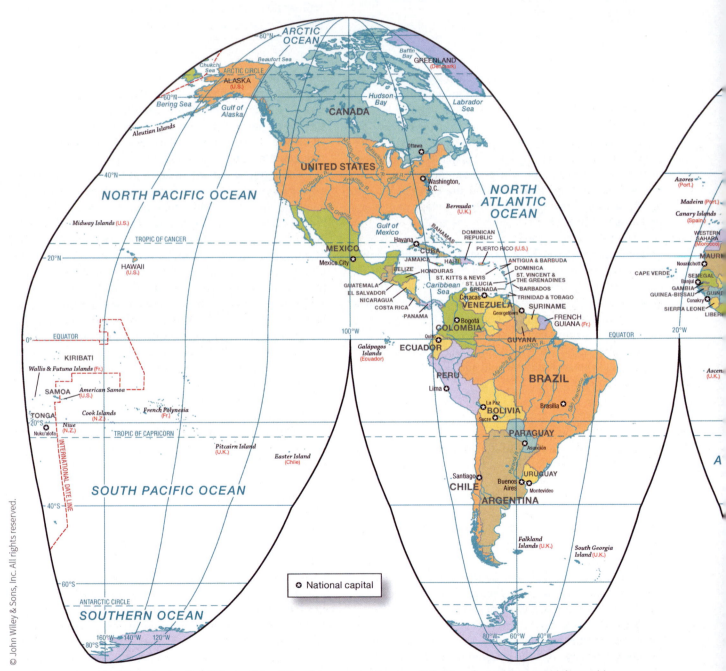

FIGURE 1.22a   Political geography of the world.   This reference map shows capital cities of countries around the world.

and environmental outcomes to current and hypothetical situations.

Remote sensing is a method of collecting data by using instruments, including sensors on satellites, which are far from the area of study. It enables geographers to study environmental and human phenomena at a variety of scales. Through remote sensing, geographers can analyze change in places over time—to the vegetation, landscape, water, or climate. Geographers can also use remote sensing to study migration flows and political maneuvers. Remote sensing and GIS can be combined by overlaying satellite photographs with GIS spatial and attribute data.

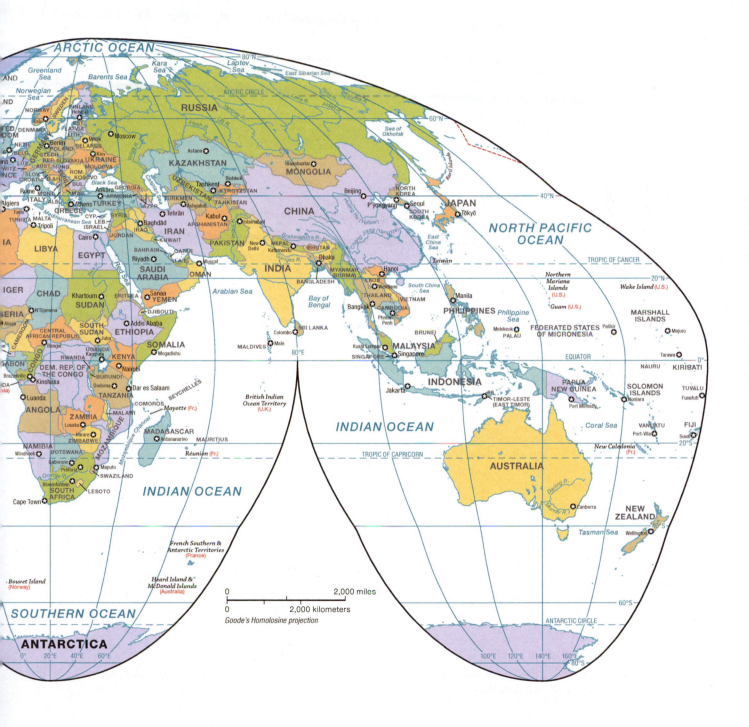

Data from CIESIN—Columbia University; United Nations Food and Agriculture Programme—FAO, and Centro Internacional de Agricultura Tropical—CIAT. 2005. Gridded Population of the World, Version 3 (GPWv3): Population Count Grid, Future Estimates. Palisades, NY: NASA Socioeconomic Data and Applications Center (SEDAC). http://dx.doi.org/10.7927/H42B8VZZ. Accessed December 2014.

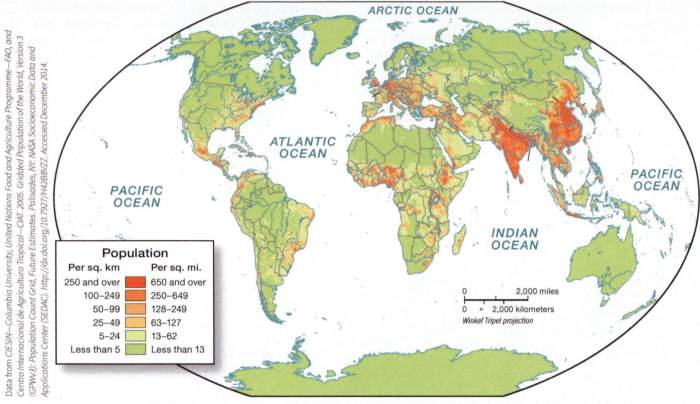

FIGURE 1.22b   Population density of the world.   This thematic map shows the general pattern of people across the world.

The power of GIS and remote sensing is not in the tools themselves but rather in how geographers use the tools to deepen their analysis.

LANDSCAPES

When geographers travel, they see landscapes—both physical and cultural. Each subdiscipline in physical geography studies a different aspect of the physical world. For example, geomorphology analyzes the topography, landforms, and physical landscape of Earth (Figure 1.23). Climatologists analyze weather over time to explain the climatic conditions and variances of regions and places in the world, along with climatic phenomena such as floods and climate change. Geographers who study hazards analyze the potential changes of physical geography processes and the likely impact of hazards on places humans have altered.

The Physical Landscape

By examining the **physical landscape**, a geographer can see how processes, whether glacial, fluvial (river or stream),

or tectonic, have shaped it (Figure 1.24). For example, traveling through East Africa, a physical geographer would immediately notice the signs of tectonic activity shaping the physical landscape.

The Cultural Landscape

No place is untouched by people. People have shaped each place on Earth in a multitude of ways, and as a result we can read the world around us to see human imprint on the landscape. A **cultural landscape** is the visible imprint of human activity. Cultural landscapes can show layers of history as different people who lived in a place over time each made their own imprint in a process known as *sequent occupance* (Figure 1.25).

Geographers have developed several ways to interpret cultural landscapes by focusing on the layers of history and on the changes that events bring to

---

cultural landscape   the visible human imprint on the landscape.

(a)

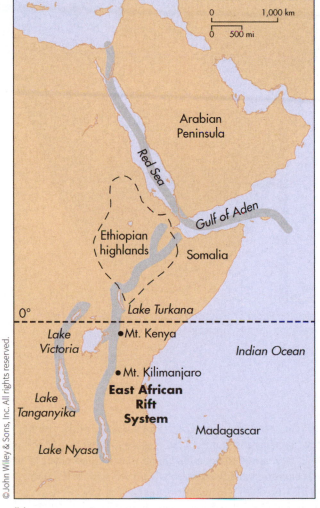

(b)

landscapes (Lewis 1979). Cultural geographers study ordinary landscapes in addition to cataclysmic landscapes in order to understand people, place, and culture (Groth and Bressi 1997).

The images your imagination has gleaned from magazine covers, movies, and books are permanently in your mind, but the people in these places have changed since you first glimpsed those images. People and places are always changing, making world regional geography an intensely dynamic subject. Understanding our increasingly complex and interconnected world requires continually building our schemata. For example, the Afghani woman in the opening image of this chapter represented for Americans the human face of the war between the Soviet Union and Afghanistan when she appeared on a 1985 cover of *National Geographic*. Today, Gula still represents a war-torn country.

FIGURE 1.23   East African Rift Valley. (a) In the East African Rift Valley, the continental plate is thinning, and the elevation of lands in the valley is lowering as the tectonic plate thins through the rift valley. (b) The map of the East African Rift Valley confirms the thinning or spreading of the plate, because the major lakes in East Africa are in the rift valleys, where elevations are lowest.

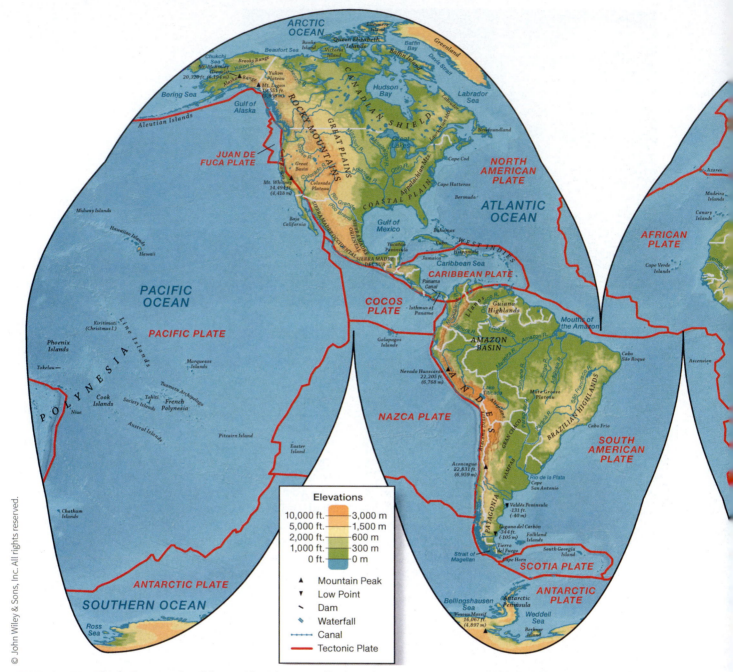

FIGURE 1.24 **Physical geography of the world.** The areas of highest elevation appear orange on this map, and other areas of relatively high elevation are a lighter orange to beige. The areas of lowest elevation are the darkest green.

The 1985 photograph by Steve McCurry was so well known that it came to symbolize the *National Geographic*. The magazine used the photograph so frequently after 1985 that it almost became a brand for the magazine, making people think of *National Geographic* when they saw the photograph. Millions in the United States and around the world had seen her photograph, but when the photographer, Steve McCurry, sought out Gula again to photograph her in 2002, he learned to his amazement that she had never seen it. Technological advances between 1985 and 2002 enabled the FBI, through iris mapping, to say with 100 percent certainty that Gula was the woman photographed in 1985.

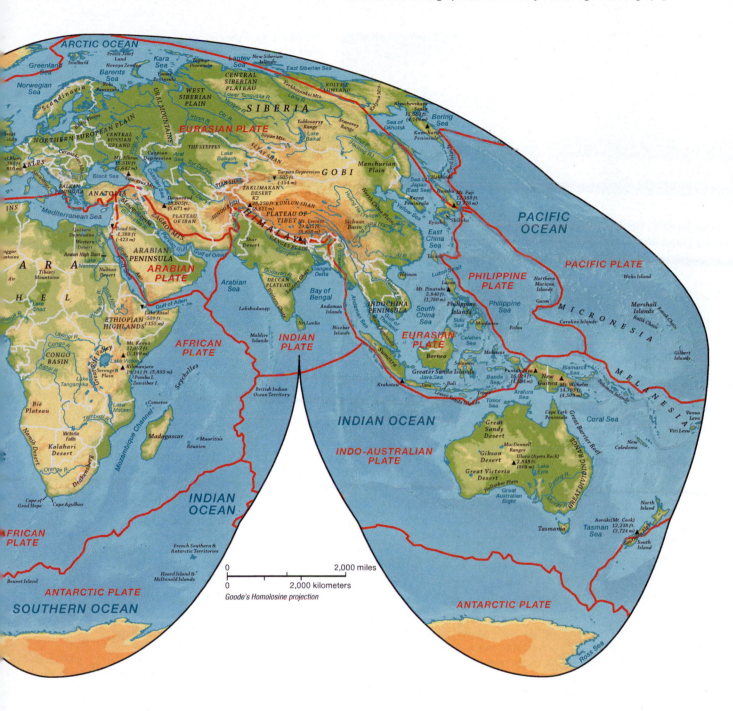

The world felt closer to Gula and the plight of her home country through her 1985 photograph, but the same photo, which she did not know existed, meant little in her life. The world perceived a region through her eyes, but at the scale of her family and locality, the same photograph meant nothing. Her perception was drawn from her experience and concentrated on enduring the ongoing conflict in her country.

CONCEPT CHECK

1. **How** do geographers use GIS and remote sensing to study geographic phenomena?

2. **What** is a physical landscape and how do geographers analyze physical landscapes?

3. **What** is a cultural landscape and **how** do geographers analyze cultural landscapes?

© Michael Siluk/The Image Works.

FIGURE 1.25   The cultural landscape: Indian Mounds Park in St. Paul, Minnesota.   If you were in the Dayton's Bluff neighborhood of St. Paul, Minnesota, you might wander into Indian Mounds Park and look down on the Mississippi River as it flows between Minneapolis and St. Paul. Standing on this hill, you would see the concept of sequent occupance in the cultural landscape. The homes in the nearby neighborhood that date mostly to the 1800s and early 1900s are one layer of the landscape. The hill you are standing near is a remnant of two other occupants. The Hopewell Indians, who lived in the area beginning between 1,500 to 2,000 years ago, built the hills as burial mounds. The Dakota Indians, who arrived around the 1600s, added to the Hopewell mounds and likely built mounds of their own. Between the houses of the early 1900s and the 2,000-year-old burial mounds, the arrival of European Americans brought layers of intense agriculture, river barge traffic, a fish hatchery, an asylum, and, in 1893, the establishment of a park to preserve these mounds and green space. On this land where 37 Indian burial mounds once stood, only six stand today. Some mounds were razed for research, some for agriculture, and 11 mounds were reportedly razed to landscape the park when it was established.

SUMMARY

What Is Geography?

1. **Geography** can be thought of in simple terms as the study of people, place, environment, and space.

2. Geography incorporates two major fields of study: **physical geography** and **human geography**. Physical geographers take a geographic approach (analyzing people, place, and space) to the study of earth, environment, and human–environment interactions. Human geographers take a geographic approach to the study of peoples, cultures, and their shaping of place and environment. Geographers also study the reciprocal relationship between the physical and the human, in a field of study called **human–environment** relations.

What Are Regions, and Why Do Geographers Use Regions?

1. Traditionally, geographers used regions as the basis of their analysis—as the basic units whereby they studied the world. More recently, geographers researching in the field of new regional geography have viewed regions as imagined entities and now question how regions are drawn, what they represent, and how they should be used. The regions used in world regional geography books are created by humans and are usually determined based on a combination of physical and cultural criteria. Governments and international organizations define regions and build their institutions using these regions.

2. Geographers use **scale** to describe the areal extent of a study, for example, countries of the world or counties in a state.

3. Regions are powerful concepts because each of us uses them to order and interpret our world and to comprehend the vastness of the world and our place in it.

4. Regions reflect the social, political, and economic processes weaving through places and across peoples.

What Regions Are Used in *Understanding World Regional Geography*?

1. A **formal region** distinguishes an area of land by cultural or physical criteria. A **functional region** is an area that has a shared political, economic, or social purpose. **Perceptual regions** are based on what individual people see in their minds. We perceive the region where we live, and we also perceive the regions where others live.

2. The regions in this book are drawn based on two factors: historical spatial interactions and physical and cultural processes that occur globally and locally and are manifested regionally in the world.

3. The map of world religions demonstrates the two major types of diffusion: **expansion diffusion** and **relocation diffusion**. Ideas and innovations spread across people and places. With expansion diffusion, people remain stationary and ideas spread across places. With relocation diffusion, people move and take ideas and innovations with them.

4. Between 1500 and 1950, growing connections among people and places around the world established **globalization**. Civilizations reached beyond regional seas and coastlines to cross oceans and interconnect the world through trade, culture, and physical processes. With globalization, interactions among people and places continue to increase across the world, often ignoring state borders.

What Tools Do Geographers Use to Study World Regions?

1. Geographers use two key visual tools: cartography and landscape. **Cartography** is the art and science of making maps. Geographers use maps as tools to present data spatially. Cartographers design maps with a certain purpose in mind—such as showing data, swaying an opinion, or conveying an insight.

2. Cartographers divide maps into two basic types: *reference maps* and *thematic maps*. A reference map shows locations of places and geographic features. A map in a road atlas is a good example of a reference map. The goal of a reference map is to convey **absolute locations** of places. A thematic map tells a story about the degree of an attribute, the pattern of its distribution, or its movement. A thematic map shows **relative locations** of places. Thematic maps are used to tell the story of data—whether by location, distribution, or movement.

3. Most maps today are created on computers in the field of geographic information sciences (GISc) and remote sensing. The technologies in these fields have revolutionized the process of mapmaking. A **geographic information system (GIS)** is a combination of computer hardware and software designed to show, analyze, and represent geographic data.

Geography in the Field   YOUR**TURN**

This sandstone block is located in Fredericksburg, Virginia, a city sited on the Rappahanock River and situated halfway between the North's capital of Washington, D.C,, and the Confederate capital of Richmond, Virginia, during the Civil War. The piece of

Reza A Marvashti/The Free Lance-Star/AP Photo.

sandstone served as a slave auction block prior to the war. The plaque on the block simply states "Auction Block: Fredericksburg's Principal Auction Site in Pre-Civil War Days for Slaves and Property 1984 HFFI." In 2005, the slave auction block was vandalized, with two large pieces and several smaller pieces broken off the top of the sandstone block. After the vandalism, the city council debated whether to leave the block as it was or to repair it. The city council voted to repair the slave auction block.

Thinking Geographically

- Why do you think this slave auction block was dedicated in Fredericksburg, Virginia, in 1984?

- What does the slave auction block mean in the cultural landscape of this place, and why do you think the city council voted to repair the block after it was vandalized in 2005?

Read:

Moyer, Laura. Slave Auction Block Vandalized. *Fredericksburg Freelance Star*. May 6, 2005. http://fredericksburg.com/News/FLS/2005/052005/05062005/1757761

Hanna, Stephen. 2008. "A Slavery Museum? Race, Memory, and Landscape in Fredericksburg, Virginia." *Southeastern Geographer*, 48(3): 316–337.

KEY TERMS

geography	context	perceptual region	relocation diffusion
people	contiguous	identity	globalization
place	scale	toponym	fieldwork
environment	modifiable areal unit problem	perception of place	cartography
space	region	sense of place	absolute location
physical geography	relative humidity	spatial interaction	relative location
human geography	Institutionalize	time–distance decay	geographic information system (GIS)
remittances	schemata	hearth	remote sensing
human–environment	formal region	diffusion	physical landscape
threshold concepts	functional region	expansion diffusion	cultural landscape

CREATIVE AND CRITICAL THINKING QUESTIONS

1. Read about Sharbat Gula, the Afghan woman photographed in the chapter opener, and her life now at: http://news.nationalgeographic.com/news/2002/03/0311_020312_sharbat.html

   How has her status as a global symbol of Afghanistan affected her **identity** at the local scale? At the global scale?

2. How has the **situation** of Afghanistan changed in Sharbat Gula's lifetime?

3. Education is a key component of combating HIV/AIDS. How can geographers use their understanding of regional differences to advise nonprofits and governments on the best approach to tailor education programs that slow the **diffusion** of HIV/AIDS in each world region?

4. Read about the rice terraces of the Ifugao Province of the Philippines on the UNESCO website and determine whether the terraced region best fits the definition of a functional, formal, or perceptual **region**.

5. How have your perceptions of the world been formed? What experiences have you had traveling or talking with family and friends that have shaped your perceptual **regions** of the world? What movies, television shows, and books have shaped your perceptions of the world?

6. What cultural and physical geographic criteria would you use to define the formal **region** where you live? How many functional regions is your hometown in (list as many as you can)?

7. The Great Plains is one of many formal regions in North America. Name the region where your hometown is located and describe either the physical or **cultural landscape** and

how the landscape distinguishes your region from surrounding regions.

8. Choose one religion mapped on the map of world religions. Read about the religion in this book (use the index) and online (try BBC), and find out where the religion's hearth is and how it **diffused**. Look at the map to see how far the religion has spread and hypothesize what kinds of diffusion brought this religion to these places in the world.

9. Choose a feature in the **cultural landscape** on your campus or in your hometown. Describe what the feature looks like and explain how it fits into the surrounding landscape and the context of the place.

SELF-TEST

1. Geography is divided into two subdisciplines, broadly called _____.
   a. environmental geography and political geography
   b. remote sensing and cartography
   c. physical geography and human geography
   d. place geography and space geography

2. In both of the subdisciplines, geographers _____.
   a. memorize capitals of countries
   b. analyze people, place, and space
   c. use timelines to analyze why places are different
   d. study how environments determine world cultures

3. The world region in which HIV/AIDS impacts women more than men is _____.
   a. East Asia
   b. North Africa and Southwest Asia
   c. Latin America
   d. Subsaharan Africa

4. To understand the diffusion of HIV/AIDS or other diseases, geographers explore how disease transmission can vary by the demographic characteristics of different regions. Recent studies report that people infected with this disease may be "supercontagious" when also infected with HIV/AIDS.
   a. chicken pox   c. swine flu
   b. smallpox   d. malaria

5. When we change scales of analysis, the distribution of an attribute _____.
   a. changes
   b. stays the same

6. The structure in the mind through which humans build knowledge and filter information is _____.
   a. metacognition   c. region
   b. scale   d. schema

7. The humid subtropical climate is humid because _____.
   a. high-pressure systems in the ocean bring warm, moist air to the east coast of continents
   b. high-pressure systems in the ocean bring warm, moist air to the west coast of continents
   c. low-pressure systems in the ocean bring cool, moist air to the east coast of continents
   d. low-pressure systems in the ocean bring cool, moist air to the west coast of continents

8. The German-speaking area of Europe is a good example of a _____.
   a. formal region   c. contiguous region
   b. functional region d. continental region

9. In the map of the distribution of diseases circa 300 CE (during Roman times), diseases were _____.
   a. concentrated in tropical Africa
   b. diffused mainly along trade routes
   c. diffused mainly around military bases
   d. concentrated along the equator

10. In the map above of the distribution of diseases, Subsaharan Africa is shaded gray because during this time _____.
    a. no one lived in Subsaharan Africa
    b. the people in Subsaharan Africa did not spatially interact with people in North Africa
    c. the Sahara covered all of Africa
    d. people in Subsaharan Africa were immune to these diseases

11. With this type of diffusion, ideas spread from the most important place to the next most important place.
    a. contagious   c. hierarchical
    b. relocation   d. stimulus

12. A map in a road atlas is a good example of this type of map.
    a. reference   c. physical
    b. thematic   d. cultural

13. Sequent occupance shows _____.
    a. the layers of cultures that have lived in a place
    b. how rivers have changed course over time
    c. how shipping methods have changed with globalization
    d. the layers of climate change in a place

14. Geographers call the visible human imprint on the landscape the _____.
    a. physical landscape
    b. environmental landscape
    c. cultural landscape
    d. historical landscape

15. Globalization is a set of processes that are _____.
    a. deepening the relationships of people and places around the world
    b. creating a more even exchange of global goods around the world
    c. redefining country borders in Africa and Asia
    d. leading to European colonization of the world

ANSWERS FOR SELF-TEST QUESTIONS

1. c, 2. b, 3. d, 4. d, 5. a, 6. d, 7. a, 8. a, 9. b, 10. b, 11. c, 12. a, 13. a, 14. c, 15. a

GLOBAL CONNECTIONS

The man on a motorcycle looks dressed and ready to go to work in the HITEC City (Hyderabad Information Technology and Engineering Consultancy) area of Hyderabad, India. He may work at a job providing technical support for a major information technology company in the United States. He rides his motorcycle past a small settlement of tents covered by blue tarps and corrugated metal where a dog seems to be standing guard. The people who live in the tent settlement likely migrated to the city from the rural area of Telangana, the state in India where Hyderabad is located, or the neighboring state of Andhra Pradesh. The migrants who live in this tent settlement work to construct the new buildings going up all over the HITEC City, including the one being constructed in the background of this photograph.

While the tech worker and migrants in the tent settlement live and work near each other, and both are intelligent and entrepreneurial, the worlds they inhabit are not the same. They have different friends, professional networks, value systems, and world views, and they might not even know what to talk about if they were to speak to one another.

This situation would seem to seriously contradict geographer Waldo Tobler's first law of geography, that is, "everything is related to everything else, but near things are more related than are distant things." Whether the scene contradicts this maxim depends a lot on how we conceptualize "nearness." Most of us think of nearness in terms of physical distance or proximity. If we limit ourselves to this definition, then we clearly have a contradiction. However, many geographers also think of nearness in terms of connectivity, the ability of individuals to interact with one another via telecommunications, the Internet, trade, or other means. Under this scenario, this tech worker in Hyderabad is "nearer" to other information technology workers and consumers around the world than he is to the migrant who lives in the tent settlement outside his office.

Hyderabad, India. Late on a summer afternoon, a car and motorcycle drive through a neighborhood near HITEC City, an area of the city home to Tata Consultancy Services, HSBC Global Resourcing, Oracle Software, Dell Computers India, and Deloitte Consulting India. The tent settlement just beyond the motorcyclist is one of the city's hundreds of slums, some small like this and some covering huge expanses of land. The 2011 census of India found that one in three people in Hyderabad lives in a slum, including the tent slums, which was a 264 percent increase since the 2001 census. Migration from rural to urban areas is the main source of the growth in slums over the last 10 years.

Courtesy of Erin Fouberg.

CHAPTER OUTLINE

What Is the Nature of Globalization, and How Is It Impacting the World System?
- Different Understandings of Globalization
- Colonialism
- Shifting Regional Configurations

How Do International Organizations Create and Shape Global Connectedness?
- International Financial Institutions
- United Nations
- Bilateral Donors, Nongovernmental Organizations, and Multinational Corporations
- Hegemon
- The Influence of Local Actors

How Are Places Connected, and Why Are Some Places More Connected than Others?
- Networks and Global Connections

- Global Processes and Local Places
- Local Factors and Global Connectivity

How Is Global Environmental Change Affecting the World?
- Global Environmental Change
- Vulnerability
- Navigating Globalization and Environmental Change

 THRESHOLD CONCEPTS in this Chapter

Anthropocene
Globalization
Networks

WHAT IS THE NATURE OF GLOBALIZATION, AND HOW IS IT IMPACTING THE WORLD SYSTEM?

LEARNING OBJECTIVES

1. Critically **discuss** the differing theories of weak and strong globalization.

2. **Describe** the enduring impact of colonialism on the world system.

3. **Explain** how patterns of exchange are continually reshaping the world, which may lead to different regional configurations.

Using regions to study the world can prevent us from seeing global and local processes happening at the same time. In a sense, a region is like a lens, and global and local forces are refracted by looking through the lens. In this chapter, we focus on the study of global processes and agents that shape, pass through, and are shaped by our regional frames of reference. In this first section, we focus on different ideas regarding the amorphous process known as globalization, the historical experience of colonialism, and the influence of globalization on how we define regions.

DIFFERENT UNDERSTANDINGS OF GLOBALIZATION

 The concept of **globalization** is frequently used to describe the character and intensity of connections between various places around the globe. International relations theorists Andrew Hurrell and Ngaire Woods (1995, 447) define it as "the process of increasing interdependence and global enmeshment which occurs as money, people, images, values and ideas flow ever more swiftly and smoothly across national boundaries." Geographers often approach globalization in ways slightly different from other disciplines (Murray 2005). Geographers are sensitive to, or aware of, differences between places and across time. While some theorists see globalization as a homogenizing force that creates a global culture blanketing the world and masking local differences, geographers see globalization as a complex set of processes that differentiate places (Figure 2.1). Geographers are keenly aware of interactions across scales and recognize the value of global, regional, and local processes.

Geographers often think of globalization as something that is far from new. This conception of globalization, often referred to as **weak globalization**, sees globalization as a long-standing process, dating back to a time when systems of

globalization processes heightening interactions, increasing interdependence, and deepening relations across country boundaries.

exchange became more global in scope (Hirst and Thompson 1996). Many geographers argue that globalization began with the expansion of the Spanish and Portuguese empires in the 1500s. The triangular trading system involving Africa, Europe, and the Americas is one example of an earlier trading system that was global in scope.

FIGURE 2.1a **Venice, Italy.** The Roman Catholic Church is a global church centralized hierarchically under the pope in Rome, Italy, and St. Mark's Basilica in Venice, Italy reflects the local design aesthetic with its Italian marble columns and gold mosaics.

FIGURE 2.1b **Andhra Pradesh, India.** The Roman Catholic Church takes on the aesthetic of the local culture in India, as this Catholic church displays bright colors and towers shaped similarly to the shikhara of Hindu temples.

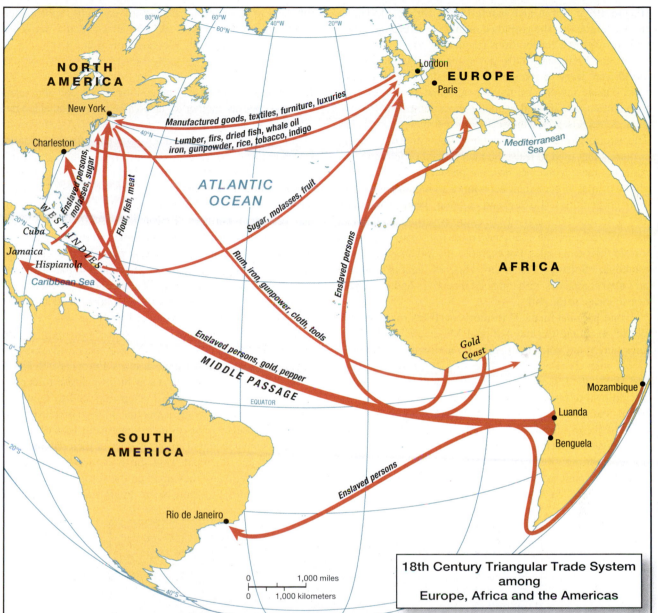

FIGURE 2.2   The Columbian Exchange.   The eighteenth-century triangular trade system among Europe, Africa, and the Americas helped establish the foundations of the capitalist world economy.

In the Columbian Exchange, the triangular trading system (Figure 2.2) that operated during the seventeenth, eighteenth, and early nineteenth centuries, Europe exported cheap manufactured goods to Africa to be exchanged for enslaved people. Europeans then forcibly exported enslaved people to the plantations of the Americas in exchange for commodities (cotton, tobacco, rice). Traders shipped the commodities back to Europe and the circuit began anew. The Atlantic Slave Trade eventually diminished in intensity following a ban by the British government in 1807. New forms of commodity production and exchange connecting Europe, Africa, the Americas, and Asia developed during the colonial era.

Once colonies gained independence, the relationships they had with former colonizers and the patterns of global trading systems changed again. Wealthy countries and companies still strongly influence, and in many cases control, production worldwide. Weak globalizationists conclude that the trade relationships operating today have their roots in the Columbian Exchange, stretching back hundreds of years. This view supports the weak globalization argument that globalization is not anything new, but rather a long-standing process.

In contrast to the weak globalization perspective, the **strong globalization** view holds that a global economy has emerged since the 1970s and is significantly different from that which existed previously. Strong globalizationists contend that the economy is much more global and that states (countries) are increasingly irrelevant in world trade today. In the new global economy, the market is all important and is not constrained by state governments

(Fagan 1995; Bryan and Farell 1996). Strong globalizationists argue that the global economy is increasingly controlled by global corporations (Fagan 1995).

In the 1960s and 1970s, governments in the developing world were leery of foreign direct investment and discouraged outside investment. During this time period, governments often nationalized or took over a number of foreign-owned enterprises, including banks and petroleum companies. The strong globalization viewpoint holds that developing countries that used to resist the intervention of foreign companies in their economies now are persuaded by the lure of global capital to open their economies to investment by foreign corporations (Loxley and Seddon 1994; Sandbrook 1997). In the 1980s and 1990s, many governments that had nationalized companies adopted a more free-market approach toward their economies. For example, the Indian government established Software Technology Parks of India in 1991 and provided companies with financial incentives and infrastructure to locate in the parks.

International financial institutions (IFIs), including the World Bank and the International Monetary Fund (IMF), have played key roles in opening up national economies to global markets (Roy 1997; Bracking 1999). In the 1980s and 1990s, the IFIs pushed economic reforms that included the removal of tariff barriers, an emphasis on export orientation, and the privatization of state-owned companies. All three of these reforms encouraged greater integration into global markets and the flow of investment capital across state borders. Strong globalizationists contend that a combination of economic policies that increased the flow of capital across borders, the emergence of a global culture and the

diffusion of common ideas, and increasingly sophisticated communication networks and globalized media marked a significant shift to strong globalization after 1970.

COLONIALISM

Colonialism is a physical process by which one area takes over and controls another for its own benefit. Colonialism has occurred at various times throughout history. As empires or kingdoms grew and the need for resources expanded, powerful leaders would often annex other territories through the use of military force. Both the Greeks and Romans were colonial powers who claimed large areas from which they extracted resources and tribute. For example, during the Roman colonial period in northern Tunisia (146 BCE–439 CE), the authorities took land from local people and redistributed it to Roman army veterans. The Romans wanted army veterans to farm the land and then link their production to their empire's system of trade. The intensity of agricultural production on these farms eventually led to soil degradation (Jeddou 2008).

Modern colonialism began in fifteenth-century Europe. It was different from Roman or any other preceding empire because the focus in the modern era was specifically on gaining territory, not simply tribute or taxes. European colonizers were willing to go long distances in their desire to control territories. Colonizing citizens often migrated to a colony, creating a place of not only economic but also political privilege for themselves. The colonizers physically took control of the government, economy, and territory of their colonies.

Spain and Portugal were the first modern European colonial powers, and they sparked the **first wave of colonialism** (Figure 2.3). Both Spain and Portugal experienced

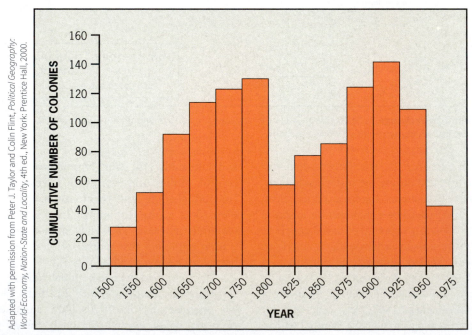

Adapted with permission from Peter J. Taylor and Colin Flint, *Political Geography: World-Economy, Nation-State and Locality*, 4th ed., New York: Prentice Hall, 2000.

FIGURE 2.3 **The two waves of colonialism between 1500 and 1975.** The first wave lasted from 1500 to 1825. Europeans colonized the Americas and the coast of Africa to fuel the triangular trade network and an economy based on agriculture. The second wave lasted from 1825 to 1975. Colonizers took control over much of Africa and Asia to fuel the capitalist world economy based on industry and manufacturing goods.

periods of internal political stability in the fifteenth century. The countries gained wealth through exploration and they expanded their influence to far-flung parts of the world during the late fifteenth century. In the sixteenth century, Britain, France, the Netherlands, and Belgium became colonial powers primarily in North and South America but also in coastal Africa and portions of India and other Asian countries (Figure 2.4).

Colonizers focused on gaining lands to establish large-scale agricultural production in the Americas and to establish trading posts for the Atlantic slave trade in Africa and ports for the spice trade in Asia, creating a nearly global system of trade. The first wave of European colonization gave way to the first wave of independence when several colonies in the Americas obtained independence in the late 1700s and 1800.

In the second half of the nineteenth century, the diffusion of the Industrial Revolution initiated a **second wave of colonization** (Figure 2.3) as Europe's industrial economies stopped at nothing to secure the resources required to grow their factories. The major colonial powers during this time were Britain, France, the Netherlands, Belgium, Germany, and Italy, all motivated by profit, competition, and proselytizing of their religions. At the Berlin Conference in 1884 and 1885, European powers arbitrarily divided up the world's second-largest continent among themselves in the **scramble for Africa**. Many European powers were ruthless in the exploitation of their colonies, often taking the best agricultural land for themselves, assuming control of mining operations, and building infrastructure, especially railways and ports, only to facilitate the extraction of raw materials.

One of the most poignant examples of the commercial (or profit) interests involved in colonialism was the first Opium War (1839–1842) between Great Britain and China in the fertile Pearl River Delta north of Hong Kong. The dispute originated when the Chinese banned opium sales in China, a trade the British controlled. China banned the trade because of the public health problems created by opium addiction in the country. Britain's main concern was the huge trade deficit it was running with China. Britain had high demand for Chinese goods (for tea and silk especially), but China had little to no demand for British goods. Controlling the opium trade enabled Britain to export to China. The Chinese seized opium from British traders in 1839, sparking war. The British prevailed with their superior maritime power and garnered considerable concessions in an 1842 treaty, including Hong Kong Island (which they did not return to China until 1997); freedom to sell opium to Chinese merchants; and the opening of five treaty ports (before this only one port, Shameen Island, was open for trade).

Religious proselytizing as a rationale for colonization was less prominent during the second wave of colonialism than the first, but it continued to be important. During the second wave, colonial administrators and religious missionaries in France often seemed to be in tension over the goal of colonialism. France produced more Catholic missionaries during the second half of the nineteenth century than any other country in the world, and they were sent to work in French colonies in Southeast Asia. A group of French Freemasons accused Catholic missionaries in present-day Vietnam of undermining France's "civilizing mission" (Daughton 2006). French missionaries eventually realized they needed to emphasize their patriotism and their commitment to the work of French empire in order to be allowed to continue their activities; as a result, they performed missionary work and advocated for the French colonial administration at the same time.

The second wave of colonialism lasted until just after World War II, when colonies demanded independence and Europe's devastated economies could no longer afford to control their colonies. The hearth of the second wave of decolonization was South Asia where the Indian National Congress began working for independence in 1885 and Pakistan and India gained independence in 1947 (Figure 2.5). From South Asia, decolonization diffused throughout Asia and into Africa. Between 1947 and 1970, a total of 60 countries became independent (Figure 2.6).

SHIFTING REGIONAL CONFIGURATIONS

Over the centuries, global trade and exchange has ebbed and flowed between different areas of the world. The shifting patterns and intensities of trade often alter and blur the boundaries between regional configurations. During the Roman Empire, the Mediterranean region was a particularly meaningful world region, as trade between Europe and Africa was concentrated in this region.

During the Age of Exploration, stretching from the 1400s to the 1900s, the **Atlantic World** was the focal point for global trade and exchange of culture. Growing research on the Atlantic area (e.g., Carney 2001; Carney and Voeks 2003; Egerton, Games, Lane, and Wright 2007) has focused on the region as a conduit for the transfer of people and ideas.

The Columbian Exchange brought numerous plant materials from Africa to the Americas, including rice, coffee, okra, and cotton. Traders brought American crops to Africa, including corn (maize) and cassava. The large numbers of brutally enslaved Africans brought to the Americas had as powerful an impact in the New World as the Europeans who transported them. Africa's profound cultural influences on the Americas range from agricultural techniques to music and dance traditions to cooking practices.

Some of the best transnational work has actually been done by scholars who have a deep understanding of, and a history of fieldwork in, a particular region. For example, Judith Carney, a geographer at UCLA, is the author of *Black Rice: The African Origins of Rice Cultivation in the Americas* (2002), which won the Herskovits Book Award of the African Studies Association in 2002. Carney effectively demonstrated how African knowledge of tidal

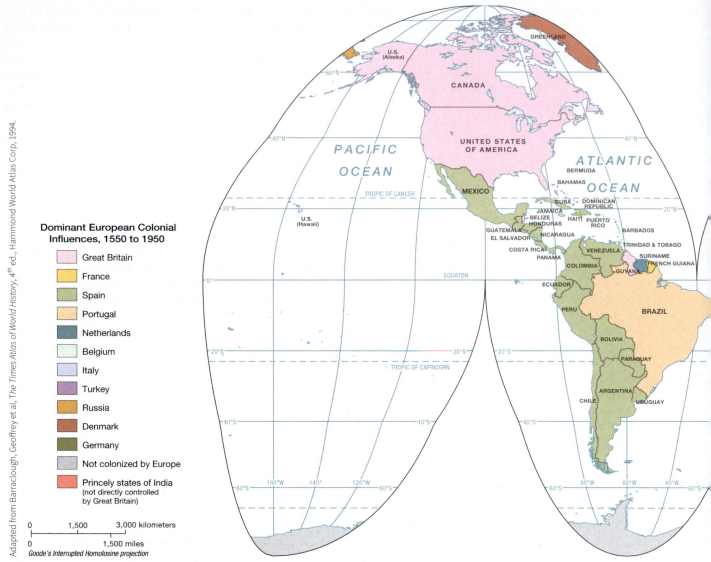

Adapted from Barraclough, Geoffrey et al, *The Times Atlas of World History*, 4th ed., Hammond World Atlas Corp, 1994.

Dominant European Colonial Influences, 1550 to 1950

- Great Britain
- France
- Spain
- Portugal
- Netherlands
- Belgium
- Italy
- Turkey
- Russia
- Denmark
- Germany
- Not colonized by Europe
- Princely states of India (not directly controlled by Great Britain)

0   1,500   3,000 kilometers

0   1,500 miles

Goode's Interrupted Homolosine projection

FIGURE 2.4   Dominant colonial Influences, 1500 to 1975.   Across the two waves of colonialism from 1500 to 1975, Europeans colonized nearly the entire world. The Japanese colonized Korea, parts of China, and several Pacific islands during the second wave of colonialism. Some countries and regions were colonized by more than one power, but this map reveals the most dominant colonial influences throughout the world across these centuries.

mangrove rice cultivation in West Africa was vital for the establishment of highly productive rice plantations in the Carolinas in the seventeenth and eighteenth centuries—an insight that has changed our understanding of African American history (Figure 2.7).

Interestingly, Carney was an Africanist first and only later became interested in the U.S. Southeast. It was her

FIGURE 2.5   Karachi, Pakistan.   Independence ceremony for Pakistan on August 14, 1947. Muhammad Ali Jinnah, the second from the left, was the founding father of Pakistan. He advocated for a Muslim state separate from India, which became a Hindu state. On the far left is Jinnah's sister, Fatima Jinnah. They shared the dais where the formal independence ceremony occurred with the British Lord Mountbatten and his wife, Lady Edwina. The British held an independence ceremony in India the next day and then showed the two countries the map of where the border between them would be on August 17, 1947. A chaotic migration of 7 million Hindus from Pakistan to India and another 7 million Muslims from India to Pakistan followed. One million people died in the resettlement.

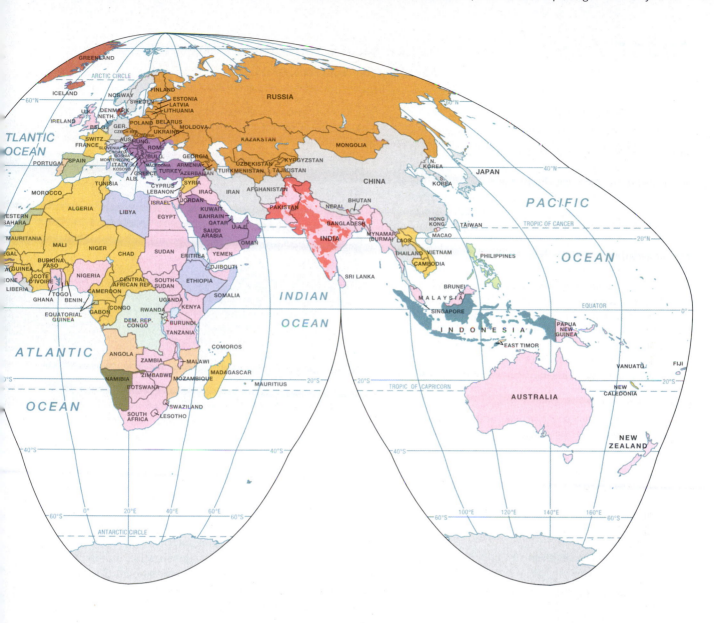

detailed understanding of rice cultivation in West Africa that allowed her to recognize the agency of Africans in the historical development of rice farming in the Carolinas. Without a previous background in African agriculture, Carney's exploration of the African connections to the historical development of rice farming systems in the Carolinas could never have been made as forcefully and as convincingly. A deep understanding of a particular place gave her the foundation on which to build her fine transnational scholarship.

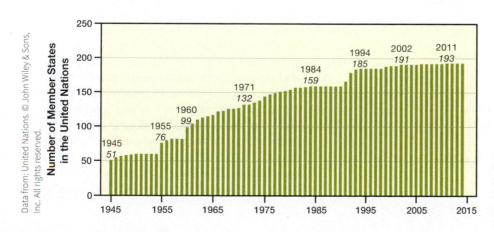

FIGURE 2.6 **United Nations member states 1947 to 2014.** The number of member states in the United Nations has grown from 1947 to the present.

GUEST *Field Note*   Agricultural Connections between West Africa and South Carolina

DR. JUDITH CARNEY
University of California, Los Angeles

More than 3,000 years ago Africans independently domesticated a separate species of rice, *Oryza glaberrima*. African rice arrived in the Americas during the trans-Atlantic slave trade (circa 1450–1808), when ship captains purchased African foodstuffs as provisions for slave voyages. Supplies occasionally left over from Atlantic crossings created opportunities for enslaved West Africans to access seeds and plant them as a subsistence preference. In some tropical and subtropical regions such as the colony of South Carolina, which shared similar lowland ecosystems with those where the grain thrived in West Africa, rice eventually became the basis of a prosperous plantation economy. Elsewhere, *O. glaberrima* has been recovered from historical botanical collections, whereas African rice is still grown in remote areas of Suriname by the Maroon descendants of runaway slaves. My fieldwork along the Gambian River in West Africa in 1989 helped me make the connection between rice cultivation in Africa and North America. The Mandinka women in the photograph plant, harvest, and mill rice along rivers and their tributaries as their forebears did centuries ago. Rice growers from West Africa's rice region transferred these techniques to South Carolina, Brazil, the Guianas, and Mexico.

Courtesy of Judith Carney, University of California, Los Angeles.

FIGURE 2.7   Gambia River, The Gambia.   Mandinka women weeding a tidal rice field along The Gambia River.

CONCEPT **CHECK**

1. **Explain** the difference between strong and weak globalization.

2. **Compare** and contrast the first and second waves of colonialism.

3. **Describe** how flows of trade and people have influenced ideas about regional configurations.

HOW DO INTERNATIONAL ORGANIZATIONS CREATE AND SHAPE GLOBAL CONNECTEDNESS?

LEARNING OBJECTIVES

1. **Describe** the major international financial institutions.

2. **Explain** how international institutions, including nongovernment organizations, and actors shape global connectedness.

3. Critically **discuss** the role that local actors play by engaging with global entities.

People create governments and organizations and these institutions establish connections and affect change globally and locally. Increasingly, the international institutions people have made are a force for globalization. In this section, we explore a broad range of international organizations, their influence on the global system, and the impact of local actors on global processes.

INTERNATIONAL FINANCIAL INSTITUTIONS

In the closing days of World War II, representatives of the 44 Allied powers met in Bretton Woods, New Hampshire, to discuss ways to provide for stability and growth in the postwar era. A major concern was rebuilding industrialized economies that had been ravaged by war. World leaders were also keen to prevent another global depression like the one that had occurred in the 1930s. At Bretton Woods they established two international financial institutions, the World Bank and the International Monetary Fund (IMF), and negotiated an agreement to eventually create a third, the General Agreement on Trade and Tariffs (the predecessor to the World Trade Organization), to play a major role in development and trade policy. By informal agreement, the president of the World Bank is always an American and the president of the IMF is always a European.

The **World Bank**, also known as the International Bank for Reconstruction and Development, was initially created to help rebuild Europe following World War II. Its major purpose was to provide loans for reconstruction and development purposes. The Bank has since gone on to be a major source of development funding in poorer countries. While the Bank does provide loans at concessionary (or below-market) interest rates to the poorest countries, these loans must be paid back or recipients face the serious prospect of a reduced credit rating and no future loans. In the 1950s, the Bank often gave loans for large infrastructure projects, such as building dams and highways.

The main function of the **International Monetary Fund** (IMF) is to promote stability of the global financial system by monitoring national policies that influence exchange rates and balance of payments. Exchange rates are the conversion rates between different currencies. A balance-of-payments sheet is the accounting record of all transactions between a country and the rest of the world. The IMF provides loans for short-term balance-of-payments problems. If a country is importing far more than it is exporting, it may grow short on cash. While wealthier countries may borrow funds from commercial banks or sell bonds, poorer countries often do not have such recourse. By providing short-term loans to poor countries, the IMF helps prevent them from defaulting on loans and keeps the international financial system running smoothly.

Both the World Bank and the IMF are largely controlled by their major shareholders, the world's wealthiest countries. Even after voting powers at the World Bank were revised in 2010 to increase the voice of developing countries, the biggest contributors to the Bank still have the majority of voting power. The countries with the greatest voting power are now the United States (15.85 percent), Japan (6.84 percent), China (4.42 percent), Germany (4.00 percent), the United Kingdom (3.75 percent), and France (3.75 percent).

Although the **World Trade Organization** (WTO) was not always considered an international financial institution, it was also an eventual outcome of the 1944 Bretton Woods conference. The General Agreement on Trade and Tariffs (GATT), the predecessor to the WTO, emerged from Bretton Woods. Based on a belief that trade restrictions led to the global recession of the 1930s and that free trade goes hand in hand with economic growth, the GATT promoted free trade. The WTO came into being in 1995 following a set of discussions undertaken to elevate GATT from an agreement to a functioning international institution. The major functions of the WTO are to promote free trade and adjudicate trade disputes.

In addition to major international financial institutions, a number of regional development banks function in much the same way as the World Bank. These include the African Development Bank, Asian Development Bank, European Bank for Reconstruction and Development, and the Inter-American Development Bank Group. The financial capital for these banks is provided by countries in the respective region as well as by some wealthy countries outside the region. Loans are provided solely to countries in the respective region.

International financial institutions (IFIs) exert considerable influence on the global economic system. One of the best examples of their power is the series of policy reforms pushed by these institutions over a 20-year period following the Third World Debt Crisis of the late 1970s. A number of factors brought on the crisis. Many developing countries had become highly indebted because commercial banks had lots of **petrodollars** to loan out owing to the increasingly high profits being generated by the oil exporters of the world.

In the 1970s, the U.S. government no longer tied the value of the dollar to gold. Instead, it entered into an agreement with Saudi Arabia and other oil-producing countries to sell oil in U.S. dollars. Saudi Arabia sold every barrel of oil for U.S. dollars, which created an artificial demand for U.S. dollars. The dollars spent globally for purchasing oil are called petrodollars. When the price of oil rose in the 1970s, more petrodollars moved through the global economy, and commercial banks had excess petrodollars to lend out. Because many countries in the developing world experienced unprecedented growth rates in the 1960s, banks were eager to lend them money.

In the early 1980s, commodity prices dropped and revenue slowed for developing countries. At the same time, inefficient state-run companies had drained state resources, and it proved to be increasingly difficult for developing countries to pay off their loans. The straw that broke the camel's back was rising energy costs in the late 1970s and early 1980s, which further strained the economies of these countries (the majority of whom were oil importers).

The rich countries of the world were gravely concerned by the growing Third World debt, fearing that it would lead to widespread loan defaults. The World Bank and the IMF stepped in with stiff medicine. In exchange for sweeping economic reforms, which came to be known as **structural adjustments**, the IFIs provided loans to the poorest countries and helped reschedule debts. The structural adjustment reforms included dramatic cuts in government spending and reductions in social services, the selling off (or privatization) of state-run companies, a renewed prioritization of the export-oriented sectors of national economies, the slashing of subsidies and tariffs, and the devaluation of many national currencies (the goal of which was to encourage exports and reduce imports).

By the late 1980s, commentators were referring to this reform package as the **Washington Consensus** (named for the headquarters of the World Bank and IMF), as there was widespread agreement among the major donors to the IFIs that structural adjustments would promote economic growth. The term *Washington Consensus* subsequently came to be used much more broadly to refer to "market fundamentalism," or the imposition of an unregulated (or liberal) economic philosophy on much of the developing world. It was difficult for developing countries to reject policy advice from the World Bank and the IMF (which worked closely with the U.S. Treasury Department), because not only did these countries often desperately need the loans offered in exchange for policy reform, but many donors would deny them assistance without the World Bank's and IMF's "seals of good housekeeping." For promoting more open economies (that is, fewer restrictions on trade), many commentators considered the World Bank, IMF, and WTO to be in the front lines of promoting globalization in the developing world.

The orthodox economic positions of the IFIs began to soften by the late 1990s due to their mixed track record of promoting economic growth and to concerns about the social costs of structural adjustment policies (Ashiabi and Arthur 2004). The global financial slowdown of 2008–2010 also created more philosophical space for a less free-market-oriented approach as governments, even in wealthy countries, felt compelled to intervene in national economies.

UNITED NATIONS

In 1945, a total of 51 countries formed the **United Nations** (UN) in the aftermath of the world wars that marred the first part of the twentieth century. The UN Charter stated that its mission was to "maintain international peace and security," "to develop friendly relations among nations," to "achieve international cooperation" in order to solve economic and social problems, and to "be a center for harmonizing the actions of nations in the attainment of these common ends."

Today the 193 member states of the UN deal with a variety of issues that affect world peace, including facilitating cooperation in international law, international security, economic development, social progress, and human rights. The UN tends to be more democratic than the World Bank and IMF because every sovereign state has a vote in the UN General Assembly. The UN also has a Security Council, dominated by only five countries with permanent seats and veto power (the United States, Russia, China, France, and the United Kingdom). The Security Council makes some of the most politicized decisions, establishing peacekeeping operations and international sanctions and authorizing military action.

The UN oversees dozens of agencies that work directly with the governments of member states to promote peace, humanitarianism, and development. Examples of these agencies include the Food and Agriculture Organization (FAO), the UN Development Program (UNDP), the UN World Food Program (WFP) and the UN Children's Fund (UNICEF). These agencies provide technical advice to national governments on development questions (as opposed to loans for development projects), and they report on and track issues of importance under their mission. The UN also runs the International Court of Justice, which settles disputes between countries, and the UN High Commission on Refugees (UNHCR), which protects and supports refugees.

UN institutions are influential for quite different reasons from the IFIs. The UN plays an important role as a forum for international dialogue and the promotion of peace. For example, prior to the UN Security Council voting to send in peacekeeping forces, a number of critical negotiations and discussions take place between member states. Furthermore, UN peacekeeping forces are composed of troops from a variety of countries who must learn to work together. The UN sponsorship of peacekeeping missions is critical as it offers some semblance of neutrality (Figure 2.8). The UN has conducted 69 peacekeeping operations since 1948, with 16 ongoing missions in 2014 (in countries including Sudan, Lebanon, Kosovo, and Syria).

Commentators criticize the UN for being inefficient and bureaucratic. Supporters of the UN contend that any organization attempting to get multiple countries to work together is bound to be slow and deliberative. The UN uses hiring quotas to ensure that any one nation is not overrepresented in a particular agency. Some critics suggest that these quotas deter UN agencies from hiring the most qualified individuals and lead to inefficiency. Another criticism of the UN is that it reinforces state, or country, governments as lead actors in global issues. Nations within countries seeking autonomy or independence, such as the

© Antonio Dasiparu/Epa/Corbis.

FIGURE 2.8   Dili, East Timor.   Australian forces, working as part of the UN peacekeeping force, march during a handover ceremony. The UN peacekeeping force was comprised of forces from 36 different countries, including Australia, Brazil, Bangladesh, Sri Lanka, Philippines, Turkey, Uganda, and Zambia. East Timor is one of the poorest countries in Asia. Portugal colonized the country from the sixteenth century to 1975, making little investment in the place and leaving unilaterally. Indonesia invaded East Timor immediately after its 1975 independence and occupied the country until 1999, when international pressure led to a referendum where East Timorese voted for independence. UN peacekeeping forces helped build the country after its official independence in 2002 and returned in 2006 to keep peace until the end of 2012 in a country where over 100,000 Timorese have died in the violence since 1975.

Kurds of Iran, Iraq, Turkey and Syria, the Tibetans of China, or the Basques of Spain and France, have no representation in the UN, which makes it quite difficult for these groups to make their case for rights or independence.

BILATERAL DONORS, NONGOVERNMENTAL ORGANIZATIONS, AND MULTINATIONAL CORPORATIONS

In addition to the UN and international financial institutions, the world of development is populated by a host of governmental foreign-assistance agencies, **nongovernmental organizations** (NGOs), and **multinational corporations**. Government agency donors include the U.S. Agency for International Development (USAID), the United Kingdom's Department for International Development (DFID), and the Canadian International Development Agency (CIDA). These government agencies fund development projects and programs in other countries, either by working directly with governments or through NGOs.

NGOs are private organizations that do policy, relief, or development work. They may be international, national, or local. International NGOs are typically headquartered in wealthy countries and work in multiple countries. They may be funded by private donations or receive grants from government agencies. These NGOs may do policy or advocacy work, such as Oxfam's efforts to establish better prison conditions for women in Yemen; emergency relief, such as Save the Children's work in southwestern China to help children affected by back-to-back earthquakes; or development projects, such as CARE's efforts to improve the water supply and sanitation in Honduras. National-scale NGOs (such as Teach for America) work within one country and are often staffed by people from that country. Finally, local NGOs or grassroots organizations work in a much more circumscribed area or set of communities, typically focusing on development projects and community organizing, and often funded by larger NGOs or foreign governments. For example, the Sudanese Women Empowerment for Peace, which operates to empower women to advocate for themselves and build capacity to act in government, has received funding for its grassroots movement from the Netherlands since 1997.

Since the early 1980s, the number of NGOs worldwide and their influence have grown. These nonstate actors give voice to otherwise unheard constituencies and lobby for issues that are not a priority for state governments. For example, while 172 governments participated in the UN Conference on Environment and Development held in Rio de Janeiro, Brazil, in 1992, approximately 2,400 representatives of NGOs attended a parallel NGO global forum. The NGOs lobbied hard for the governments to include certain provisions in agreements made at the meeting. From that point forward, NGOs have played a much more active role at UN conferences, such as the 2002 Conference on Environment and Development in Johannesburg, South Africa, or the 2009 United Nations Climate Change Conference in Copenhagen, Denmark.

NGOs are also credited with bringing about reforms at the World Bank. In the 1990s, the World Bank began to require environmental impact statements for new projects as a result of years of NGOs criticizing the adverse environmental effects of World Bank-sponsored projects. NGOs also took the lead in pressuring the World Bank to establish an independent inspection panel to review and investigate complaints related to World Bank-sponsored projects and programs. The Bank established the panel in 1994.

Multinational corporations are also major international actors. Large private banks are a major source of credit for national governments. International companies provide inputs (seeds, fertilizers, and pesticides) for farming and buy agricultural commodities, operate extractive industries (mines, timber operations, and oil wells), contract to build infrastructure (roads, dams, and power plants), and manufacture goods.

Some watchdog groups are concerned that the power of private sector, for-profit actors is too strong. Watchdog groups have set up campaigns against Monsanto, a $44 billion global agriculture company that produces genetically modified crops and the chemicals that help them grow. One group even produced a film entitled *Stop Monsanto*. Monsanto and many farm groups contend that the company's engineering is a necessary, effective, and safe way to feed a growing planet. For example, Monsanto is genetically engineering a soybean to provide Omega 3 fatty acids, which in turn will help avoid the depletion of fish stocks that are currently used to create Omega 3 fatty acid supplements.

HEGEMONS

The World Bank and IMF are based in Washington, D.C., and the headquarters of the UN is in New York City; thus, U.S. interests have played a crucial role in shaping global cooperation on trade, economies, politics, and human relations since Bretton Woods in 1944. Political geographers describe the dominance of one state on the global stage as a **hegemon**. Hegemons are world powers that define and enforce the prevailing rules of the game in a particular era (Agnew 1998). Political scientist George Modelski (1987) outlined four hegemons between 1500 and 2000: Spain in the 1500s, the Netherlands in the 1600s, Great Britain in the 1700s and 1800s, and the United States in the 1900s. The global institutional architecture that influences international relations and economic interconnections is in no small measure tied to the hegemonic power of the day. It is as yet unclear who will be the superpower of the twenty-first century. The United States may continue its dominance for a second century, but many scholars believe that China is most likely to assume this role.

Although the Chinese economy has grown and prospered as a global epicenter of manufacturing, it has required increasing levels of resources from and engagement with other areas of the world to acquire materials. Today, companies owned by Chinese entrepreneurs are some of biggest investors around the world, especially in energy and resource-related industries. To protect their interests, the Chinese government has made large investments in and provided vast amounts of foreign assistance to certain African, Latin American, and Asian countries.

Some commentators view Chinese investments in the developing world as problematic, seeing them as just another round of foreign imperialism (Carmody and Owusu 2007), while others see this engagement as benign, if not positive. These advocates see the Chinese as providing an alternate source of investment and a counterbalance to traditional Western powers. Some scholars suggest that China's willingness to build infrastructure in the developing world is critical for fostering development (Zafar 2007). In contrast, some Americans see China as competing with the United States for resources in the developing world and turning a blind eye to the human rights records of many of these countries.

THE INFLUENCE OF LOCAL ACTORS

Scholars of globalization sometimes tend to think international actors are more important than local actors. Such discussions stem in part from twentieth-century theories that emphasize the long-running global system dating back to the first wave of colonialism, in which the world's wealthiest countries extracted resources from and dominated poorer countries (first through colonial control and then through unfair trading relationships). Although such a lens is useful for understanding the structures of world trade and international finance that privilege wealthy countries, geographers are careful not to attribute every phenomenon to big structural factors.

Geographers recognize the power of local actors to influence events. The **agency** of individuals affects a wide range of global issues, such as the content of national policies or the behavior of multinational companies. For example, at first glance it appeared that the government of Côte d'Ivoire, the Franco-Ivoirian Cotton Company, and global markets had established successful cotton production in Cote d'Ivoire. But upon closer inspection, geographer Tom Bassett (2001) pinned the cotton industry's spectacular growth on the agency of small cotton farmers who, acting individually and collectively, influenced and shaped the government's cotton policy.

Selective adoption is a powerful way for local actors to assert their agency by choosing what global options suit them. For example, in China large-scale supermarkets are quickly entering the city of Shanghai (Figure 2.9). A survey showed that residents selectively adopted supermarkets for the majority of their purchases of canned and dry goods but still relied on traditional vendors to buy fresh foods (Goldman 2000). Similarly, it may seem that World Bank, IMF, or UN programs are introduced to a local community and then adopted without question. The reality is often much more complicated. Local people in many parts of the world have been exposed to outside ideas for centuries. Most individuals or communities examine a new technology or idea to see how appropriate it is for their particular context. Good ideas persist and spread through contagious diffusion; bad ones are discarded. In response to selective adoption, outsiders have sometimes disparaged local communities who resist change or do not adopt new ideas.

In some cases, the external and internal forces at play over a particular issue cannot be neatly sorted into two separate columns. Throughout the world local actors appropriate and reformulate outside ideas (sometimes to very different ends). The interplay between internal and external ideas is sometimes referred to as **glocalization**. A classic example of glocalization is McDonald's adaptation to local customs, for instance, serving wine in France or lamb in the Middle East.

A fascinating example of glocalization occurred during the Arab Spring of 2011, as protesters put global social media sites to local political use. In Libya, the secret police monitored Facebook and Twitter, making it very dangerous to use the social media sites as platforms for organizing the rebellion. To circumvent this risk, revolutionaries shifted their communications over to a Muslim dating site, Mawada. They created fake profiles and communicated via coded messages. For example, the phrase "I LLLLLove you" meant that the person had five other rebels with him, one for each L in LLLLLove. The rebels in Libya created a new use and language for the dating site in response to the specific situation in their country (Kofman and Heussner 2011).

Imaginechina/AP Photo.

FIGURE 2.9   Shanghai, China.   Auchan is a French grocery company that operates in more than ten countries globally, including China. The first store Auchan opened in China was in Shanghai, and today the company reports that it has 59 Auchan hypermarkets, including this branch in Changyang, and 264 RT Mart hypermarkets, employing over 115,000 people in China.

CONCEPT CHECK

1. **Why** were the Bretton Woods institutions created in the wake of World War II?

2. **How** have international institutions shaped global connectedness?

3. **How** does local agency temper and change international initiatives?

HOW ARE PLACES CONNECTED, AND WHY ARE SOME PLACES MORE CONNECTED THAN OTHERS?

LEARNING OBJECTIVES

1. **Describe** how networks influence connectivity.
2. Critically **discuss** how global processes connect places.
3. **Explain** how local factors may influence connectivity.

The production of goods and services feeds off and creates connections. Whether and how two places are linked or networked depends on a number of factors when looking at economic production. The **hearth** of a new good or service is the origin or the place that initially produces the good. Ideas and goods will diffuse first to the places most connected to the hearth.

NETWORKS AND GLOBAL CONNECTIONS

A **network**, in the most general sense, is the system of links among places, institutions, or people. A traditional way to think about networks in geography is in terms of transportation infrastructure, such as roads, rails, and canals, and scheduled transit services, such as buses, trains, and airplanes. The form and existence of these networks exert a major influence on the connectivity between different places and the flow of resources from one place to another.

A network's orientation may affect movement in some directions over others. For instance, many transportation networks established in colonies simply went from a point of resource extraction (such as a mine) to a port city. Argentina's railway network established by 1948 is a case in point (Figure 2.10). According to geographer David Keeling,

---

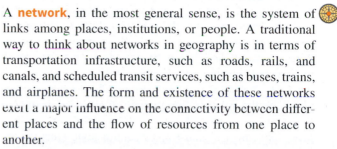 **network** a system of connections among people and places.

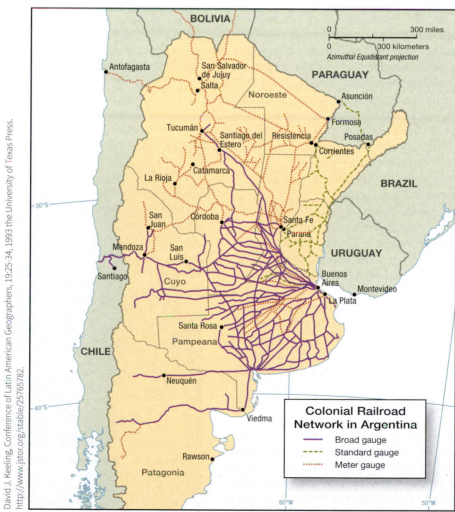

David J. Keeling, Conference of Latin American Geographers, 19:25-34, 1993 the University of Texas Press. http://www.jstor.org/stable/2576782.

FIGURE 2.10 **Railroad networks in Argentina.** Designed to move goods from the interior to coasts for export, rail lines are concentrated in port regions. Argentina built railways into all areas of the country with the goal of helping them develop. Despite the presence of rail lines in northwest Argentina, the towns in the northwest "remained isolated, connected by poor to non-existent roads and terminally slow overland travel to railheads" (Keeling 1993, 28).

Argentina's railroad network established an "overreliance on export trade" for the country. The railroads connected cities in the northwest to the ports to help export resources and import consumer goods. Port cities, including Buenos Aires, Rosario, and Bahia Blanca, are zones of export, import, and consumption (Keeling 1993).

Transportation networks often reflect *urban hierarchies*, or the relative importance of cities over smaller towns and rural areas. For example, a *radial network* may favor movement to and from a central node (such as the national capital or an important port city) over circumferential movements between nodes (such as between provincial seats of government). At the scale of (especially pre-World War II) cities, one often sees a radial network of commuter trains that goes out from the **central business district** (or main concentration of businesses) of a city to residential neighborhoods and suburbs. Such public transportation networks are efficient if one wants to go to and from the central city, but they are challenging to use between suburbs. The growth of suburb-to-suburb commuters in some cities (such as Washington, D.C.) has led to massive traffic and congestion.

A variation on the radial network is the *hub-and-spoke system* with a handful of central nodes (or hubs) connected to smaller points. Airlines use this system (Figure 2.11). For example, a resident of St. Paul, Minnesota, can fly directly to most areas of the country from Delta Airlines' hub airport of Minneapolis-St. Paul. But a resident of Aberdeen, South Dakota, has to fly from her spoke airport to the hub at Minneapolis-St. Paul, on one of only two flights a day, before flying on to her final destination. Because the costs of flying to and from spoke destinations are often higher in terms of time and money, people at these locations are at a disadvantage.

Since governments construct transportation networks to respond to market demands to carry goods and people between certain nodes, they generally reflect hierarchies of economic power, whether globally or within a country. Transportation networks may also be intentionally designed to enhance connectivity for some communities and to marginalize others. A classic example of this occurred during the high-Apartheid years in South Africa between 1946 and 1994, when the white minority government moved blacks out of white-only areas and invested in transportation networks that favored white over black areas (Figure 2.12).

Globally, **accessibility**, the travel times to urban centers or key services, has been improving over time. A study by the European Commission (2008) showed that only 10 percent of the world is more than two days' travel from a city of 50,000 people or more (Figure 2.13). According to this analysis, the Tibetan Plateau is the most remote place in the world (in terms of the number of people who live far from a city of 50,000 or more).

Access to global transportation networks is a crucial factor in any country's overall economic condition. The

© Erin H. Fouberg.

FIGURE 2.11a   Aberdeen, South Dakota.   The departures and arrival board at the Aberdeen regional airport in South Dakota displays one departure and one arrival, connecting the spoke of Aberdeen only to the hub of Minneapolis. Delta Airlines is the only carrier that serves Aberdeen, and the only place a passenger can fly commercially out of Aberdeen is to Minneapolis.

© William Moseley.

FIGURE 2.11b   Minneapolis, Minnesota.   The departure board in the Minneapolis Airport shows hundreds of flights departing directly to locations throughout the United States and globally. The Minneapolis Airport is a hub in Delta's hub and spoke system, so flights from dozens of feeder regional airlines arrive and passengers depart from this airport for their final destinations or larger airline hubs.

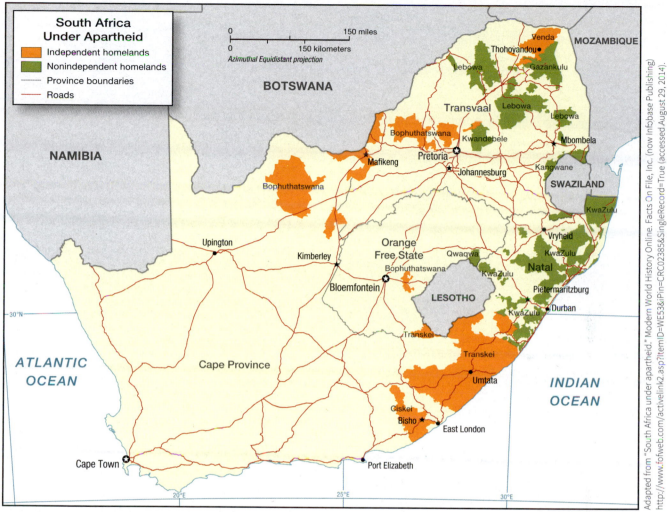

FIGURE 2.12 **Homelands and transportation networks in Apartheid South Africa.**   During Apartheid, the South African government restricted blacks to living in homelands designated by the government (colored orange and green on this map). Transportation networks largely skirted around the homelands, instead connecting major cities and areas where whites lived with one another. The design of South Africa's transportation system disconnected black homelands from the rest of the country.

Adapted from "South Africa under apartheid." Modern World History Online. Facts On File, Inc. (now Infobase Publishing) http://www.fofweb.com/activelink2.asp?ItemID=WE53&iPin=CRC02385&SingleRecord=True (accessed August 29, 2014).

cheapest way to transport heavy goods and materials is by sea, so a country's access to a seaport can be critical to its success in the global economy. In the world's 43 **landlocked** countries (Figure 2.14), exporters have to transport goods by rail or road through another country before arriving at a seaport. Not only distance but also the time and money needed to cross country borders make such transportation expensive.

Information technology networks are an integral part of contemporary globalization. While the Internet has connected the world's far-flung places, considerable disparities in bandwidth mean that some people have better access to Internet and communication technologies than others (Figure 2.15), a problem known as the **digital divide**. While commentators have argued that mobile phones allow the poorest people to leapfrog over intermediate technologies, such as land lines for phones, and help narrow the digital divide, others have suggested that these technologies may reinforce existing power relations (Otiso and Moseley 2009).

When Internet and communications technologies are only accessible to a society's wealthier members, and when the languages, software, and hard technologies reflect the elite's skills and desires, new technologies may reinforce existing power dynamics. The development of the telegraph and rail in the late 1800s and information technology networks in the late 1900s brought improved transportation and communication technologies that deepened connectivity among some and bypassed others.

Information technology networks contribute to **time-space compression**, a term geographer David Harvey (1990) coined to describe processes that accelerate the experience of time and reduce the significance of distance. Harvey saw time–space compression intensify at certain moments in history when rapid changes in technology and economic systems took place in the late 1800s and late 1900s. During these periods, improved transportation and communication technologies deepened connectivity among some and bypassed others.

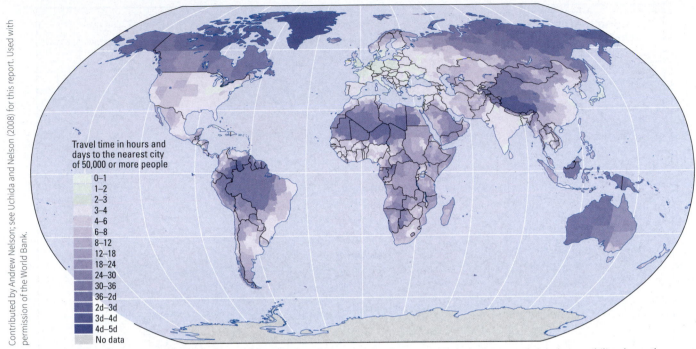

Travel time in hours and days to the nearest city of 50,000 or more people

0–1
1–2
2–3
3–4
4–6
6–8
8–12
12–18
18–24
24–30
30–36
36–2d
2d–3d
3d–4d
4d–5d
No data

FIGURE 2.13   Global accessibility.   Created by the European Commission and the World Bank, the map of global accessibility shows the travel time needed to reach cities of 50,000. Globalization has most increased accessibility to the cities in the lightest colors on the map, while cities and areas in the darkest blue are distanced by lack of accessibility to the world economy.

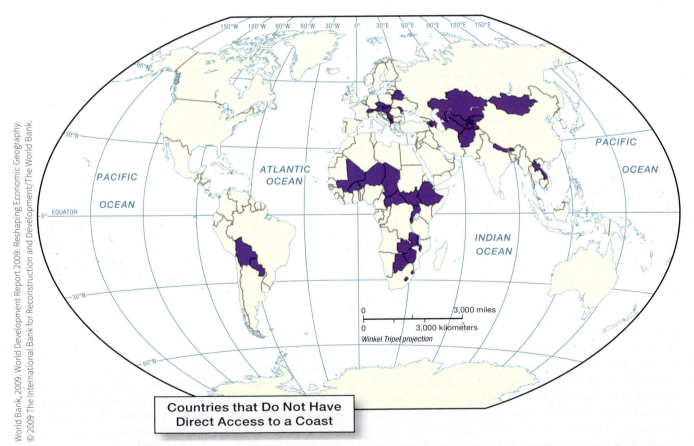

PACIFIC OCEAN

ATLANTIC OCEAN

PACIFIC OCEAN

INDIAN OCEAN

EQUATOR

0   3,000 miles
0   3,000 kilometers
Winkel Tripel projection

Countries that Do Not Have Direct Access to a Coast

FIGURE 2.14   Landlocked countries.   Forty-three countries have no direct access to a coast. Being landlocked does not alone create poverty in a country. Switzerland and Luxembourg are both landlocked and are two of the wealthiest countries in the world. The World Bank claims that being landlocked deters development for countries "with poor neighbors" or for countries that are "far from markets."

FIGURE 2.15 **The Geography of the Internet.** TeleGeography mapped the geography of the Internet, examining connections, bandwidth, and traffic among cities and regions across the world. The Internet functions as a hub-and-spoke system, with the greatest amount of inter-regional traffic flowing through major world cities. The lines on the map are proportional to the bandwidth of flow between cities. The pie graphs on the map show the proportion of traffic within and between regions. The gray is always the traffic within a given region. Reading the pie charts, Europe has 77 percent of its Internet traffic within Europe and 15 percent with the US and Canada (blue). The US and Canada has the most interregional connection with Europe (blue) and second most with Latin America and the Caribbean (brown), and 25 percent with Asia and the Pacific (yellow). Latin America and the Caribbean show an interesting pattern of 85 percent of their Internet traffic with the US and Canada. In fact, 60 percent of all of Latin America and the Caribbean's interregional Internet traffic is with the world city of Miami, Florida in the United States. Asia and the Pacific has more than 40 percent of its interregional traffic with the US and Canada (yellow) and 30 percent with itself (gray) and 26 percent with Europe (purple). Africa has more than 80 percent of its traffic with Europe (green), 9 percent with Asia and the Pacific (orange), and 3 percent with the US and Canada (red).

Contemporary communication technologies, including cell phones, Facebook, YouTube, blogs, Twitter, Skype, e-mail, and e-mail listservs, allow social networks to traverse great distances at very low cost. Artists, authors, intellectuals, performers, and scientists are able to share their work via Internet video, chat rooms, and podcasts, changing the way that artistic and scientific achievements are disseminated, discovered, and discussed. NGOs and grassroots movements rely on contemporary communication technologies to mobilize political support at the international level in an unprecedented way.

Social networks and technologies have had a revolutionary effect on the diffusion of ideas and ideals. In an interview with the Canadian Broadcasting Company in 1997, Nobel Peace Prize winner Jody Williams described sending out hundreds of faxes around the world, each day, over several years, to garner support for her campaign to ban and clear antipersonnel mines. Affordable faxing was a revolutionary form of communication in the late 1980s and early 1990s, and

it was critical in helping Williams mobilize her campaign. Now faxing is seen as expensive and antiquated.

A decade later political movements have a whole new set of technologies at their disposal. For example, pro-democracy forces in Iran launched an amazing grassroots protest following that country's disputed presidential election in June 2009. In the midst of a government crackdown against protestors, ordinary people using cell phone technology were able to send photos, videos, and Tweets from the street, mobilizing international awareness and support. Still, the potential democratizing effects of Internet and communication technology networks are limited in many places by bandwidth, government censorship, and the underlying power structures of cell phone access.

GLOBAL PROCESSES AND LOCAL PLACES

Local phenomena are rarely just local. They are connected to processes operating at a variety of scales. Global environmental change, international markets, regional trade relations, national policies, and local community dynamics all interact to produce a local outcome. A case described by the geographer Tom Bassett (1988) for Côte d'Ivoire in West Africa is a good example. Bassett studied conflicts between farmers and herders in the northern parts of the country, a problem that might seem to be quite local. Bassett came to learn that, in order to avoid importing costly European meat, the government instituted policies to encourage self-sufficiency in meat production that were, in part, responsible for encouraging Fulani herders from Mali and Burkina Faso to resettle in Côte d'Ivoire. A drought in the Sahel region of West Africa to the north of Côte d'Ivoire also facilitated the migration of herders into the country. The drought-related loss of livestock had been leading herders to push further south in search of wetter conditions. These two causal factors (one national and one regional) meant that more and more herders settled in areas of northern Côte d'Ivoire dominated by Senufo farmers. The farmers, who were working under very slim profit margins for the cotton they produced (as a result of national and international commodity pricing policies), came into conflict with growing numbers of herders looking for cattle pasture. The concept of scale helps us understand local phenomena by analyzing linkages in space and time.

In many cases, global economics, national policies, and local behavior come together to produce particular landscapes that in turn influence behavior, policy, and economics. The relationship between energy consumption patterns in the United States and its post–World War II development patterns is a good example of this. The United States has the highest level of per capita energy consumption in the world. How did it get this way? A petroleum-based form of development emerged in the United States in the 20th century. Following the rise of the auto and petroleum industries, the government invested massively in the federal highway system, which greatly encouraged auto-based travel within cities. Because the new systems included ring roads around cities, the transportation system also facilitated an increasingly diffuse or low-density urban form on the outskirts of cities known as suburbs. Americans built suburbs with federally backed housing loans for largely white middle-class families. Public transportation within urban areas also declined significantly in all but the largest American cities between 1950 and 1970. The degree to which U.S. auto companies were complicit in the demise of public transit options is contested. What is clear is that the federal government heavily favored autos by funding road building and reducing subsidies for trains and trolleys.

Cheap oil changed American society to the point where U.S. citizens felt they could not survive without it. Suburbs, large roads, limited sidewalks, and little to no public transportation characterize the country (Figure 2.16). Most Americans literally cannot function without a car and access

FIGURE 2.16a   Mesa, Arizona.   Mesa is a suburb of Phoenix and is home to 452,000 people. People rely primarily on automobiles to navigate the greater Phoenix metropolitan statistical area, which expands over 16,500 square miles and includes 4.5 million people.

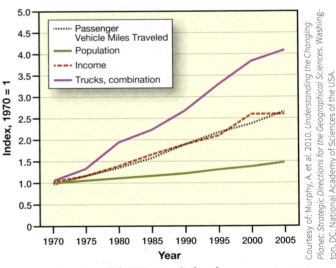

FIGURE 2.16b   Growth in U.S. Population, income, passenger vehicle miles traveled, and combination truck miles traveled, 1970–2005.   The growth in income and vehicle traffic correlates well, while both outpace the growth in population between 1970 and 2005.

to cheap gasoline. As a result, American society has been rocked every time global energy prices rise significantly. The formation of a cartel of petroleum exporting states (the Organization of Petroleum Exporting Countries or OPEC) in the 1970s and the Iran–Iraq War in the early 1980s both sent oil prices soaring and Americans howling.

In 2008, a booming global economy and a problem known as "peak oil," where declining global oil reserves limit the ability of oil economies to keep producing more oil, also sent energy prices skyrocketing and some Americans clamoring to reduce gasoline taxes. This addiction to low-cost fossil fuels has had grave consequences for the United States and the rest of the globe. For decades, the American government has supported undemocratic regimes around the world (such as Saudi Arabia, Nigeria, and Equatorial Guinea) because of their need to maintain cheap supplies of energy. That said, each time a gasoline price shock has occurred, Americans have also changed their behavior by increasing their use of public transportation and purchasing more fuel-efficient vehicles. Policy analysts question why the U.S. government has not done more to support Americans in their periodic, crisis-driven attempts to become more energy efficient.

LOCAL FACTORS AND GLOBAL CONNECTIVITY

Local factors have a significant influence on a place's degree of connectedness to larger, even global, systems. **Site** refers to the basic physical attributes of a location that make it attractive for human habitation or economic activity. For example, a good site might have fresh water, arable soils, and a disease-free climate. **Situation**, on the other hand, is the position of a place relative to others in a network. Venice, Italy, is a classic case of city that once prospered because of its situation rather than its site. Venice rose to prominence as a major maritime and trading power in the eleventh century. While its site is a series of undesirable swampy islands, it had an excellent situation on the Adriatic Sea in northeast Italy between the mouths of the Po and the Piave rivers, for capitalizing on the trade between the Middle East and Europe (Figure 2.17).

A city's situation may be a product of decisions at the local level. For example, Atlanta, Georgia, is an important city today in part because it built a major international airport just as flight was becoming a critical mode of transportation. The city's airport has been the busiest passenger airport in the world since 1999, giving Atlanta an excellent situation in many trading and business networks.

The situation of a city in global networks may also be a function of the businesses that cluster there. For example, partly because a number of prestigious and powerful fashion houses, including Chanel, Louis Vuitton, and Dior, are based in Paris, the city is considered the fashion capital of the world. Cost advantages accrue to individual firms that cluster together in the same location. These **agglomeration effects** include sharing ideas and talent with one another, as well as attracting related firms that supply them with inputs or buy and help them sell their product. Fashion houses minimize transport and communication costs by sourcing cloth, thread, leather, or buttons from nearby businesses. In turn, Parisian specialty boutiques are the first to market the houses' creations to high-end customers, and local fashion magazines (like *Vogue*) diffuse and market the houses' ideas. The relationship between a company and its suppliers is a **backward linkage**, whereas the relationship a firm has with purchasers or disseminators of its output is a **forward linkage**. Enough fashion houses clustered together may also attract ancillary services. These are more general services, including maintenance, secretarial services, and security, which may also supply industries other than fashion houses.

Not just cities but also regions may be known for their economic activity or affluence. Historic manufacturing belts have included the northeastern and southern Great Lakes regions of the United States, northern Europe, central Japan, and the Sao Paulo and Rio de Janeiro region of Brazil. A combination of external or structural factors (such as dualism, discussed in Chapter 3) and local factors together created these regions of high economic activity.

An early start in economic development can yield advantages, known as **initial advantage**. The basic idea is that areas that decide to do something first, such as manufacture computers, face little to no competition and may gradually build an environment (such as the agglomeration economy) that is conducive to further success. As places with initial advantage prosper, it becomes increasingly difficult for new places to compete.

De Agostini/Publiaer Foto/De Agostini Picture Library/Getty Images.

FIGURE 2.17   **The site of Venice, Italy.**   Venice is a group of swampy islands in the Venetian Lagoon, protected by a series of barrier islands from the Adriatic Sea. The site, basically at sea level, is not ideal for a city, as it makes it prone to flooding. However, the situation of Venice, between the Middle East and western Europe, helped Venice grow as an important trading port during the Middle Ages.

Reading the **CULTURAL** Landscape

Agglomeration Effect in Silicon Valley, California

FIGURE 2.18   Menlo Park, California. Facebook employees enjoy lunch in the courtyard of the Facebook campus. In the background, the Smokehouse Barbecue shack offers fresh barbecue with all the fixings. The 57-acre campus was once home to Sun Microsystems. Facebook transformed the offices with the help of Gensler architecture firm to create a balance of open spaces that encourage collaboration and private spaces where employees can focus and complete work.

Jessica Brandi Lifland/Polaris/Newscom.

Silicon Valley, California, experienced initial advantage in computer and hardware software engineering during the twentieth century. The region was the hearth of Hewlett-Packard and Apple. Working in concert with researchers at Stanford University and drawing from graduates of the prestigious university, start-up computer companies clustered in Silicon Valley in the 1960s and 1970s. Silicon Valley still experiences an agglomeration effect, as designers, parts companies, marketing companies, and head hunters cluster in the region to serve major corporations. Now home to Google and Facebook, Silicon Valley continues to attract new service industries designed to simplify the lives of busy employees in the region and to sell them high-priced consumer goods they can afford when their company's stock is doing well. Campuses such as Facebook's, pictured here, offer restaurants like the barbeque smokehouse in the background, to make staying at work easier on employees. Although houses in Silicon Valley may look like average homes in any other neighborhood in U.S. suburbs, they sell for a minimum of $1.5 million. Even bike shops are deceptively expensive. VeloTech Cycles, which stands behind a nondescript façade, sells Italian bicycles that retail for upwards of $10,000.

When Silicon Valley, California, developed as a region of high-tech innovation in the 1980s, it drew talent from all over the world, including large numbers of software engineers from places like India. Even today, despite the presence of other high-tech hubs in the United States, Silicon Valley continues to draw one-third of all the country's venture capital investment (Figure 2.18). As such, it could be argued that some parts of the United States and the world felt Silicon Valley's **backwash effects** when it drew capital and talent away from potential competitors. The concept of backwash effect, or the negative impact on a region from economic growth in another region, helps to inform the core–periphery dichotomy in Chapter 3.

Sometimes economic activity in one region slows growth in another, and at other times it spurs it. China is an example of a country that has catapulted itself along a fast track of development by producing cheap manufactured goods for wealthy regions in the world, including the United States and Europe. Economic growth in the first region increases demand for food, consumer products, or manufactures. When demand cannot be met internally, it spurs economic activity elsewhere. Cheaper land and

labor in another region may inspire entrepreneurs to invest there. Growth in China has now progressed to the point that its vast internal market of middle-class consumers can sustain many Chinese firms, not to mention firms in others parts of the world. As a result, agriculture and fossil fuel production in Central Asia are rising to meet China's demands. Sometimes the process of economic growth in one region spurring activity in another, also known as **spread effects**, is so strong that it creates self-sustaining economic growth in the other.

Development can jump-start in a new area when a novel investment opportunity, or cluster of technologies, emerges and a particular region is first to engage this opportunity, thereby gaining an initial advantage over areas that try to do the same later. For example, Detroit became the leading auto manufacturing center in the United States by being first to seize on this new industry in the early 1900s, just as Mumbai, India (also known as Bollywood) became the center of Indian filmmaking in the 1930s. During these windows of locational opportunity, a combination of willing investors and entrepreneurs made the establishment and growth of the industries in these regions possible.

When production of a good diffuses beyond its hearth and the window of locational opportunity closes, hearth areas decline and become gradually less connected to the global economy. Flint, Michigan, was a prosperous American auto manufacturing city until the 1980s. In the face of stiff competition from Japanese firms, auto manufacturers started moving their plants to other parts of the world where costs were lower, such as Mexico, precipitating Flint's rapid decline. Flint and other established industrial areas often have higher wages, property values, levels of congestion, and/or taxes. Firms can only afford these higher costs if they can charge a higher price for their product, and competition from other areas often makes this difficult. The higher costs then push firms to go elsewhere, leading to declining postindustrial landscapes.

CONCEPT CHECK

1. **How** do transportation and communication networks influence connectivity?

2. **What** types of global-scale phenomena connect places?

3. **Why** are local geographical factors able to influence connectivity at broader scales?

HOW IS GLOBAL ENVIRONMENTAL CHANGE AFFECTING THE WORLD?

LEARNING OBJECTIVES

1. Critically **discuss** different dimensions of global environmental change.

2. **Explain** vulnerability to environmental change.

3. **Explain** the problem of double exposure.

Long before the production of goods was globally networked, environment connected far-flung parts of the world. Earth's atmosphere, oceans, and lithosphere are globally linked so that change in one area can affect other areas of earth. For example, air pollution created in one state does not stay within the borders of that country. Global winds move air pollution from its origin across continents and oceans. In 2014, scientists estimated that between 12 and 24 percent of the air pollution in Los Angeles, California, originates in China.

GLOBAL ENVIRONMENTAL CHANGE

Global environmental change has many different dimensions, and one of the most significant is **climate change**.

Earth's climate changes naturally over time, and the planet goes through warm and cool periods. Scientists are concerned about human-induced (or anthropogenic) changes that go above and beyond Earth's natural fluctuations. The difficulty in distinguishing between natural and anthropogenic climate change has led to considerable debate. However, an increasingly global consensus has emerged since 2001 that humans are the cause of the most recent climate change. In 2001, the Intergovernmental Panel on Climate Change, an international body of climate scientists, published its position, stating: "An increasing body of observations gives a collective picture of a warming world and other changes in the climate system. There is new and stronger evidence that most of the warming observed over the last 50 years is attributable to human activities" (IPCC 2001).

Earth's lower atmosphere is warming at unprecedented rates over the last 1,000 years. In fact, the level of climate change is predicted to be so significant that this warming, along with a number of other human-induced environmental modifications, have led some scholars to refer to the current era as the **anthropocene**. The influence of human behavior on Earth since the Industrial Revolution in the late 1700s is so significant that it constitutes a new geological era (Figure 2.19). The current geologic era, the Holocene, which began 10,000 years ago has been replaced by the Anthropocene, with "anthropo" meaning human and "cene" meaning new.

Earth's atmosphere regulates the flow of energy between space and Earth's surface. Sunlight passes through the atmosphere relatively unimpeded. Earth absorbs solar radiation and then releases it at longer wavelengths. In the lower layer of the atmosphere, the troposphere, greenhouse gases cluster and absorb and reemit the energy as longer waves. Some of the energy reemitted by greenhouse gases comes back to Earth's surface, creating a natural process known as the greenhouse effect (as glass on a greenhouse keeps the heat in). The greenhouse effect keeps Earth at a livable temperature, about 59 degrees Fahrenheit (33 degrees Celsius) warmer than earth would be without the greenhouse effect.

The beneficial, natural greenhouse effect becomes a problem when it is amplified. Several components in the atmosphere are responsible for the greenhouse effect, including water vapor, methane, ozone, nitrous oxides, carbon dioxide, and trace gases. These are collectively known as greenhouse gases. Many of these gases have natural as well as human sources. Fossil fuel combustion, some industrial processes, deforestation, and livestock-raising increase greenhouse gas concentrations, which lead to amplification of the greenhouse effect and facilitate global climate change (Figure 2.20). The increases in greenhouse gas emissions (most notably carbon dioxide) have led to appreciable increases in temperatures in the lower atmosphere (Figure 2.21).

 anthropocene the current geologic era in which humans play a major role in shaping Earth's environment.

Reading the **PHYSICAL** Landscape

The Anthropocene in the Urban Realm

The anthropocene is not confined to cities such as Washington, D.C., where the impact of humans on the environment is clear. Nearly every inch of the city has been purposely built, shaped, or paved by humans. What looks "natural" is not. Bodies of water are lined with concrete, river flow is constrained, and nearly every tree is planted. The Capitol Reflecting Pool (Figure 2.19a) was built in the 1960s and 1970s and designed by Skidmore Owings and Merrill to reduce traffic around the U.S. Capitol. At the same time, the city built the Third Street tunnel, which moves traffic under the mall.

The rural landscape of the Washington metropolitan area looks more "natural" (Figure 2.19b), but humans have shaped this physical environment as much, and perhaps even longer, than they shaped the city. Iroquois Indians farmed the area of Leesburg, Virginia, for centuries before 1722 when they ceded the land to the Virginia colony. The situation of the town changed over time from trade route between the Shenandoah and Alexandria in the early years to battleground during the Civil War to breadbasket for Washington, D.C., to a draw for Washingtonians seeking a rural or country life. The fertile soils of the Leesburg area are now home to small hobby and horse farms. In each situation, the people living here shaped and molded the land and environment.

© Stuart Pearce/AGE Fotostock.

FIGURE 2.19a   Washington, D.C. The Capitol Reflecting Pool covers 6 acres on the west side of the U.S. Capitol and is the eastern end of the Mall in the city.

© Jeff Greenberg "0 people images"/Alamy.

FIGURE 2.19b   Leesburg, Virginia. While the landscape of this horse farm may look "natural," it too is an example of the anthropocene as people living here have changed the environment multiple ways over time to suit their purposes.

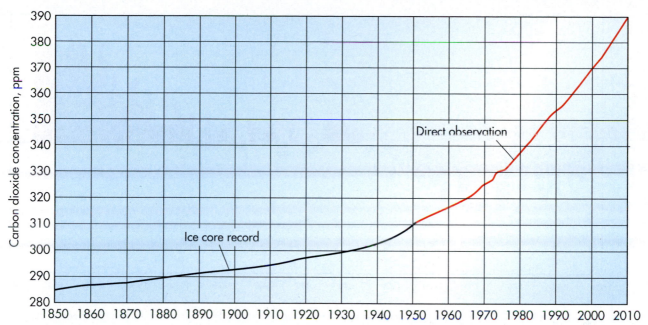

Sources: Pre-1958: A. Neftle et al., Historical CO2 record from the Siple Station ice core, in Trends: A Compendium of Data on Global Change, 1994, U.S. Department of Energy. Post-1958: Earth Systems Research Laboratory (ESRL) National Oceanic and Atmospheric Administration (NOAA).

FIGURE 2.20 **Change in atmospheric carbon dioxide, 1850 to 2010.** The amount of carbon dioxide in the atmosphere is calculated from two sources: ice core samples and direct observation. The line on this graph charts the change in the concentration of carbon dioxide in the atmosphere between 1850 and 2010. The release of carbon from storage (in rocks) into the atmosphere through burning fossil fuels is the primary source of the increase in atmospheric carbon dioxide since 1850. Compare this chart with Figure 2.21 and note how carbon dioxide and surface temperatures correlate.

Increasing global temperatures are a problem for several reasons. First, as water expands with heat and polar ice caps melt, rising sea levels threaten the 25 percent of the world's population who live within 1.1 meters of sea level in coastal areas. Sea-level rise will inundate low-lying areas, submerge coastal wetlands, increase coastal erosion, increase flood and storm damage from coastal storms, and increase the salinity of groundwater in low-lying areas (threatening drinking water supplies). Figure 2.22 shows coastal vulnerability to sea-level rise in the United States. Other potential effects of global climate change include: decreases or greater variability in rainfall; expanding deserts; shifting vegetation communities; disruption of agriculture; and spread of tropical disease such as malaria. Some research also suggests that climate change may be linked to the increasing frequency of severe storms and El Niño, which is a climate pattern that occurs across the tropical Pacific Ocean on average every five years. El Niño is associated with floods, droughts, and other weather disturbances in many regions of the world.

VULNERABILITY

Not only will the effects of global climate change be distributed unevenly across the planet, but climate change will also not affect all social groups in the same way. The reality is that some groups of people are simply more vulnerable to climate change than others. Part of this reality has to do with where you live and the physical characteristics of that place. What people do for a living (or their livelihood) and their ability to recover from fluctuating environmental conditions are also important factors to consider.

Three factors contribute to **vulnerability**. The first component is the risk (or probability) that a biophysical disturbance, such as a drought or a severe storm, will occur. Clearly, some places are at greater risk from climate change than others. For example, people living in coastal areas of low-lying islands (such as Micronesia) are much more likely to experience the effects of sea-level rise than those living on higher ground.

The second component is the sensitivity of a livelihood to an environmental change. What people are doing to make a living, and their mix of different economic activities, will affect sensitivity. Farmers, for example, are more likely to be directly affected by declines in rainfall than coal miners. Many households, however, don't just farm, but engage in a variety of activities. In addition to farming crops, they may keep cattle, work in town, and run a small store. These more complex or diversified livelihoods tend to be less sensitive to any given environmental perturbation. In fact, households

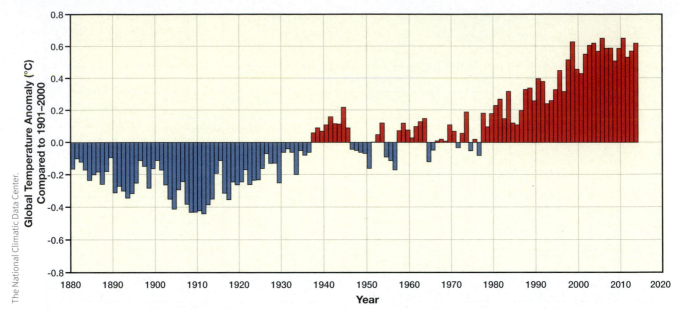

The National Climatic Data Center.

FIGURE 2.21   Trends in global average surface temperature.   The Intergovernmental Panel on Climate Change unequivocally holds that warming surface temperatures since the 1930s are a result of anthropogenic sources of greenhouse gases, especially carbon dioxide.

may intentionally diversify their livelihoods in order to manage the risk of environmental distress. Sensitivity varies considerably within individual livelihood strategies. For example, farmers may plant a variety of crops, some of which are more drought resistant than others.

The third component of vulnerability is a household's ability to recover from an environmental shock, or its resilience. More resilient households may have grain or cash reserves to fall back on in times of difficulty. They may also have a variety of networks they rely on in times

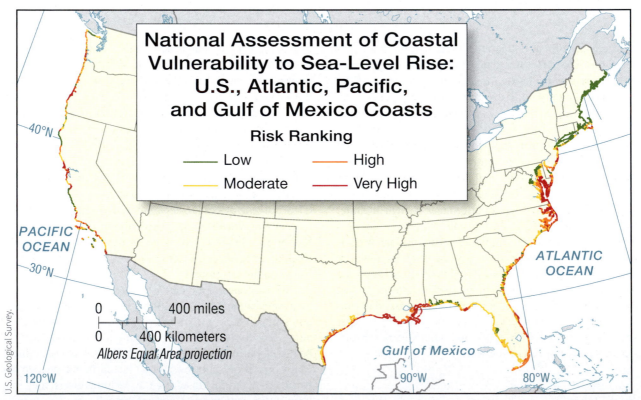

U.S. Geological Survey.

FIGURE 2.22   Coastal vulnerability to climate change.   Working with the Coast Guard, the U.S. Geological Survey derived and mapped the coastal vulnerability index for the United States in order to assess the likelihood of coastal inundation by sea-level rise. The shape of the coastline, history of flooding, change in shoreline over time, and average wave height were among the variables they considered.

of crisis, including extended family, the community, or the national government. Communities or governments, for example, may have grain reserves that they store and save for distribution following severe droughts.

These various components of vulnerability are not entirely determined by locale. In many instances, broader political and economic forces shape or influence where people live and what types of livelihood activities they pursue, as well as their ability to recover from environmental distress. For example, poverty may force some households to live in more vulnerable environments such as unstable hillsides or floodplains. Market prices or private companies may encourage households to grow crops that are not

drought resistant. Finally, municipal or national governments may choose whether to provide social safety nets for their citizens.

NAVIGATING GLOBALIZATION AND ENVIRONMENTAL CHANGE

Rural households around the world are trying to manage their livelihoods and navigate the twin challenges of global environmental change and globalization. Geographers Robin Leichenko and Karen O'Brien (2008) uncovered patterns of advantage and disadvantage among agricultural communities in India that faced **double exposure** to

---

 ## USING GEOGRAPHIC TOOLS

Geographic Information Systems (GIS) Analysis

Any data with locations attached to it can be integrated into a geographic information system (GIS). Geographers collect, assign attributes, and enter data into a GIS. Once each factor in a problem is represented as a layer in the

GIS, geographers can use analysis tools built into the GIS to pinpoint locations with certain attributes. In order to make predictions about what places are the most vulnerable to globalization (Figure 2.23a) and what places have the

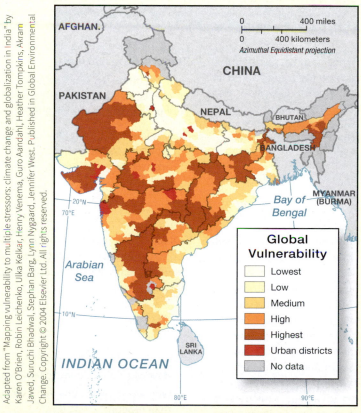

FIGURE 2.23a   **District-level vulnerability to globalization.**   The places with the highest levels of vulnerability to globalization are sensitive to imports and have low adaptive capacity.

(Continued)

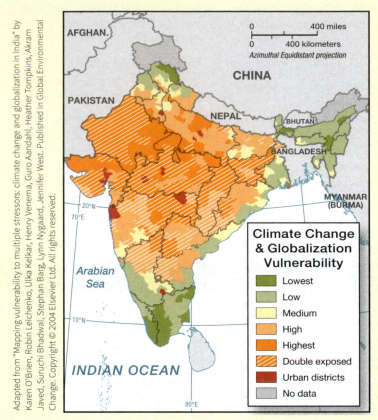

FIGURE 2.23b **District-level double exposure.** Places with double exposure are vulnerable to both climate change and globalization. The background colors of green to orange on this map show vulnerability to climate change. Orange areas that are not cross hatched as 'double exposed' are vulnerable to climate change but not to globalization.

greatest sensitivity to climate change in India, geographers had to collect all the variables that they considered in measuring "vulnerability", starting with elevation and proximity to rivers and coast. Once they created their two vulnerability maps, they used GIS to analyze where two maps overlapped to distinguish the regions with double exposure (Figure 2.23b).

globalization. Faced with both trade liberalization and climate change, some Indian farmers have become more vulnerable to change and some less.

Drawing together data on the effects of climate change on crop yields, changes in plant pollination and competition, and human vulnerability to climate change, O'Brien, Leichenko, et al. (2004) constructed maps revealing the regions that are most vulnerable to predicted climate changes across the country. A geographic information system (GIS) analysis of the relationship between that map and patterns of agricultural export advantage, sensitivity to imports, and the resilience of farmers to socioeconomic change allowed them to identify places across India that are double exposed to climate change and trade liberalization and places that are less exposed, and therefore likely less vulnerable in the face of these processes (Figure 2.23).

Thinking Geographically

1. In addition to elevation and proximity to rivers and coasts, what other factors would you map in assessing vulnerability to global climate change?

2. O'Brien and Leichenko applied the concept of double exposure to India. Is the region where you live double exposed to global processes?

Simultaneous processes of broad-scale environmental and economic change are doubly challenging some rural communities, making their ability to thrive all the more unlikely.

CONCEPT **CHECK**

1. **How** will global climate change likely impact communities?

2. **Why** are some groups more vulnerable to climate change than others?

3. **What** twin challenges does "double exposure" refer to, and how best, in your opinion, might this problem be managed?

SUMMARY

What Is the Nature of Globalization, and How Is It Impacting the World System?

1. Globalization is a long-standing process of economic and cultural exchange that has unevenly influenced connectivity between different places over time.

2. The view of weak globalization tracks the process of globalization back to the 1500s and the first wave of colonialism. The strong globalization viewpoint holds that the level of corporate involvement in the state since the 1970s is a fundamental shift and that modern globalization began then.

3. Colonialism is the process of one territory physically taking over and controlling another territory. Colonialism had a chronology of two waves and a geography. The first wave of colonialism took place between 1500 and 1850, with European colonies largely in the Americas. The second wave of colonialism took place between 1850 and the 1960s, with European colonies largely in Africa and Asia.

How Do International Organizations Create and Shape Global Connectedness?

1. International financial institutions, including the World Bank and the IMF, have shaped economic development through policies including structural adjustment, whereby governments of developing countries were required to open their economies to foreign direct investment and privatize state-run institutions in exchange for loans.

2. Fifty-one countries formed the United Nations (UN) in 1945 to promote peace and dialogue among countries. The UN Security Council can deploy peacekeeping forces in crisis areas, and the peacekeeping forces are composed of forces from across the countries that are members of the UN.

3. A hegemon is a state that is a world power in a given era. As the hegemon, a state can shape the laws and norms of trade and politics globally. The United States was the hegemon in the twentieth century. The hegemon for the twenty-first century is still debatable, although many scholars look to China.

How Are Places Connected, and Why Are Some Places More Connected than Others?

1. Globalization is made possible and shaped by a multitude of networks that operate to connect places and peoples. Networks include transportation and communication routes and organizations that operate globally such as nongovernmental organizations (NGOs).

2. Location of places on networks helps determine their accessibility to other places, the flow of information, and the hierarchy of power. In a networked, globalized world, local problems often have global, regional, and state contexts that, when understood, help us analyze the local issue.

How Is Global Environmental Change Affecting the World?

1. The current era in geology is known as the anthropocene because of the central role humans play in shaping earth's environment, from land use to global climate change.

2. Geographers employ the concept of vulnerability to analyze the potential impact of global processes on local places, whether rising sea levels or trade liberalization. How a household or country bounces back from environmental shock is a measure of resilience.

Geography in the Field   YOUR**TURN**

This photo of a soccer fan in Johannesburg, South Africa, at the 2010 World Cup marks the first time an African country had hosted this global sporting event. Up until June 2010, many people might not have known the name of the horn in this photo (the

Johannesburg, South Africa. A fan blows a vuvuzela in the stands prior to the start of the opening match of the 2010 FIFA World Cup between South Africa and Mexico.

REUTERS/Kai Pfaffenbach/Landov.

vuvuzela) or recognized the South African flag. Was this event good for South Africa? Most commentators contend the 2010 World Cup was a great success for the host country. The country received a throng of international tourists and was covered positively by the media, and concerns about crime never materialized. Within South Africa itself, intense debates considered the pros and cons of hosting this event, although little was heard of this controversy once the euphoria of the World Cup gripped the nation in 2010. Hosting the event required enormous expenditures to build stadiums and infrastructure in a country that lives with great disparities between its rich and poor. At times South Africa made efforts to direct investment into lower income areas, but these attempts were often thwarted by higher level decision makers, including the World Football Federation's governing body (FIFA), which was concerned about crime.

Thinking Geographically

- What has happened to the massive stadiums built throughout South Africa since the 2010 World Cup?
- What are some of the services and businesses that likely agglomerated in Johannesburg during the World Cup?

Read:

Fortune, Quinton. 2014. South Africa spent 2.4bn to host the World Cup. What happened next? *The Guardian*. 23 September.

KEY TERMS

globalization
weak globalization
strong globalization
colonialism
first wave of colonialism
second wave of colonization
scramble for Africa
Atlantic World
World Bank
International Monetary Fund
World Trade Organization

international financial institutions
petrodollars
structural adjustments
Washington Consensus
United Nations
nongovernmental organizations (NGOs)
multinational corporations
hegemon
agency

selective adoption
glocalization
hearth
network
central business district
accessibility
landlocked
digital divide
time–space compression
situation

agglomeration effect
backward linkage
forward linkage
initial advantage
backwash effect
spread effect
climate change
anthropocene
vulnerability
double exposure

CREATIVE AND CRITICAL THINKING QUESTIONS

1. Which perspective on **globalization** do you find more compelling, the weak or strong globalization thesis, and why? In other words, what is the nature of globalization and how long has it been going on?

2. How could European powers use the concept of **identity** to rationalize the second wave of colonialism? Is colonialism dead, or does it persist in different forms today?

3. How have the Bretton Woods institutions (the World Bank, IMF, and WTO) helped establish the **situation** of the United States in the global system? What do you perceive to be the strengths and weakness of the IFIs?

4. What is hegemon? How does **globalization** help create hegemons?

5. How do **networks** apply to your everyday life? How do the form and orientation of transportation networks influence how you move about your community, your country, or the world?

6. How do cyber **networks** affect your life, for example, do you subscribe to e-mail lists, use Facebook, or send and receive tweets? What are the implications of the spatial patterns of these networks and the type and quality of information these provide you (and how might this compare to social networks and information exchange in a pre-Internet era)?

7. Why is it that some places seem to benefit from **globalization** and others do not? Think about the global and local factors that may contribute to a place's "success." How might these patterns (of success or decline) change?

8. Do you think globalization and global climate change are two completely autonomous processes, or might they somehow be connected in the **anthropocene**? If you do see connections among these processes, what might they be?

SELF-TEST

1. Those who subscribe to the "weak" view of globalization believe that:
   a. a global economy has emerged since the 1970s that is significantly different in several ways
   b. the nation-state is increasingly irrelevant in the new global economy
   c. this is a long-standing process that dates back to a time when systems of exchange became more global in scope, starting with the expansion of the Portuguese empires in the 1500s
   d. global corporations are the key actors in the new economy, and they are unencumbered by the constraints of national governments

2. The Atlantic World is all of the following except:
   a. Different from traditional world regions, which are often defined by continental boundaries
   b. An alternative type of region that focuses on the Atlantic basin as an area drawn together by the exchange of people, ideas, and plant material across the ocean

   c. A useful lens for understanding Africa's cultural influences on the Americas
   d. Began with the start of regular transatlantic aircraft flights

3. All of the following are Bretton Woods institutions except:
   a. FAO
   b. World Bank
   c. IMF
   d. GATT/WTO

4. Structural adjustment refers to:
   a. policies pushed by the United Nations to decrease infant mortality rates
   b. policies pushed by the World Bank and IMF to raise revenues and cut government costs in developing countries
   c. policies pushed by NGOs to redress structural inequalities in the global economic systems
   d. policies pushed by UNICEF to attain zero population growth

5. All are true of the United Nations (UN), except:
   a. More countries tend to participate in the running of the UN than the World Bank
   b. Many UN agencies (such as the FAO or UNICEF) work directly with governments by placing technical experts in relevant national agencies
   c. UN agencies mainly provide loans for development projects
   d. The UN provides an important forum for international dialogue and the promotion of peace

6. Within the context of this chapter, agency refers to:
   a. the hegemonic World Bank
   b. the role and power of local actors to influence local and broader scale events
   c. the power of broad structural processes and relationships to influence local events
   d. an aspect not accounted for in glocalization

7. All of the following might be considered a type of network except:
   a. a country's pattern of intercity railways lines
   b. airline flight routes around a continent
   c. the pattern of Internet traffic around the world
   d. the position of a place relative to others

8. The initial locational advantage of a place for economic activity often spirals into a continuous buildup of advantages that attract people and investment to that place. This effect is called:
   a. backwash effects
   b. agglomeration diseconomies
   c. deindustrialization
   d. cumulative causation
   e. creative destruction

9. Which of the following cities likely has the worst "situation"?
   a. Venice in the eleventh century
   b. Atlanta, Georgia (USA), in the late twentieth century
   c. Aberdeen, South Dakota (USA), in the current period
   d. Rome in the second century CE

10. The high-energy consumption patterns in the United States are likely facilitated by all of the following except:
    a. the high-density urban living found in New York City
    b. low-density suburbs
    c. limited public transit options in all but older and larger American cities
    d. pricing structures that favor automobile-oriented transit

11. Which of the following is not a component of vulnerability discussed in the chapter:
    a. the risk (or probability) that a biophysical disturbance will occur
    b. the sensitivity of a livelihood to an environmental perturbation
    c. a household's ability to recover from an environmental shock (aka resilience)
    d. a phenomenon that affects all segments of society equally

12. Within the context of this chapter, "double exposure" refers to:
    a. a problem experienced only in the Southern Hemisphere
    b. households or areas that face the twin challenge of coping with globalization and climate change
    c. cities that have to deal with poor site and situation
    d. an analytical approach that recognizes the influence of broad structural forces and local agency

13. Some scientists have argued that we should call the current era the Anthropocene because:
    a. it is the era in which we discovered global warming
    b. it is time we recognized the importance of anthropology
    c. the influence of human behavior on earth since the Industrial Revolution in the late 1700s is so significant that it constitutes a new geological era
    d. it is the era in which humans discovered how to manage earth's resources sustainably

14. Which of the following is not believed to be a known or potential impact of global warming:
    a. a thinning of the stratospheric ozone layer
    b. rise in sea level
    c. an increase in tropical diseases within regions that are currently temperate in nature
    d. a disruption of agriculture

15. All of the following statements are true about nongovernmental organizations (NGOs) except:
    a. NGOs may be international, national, or local
    b. NGOs never receive funding from governments
    c. NGOs rose to prominence beginning in the 1980s
    d. NGOs have been important for giving voice to certain constituencies and for lobbying on behalf of particular issues that are not a priority for state governments

ANSWERS FOR SELF-TEST QUESTIONS

1. c, 2. d, 3. a, 4. b, 5. c, 6. b, 7. d, 8. d, 9. c, 10. a, 11. d, 12. b, 13. c, 14. a, 15. b

GEOGRAPHY OF DEVELOPMENT

V ery few people live as hunter-gatherers in the world today. The San people of the Kalahari Desert in southern Africa are one such group. What thoughts come to mind when you see a picture of hunter-gatherers? Most Westerners see such groups as primitive, backward, or underdeveloped. We may think of hunter-gatherers as "less developed" than city dwellers in New York or London. Whether we are conscious of it or not, we likely place people on a continuum of development, a scale typically linked to indicators of material well-being.

Ariadne Van Zandbergen/Africapictures.net/Newscom.

Kalahari Desert. The San people are one of the best known hunter-gatherer groups living on the continent of Africa today. Archaeologists estimate that the ancestors of the San hunted and gathered in this same region over the last 20,000 years (Marshall 2003). The two hunters in this photograph are inspecting a set of animal tracks. The government of Botswana has worked over the last few decades to encourage the San people to settle in villages and live sedentary lives. The pros and cons of this policy are hotly debated.

What criteria do we use to measure development in our mind, and why do we use these criteria? Development implies progress, but progress in what? Does development mean amassing wealth? Does development mean access to clean water and a steady food supply? Can people be poor and developed at the same time?

We perceive hunter-gatherers as primitive or underdeveloped, but are hunter-gatherers necessarily worse off than we are? Studies suggest that one group of San spent 12 to 19 hours per week working to obtain food as compared to the 40-some-hour workweek of most people in the so-called developed world. The same study found that the San had more time for leisure, slept more, ate a more balanced diet, and worked less than their "more developed" farming neighbors. The San and other hunter-gatherers around the world know where they can find different resources, including food, shelter, and water, during the course of the year, and they migrate seasonally and purposely to find resources necessary for survival.

In this chapter, we explain what development is, where the idea of development originated, how theories about the causes and effects of development have changed over time, and how geography helps us understand the uneven development of our world.

What Is Development and Where Did the Idea of Development Originate?
- Origins of the Concept of Development
- Measures of Development
- Development at What Scale?

How Does Geography Help Us Understand Development?
- Development as Modernization
- Structuralist Theories of Development
- New Economic Geography
- Geographies of Difference

Why Is Development Uneven?
- Environmental Determinism
- Dualism as a Concept for Understanding Uneven Development

What Role Does Development Assistance Have in the World Economy?
- International Development Assistance
- Participatory Development
- Development and Primary Production
- Sustainable Development
- Fair Trade

 THRESHOLD CONCEPTS in this Chapter

Development
Unequal Exchange
Commodity Chain
Mental Map

WHAT IS DEVELOPMENT AND WHERE DID THE IDEA OF DEVELOPMENT ORIGINATE?

LEARNING OBJECTIVES

1. **Understand** the origins of the concept of development.
2. **Critically discuss** different methods of measuring development.
3. **Analyze** development at different scales.

Open almost any book dealing with global politics, geography, anthropology, or economics and these texts will almost immediately begin discussing the globe in terms of varying levels of development. Authors refer to development using dozens of different terms, including: traditional versus modern; least developed, developing and developed nations; low income versus high income; Global South versus Global North; Fourth, Third, Second, and First Worlds; and nonindustrial versus industrialized. Although each of these terms (or sets of terms) has a slightly different history and connotation, they all get at some notion of development. To understand development, we must first grasp the origins of the concept.

ORIGINS OF THE CONCEPT OF DEVELOPMENT

Countries have long compared themselves to one another in terms of wealth. For example, during the era of mercantilism from the sixteenth to eighteenth centuries, European states competed with each other to see who could accumulate the most gold reserves. Countries also drew distinctions between themselves and others based on who had superior technology. Gauging development based on technological advances seems like an impartial way to look at which countries are developed and which are not. However, even the most essential technology to us, a flush toilet, may not be the best technology in every environment. While flush toilets might appear to be more advanced than pit latrines (holes dug in the ground), flush toilets are best suited for places where water is abundant. Instead of looking at the presence or absence of a technology, a better way to gauge development may be to look at sanitation and health. If children are dying from diarrheal diseases caused by unsanitary conditions, we have a development issue whether the toilet flushes or is a pit latrine.

Such awareness of context, however, was not even a question during the colonial era when all things European were deemed superior. Within the British Empire, for example, the English framed the conversation in terms of "progress," implying that all colonies were on a linear path to becoming civilized, that is, British. The irony is that while the British justified their colonial occupation in terms of "progress," they extracted resources from their colonies at a ferocious rate, impeding economic development for the colony itself and leaving behind serious social and environmental consequences. Bishop Desmond Tutu, a South African religious and resistance leader, once summarized the colonization of Africa by humorously quipping, "When the missionaries came to Africa they had the Bible and we had the land. They said 'Let us pray.' We closed our eyes. When we opened them we had the Bible and they had the land."

What is **development** and how do we measure it? Economists may define development in terms of economic growth, or political scientists in terms of good governance or free and fair elections. In contrast, geographers have tended to describe development as a process of change in the composition of an economy of a particular region and the well-being of its inhabitants, relative to other areas. Discussion of development typically involves describing how the nature of an economy (agricultural, industrial, or service-based) changes, but another way to consider development is to measure the variability of human well-being (education, health, gender equality) in space.

Colonizers certainly used the idea of development, referring mostly to how "civilized" a place is compared to another, but the idea of development in the current context came in the aftermath of World War II. As we discussed in Chapter 2, in the closing days of World War II, the major powers met in Bretton Woods, New Hampshire, to plan for peacetime. They proposed several global institutions, including the World Bank—also known as the International Bank for Reconstruction and Development. The primary purpose of this organization was to provide loans for the reconstruction of Europe and Japan following the devastation of the war. A few years later in 1947, U.S. President Harry Truman launched the Marshall Plan, a massive initiative to rebuild and develop Europe. The United States launched the Marshall Plan near the beginning of the Cold War because it feared that, in the absence of a speedy economic recovery, communism would spread to Western Europe. The Marshall Plan was a remarkable humanitarian gesture through which the United States gave Europe $15 billion to rebuild itself with few strings attached. The new institutions and development programs after World War II set the tone for how we have thought about development since: as first and foremost economic growth stemming largely from construction of transportation, communication, and production facilities.

MEASURES OF DEVELOPMENT

Concerns about development mounted in the wake of World War II not only because of the need to rebuild Europe but also because with decolonization in Africa and Asia, the number of independent countries in the world grew quickly (Figure 3.1). The newly independent states were called the Third World or the Global South. These

 development improvement in the economy and well-being of a place relative to another place.

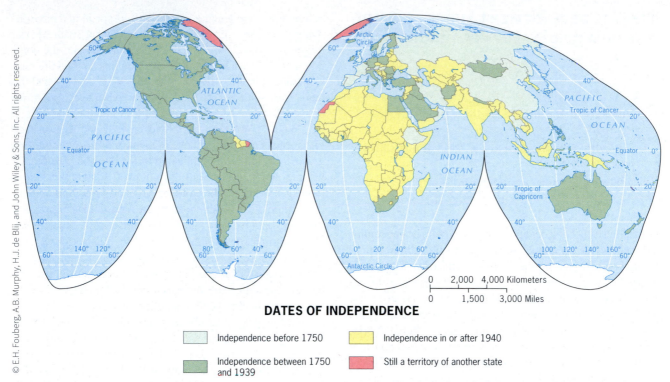

DATES OF INDEPENDENCE

- Independence before 1750
- Independence between 1750 and 1939
- Independence in or after 1940
- Still a territory of another state

FIGURE 3.1 **Decolonization after World War II.** The countries that went through decolonization to become independent states are highlighted in green and yellow on this map. The areas in green were colonized and then decolonized during the first wave of colonialism. Africa and Asia (places colonized during the second wave of colonialism) gained independence after World War II.

two terms have different origins. The term **Third World** arose during the Cold War to define countries that were not aligned with capitalism, NATO and its allies (which represented the First World), nor with communism, the Soviet Union and its allies (which represented the Second World). While it was originally a political category, the term *Third World* came to represent the developing countries of the world. The divide between the Global North and the **Global South** is a socioeconomic and political categorization between wealthy developed countries (the former) and poorer developing countries (the latter). Most states comprising the North are in fact located in the Northern Hemisphere (with the notable exceptions of Australia and New Zealand). As states become developed, they may become part of the North, regardless of geographical location.

International financial institutions, including the World Bank, looked at development as economic growth and devised development measures based on this perspective. The most common measurements of wealth are gross national product (GNP),[1] gross domestic product (GDP),[2] and **gross national income (GNI)**.[3] Each may be expressed on a per capita

(or per person) basis, which allows comparisons between countries regardless of population size (Figure 3.2).

Is development synonymous with economic growth? One limitation with each way we measure development is that wealth generated by economic growth is not usually spread across a population evenly. If a country's GNI increases, it may mean the wealth of the richest segment of the population has increased, which in turn has increased the total GNI and the GNI per capita. However, the poorer segment of population in the country may see little to no improvement in their condition, despite the state's increased GNI.

In order to gauge how wealth is distributed within a country, development analysts use the Lorenz curve (Figure 3.3). The 45-degree line of the Lorenz curve represents perfect equality, where equal portions of the population control equal amounts of income. When area A in the diagram becomes larger, it shows greater inequality.

A number of social scientists, including geographers, suggest that economic growth may not always lead to broad-based improvements in the human condition. They argue that if we are really focused on development, we should measure more than wealth; we should also assess parameters such as literacy, life expectancy, and infant mortality. The **Human Development Index (HDI)** provides one such alternative measure of development and is an unweighted average of life expectancy, literacy, the average years of schooling, and per capita GDP. Figure 3.4 shows how the HDI index varies across the planet.

---

[1] Gross national product (GNP) is the total value of goods and services produced by residents of a country, including domestic and foreign production.

[2] Gross domestic product (GDP) is total value of all final goods and services produced within the borders of a country.

[3] Gross national income (GNI) is GDP plus income received from investments outside of the country.

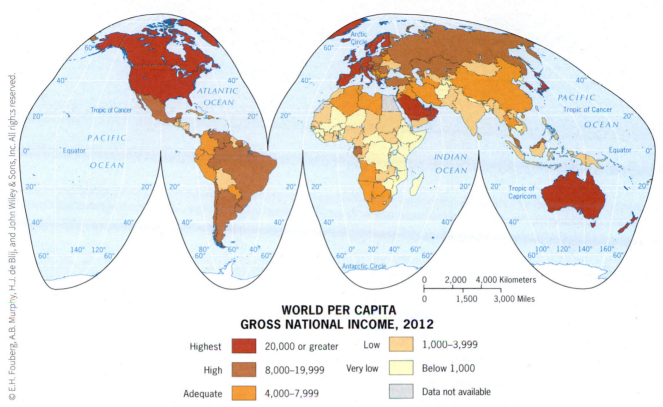

**WORLD PER CAPITA
GROSS NATIONAL INCOME, 2012**

Highest	20,000 or greater	Low		1,000–3,999
High	8,000–19,999	Very low		Below 1,000
Adequate	4,000–7,999			Data not available

FIGURE 3.2 **Gross national income (GNI) per capita, 2012.** The gross national income (GNI) per capita is the most widely accepted way of measuring wealth by country in the world today because the statistic tries to account for the flow of wealth in our globalized world. GNI per capita incorporates wealth held by the country's population that is located outside of the country.

By comparing and contrasting the map of per capita GNI with the map of HDI, we can see that the wealthiest countries in the world are generally also high on the HDI. The advantage of HDI is that it reveals more variation in development in the poorer regions of the world.

DEVELOPMENT AT WHAT SCALE?

Geographers use scale conceptually to show the differences across geographic areas (see Chapter 1). Geographers can use scale to help them understand how development policies operate and how what is happening at one scale— whether individual, family, local, national, regional, or global—affects development at another scale. We can look at a development policy or program operating and ask whether it should operate at a different scale (Figure 3.5) or consider the extent to which policies or programs operating at one scale influence development processes operating at another scale.

Many development programs are created as local scale initiatives. For example, a village may come together to build a new school, water pump, or community garden. Other types of development programs or policies, such as educational reform or disease eradication programs, may be implemented at the national or regional scale. Ideally, the scale of a development initiative will correlate to the scale of the development issue or challenge. For example, a communicable disease that is spread over a geographically

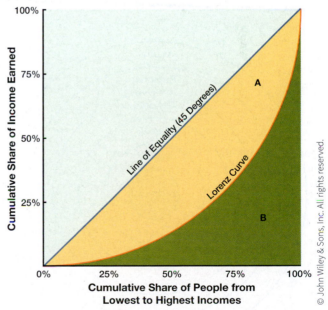

FIGURE 3.3 **Lorenz curve.** A key measure for understanding the distribution of income within a country is the gini-coefficient, which is derived from the Lorenz curve. The gini-coefficient is measured with a score of 0 signifying perfect equality and a score of 1 denoting perfect inequality. The 45-degree line of the Lorenz curve represents perfect equality, where equal portions of the population control equal amounts of income. The gini-coefficient is the ratio of area A divided by area A + B. So, as area A decreases in size, the gini-coeffient gets closer to zero, representing perfect equality in population and income.

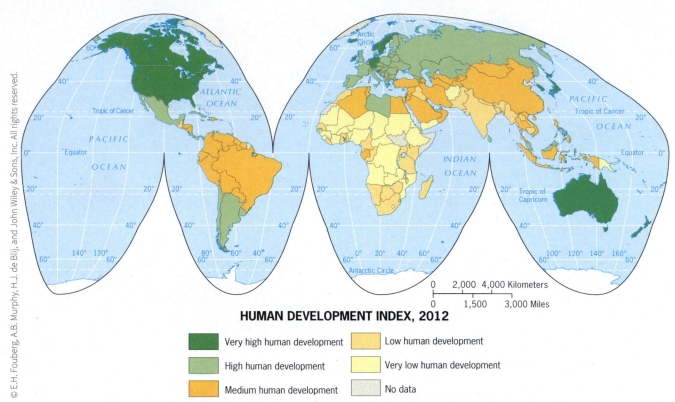

HUMAN DEVELOPMENT INDEX, 2012

- Very high human development
- High human development
- Medium human development
- Low human development
- Very low human development
- No data

FIGURE 3.4   The Human Development Index.   The United Nations Development Program derives the Human Development Index annually by creating a composite statistic covering three different areas of human development, including lifespan and health, education, and standard of living. The measurement is not perfect, but the HDI does give development another dimension beyond income alone.

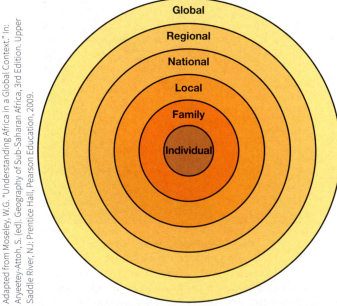

FIGURE 3.5   Schematic representation of scale.   One way geographers have thought about scale is by using concentric circles to represent the finest resolution of scale, from the individual up to the global scale. This schematic is a helpful way to begin to think about scale. Each circle does not exist independently of the others. Each scale interplays with the other scales, all at the same time.

broad area will likely demand a similarly scaled eradication program. The campaign to eliminate the global scourge of smallpox was successful because of the global-scale approach it took, and the campaign likely would have failed if it had only launched in some countries or regions. In contrast, a global campaign to situate a community garden in every village would appear overly top-down and insensitive to local variation and needs.

It is also tremendously important to understand how policies or programs operating at one scale may be influenced by (or influence) processes operating at another scale. Being oblivious to connections across scales may lead one to miss conflicting forces in the development process. For example, geographer Rick Schroeder found that in the 1970s, development agencies helped women establish community gardens in the lowlands of the Gambia River. The program successfully helped women produce goods that could be sold. In the 1990s, aid agencies in The Gambia initiated environmental projects but failed to take account of local land ownership practices and the motivations of males who participated in their programs (Schroeder 1997). As a result of the programs, females lost their gardening space.

In another case study of development and scale, soil conservation efforts among poor hillside farmers in upland Nepal failed to take account of the actions of wealthy farmers who expanded their fields in relation to export markets, which drove poorer farmers onto hillsides (Blaikie and Brookfield 1987). No amount of conservation could stem soil erosion if the real driver of the problem was cash crop production in the valley bottoms, driving subsistence farmers onto marginal and highly erodible slopes.

CONCEPT CHECK

1. **Why** is it important to understand the history of the development concept?

2. **What** do you think is the best way to measure development and why?

3. **How** does the concept of scale inform our understanding of development?

HOW DOES GEOGRAPHY HELP US UNDERSTAND DEVELOPMENT?

LEARNING OBJECTIVES

1. Critically **discuss** problems with modernization theory.

2. **Understand** different development theories.

3. **Explain** how global structures, physical geography, and social difference influence development.

Geographers who study development generally avoid narrowly defining development as economic growth. Instead, they focus on how economies change over time, the implications of these changes for human well-being, and the connections between development in one place and another. This broad conception of development in geography encompasses a lot of different perspectives within the discipline. This section of the chapter examines ideas about development within geography and how ideas have changed over time.

DEVELOPMENT AS MODERNIZATION

After World War II, colonies in Asia and Africa gained independence, and the number of states in the world quickly grew. Scholars in geography and other disciplines, in an attempt to advise newly independent countries about how to develop, sought to explain why some countries developed economically while others did not. Scholars looked for steps or stages that the newly independent countries could follow. This suite of theories, often collectively referred to under the umbrella term **modernization**, suggested that the European industrial economy was the ideal or pinnacle stage of development. These theorists argued that with the right combination of capital, know-how, and attitude, economic growth

would proceed down a certain path already forged by the wealthy countries of the world. They posited that countries would make a transition from traditional to modern states. While these theories were most popular in economics, they influenced thinking in geography and other disciplines as well.

W.W. Rostow, a macroeconomist writing in the 1950s and 1960s, outlined a model of the stages of economic growth (1960) and became the best-known of the modernization theorists. Rostow hypothesized that a country went through five stages, starting with a traditional agricultural economy (Figure 3.6). Once the society commercialized agriculture, it would accumulate a surplus of funds. The society used the surplus as critical investment capital for a take-off phase for industrialization. Traveling through industrialization and maturity, the society eventually reached the fifth stage, high mass consumption. In this stage, the society was capitalist and the economy grew through mass production and consumption of goods. Rostow used the metaphor of an airplane lumbering down the runway for take-off. The biggest challenge was gathering enough speed to get off the ground.

Modernization theory shaped development policy over the second half of the twentieth century. In the 1950s and 1960s, development agencies, including the World Bank, pushed developing countries to industrialize agriculture and to build infrastructure such as dams and roads (Figure 3.7). The idea was that such big investments in infrastructure—especially dams—could jump start an economy and put it on the path to industrialization by providing irrigation for commercial agriculture and cheap

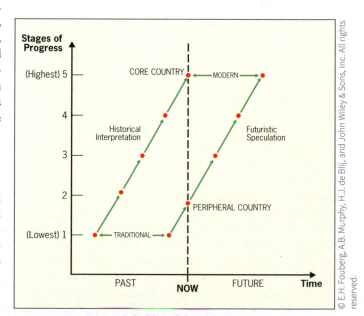

FIGURE 3.6   Rostow's ladder of development.   One way to represent Rostow's modernization theory is as a ladder of development, with each rung representing one of his five stages: (1) Traditional or preindustrial, (2) preconditions for take-off, (3) take-off into self-sustaining growth, (4) drive to maturity, and (5) the age of mass consumption.

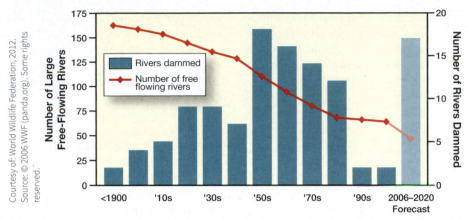

FIGURE 3.7 **Global dam development.** This graphic shows the growth in dam development around the world since 1900. In the developing world, the vast majority of dams were built as development projects, in an effort to help launch newly independent countries onto the ladder of development. Today geographers who study development recognize that this wave of dam developments had major environmental implications: The total number of free-flowing rivers has declined, affecting wildlife and plant species. Moreover, building dams has often failed to foster development as predicted.

electricity for manufacturing. Roads could open up new areas of the country and connect them to the national market economy.

Modernization theory employed a "pull yourself up by the bootstraps" philosophy. Followers of modernization theory argued that if a country adopted the right type of policies and made the right type of investments, development would automatically follow. The idea that a country could choose to develop autonomously by making certain policy decisions was conceptually appealing because of its tidiness. However, many geographers and economists now recognize that countries do not develop autonomously in the real world.

Multiple critiques of modernization theory emerged in subsequent decades. One of the first critiques was of the ladder of development itself (Figure 3.6). Rostow created the five steps based on the example of Great Britain's economic development. Britain was the hearth of the Industrial Revolution (see Chapter 9), so it climbed the ladder first. If we think through how Britain climbed the ladder and if we then consider how newly independent countries in the 1960s and 1970s, such as Nigeria, were supposed to climb this same ladder, we can come up with several critiques of modernization theory.

First, if the ladder were correct, the global context in which Britain climbed the ladder in the 1700s was vastly different from the global context for Nigeria's climb in the 1960s. No countries were ahead of Britain on the ladder when it climbed, whereas Nigeria and other developing countries had to climb with dozens of countries on the rungs above them. Take this visual a step further and recognize that in Britain's climb up the ladder, it stepped on the heads of dozens of other countries. Through colonization, Great Britain exploited the lands, peoples, and resources of one quarter of the world between the 1700s and 1960, including Nigeria. Surely, the wealth returning to Great Britain through colonialism helped its quick advancement up the ladder (see Figure 3.6).

To understand the role of context in development, geographers, including Peter J. Taylor (1992), draw from R.H. Tawney's tadpole philosophy (Wright 1987), which considers that a small percentage of tadpoles develop into frogs in a pond. A pond may have a thousand tadpoles, but only small percentage become frogs because all of the tadpoles compete in the same pond for food, sun, and other resources. Likewise, in the world economy, not all countries can become wealthy. For a few countries to become wealthy, they must exploit the others and the global resources. Taylor and academics who study the capitalist world economy argue that the capitalist structures make it impossible for all countries in the world to become wealthy, thus making development a futile task in many ways.

Several of these critiques of modernization theory have been incorporated into what we might loosely call the geographic perspective on development. Below we explore three such perspectives: (1) structuralist theories of development, (2) new economic geography, and (3) geographies of difference.

STRUCTURALIST THEORIES OF DEVELOPMENT

In reaction to modernization theory, a new set of development ideas began to arise in the 1960s that emphasized structure, or the global framework under which countries operated. This emphasis on structure meant that the relationships among countries were as or more important than internal policies for determining the future development of a country.

Dependency theory, originally conceived by Andre Gunder Frank (1979), is one example of such a theory. This approach suggested that during the colonial era the economies in the tropics (Africa, Latin America, and Asia) were essentially "underdeveloped." The European colonizers set up the economies of the colonies to benefit Europe. According to Frank, European colonizers sought primary goods, including crops and resources, from colonies, and the colonizers teamed with elites in developing countries to gain wealth and power. In many instances, colonizers

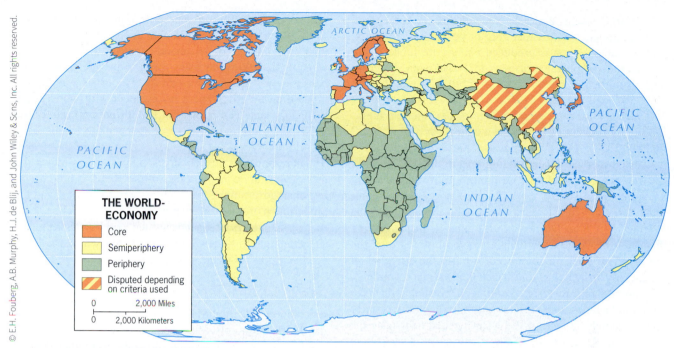

THE WORLD-
ECONOMY

- Core
- Semiperiphery
- Periphery
- Disputed depending on criteria used

0   2,000 Miles
0   2,000 Kilometers

FIGURE 3.8   Core, periphery, and semiperiphery.   One classification of the world is into core, periphery, and semiperiphery. Wallerstein saw the semiperiphery as a place, but he considered core and periphery to be processes. This map marks the core and the periphery as places where, respectively, primarily core or periphery processes are taking place.

took over ownership of farmland, pushing farmers in tropical countries to pursue wage labor on large plantations or in emerging mining sectors. Where local farmers continued to operate, they were encouraged or forced to produce commodity crops for the European market (instead of producing subsistence to feed themselves). Colonies traded primary commodities for manufactured goods from Europe and other industrialized countries in a global system of unequal exchange that favored Europeans. Taylor (1992) described the creation of the modern world economy as being "made by Europeans for Europeans as one great functional region centered on Europe."

Wallerstein (1979) expanded on Frank's ideas though his **world-systems theory**, which offers a geography for dependency theory by depicting the world in terms of the core, semiperiphery, and periphery (Figure 3.8). Under this schema, core countries are the most developed states where production of goods requires high levels of education, research and development, and offers high wages and benefits (Figure 3.9). The semiperiphery is an emerging group of states where both core and peripheral processes are involved in the production of goods. Finally, the periphery represents

those countries where economic production requires low education, involves little research and development, and offers relatively low wages and few benefits.

FIGURE 3.9   Hong Kong harbor in China.   Some countries have advanced from periphery to semiperiphery to core. China is a good example of a country that moved from the periphery to the semiperiphery as it increasingly engaged in manufacturing and became a major global exporter, as exemplified by the heavy freight traffic in Hong Kong harbor. Some scholars believe that China is poised to join the core and to become a preeminent global power.

World systems theorists contend that the same good can be produced with either core or peripheral processes. For example, cotton is produced using core processes in Texas and peripheral processes in Uzbekistan. Purchasing a John Deere cotton picker combine costs more than $600,000, but the picker reduces "the need for four to six workers, two to three tractors, one or two module builders, and a boll buggy as part of a cotton harvesting crew" (Cline 2008). Farmers who use these technologically advanced methods of harvesting are producing cotton in a core way, which churns up more activity in the world economy, along every step of the production of the combine. The implements dealer who sells the combine, the engineers at John Deere who designed it, the marketing firm that designed advertising campaigns for it, and the mills where steel was produced to build the combine all financially benefit from a farmer purchasing a combine. Meanwhile, in Uzbekistan, cotton is harvested by hand or simple machines, and agricultural policies in the country have drained the land of the water needed to irrigate the cotton (see Chapter 10). Both countries produce cotton, but the core production processes generated much more wealth than the peripheral processes did.

All three of these regions operate as a world system, with deep connections existing among each sphere. While Figure 3.8 is just one representation of core, periphery, and semiperiphery, it suggests that most countries in the tropics still play the role of the peripheral producers in the global system. The problem, as world systems theorists see it, is that many tropical countries will find it difficult to break out of their role of producing goods with peripheral processes, including relatively low wages, low levels of technology, and low levels of education. In other words, Rostow's stages of economic growth will not occur because the least developed countries are locked into a set of exploitative relationships with more developed countries.

Dependency theory and world-systems theory did lead to some real policy changes on the ground in developing countries. Probably the most significant of these was an approach known as import substitution, which is often associated with the Argentine economist Raúl Prebisch (1959), who argued that tropical countries would be forever stuck as producers of primary products (and therefore would fail to develop) unless they took proactive steps to change the nature of their economies in relation to those of others. The idea behind import substitution was that manufactured goods needed to be produced at home rather than imported from the core countries (hence the name "import substitution" for such policies). Given the lack of private capital available for industrialization and stiff competition from producers in the core countries, the governments of peripheral states often became directly involved in creating such enterprises, establishing state-run enterprises known as parastatals. Governments typically also erected tariff barriers to protect the new industries until they could stand up to international competition.

Import substitution was quite popular until the late 1970s when the Third World debt crisis struck. This crisis involved a number of peripheral countries that were on the verge of defaulting on loans owed to private commercial banks as well as public creditors. The crisis was largely brought on by government involvement with increasingly inefficient, state-run enterprises and the energy crisis of the 1970s when high oil prices were particularly challenging for oil importers.

In response to the debt crisis, public lenders imposed a new neoliberal economic order in the 1980s–2000s, stressing small government, free trade, export orientation, and the privatization of state-run enterprises. Critics of neoliberalism (largely supporters of dependency and world-systems theory) suggested that this "new" order was actually a return to the economic policies of the colonial era.

NEW ECONOMIC GEOGRAPHY

In addition to structuralist theories of development, by the 1990s economists and economic geographers developed a growing critique of neoliberal development policy known as new economic geography. Most new economic geographers continued to believe free trade was the best route to development, but they recognized the challenges faced by some countries in terms of physical geography, including tropical diseases, poor agricultural soils or inadequate roads, and lack of rail and shipping connections to global trading networks (Hausmann 2001; Sachs 2006). In terms of policy prescriptions, new economic geographers have tended to emphasize public investments in transportation infrastructure or health initiatives to reduce disease burdens. While many development geographers are supportive of these investments, others are skeptical of the continuing emphasis on free trade within new economic geography. Some are concerned that new economic geography deploys a simplistic vision of geography in an overly deterministic fashion (see the discussion of environmental determinism in the next section).

GEOGRAPHIES OF DIFFERENCE

Another significant evolution of thought within development geography during the 1980s and 1990s was a growing acknowledgment of the importance of recognizing power differences within populations across the developing world. Until the 1970s, much development assistance had been focused on improving the welfare of a generic population—undifferentiated by wealth, **gender**, race, or ethnicity. As a result, the vast majority of development aid tended to favor the most powerful members of a community (namely, men, the wealthy, and other dominant groups) who were better at articulating their wants and capturing new resources. Scholarship by geographers Katz (1991), Rocheleau et al. (1997), and Schroeder (1997) showed how important it was to recognize such power differences in development work.

Geographer Holly Hapke examined the impact of mechanization and commercialization on small-scale fish traders in Kerala, India, paying special attention to gender and the impact of economic transformation on female fish traders (2001). She argued that because women's roles in the fish economy had been overlooked, they experienced economic marginalization at the same time that their labor had become increasingly important for household survival. Market-oriented development of the fish trade changed women's relationship to production and marketing, marginalizing them in the process.

CONCEPT **CHECK**

1. **How** was modernization theory an ethnocentric approach to development?

2. **What** do you perceive to be the strengths and weaknesses of new economic geography?

3. **Why** is it important for development agencies to understand power differences within counties and regions when planning for development?

WHY IS DEVELOPMENT UNEVEN?

LEARNING OBJECTIVES

1. **Understand** environmental determinism.

2. **Explain** how the concept of dualism helps us understand uneven development.

3. **Analyze** development patterns within your own state, province, or region for signs of uneven development.

From a geographical perspective, one of the most striking aspects of development is its unevenness. We are perhaps most familiar with such differences from country to country, but the gulf between the rich and poor is often just as wide within countries. Why some countries and areas are more prosperous than others can be explained in numerous ways. Some attribute the uneven outcomes of development policies to differences in natural endowments or natural resources. Others focus on political leadership that influences the type of economic policies put in place. Some theories use historical explanations, especially past histories of colonialism, neocolonialism, and imperialism, or economic and political interaction between countries that may help some countries and hurt others.

ENVIRONMENTAL DETERMINISM

As students learn more about world geography, many are tempted to use physical geography, including location and climate, to explain development. Geographers went through their own period where they used environmental differences to explain everything from intelligence to wealth. This set of theories is called **environmental determinism**. Although geographers no longer espouse environmental determinism, seeing it as a vast oversimplification of the world, other theorists and political analysts propose environmental deterministic theories regularly. For this reason, a student who is well versed in geography needs to recognize environmental determinism and understand how to break down and analyze environmentally deterministic theories and policies.

Social, political, or intellectual context often has a bearing on the theories and ideas that are popular at a certain time. For example, many would argue that Rostow's modernization theory was influenced by decolonization and the Cold War after World War II. Similarly, we must consider the Darwinian and colonial context in which environmental determinism emerged.

Charles Darwin's theory of evolution, published in 1859, holds that through a process of natural selection, species evolved over time in relation to changes in the physical environment and biotic community. While Darwin's ideas eventually gained acceptance in mainstream science, much more controversial were a group of thinkers who tried to apply Darwin's ideas to the human or social sphere—an approach known as social Darwinism. One such social Darwinist thinker was Fredrich Ratzel who argued that differences between human societies could be explained in relation to the varying climates in which they evolved. Ratzel asserted that warm tropical climates produced (or environmentally determined) certain types of human societies that were quite distinct from those that evolved in temperate or arctic climates. Ratzel, the first environmental determinist, trained a number of geographers, including Ellen Churchill Semple and Ellsworth Huntington, who in turn were quite influential within the discipline of geography.

Environmental determinism became quite popular in the social sciences by the early twentieth century, particularly in geography and anthropology. The appeal was in part related to the desire of social scientists to have a grand theory for explaining social difference (similar to theories or laws in the natural sciences). The problem was that scholars began to use these ideas to try to "explain" why one group was superior or more advanced than another—an inherently problematic exercise. The following quote is one example of such an explanation: "[T]he well-known contrast between the energetic people of the most progressive parts of the temperate zone and the inert inhabitants of the tropics and even the intermediate regions, such as Persia, is largely due to climate" (Huntington 1915, 22). In other words, the people of the tropics, coddled by lush warm environments, were slow moving, primitive, and unsophisticated. In contrast, people of the temperate regions (forced to plan ahead for long winters) developed complex civilizations and were energized by their cooler climates.

Geographers eventually, especially after World War II, saw environmental determinism as a deeply flawed, if not highly racist, concept. Scholars began to question the ideas of environmental determinism because they found too many exceptions to the generalizations being made about the tropics. How does one explain Egyptian pyramids, Mayan temples, or the Great Zimbabwe if people in the tropics were supposed to be primitive and lacking a history of great civilizations? Furthermore, this set of ideas was developed at the height of the colonial period, and many came to realize that theories such as environmental determinism helped justify the behavior of colonializers by supporting the notion that people from the temperate regions were superior to those from the tropics. Using environmental determinism, a theory that explained differences in

wealth solely in terms of climate, ignored larger relevant political processes such as colonialism.

DUALISM AS A CONCEPT FOR UNDERSTANDING UNEVEN DEVELOPMENT

Not all explanations of uneven development are as problematic as environmental determinism. One look through the table of country statistics in Appendix A (found on the book's website) reveals undeniable differences in human well-being, from literacy rates to infant mortality rates. The infant mortality rate (number of deaths under age 1 per 1,000) in Niger was 71 in 2013 versus 2.4 in Norway (Population Reference Bureau 2013). The images from the 2010 earthquake in Haiti (Figure 3.10) helped North Americans visualize that

Reading the **PHYSICAL** Landscape

Haiti Earthquake, 2010

Port-au-Prince, Haiti. View of a neighborhood devastated by the March 2010 earthquake. The earthquake left one million people homeless in Port-au-Prince and its surrounding area.

The devastating earthquake in Haiti in 2010 helped open the eyes of many North Americans to the crippling poverty in the country. In Figure 3.10, you can read the devastation visited on the landscape in the way the buildings crumbled under the magnitude 7.0 earthquake on the Richter scale, killing approximately 230,000 people. In the same year, an 8.8 magnitude earthquake hit Chile, but the death toll was significantly lower than in Haiti: Only about 700 people perished in Chile.

Haiti is the poorest country in the Western Hemisphere and has several major strikes against it that make development difficult: significant deforestation, a dry climate that deters crop agriculture, an economy exploited by the French during the colonial era, and a series of corrupt Haitian dictators since independence. Chile, on the other hand, has the highest GNI per capita in South America, Central America, and the Caribbean. Chile experienced the strongest earthquakes ever recorded in 1960 (9.5 on the Richter scale), and in 1985 the government enacted strict building codes to better handle future earthquakes. In contrast, Haiti has no building codes to prepare for earthquakes.

The greater amount of destruction caused by the Haiti earthquake was also a result of the physical geography of the earthquake itself. According to *The Week*, the epicenter of the Haiti earthquake was 10 miles (16 km) from Port-au-Prince (population 3 million), while the epicenter of the Chile earthquake was 70 miles (112 km) from Concepción (population 200,000). The epicenter of the Chile earthquake was also deeper, 22 miles (35 km) below the surface, but the epicenter of the Haiti earthquake was about 11 miles (18 km) below the surface.

FIGURE 3.10   Port-au-Prince, Haiti. A major street is filled with debris from the destruction caused by an earthquake in 2010.

50 percent of the population in Haiti were undernourished in 2012 versus 5 percent in Japan (World Bank 2012). One useful concept for understanding varying levels of development is **dualism**. Dualism refers to situations in which two areas are in relationship with one another (through trade, for example), and one area is developing at the expense of the other. This concept may be examined at a variety of different scales, including global, national, local, urban, and rural.

Global-scale dualism refers to different countries that are in relationship with one another via trade or **unequal exchange**. Unequal exchange occurs when laborers in one country produce a good for low wages and then the good is processed through a **commodity chain** incorporating shipping and marketing and sold at a relatively high value. Two theories that take dualism at the global scale, or unequal exchange, into account are dependency and world-systems theory, both of which are discussed in the previous section. The classic case of unequal exchange is the coffee farmer in Colombia who receives very little for his crop and then sees it sold in the store down the street under a Nescafé label for hundreds of percentages more per pound than what he was paid. Unequal exchange is built into capitalism, as branding and marketing help companies demand much higher prices for their goods than they pay for the production of the good.

What is important to understand is that dualistic relationships, like those found in unequal exchange, also exist at more local scales. Within most countries, dualisms often exist between urban and rural areas. In the United States, for example, many would argue that core areas exist where a disproportionate amount of investment and wealth accumulate. These areas, typically cities, have historical relationships with more rural zones of resource extraction that function as peripheries to the urban cores.

In *Nature's Metropolis* (1992), environmental historian and geographer William Cronon wrote of the relationship between Chicago and vast interior regions of the Midwest. Timber, iron ore, livestock, and grain historically came to Chicago from the Upper Midwest in exchange for manufactured items. Cronon explained that while Chicago became fabulously wealthy during the nineteenth and twentieth centuries, the regions of extraction (such as northern Minnesota, southern Illinois, and northern Wisconsin) have little to show for their exports today.

Geographers are drawn to dualism in order to understand development because the theory works at different scales, much like world-systems theory, to explain the uneven development of the world. Geographers apply dualism both globally and more locally, looking at differences within rural and urban areas.

Dualism is apparent in rural areas where one may have an export-oriented commercial agricultural sector that exists alongside smaller-scale, mostly subsistence farming. The classic situation is one wherein small subsistence farmers spend the day laboring on nearby large plantations and return to work on their own farms in the evenings and during the weekends. The linkage between the two sectors is often disadvantageous for the small farmer or laborer.

For commercial farms, the advantage of laborers who moonlight as small farmers is that they need not be paid a living wage because they have their own production to cover a portion of their annual food needs. Many small farmers who end up working on larger farms may initially take on such employment in order to bridge a food production shortfall. In so doing, however, they often embark on a slippery slope of declining production on their own small farm, as they spend critical time working for the wage-paying farm. In turn, the small farmer may compromise her own ability to produce food.

A similar set of dualistic relationships exist in urban areas between formal and informal employment sectors. An employer in the formal economy, such as a large export-oriented company, may hire temporary laborers who are otherwise employed in the informal economy to ramp up production. At the same time, work in the informal economy includes a wide array of undocumented full- and part-time employment activities such as petty commerce and artisanal production that will in turn be bought, sold, and traded in the formal economy. Do exchanges between the informal economy and formal economy actually hurt the informal sector in an urban area? Or does such a system simply allow for a supply of cheap labor?

The impacts of dualism are contested and views vary on both sides of the exchange. Dualism highlights processes of unequal exchange that development theorists see as slowly undermining one group or sector in favor of the other.

CONCEPT **CHECK**

1. **Why** is environmental determinism a problematic theory?

2. **How** could development in one area lead to underdevelopment in another?

3. **Do** uneven patterns of development exist in your own area or region and why?

WHAT ROLE DOES DEVELOPMENT ASSISTANCE HAVE IN THE CONTEMPORARY WORLD?

LEARNING OBJECTIVES

1. **Understand** patterns of poverty across the world.
2. Critically **assess** the future of development.
3. **Analyze** different development alternatives.

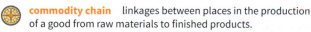

unequal exchange uneven relationship between low labor costs and high-value products.

commodity chain linkages between places in the production of a good from raw materials to finished products.

What role does development assistance have in the contemporary world? The answer to this question may depend on your political philosophy. If you subscribe to political realism, a school of international relations that prioritizes national interest and security over ideals or ethics, then you may be more likely to believe that foreign aid is always given to advance the agenda of the country that is donating the money or that assistance is given to advance the agendas of international institutions controlled by powerful countries. In contrast, if you espouse liberal international relations theory, based on a belief that absolute gains can be made through cooperation and interdependence, then you may be more apt to believe that international development assistance could actually help some countries.

INTERNATIONAL DEVELOPMENT ASSISTANCE

International development assistance comes in all shapes and sizes. It varies in terms of the provider (e.g., intergovernmental organizations such as the United Nations, a country's government, or a nonprofit organization) as well as in terms of its form, whether a loan for a dam, emergency food relief, technical assistance to a

government agency, or funds to dig a community well. As measured by the Organization for Economic Cooperation and Development (OECD), in 2005, countries and development organizations contributed $78.6 billion in official development assistance to poorer countries. The United States contributes the most total dollars in foreign aid, but as a percentage of the country's GNI, the United States falls behind Canada and several countries in Europe (Figure 3.11).

After hundreds of billions of dollars given in international development aid over the last 20 years, academics, donors, and donor countries are questioning the value of international development assistance. Let us first examine the arguments against international development assistance and then explore arguments in favor of it.

Ample evidence supports the view that international development assistance is problematic. The vast majority of foreign aid provided by countries tends to go to close allies rather than to the countries most in need of assistance. Some classic critiques of foreign assistance (e.g., Hancock 1994; Maren 2002; Moyo 2009) document the inherent self-interest in much foreign assistance. Donors, for example, have often required that goods and services be

World Bank, World Development Indicators: Aid Dependency, 2012.

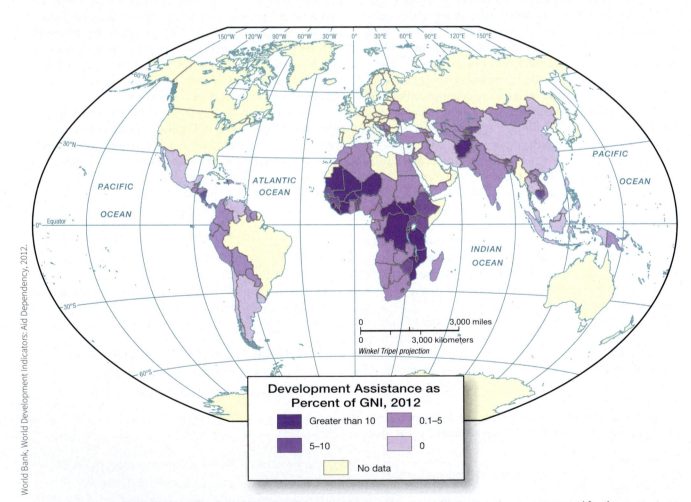

FIGURE 3.11   Development assistance received as a percent of GNI, 2012.   Development assistance accounted for the greatest proportion of gross national income in 2012 in the countries shown in the darkest color on the map.

purchased from suppliers in their own countries. A case in point is how the U.S. Food for Peace Program traditionally purchased surplus food stuffs in the United States for distribution in other countries. Although this program may have been providing needed assistance in food-deficit areas, it also served to support American farmers.

What at first seems like a win–win situation unfortunately has negative consequences for the country receiving aid. Emergency relief efforts using American grains often hurt farmers in the target areas by depressing demand for their crops or encouraging locals to develop tastes for nonlocal grains such as wheat. Development experts argue that a better course of action in most cases is to purchase grains locally if they are available, in order to bring relief and economic development to the receiving country.

One of the most troublesome aspects of development programs is the unintended consequences of any policy. In the case above, undermining local grain farming is an unintended consequence. Likewise, the documentary "T-Shirt Travels" and the book *Travels of a T-Shirt in the Global Economy* demonstrate how a simple act of kindness, donating your old t-shirts to the Salvation Army or Goodwill, has resulted in undermining the textile industry in Zambia and other African countries where secondhand clothing is more popular than locally made clothing (Figure 3.12).

The secondhand clothing trade in Africa grew rapidly after 1990, partially in response to structural adjustment[4] loans from the World Bank and IMF that began flowing into Zambia in 1983. Zambia and other countries that receive structural adjustment loans must open their economies to foreign direct investment, privatize publicly owned enterprises, and establish free trade practices in order to receive large loans. Through these kinds of loans, Zambia altered its economy and took on a great deal of debt. The consequences for Zambia have been overwhelmingly negative. What were well-established industries, such as the textile industry, were undercut by the flow of secondhand clothing into the country. Today, less than 10 percent of the population has a job in the formal sector of the economy. Instead, the informal sector, including the secondhand clothing trade, employs the vast majority of working people in Zambia.

Given the history of troubled development efforts and the many negative, unintended consequences of development, you may want to write development off as a failed "project." Today, development specialists are acknowledging past problems with development and rethinking this process. Is there a way to approach development that might

---

[4] The literal sense of structural adjustment is that it is a set of policies that will restructure or rebalance a nation's economy. See Chapter 2 for a discussion of these policies.

Reading the **CULTURAL** Landscape

T-Shirt Travels

Stands of used Western-style clothing are a common sight in marketplaces in Subsaharan Africa. The Salula or "dead white people's clothing" trade has replaced much of the local textile industry in the region. An Independent Lens documentary traced a t-shirt donated at a Salvation Army store in Fredericksburg, Virginia, and found that it made stops in New Jersey and Mombassa, Kenya, before reaching the marketplace in Mongu, Zambia. T-shirts donated throughout North America and western Europe are compressed in large bales and placed on container ships in New Jersey. Once in Kenya, Indian businessmen take the large bales and break them into smaller bales (about the size of a hay bale you might find at a fall festival), and then African entrepreneurs travel to city centers to purchase bales and take them home to sell them in marketplaces like the one shown in the photograph.

FIGURE 3.12　Bangula, Malawi. Secondhand clothing stalls are commonplace in markets throughout Subsaharan Africa.

© Jenny Matthews/Alamy.

be less problematic? Below we outline four alternative approaches to development that geographers have studied: (1) participatory development; (2) producing primary goods, including nonrenewable resources, for the benefit of local people; (3) sustainable development; and (4) fair trade.

PARTICIPATORY DEVELOPMENT

One theme that has received considerable attention in foreign assistance circles since at least the early 1990s is **participatory development**. The emphasis on participation came about in reaction to a history of top-down development projects that left local people with little to no sense of ownership of the projects undertaken by outsiders in their communities. Besides being problematic in its own right, a lack of participation often inhibited the longer-term sustainability of programs because local people had little interest in maintaining projects they did not initiate or request. In order to foster community participation, some development actors have employed a range of techniques to facilitate community problem identification and group problem solving. Geographers and others social scientists

have devised a number of techniques to enable greater community input. One such technique is creating **mental maps** (or cognitive maps) (Figure 3.13).

Cognitive or mental maps are powerful tools in development because they reveal indigenous knowledge of soils, resources, seeds, political relationships, social relationships, climate, and more. Through examining mental maps, geographers gain a sense of how people understand their locality and how a particular place functions in the larger context. By working with mental maps, development experts can better mesh development efforts with the way people in a place live, work, and function.

Geographer Peter Herlihy (2003) facilitated a participatory research mapping (PRM) project to document the lands used by the indigenous populations of the Darién Province in eastern Panama. Having fought for recognition of their land rights in the face of encroaching outsiders, indigenous leaders were well aware of the power and importance of cartographic information. Indeed, the Darién

---

mental maps   (also known as cognitive maps) maps of an area made from memory and experience by individuals or groups.

USING GEOGRAPHIC TOOLS

Using Cognitive or Mental Maps to Understand Local Development Priorities

A cognitive or mental map is a hand-drawn map of an area produced by an individual or group. Such maps may be contrasted with more formal maps produced by professional mapmakers or cartographers. Geographers have long used cognitive maps with study participants in order to gain insight into a person's understanding of and relationship to an area. These maps are also extremely useful in development project research and planning as a participatory information-gathering tool. Let's say, for example, that you wanted to better understand the relationship between a community and the physical landscape. By asking people to draw cognitive maps of their village and surrounding lands, development workers can begin to understand how different groups depend on the resource base. For example, Batterbury (1996) describes how cognitive maps drawn by women in rural communities in West Africa emphasize forests and sources of water, which makes sense given their traditional responsibilities as collectors of firewood and water.

Thinking Geographically

1. What would be included in a mental map and what would be left out?

2. How can mental maps be interpreted in order to help inform development projects in a country?

Source: Herlihy, P. 2003. "Participatory Research Mapping of Indigenous Lands in Darién, Panama." *Human Organization*, 62(4): 315–331.

Courtesy of William Moseley.

FIGURE 3.13   Central Mali.   Locals in a village near the town of Mopti work together on a cognitive map of agricultural and natural resources.

One of the great innovations of the Six-S network was its emphasis on flexible funding, an approach allowing the village-based groups to propose their own projects that were then selected by a committee of village representatives, rather than by a donor. By all accounts, Six-S was amazingly successful over a 15-year period, serving several hundred thousand people organized into 30,000 groups in 150,000 villages (Lecompte and Krishna 1997).[5]

DEVELOPMENT AND PRIMARY PRODUCTION

Geographers often describe the nature of economies in terms of primary (mining, forestry, and agriculture), secondary (manufacturing) and tertiary (service and financial sector) economic activities. A major obstacle that developing countries often face is the problem of being stuck as a producer of raw materials (or stuck in an economy dominated by the primary sector). If you examine the poorest countries in the world today, often called the Global South, many of them have economies that are overly dependent on the export of one or two primary commodities. To use Andre Gunder Frank's terminology, these are economies that were underdeveloped during the colonial period by being made to produce a few commodities for the core, today's Global North. In many cases, this orientation has only been reinforced by the policies of global financial institutions such as the World Bank and the International Monetary Fund (IMF). Consequently, we need to consider development alternatives for economies that are dependent on the export of one or two primary commodities.

One alternative is the trust fund approach. This approach has been tried in a few countries, both less and more developed. A good example of a less developed nation that has pursued this approach is the Micronesian island state of Kiribati. With a population of 90,000 spread over 34 islands, Kiribati had a per capita income in 2013 of $2,620. While most of the population (80 percent) is engaged in a subsistence-based farming and fishing economy, the country also has significant phosphate deposits, which it is mining and exporting. Since 1956, the proceeds from phosphate extraction have been placed in a trust fund that is invested offshore by two London-based account managers. The returns on this fund are used to finance government services, including health care and a communication and transportation infrastructure between the islands (Figure 3.15). Most residents in Kiribati are therefore free to continue living a subsistence lifestyle, yet still have access to sustainably financed government services (Gibson-Graham 2004).

Another positive example of revenues from mineral resources being used to develop a country is the use of

FIGURE 3.14   Falagountou, Burkina Faso. Villagers repair a water pump in Burkina Faso.

was the most inaccurately mapped province in the country, and indigenous leaders embraced the idea of a mapping project to document their expanding settlements and natural resources. Community representatives were trained to complete land-use assessments using questionnaires and sketch maps. They worked with a team of specialists, including Herlihy, to transform this information into standard cartographic and demographic results. The methodology shows how indigenous peoples can work with researchers in data collection and interpretation to transform their cognitive knowledge into standard maps, produce excellent scientific results, and enhance their ability to manage their own lands.

The problem, and a latent contradiction, is that communities may identify problems that many development agencies are not prepared to solve, at least in the quick and efficient manner that their funders expect. The result is that communities often identify problems and select programmatic solutions they think a development agency can deliver. Some development projects have been successful, and the most successful address development at multiple scales at the same time.

One of the most acclaimed nongovernmental organization (NGO) initiatives in Africa was the Six-S network in Burkina Faso, Mali, and Senegal (Figure 3.14). Six-S (a French acronym meaning "Making Use of the Dry Season in the Savannah and the Sahel") was founded in 1977 by Bernard Quedrago, a teacher and school inspector from Burkina Faso, and Bernard Lecomte, a French national with considerable development experience. Both Quedrago and Lecompte were frustrated with traditional development approaches that tended to be "top-down" and that emphasized projects. Six-S was an NGO created to support traditional, village-based groups in a variety of locally inspired initiatives, from micro-credit schemes to soil conservation.

[5] While elements of this program continue to function autonomously today, the larger program has ended. The reasons for the end of the program are too complicated to adequately explore here, but too rapid expansion and too much external funding may have led to its eventual demise.

FIGURE 3.15   Tebunginako, Kiribati Islands. Villagers on the island of Abaiang have chosen to relocate their village due to rising seas and erosion. The new village is now also under threat of rising waters from climate change, and the villagers are constructing sea walls in response.

Fairfax Media via Getty Images.

diamond mining in Botswana (Figure 3.16). These examples stand in stark contrast to some other countries where the proceeds of resource extraction largely flow into the pockets of business interests and/or to a few corrupt leaders.

SUSTAINABLE DEVELOPMENT

Another development approach that has received considerable attention since the late 1980s is sustainable development. Up until this point, taking care of the environment (environmental stewardship) and developing the economy were largely seen as conflicting objectives. Sustainable development seeks to reconcile these goals. Although the term *sustainable development* was first used in 1980, it was the 1987 Brundtland Commission Report, or World Commission on Environment and Development, entitled *Our Common Future*, which popularized the concept. It defined **sustainable development** as "development that meets the needs of the present without compromising the ability of future generations to meet their own needs." This sufficiently broad and vague definition of the term, not uncommon for a United Nations document, was both a strength and a weakness. The advantage is that it allows several diverse constituencies from pro-growth advocates to strict environmental conservationists to support the approach because each group can interpret the term in a slightly different manner. The disadvantage of this ambiguity is that it has often resulted in different, and sometimes diverging, approaches to development (Williams and Millington 2004).

Today, development specialists view sustainable development in two different ways. The minority perspective, sometimes referred to as *strong sustainable*

GUEST *Field Note*   Development in Botswana

DR. ABDI SAMATAR
University of Minnesota

Durban, South Africa. Former President of Botswana Ketumile Masire, on the right, received the Mahatma Gandhi International Peace and Reconciliation Award in 2004 from Ela Gandhi, the granddaughter of Mahatma Gandhi. In the center is Vasu Gouden, of the African Center for Constructive Resolution of Disputes. Masire and Gouden won the award for their work in resolving disputes in Africa.

Botswana was an unlikely candidate for an African development miracle. Most policymakers and scholars thought it would remain indefinitely impoverished and weak at the time of its independence in 1966. It was also wholly surrounded by then white-ruled, potentially aggressive neighbors. Botswana used its diamond wealth to build a first-class education system, develop a desperately needed transportation and health infrastructure, and expand other segments of the economy. The result is that Botswana is arguably the world's greatest economic success story of the last third of the twentieth century. Furthermore, it achieved this development as a respected liberal democracy, unlike the undemocratic newly industrialized countries (NICs) of Asia. This story has been chronicled many times, but what I tried to do

RAJESH JANTILAL/AFP/Getty Images.

FIGURE 3.16   Durban, South Africa. Mahatma Gandhi International Peace and Reconciliation Award ceremony.

was explain this in my book, *An African Miracle: State and Class Leadership and Colonial Legacy in Botswana Development* (1999), and draw lessons from that explanation for development policy in other countries.

development, suggests that we need to fundamentally change development so that it is more compatible with the environment. This involves re-conceptualizing development as a process that is different from economic growth. The idea is to improve life and to address basic human needs, without necessarily increasing consumption or harming the environment. Furthermore, it is an approach to development that respects biological limits. This means not producing more pollution than can be assimilated by the environment, or not using renewable resources (such as forests) more quickly than they are able to regenerate. If the goal is to use fewer resources in order to stay within biological limits, this not only means consuming less, but also recycling, using renewable energy, and improving efficiency of energy and material use. A common approach to ensure that biological limits are respected is government regulation. This approach to development is different enough from the traditional economic growth model that most governments do not embrace it.

The more common definition of sustainable development, sometimes referred to as *weak sustainable development*, suggests that development and environmental stewardship are complementary. According to this argument, economic growth is necessary for good natural resource management because (1) you need wealth to invest in conservation and environmentally friendly technology; and (2) wealth allows you to have a longer-term vision and to consider future environmental conditions. Furthermore, it suggests that an improved natural resource base furthers development.

Some analysts suggest that weak sustainable development was invented by and is supported by international financial institutions (such as the World Bank and IMF) and corporations. The 1992 World Bank annual report, which was devoted to sustainable development, supported weak sustainable development. A number of critiques have been levied against weak sustainable development. First, economic growth may create wealth that can be invested in environmental protection, but additional production and consumption also create waste and use up resources. It is not clear if the additional protection completely compensates for this loss. Second, while wealthier economies may appear to be cleaner, much of this gain may be the result of exporting dirty industries to other countries (hence there is no net gain for the global environment). Third, while wealth may make it possible to invest in cleaner technologies, this does not necessarily happen. For example, an extended period of economic growth in the United States during the 1990s was accompanied by a significant increase in consumer purchases of fuel-inefficient sport utility vehicles (SUVs). Finally, ample evidence from the developing world suggests that poor households are often more frugal and more apt to take care of environmental resources than wealthier households (Moseley 2001).

Despite disagreements about whether sustainable development should be conceptualized in a strong or weak framework, and thus how sustainable development plans should be implemented, the debate has allowed concerns about the environment to become a part of the development agenda around the world.

FAIR TRADE

Fair trade is both a social movement and a means of market-based exchange that emphasizes fair compensation and healthy and environmentally sound practices at the site of production. Most of this exchange has been between producers in the Global South and consumers in the Global North.

How did fair trade become an alternative model for development? In the first decade of the twenty-first century, scholars and policymakers alike recognized an ever widening disparity between rich and poor countries and questioned the veracity of traditional development models. Some suggested a complete withdrawal from the world trading system. Advocates of this "disengagement" argument (Bond 2002) suggested that involvement with the global economy has only hurt poor countries and that global trade flows are not "free" but rather are structured in a way that benefits the Global North. Others have increasingly argued for a return to protecting local production and encouraging import substitution (Bruton 1998). Still others have been pushing for fair trade as a new form of exchange that will promote development in the Global South (Raynolds et al. 2007).

The history of the fair trade movement in the Global North is often traced back to two developments. First, consumers boycotted products in the 1960s and 1970s which they saw as unfairly produced. Examples include the boycotts of grapes in the United States by the United Farm Workers Union in 1965 and the boycott against Nestlé in 1977 to challenge its marketing of baby formula in the developing world (Murray and Raynolds 2007). Second, another important precursor to fair trade, also occurring in the 1960s and 1970s, were nonprofits and church groups that ran handicraft stores that sold goods from the developing world. Their aim was to try to help producers retain a greater share of profits (Raynolds and Long 2007).

Fair trade as we know it today was established in the late 1980s with development of product labeling and certification (Figure 3.17). Fair trade has grown steadily since the 1990s, more recently recording $6.6 billion in sales in 2012, according to Fair Trade International. Over five million farmers and farm workers across the Global South now produce fair trade-certified goods. Coffee is the leading fair trade product, followed by bananas, cocoa, tea, and sugar. While fair trade sales represent a very small share of the total market, the volume of fair trade-certified goods grew by nearly 500 percent between 1998 and 2005 (Raynolds and Long 2007). Most fair trade goods are produced in Latin America, especially Mexico, but certified fair trade production is expanding rapidly in Africa and Asia.

Despite its alternative beginning and the tremendous rate of growth in recent years, many remain skeptical of the

Ian Hanning/REA/Redux.

FIGURE 3.17 Fair trade coffee. Fair trade coffee is produced in Latin America and Africa and purchased by large coffee roasting companies, including Starbuck's (based in Seattle, Washington) and Malongo (based in Nice, France). An employee of Malongo checks the packaging on coffee at a roasting plant.

fair trade movement and its potential to bring about lasting social change (Bryant and Goodman 2004; Mutersbaugh 2005; Watts et al. 2005; McCarthy 2006). For example, Bryant and Goodman are concerned that fair trade represents "caring at a distance," which leads to the "uncritical acceptance of consumption as the primary basis of action" (2004, 344). In other words, perhaps we have come to view the market as the only means of bringing about social and economic change. This belief may undermine the continued exploration of using government as an agent of such change.

Another concern relates to the great expense involved in becoming fair trade certified. McCarthy describes how the fair trade certification process may ensure premiums, but it also "imposes costs on producers, and excludes many potential beneficiaries." (2006, 808). As such, fair trade may favor relatively wealthier and more organized producers.

Development implies progress, and during the twentieth century, most development theorists, governments, nongovernmental organizations, and intergovernmental organizations thought of development as gaining economic wealth. Since the 1970s, geographers have been asking whether we should measure development in other ways, why development varies so vastly across scales, and how projects can be better set up to address development across multiple scales concurrently. By studying unequal exchange, dualism, core, and periphery, geographers are gaining a better understanding of development. Today, participatory development, the trust fund model, sustainable development, and fair trade are espoused by scholars as better models of development, and geographers are studying these projects and weighing how they differ from structural adjustment projects, or modernization-inspired dam building, which drove development in the second half of the twentieth century.

CONCEPT CHECK

1. **How** have past forms of development assistance been problematic for receiving countries?

2. **Why** might it be important for development to be a participatory process?

3. **What**, in your view, are the strengths and weaknesses of fair trade as an approach to development?

 ## SUMMARY

What Is Development and Where Did the Idea of Development Originate?

1. Countries have long compared themselves to one another in terms of wealth. For example, during the era of mercantilism in the sixteenth to eighteenth centuries, European states competed with each other to see who could accumulate the most gold reserves. Countries also drew distinctions between themselves and others based on notions of technological superiority.

2. Although we can clearly see notions of development during the colonial era, what we call development in contemporary terms was born in the aftermath of World War II. In the closing days of the war, the major powers met at Bretton Woods, New Hampshire, to plan for peacetime. They proposed several global institutions, including the World Bank, whose purpose was to provide loans for the reconstruction of Europe and Japan following the devastation of the war.

3. Is development synonymous with economic growth? One limitation with each measurement of a country's economy (GDP, GNP, GNI) is that wealth generated by economic growth is not usually spread across a population evenly. If a country's GNI increases, it may mean a lot of wealth has been generated for the richest segment of the population, thereby increasing the GNI per capita and GNI. However, the poorer members of society may see little to no improvement in their condition, despite the state's increased GNI.

4. Many development programs are created as local scale initiatives. For example, a village may come together to build a new school, water pump, or community garden. Other types of development programs or policies, such as educational reform or disease eradication programs, may be implemented at the national or regional scale. Ideally, the scale of a development initiative will bear some resemblance to the scale of the development issue or challenge.

How Does Geography Help Us Understand Development?

1. In the aftermath of World War II, the growth in the number of states led scholars to want to offer explanations of how countries became developed in order to advise the newly independent

Fieldwork in Geography YOUR**TURN**

This photo is of a shop on a crowded market street in Kuwait City, Kuwait. Kuwait is a relatively wealthy, oil-rich, and geographically small country on the Persian Gulf that has invested heavily in development. Relative to other countries in Southwest Asia and North Africa, it measures higher on many development scales, especially in terms of education and health care.

Courtesy of William Moseley.

Kuwait City, Kuwait. Proprietors and shoppers in a marketplace.

Thinking Geographically

- When you look at this market scene, what do you see? How does this image of what is "developed" in this region mesh with your idea of what development looks like?

- How is this shop similar to or different from one you might have in your neighborhood?

- Set aside your preconceived idea of development and describe what details you see in this photo that supports the idea that Kuwait is relatively developed.

Read: The Arab Fund for Economic and Social Development: http://www.arabfund.org.

Mohammed, Nadeya Sayed Ali. *Population and Development of the Arab Gulf States: The Case of Bahrain, Oman and Kuwait.* Burlington, VT: Ashgate, 2003.

---

countries on a path to development. Theorists looked for steps or stages the newly independent countries could follow. This suite of theories, often collectively referred to under the umbrella term **modernization**, posited that the European industrial economy was the ideal or pinnacle stage of development.

2. Multiple critiques of modernization theory emerged in subsequent decades. To understand the role of context in development, geographers, including Peter J. Taylor, draw from R.H. Tawney's tadpole philosophy, which considers that only a small percentage of tadpoles develop into frogs in a pond. A pond may have a thousand tadpoles, but only a small percentage become frogs because all of the tadpoles compete in the same pond for food, sun, and other resources. Similarly, in the world economy, not all countries can become wealthy.

3. In reaction to modernization theory, a new set of development ideas began to arise in the 1960s that emphasized structure, or the global framework under which countries operated. This emphasis on structure meant that the relationships between countries were as or more important than internal policies for determining the future development of a country.

4. **Dependency theory**, originally conceived by Andre Gunder Frank (1979), suggested that economies in the tropics (Africa, Latin America, Asia) were essentially "underdeveloped" during the colonial era as European countries refashioned these economies (through a combination of taxation policies and forced coercion) for their own benefit.

Wallerstein (1979) expanded on Frank's ideas though his **world-systems theory**, which basically gave a spatial face to dependency theory by depicting the world in terms of the core, semiperiphery, and periphery.

5. In addition to structuralist theories of development, by the 1990s economists and economic geographers had developed a growing critique of neoliberal development policy known as new economic geography. Most new economic geographers continued to believe that free trade was the best route to development, but they recognized the challenges faced by some countries in terms of physical geography, such as tropical diseases, poor agricultural soils, or inadequate road, rail, and shipping connections to global trading networks.

6. Until the 1970s, much development assistance had been focused on improving the welfare of a generic population—undifferentiated by wealth, **gender**, race or ethnicity. As a result, the vast majority of development aid tended to favor the most powerful members of a community (namely, men, the wealthy, and other dominant groups) who were better at articulating their wants and capturing new resources.

Why Is Development Uneven?

1. From a geographical perspective, one of the most striking aspects of development is its unevenness. We are perhaps most familiar with such differences from country to country, but the gulf between the rich and poor is often just as

wide within countries. These differences can be explained in numerous ways; why some countries and areas are more prosperous than others is a very big and long-standing question.

2. Rostow's modernization theory came out of the context of decolonization and the Cold War after World War II, and environmental determinism grew in an era of Darwinism and colonialism. Charles Darwin's theory of evolution, published in 1859, holds that through a process of natural selection, species evolved over time in relation to a changing physical environment and biotic community.

3. Geographers eventually, especially after World War II, saw environmental determinism as deeply flawed, if not highly racist. Scholars began to question the ideas of environmental determinism because they found too many exceptions to the generalizations being made about the tropics.

4. One concept that is useful for understanding varying levels of development is **dualism**. Dualism refers to situations where two areas are in relationship with one another (through trade, for example), and one area is developing at the expense of the other. This concept may be examined at a variety of different scales, including global, national, local, urban, and rural.

What Role Does Development Assistance Have in the World Economy?

1. International development assistance comes in all shapes and sizes. It varies in terms of the provider (e.g., intergovernmental organizations such as the United Nations, a country's government, or a nonprofit organization) as well as in terms of its form, whether a loan for a dam, emergency food relief, technical assistance to a government agency, or funds for a community well. After billions of dollars have been given in international development aid over the last 20 years, academics, donors, and donor countries are questioning the value of international development assistance.

2. One theme that has received considerable attention in foreign assistance circles since at least the early 1990s is **participatory development**. The emphasis on participation came about in reaction to a history of top-down development projects that left local people with little to no sense of ownership of the projects undertaken by outsiders in their communities. Besides being problematic in its own right, a lack of participation often inhibited the longer-term sustainability of programs because local people had little interest in maintaining projects they did not initiate or request.

KEY TERMS

development	modernization	dualism	sustainable development
third world	dependency theory	unequal exchange	fair trade
global south	world-systems theory	commodity chain	
gross national income (GNI)	gender	participatory development	
human development index (HDI)	environmental determinism	mental map (cognitive map)	

CREATIVE AND CRITICAL THINKING QUESTIONS

1. **Development** has often been conceptualized as economic growth or increasing wealth. Do all peoples and cultures value wealth over all other measurements of development? How can you measure development to allow for differing cultural values?

2. How does **unequal exchange** work to create uneven development?

3. How do the **population pyramids** of developing countries differ from the population pyramids of wealthy countries?

4. If an NGO is working to "help" a people develop their economy, how might the solution they choose to pursue differ if they base their solution on their own perception rather than on the **mental maps** (cognitive maps) of the people with whom they are working?

5. Where can you purchase fair trade products in your town or city? Read the label of one of the products and the company's

website to help you piece together the **commodity chain** of the fair trade product. Describe the commodity chain.

6. Using the **commodity chain** you developed for the fair trade product, determine whether the fact that you purchased this fair trade product created or avoided unequal exchange.

7. Read about Rostow's modernization theory. Describe how you have been taught about colonialism and development in other classes or settings. Determine whether Rostow's modernization theory **diffused** into the answer you previously received. Explain how Rostow's theory has been so influential on ideas of development, and explain how you can start to think about development in a different way.

8. When the newspaper in your town or city talks about a new "development," what is it usually describing? Does the idea of development privilege one kind of **cultural landscape** over another? How does this ordering of cultural landscapes impact sustainable development?

1. Which of the following measures best accounts for wealth attributed to a country but generated outside the country's borders?
   a. gross national product (GNP)
   b. human development index (HDI)
   c. gross national income (GNI)
   d. gross domestic product (GDP)

2. Ideally, the scale of a development solution should:
   a. be higher than the problem
   b. be lower than the problem
   c. be at the same scale as the problem

3. If the GNI of a country increases over time, not everyone in the country may be seeing greater wealth because:
   a. the GNI is only measured from the wealthiest 10 percent of a country's population
   b. wealth is usually distributed unevenly among a country's population
   c. increased GNI typically comes from higher salaries among a country's poor
   d. the GNI measures happiness, not wealth

4. When Rostow created the modernization theory, he used this country, the hearth of the Industrial Revolution, as a model of development:
   a. France   c. Canada
   b. United States   d. Great Britain

5. Environmental deterministic theories draw from this context of theories shaped by this person:
   a. Rostow   c. Frank
   b. Darwin   d. Wallerstein

6. World-systems theorists think about the world in terms of core, periphery, and semiperiphery. Core and periphery are determined by:
   a. what is produced
   b. how goods are produced

7. The colonial relationship between Haiti and France is a good example of dualism because:
   a. France exploited Haiti economically
   b. France favored its colony in the Dominican Republic over its colony in Haiti
   c. France required Haiti to make structural adjustments before lending it money

8. When Chicago underwent economic development during the 1800s and 1900s, it did so by:
   a. drawing wealth from a region of extraction in the Midwest
   b. building dams on the Chicago River
   c. encouraging the development of parastatal companies
   d. following Rostow's ladder of development

9. True or False:
   Geographers often use dualism or world-systems theory to study economic development because both theories can be used across scales.

10. The countries that donate the most official development aid per GNI are located primarily in:
    a. North America and Europe
    b. North America and East Asia
    c. Europe and East Asia

11. World-systems theorists would argue that:
    a. not all countries in the world can be wealthy
    b. all countries in the world can be wealthy
    c. the world is comprised of wealthy countries and poor countries with none in between
    d. the vast majority of the word's countries are middle class

12. True or False:
    Fair trade seeks to break down unequal exchange.

13. Structural adjustment loans require countries to:
    a. close their countries to foreign direct investment
    b. privatize publicly held entities
    c. pay off all debts before receiving more loans
    d. place substantial trade barriers on imports into their countries

14. Mental maps (cognitive maps) are created:
    a. using global positioning systems (GPS) and exact measurements
    b. from the memories and experiences of individuals or groups in an area
    c. by development specialists from written records

15. The secondhand clothing market in Zambia has:
    a. undermined the textile industry in Zambia
    b. helped economically develop the interior of Zambia
    c. created many more jobs in the formal sector of the Zambian economy

ANSWERS FOR SELF-TEST QUESTIONS

1. c, 2. c, 3. b, 4. d, 5. b, 6. b, 7. a, 8. a, 9. True, 10. a, 11. a, 12. True, 13. b, 14. b, 15. a

4

SUBSAHARAN AFRICA

When you look at this photo of a mother tilling a field in the heart of West Africa, what do you see? Some of us might see a downtrodden, impoverished woman forced simultaneously to care for her child and to farm with primitive tools. Others might interpret the tree-filled field in the background as a sign of backwardness, of potential farmland unused, or the fallen wood behind her as evidence of deforestation. These immediate assumptions are largely inaccurate, based more on Western views than on a grounded understanding of the African context.

In Subsaharan Africa, women produce roughly 70 percent of the food. The woman photographed here is farming peanuts that she will use in her family's meals or sell at the market. Why the simple farming implement? She uses a short-handled hoe because these are heavy clay soils, making long-handled hoes less effective. A motorized plow would destroy the soils in this area after only a few years of use.

The trees in the background have been maintained in the field because they reduce wind erosion during the dry season and provide useful tree products. Finally, the fallen branches are dead limbs trimmed from a nearby tree for firewood that she will collect, leaving the tree to survive and grow. Although the woman in this photo certainly faces challenges, she is managing the landscape with knowledge and skill.

Courtesy of William Moseley.

Falan, Mali. This mother farming in her field with a baby on her back in southern Mali is keenly aware of how to maintain the fertility of her fields in order to provide a steady food supply to her family. She carries her small child on her back while working in the fields because her husband and sons are likely tending livestock or working in nearby cities. Her older daughters are likely working close by or performing other tasks, such as fetching water, that help keep her family going.

 THRESHOLD CONCEPTS in this Chapter

Site
Situation

WHAT IS SUBSAHARAN AFRICA?

LEARNING OBJECTIVES

1. **Understand** the major climate drivers in Subsaharan Africa and their influence on human settlement patterns.

2. **Describe** how societal roles influence the way people perceive and are affected by environmental change.

3. **Realize** the tension between local people and parks.

Africa's physical geography is imposing (Figure 4.1). The continent has the world's longest river in the Nile at 4,145 miles (6670 km), the world's largest desert in the Sahara at 3.5 million square miles (9.06 million sq km), and a major chain of mountains and valleys known as the Great Rift Valley, which runs from the Red Sea through East Africa to central Mozambique and includes snow-capped Mount Kilimanjaro near the equator (Figure 4.2). Subsaharan Africa is within the Tropical climate zone, with the equator crossing through Gabon and the Democratic Republic of the Congo. Understanding environmental conditions in Africa allows us to gain significant insights into where and how people live.

CLIMATE

Westerners often think of Africa as a dry continent. The enormous Sahara and Kalahari deserts help cement an image of dry or desert lands in our minds. However, the whole region of Subsaharan Africa is not dry. The region has

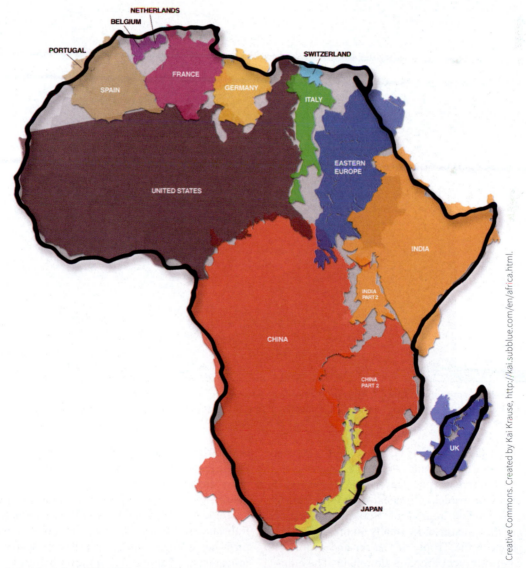

Creative Commons. Created by Kai Krause, http://kai.sublue.com/en/africa.html.

FIGURE 4.1   Imposing size of the African continent.   This map shows the area of several countries and regions of the world that can be pieced together to fit inside the continent of Africa. The map shows the 48 contiguous United States, China, India, eastern Europe, Japan, and several western European states fitting inside the continent of Africa.

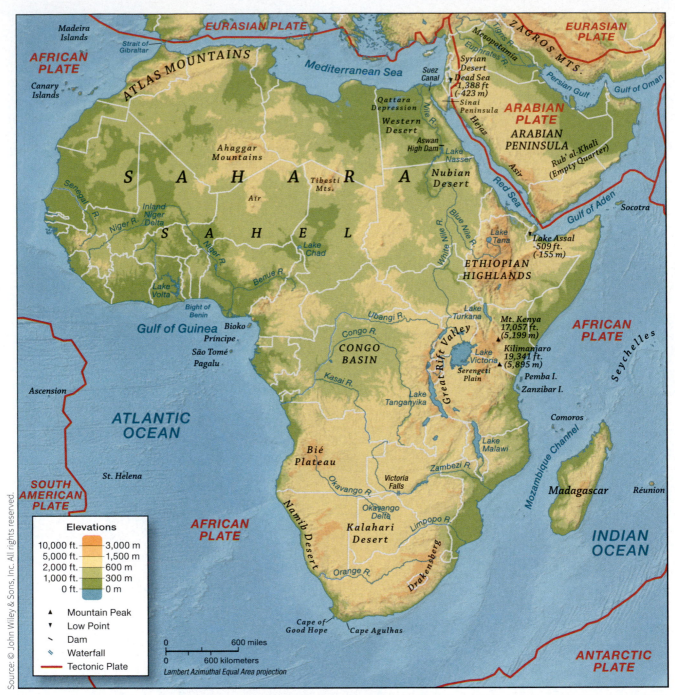

FIGURE 4.2 **Physical Geography of Africa.** The continent of Africa was once the middle of Pangaea, the massive continent that existed 300 million years ago. Africa is now being pulled in two different directions along the plate boundary that runs from the Red Sea to Mozambique.

nearly 400,000 square miles (643,000 sq km) of rainforest and another 5 million square miles of tropical savanna.

The map of climate zones in Africa (Figure 4.3) shows the variety of tropical climates, from rainforest to desert, in Africa. The region of Subsaharan Africa begins south of the Sahara, and virtually all of Subsaharan Africa lies between the Tropic of Cancer and Tropic of Capricorn, where tropical climates dominate. The equator cuts the continent in half, and the rainforest is along the equator. Tropical climates occur around the equator because this area receives direct or relatively direct

insolation (incoming solar radiation) for the entire year (Figure 4.4). The areas closest to the equator receive the most insolation, and the annual average insolation declines with each degree of latitude toward the poles. Latitude alone does not explain the climate regions of Subsaharan Africa. The large amount of insolation, in turn, has a huge impact on pressure systems and wind movements, shaping precipitation patterns.

The place on earth receiving the most insolation will absorb the most radiation from the sun. Earth then reemits the energy it has absorbed, creating a low-pressure system

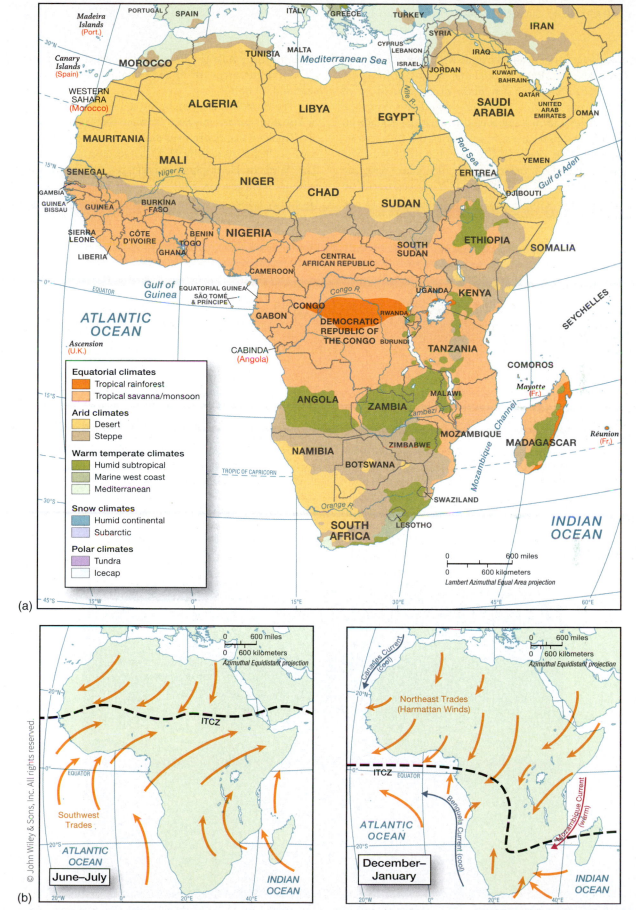

FIGURE 4.3 Climates of Subsaharan Africa. (a) Climates of Subsaharan Africa. (b) ITCZ movement in Africa. The ITCZ shifts between the Tropics of Capricorn and Cancer throughout the year, moving because earth is tilted on its axis as it revolves around the sun. The amount of insolation varies by latitude and by time of year because of earth's tilt. Differences in insolation drive seasonal wind and moisture patterns in the areas north and south of the equator.

where warm air is rising and the trade winds are converging. The low pressure along the equator draws air from the Atlantic Ocean onto the landmass of Africa. Most of Africa receives rainfall from air originating over the Atlantic Ocean and moving inland toward the equatorial zone of this low-pressure belt, known as the **Intertropical Convergence Zone (ITCZ)** (see Figure 4.3b). The location of the ITCZ migrates over the year, with the direct rays of the sun. When the ITCZ moves north (in June) toward the Tropic of Cancer, areas north of the equator in Africa receive seasonal rainfall, and when the ITCZ moves south (in December) toward the Tropic of Capricorn, areas south of the equator in Africa receive seasonal rainfall.

The latitude directly hit by the sun's rays (at a 90° angle), or **subsolar point**, migrates between the Tropic of Cancer and Tropic of Capricorn over the course of the year because earth is tilted on its axis by 23.5° as it revolves around the sun. The Northern Hemisphere experiences summer when the subsolar point reaches 23.5°N on June 21, the summer solstice. This is the rainy season in African regions north of the equator because the ITCZ draws in southwesterly winds from

Reading the **PHYSICAL** Landscape

The Serengeti

Large animals, including zebras, elephants, lions, and massive herds of wildebeest, roam the Serengeti, an area of savanna in Tanzania and Kenya (Figure 4.4a). The southern part of the region receives rainfall closer to the December solstice; so, herds of wildebeest and zebra migrate south in January, February, and March to graze on fresh grass. In June, July, and August, as the climate in the south gets drier, the herds migrate north to graze on vegetation along the Mara River. The cycle begins again in November or December, when the herds migrate south.

The relationship between earth and sun gives this region its distinct wet and dry seasons (Figure 4.4b). During the local summer months, the subsolar point hits at a latitude close by, and the ITCZ, a belt of low pressure, migrates with the subsolar point (Figure 4.4c). With the presence of the low pressure, warm, moist air is drawn in from the Atlantic because air always flows into lows. The warm air rises and cools enough to reach its condensation level, and rain falls. In contrast, the winter season is dry because the subsolar point and associated low-pressure system of the ITCZ have shifted to the opposite hemisphere. The distinct seasons of the savanna and the migration of the ITCZ have a significant impact on the migration of animals in the region.

Skip Brown/National Geographic/Getty Images.

FIGURE 4.4a   Serengeti, Tanzania.   The savanna covers much of Africa with level plains marked by thick stands of grass and isolated trees.

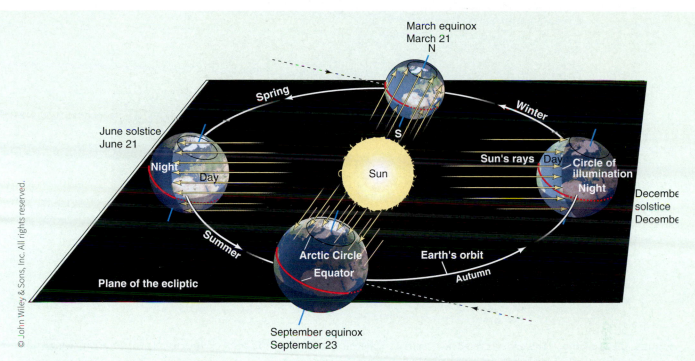

FIGURE 4.4b   Earth-Sun relationship.   The earth is tilted on its axis 23.5°. As earth rotates on its axis and revolves around the sun, the direct rays of the sun, the subsolar point, hit between 23.5°N and 23.5°S latitude over the course of the year.

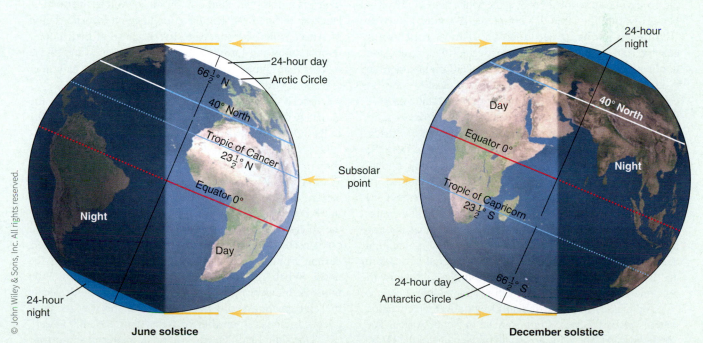

FIGURE 4.4c   Sub-solar point.   The direct rays of the sun, the subsolar point, hit at 23.5°N latitude during the June solstice, which is the northern hemisphere's summer. The subsolar point hits at 23.5°S latitude during the December solstice. The equator divides Africa in half, so that the northern part of Africa starts summer at the June solstice and the southern part of Africa starts summer at the December solstice.

Photo by William Moseley.

FIGURE 4.5   **Lesotho.**   This snow-covered hut stands on a plateau in the highlands of Lesotho. The elevation of the entire tiny country, located south of the Tropics at 30°S, is over 4,500 feet (1,400 meters).

the Atlantic Ocean. Atlantic winds penetrate as far east as Ethiopia. During the same time of year, the absence of the ITCZ gives African regions south of the equator a dry season. In the south, dry, continental winds sweep across the region.

The subsolar point migrates south of the equator as far as the Tropic of Capricorn, 23.5°S, at the winter solstice, December 21. During this time dry northeasterly winds, known as **Harmattan winds**, sweep over West Africa, north of the equator. At the same time, in East Africa north of the equator, northeasterly winds bring dry continental air originating from Asia. On the southern side of the equator, most of the west coast receives moist wind coming off of the Atlantic because the presence of the low-pressure belt of the ITCZ draws moist air from the Atlantic into southern Africa. In East Africa, south of the ITCZ, large tropical cyclones move in from the Indian Ocean, creating high winds and torrential rainfall along the coast of eastern South Africa and Mozambique.

The migration of the ITCZ creates distinct wet and dry seasons in the savanna grasslands north and south of the rainforests. The dependability of the rainy season makes these savanna grasslands and forests some of the best agricultural areas on the continent. Traditional African crop agriculture and herding practices are well adapted to the seasonality of precipitation, as are natural vegetation and wildlife.

Air masses from the Indian Ocean affect precipitation patterns in East Africa, from Somalia to South Africa. The South African Cape is located south of the range of the ITCZ, has a Mediterranean climate, and receives rainfall in the winter when midlatitude westerlies reach their northernmost point.

Two other factors shape the climate map of Subsaharan Africa: elevation and ocean currents. Higher elevations have cooler temperatures, including snow in some parts of

Africa (Figure 4.5). At an elevation of 8,000 feet, Addis Ababa in Ethiopia has lower temperatures over the year than its latitude, just 9° N, would suggest. The average temperature in Addis Ababa in July is 69°F and 77°F in May, and the average drops as low as 55°F in January. Elevated plateaus in East and southern Africa, as well as the temperate zone in South Africa, have long been a favorite region of Africans and Europeans to live because of their cooler temperatures and lower incidence of disease. Many of the insects that spread tropical disease do not live at higher elevations with cooler temperatures.

Ocean currents are the final factor in shaping the climates of Subsaharan Africa. Traveling westward from the Indian Ocean toward Africa, the North Equatorial current splits around Tanzania to flow northward as the warm monsoon drift and southward as the warm Mozambique current. Warm ocean currents keep temperatures warmer along the eastern coastal areas of the continent, and swimming is much better in the ocean on the east coast of Africa than in the cooler oceans of the west coast. On the west coast, two cold ocean currents converge near the equator: the Benguela current from the south and the Canary current from the north. When cool air masses that can only hold a little moisture hit a warmer coastline, they heat up and desiccate the land, creating a **cold water desert**. In this way, the Benguela current is responsible for West Africa's Namib Desert.

This thumbnail sketch of climate patterns and physical geography in Africa allows us to begin to understand human population distribution across the continent. Geographers love to ask the question, "Why is something where?" and maps of the distribution of human population in Africa demand these types of questions (Figure 4.6).

Perhaps the most noticeable pattern in Figures 4.6b and 4.6c is the narrow, curvaceous strip of population along the Nile River in Egypt and Sudan. A fertile river valley in the midst of the Sahara Desert is a magnet for floodplain farmers and urban centers. Approximately 90 percent of the entire population of Egypt lives within 5 miles (8 km) of the Nile River.

In contrast, the vast Sahara in the north and the Kalahari and Namib deserts in the south are sparsely populated. The densely settled coastal zone of West Africa contains port cities that have become some of the continent's major urban areas. Other populations are concentrated on the coastal zone of West Africa in the prime farming areas of the tropical rainforest and savanna zones. Along the Rift Valley in East Africa, especially near Lake Victoria, Lake Tanganyika, and Lake Malawi, rich farming and fishing opportunities attract dense concentrations of people. Finally, high population densities in much of coastal South Africa are explained by rich farming opportunities because of heavy summer rains along the southeast coast and winter rains along with a Mediterranean climate near Cape Town in the southwest.

SOCIETAL ROLES AND ENVIRONMENTAL CHANGE

Environmental conditions may influence how and where people live as a generic whole, but do not necessarily affect everyone in the same manner. People's interactions with environmental resources often vary based on gender, ethnicity, age, and relative wealth.

Take gender as an example. Many rural women in Africa are responsible for the collection of firewood and water. When firewood grows scarce, or the water table drops, it is often women who find alternative sources. The time devoted to firewood and water collection may constrain women's other activities, such as farming or commercial activities at the market.

People's daily activities also affect how they understand the environment and environmental change. Many rural men may be less aware of certain types of environmental change than women because they do not interact with some resources on a daily basis the way women do. Someone's age or ethnicity may also influence how they interact with environment. For example, geographer Cindi Katz (1991) has shown that young boys who work as herders in Sudan have very detailed knowledge of the landscape, pasture, and water conditions because of their responsibilities tending sheep. Katz asked Sudanese boys to draw their mental maps (Figure 4.7). The maps revealed the young boys' intricate understanding of their environment.

PARKS AND PEOPLE

In order to justify taking prime agricultural land from native Africans, colonial authorities argued that local people mismanaged it. During colonial times, European authorities transferred land to European settlers in East and southern Africa, and also established a number of national parks, hunting preserves, and forest reserves.

The British named the colony of Rhodesia (modern-day Zimbabwe) after the British mining magnate and politician Cecil Rhodes. In the 1930s, British surveyors zoned the entire colony in terms of agricultural potential, from natural region I (the best land) to V (the least arable land) (Figure 4.8). They then created commercial agricultural areas, mostly in regions I, II, and III, for European farmers. The British designated communal areas for African farmers, mostly in regions IV and V. Region IV has below-average rainfall, and region V is semiarid land. These communal areas are also called Native Reserves. Through the 1930 Land Apportionment Act, Britain reserved over half the land for settlers from Europe, although at that time Europeans represented less than 2 percent of Rhodesia's total population, and reserved less than a third of the land for the indigenous who comprised 98.5 percent of the population (Exploring Africa 2014).

With brute force, colonial authorities pushed Africans into the least productive land. As Africans, whom the colonists had already marginalized socially, were crowded into these drier areas, they became further impoverished, or economically marginalized. Not surprisingly, land degradation

 ## USING GEOGRAPHIC TOOLS

Mapping Population

At the scale of a continent, geographers can only represent population on thematic maps, which cannot show the exact location of every person. Thematic maps simply show the overall picture of where people live. These three maps demonstrate different ways of showing population concentrations on a regional map. The first is a map of average population density using countries as the areal unit. Maps where each area is shaded according to a quantity are called *choropleth* ("choro" means area and "pleth" refers to quantity, as in plethora). This map shows us which countries have the highest population density but not how population is distributed within any country.

(Continued)

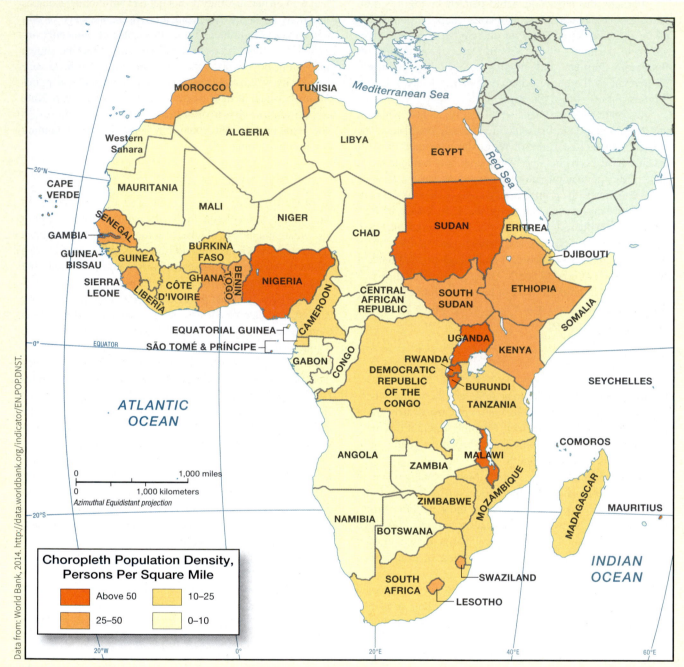

Data from: World Bank, 2014. http://data.worldbank.org/indicator/EN.POP.DNST.

Choropleth Population Density, Persons Per Square Mile

- Above 50
- 25–50
- 10–25
- 0–10

FIGURE 4.6a Choropleth map of population in Africa. In a choropleth map, data are shown by areal unit. Each country is shaded based on the population density, with the darkest color having the highest population density.

The second map, a dot density map, better reflects variations within countries, as dots represent clusters of populations within countries. One problem with dot density maps is that rural areas appear to be unpopulated, even though they certainly are not. The third map is a *dasymetric* population density map, which shows population variation within countries and perhaps better represents rural areas. Maps that show density of data according to the underlying geography are called daysmetric maps,

deriving from the Greek words for density and measurement. Dasymetric maps are like contour maps or barometric pressure maps because areas with the same data, in this case population, are connected. Areas with similar population are colored to show degrees of intensity, population concentrations on this map. Dasymetric maps work well for showing population density because they account for where people live, as well as where cities and rural areas are.

Thinking Geographically

1. What are the advantages and disadvantages of each method of mapping population density?

2. Determine which map best represents the population distribution in Subsaharan Africa.

Source: United States Geological Survey, Dasymetric Mapping: An Alternative Approach to Visually and Statistically Enhancing Population Density, http://geography.wr.usgs.gov/science/dasymetric.

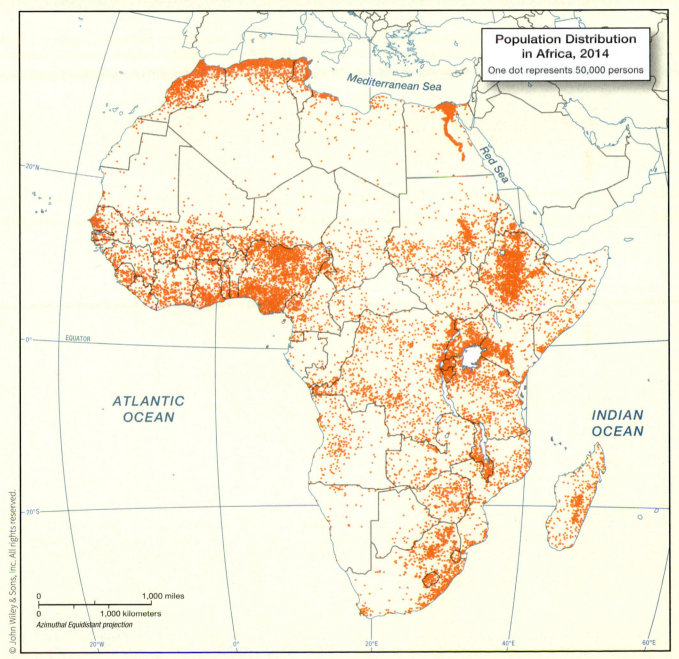

FIGURE 4.6b   Dot density map of population in Africa.   A dot density map uses a single dot to symbolize a certain number of people. On this map, one dot equals 50,000 people. The dots are placed on the map to show how people are generally distributed in Africa, with concentrations of people in cities and near freshwater sources.

(Continued)

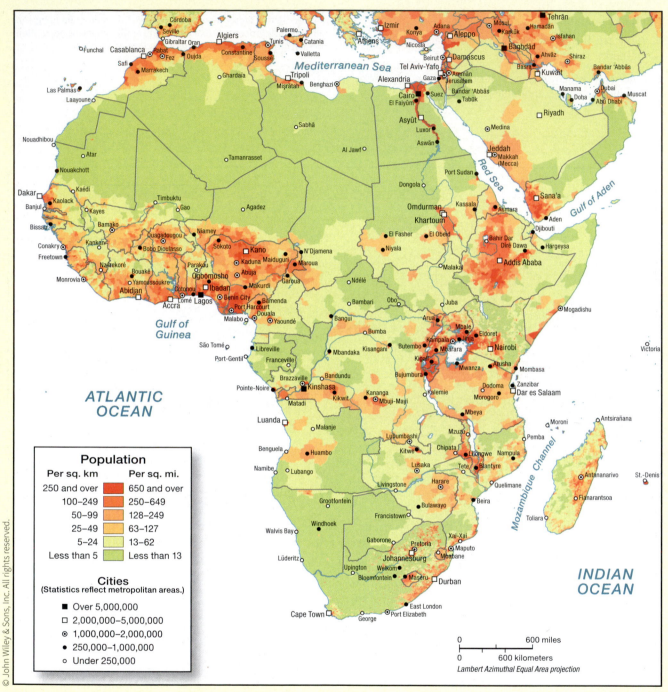

FIGURE 4.6c Dasymetric map of population in Africa. Cartographers created a way to show a theme, such as population, taking into account the landscape of the region or place. By accounting for the underlying surface and land cover, cartographers can better estimate where people live and which places are lightly populated on a dasymetric map. The population density maps throughout *Understanding World Regional Geography* are dasymetric maps.

has been high in many of these communal areas, a problem often blamed on population growth or traditional agricultural practices. However, the source of land degradation in East and southern Africa is the multiple ways, including social, environmental, and economic, that European settlers marginalized the African population. British Geographer Piers Blaikie identified the multiple forms of **marginality** in his classic text *The Political Economy of Soil Erosion* (1985).

In the postcolonial era, many national park systems have persisted and been expanded, particularly in East and southern Africa where relatively larger numbers of megafauna (big game animals) still reside (Figure 4.8). Global environmental movements spurred a burgeoning

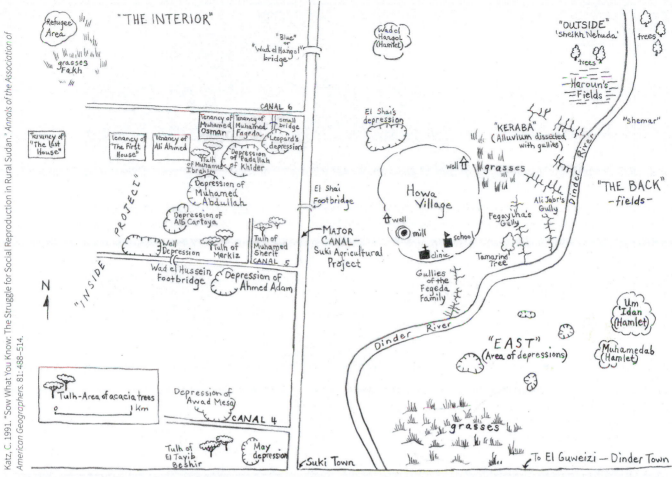

Katz, C. 1991. "Sow What You Know: The Struggle for Social Reproduction in Rural Sudan." *Annals of the Association of American Geographers.* 81: 488–514.

FIGURE 4.7 **Mental map of a young Sudanese herder.** This map shows how a young boy who worked as a Sudanese herder sees his world. Cindi Katz interviewed and traveled with a 10-year-old shepherd boy in northern Sudan to draw the map, which includes 57 different pasture areas near his village. The mental map reveals places important to the young boy, including his village, a river, wells, irrigation canals, footbridges, stands of trees, areas of low elevation (depressions), and grasslands.

ecotourism or nature-based tourism industry, which has bolstered African park systems. In countries including Kenya, Tanzania, Botswana, and South Africa, Western tourists flock to see the "Big Five," a term used to refer to the biggest, rarest, or most cherished animals traditionally sought after by trophy hunters: elephants, lions, rhinoceros, leopards, and African or Cape buffalo. Nature tourism has become big business. In Kenya, for example, it is the leading source of foreign exchange.

When national parks were established, local people often were evicted without compensation. Through land apportionment policies and establishment of parks and preserves, governments left the local people with little land to live on and work. In his 2007 book, *Transfrontier Conservation in Africa,* South African geographer Maano Ramutsindela, of the University of Cape Town, explored the history of transboundary parks in southern Africa. **Transboundary parks** are simply parks that span national borders. Governments have given multiple reasons for establishing these parks, including: the arbitrary nature of colonial borders, the promotion of peace, and the creation of larger tracts of land needed for biodiversity conservation. Since the end of apartheid (an aggressive approach to racial segregation and official state policy from 1948 to 1990) and the establishment of democracy in 1994, South Africa has helped establish six transboundary parks on its borders with each of its neighbors. Some researchers now see that the South African government's motivation to form parks was rooted in its desire to extend its political power to neighboring countries (Figure 4.9).

Land alienation, whether for purposes of securing the best agricultural lands for Europeans or developing parks and preserves, frequently led to feelings of resentment and compromised livelihoods because local people did not have the same resource base to rely on in their new locations. In 1900, Richard Leakey, the archaeologist and former head of the Kenyan Wildlife Service, established a shoot-to-kill policy to combat suspected poachers found in national parks. As a result, locals who wanted to hunt big game for sustenance could not hunt in the parks. Compressing local land ownership to smaller, marginal lands and establishing parks gave locals less land to farm.

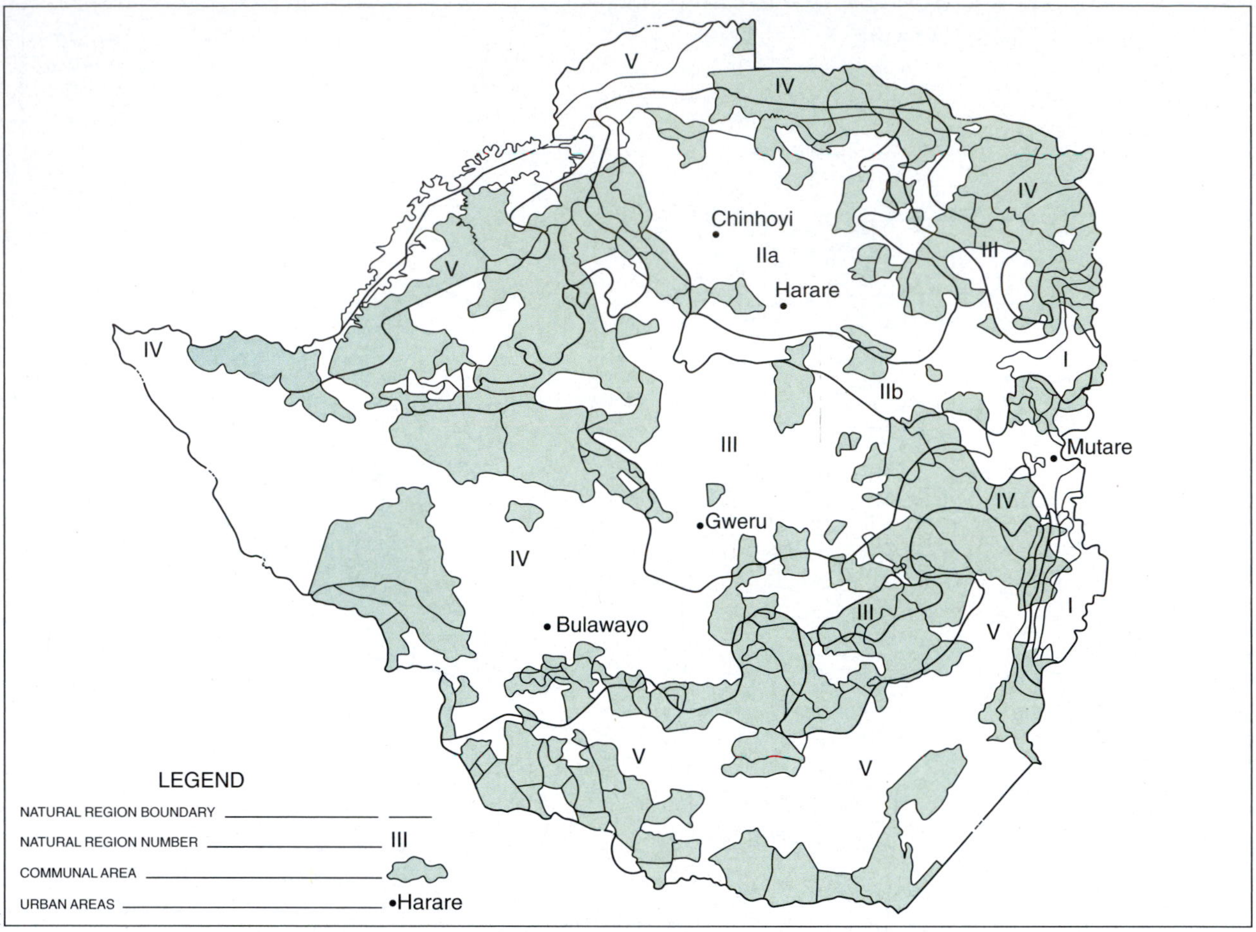

FIGURE 4.8   Colonial land classification zones and communal areas of Zimbabwe.   In 1930, British surveyors zoned the entire country in terms of agricultural potential, from natural region I, the best land, to V, the least arable land. They then created commercial agricultural areas, mostly in regions I, II, and III, for European farmers. So-called communal areas were created for African farmers, mostly in regions IV and V. The communal areas remain today.

Another source of resentment among locals is the fact that big game animals rarely respect park boundaries and may wander onto the lands of communities abutting national parks. Large animals may pose safety risks and destroy field crops. Elephants, in particular, have done a substantial amount of crop damage.

The question arises as to whether parks are *open-access resources*—resources with no particular owner—as opposed to *private property* belonging to an individual, or *common property* controlled by a group. In the absence of a legitimate controlling force or owner, open-access resources tend to be overexploited as everyone tries to maximize their personal gain with little to no regard for the overall ecological health of the resource. If a park is private property, the owner can strictly control who can access and use the park. Seeing parks as common property implies that everyone should have access to the parks' resources.

A number of conservation programs have been launched to address how parks should be viewed and to consider the needs of local people. Part of the incentive for the programs is a genuine concern for the welfare of local people, but it is also recognized that without the support of local people many conservation initiatives, including parks, are doomed to failure. Key to securing such support is to share ecotourism revenues with local communities in exchange for their participation in protecting wildlife resources. Such programs typically work with the communities' neighboring national park, and may even involve establishing buffer zones—areas surrounding national parks where the activities of local people are restricted.

CONCEPT CHECK

1. **How** does the migration of the ITCZ influence rainfall patterns in West Africa?

2. **Describe** how gender roles may influence perceptions of environmental change in Subsaharan Africa.

3. **Why** is there often tension between local people and parks in southern and eastern Africa?

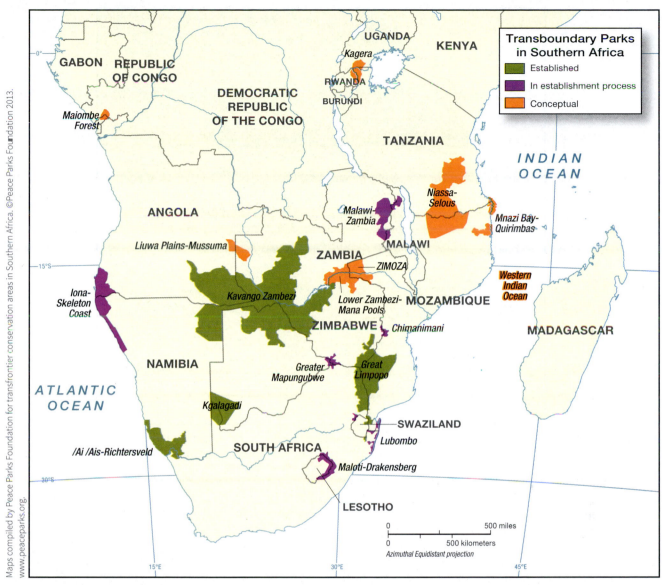

FIGURE 4.9   Transboundary parks in South Africa and Lesotho.   Maloti-Drakensberg Park straddles the border of South Africa and Lesotho, covering approximately 1,000 square miles (2,500 km²). In 2001, the governments of South Africa and Lesotho wove together Lesotho's national park with several smaller parks that had been established as early as 1903 in South Africa, to create the transboundary park, also known as a peace park.

WHO ARE SUBSAHARAN AFRICANS?

LEARNING OBJECTIVES

1. **Explain** patterns of urbanism in Africa before and after colonialism.

2. **Understand** how colonialism changed rural livelihoods in many Subsaharan countries.

3. **Discuss** how neocolonialism has operated in the postcolonial period.

Africa is the cradle of humanity. *Homo sapiens* arose about 150,000 to 200,000 years ago in Subsaharan Africa. We began to migrate out of Africa around 100,000 years ago—replacing **hominids** in other parts of the world. As such, it is no exaggeration to suggest that we are all Africans. Between that migration and today, Africans established cities of commerce including Jenne and Timbuktu in West Africa, domesticated crops and livestock between 4,000 and 5,000 years ago, and established the world's first university by the twelfth century.

PRECOLONIAL CIVILIZATIONS

Africans have been living in cities since the eighth century. They established roughly three-quarters of today's most important cities prior to the European conquest. Archaeologists mapped the cities of precolonial (before 1800)

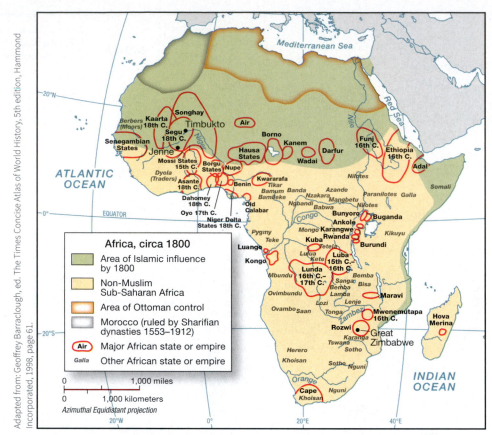

FIGURE 4.10   Major African civilizations, circa 1800.   African civilizations of various territorial size functioned throughout the massive continent. Islam had diffused through North Africa, the horn of Africa, and along the east coast. Thousands of other small civilizations existed at this time in localized areas and are not represented on this map.

African civilizations and established that most of these cities served as political capitals and as the headquarters for powerful emperors or kings (Figure 4.10). For example, Great Zimbabwe (Figure 4.11) in today's central Zimbabwe served as the capital of the Karanga Empire, which flourished between the thirteenth and fifteenth centuries.

Imperialist Europeans and Americans concocted wild theories to explain the impressive ruins at Great Zimbabwe because the sophistication reflected in the ruins did not fit Western depictions of Subsaharan African cultures as primitive and backward (ancient Egypt was another story). One such theory suggested that the grand city must have been the seat of power of King Solomon and the Queen of Sheba. In other words, the city was the work of Egyptians and Phoenicians, not Subsaharan Africans. Although Egypt is physically on the continent of Africa, Westerners differentiate Egypt, historically home of the Nile River civilization and tied to Ancient Greece and Ancient Rome, from Subsaharan African civilizations. Western depictions show Egypt as "civilized" and Subsaharan Africa as "primitive." Great Zimbabwe, however, was not tied to Egypt. Radiocarbon dating has revealed that Great Zimbabwe flourished after 1000 CE, far too late for the King Solomon and Queen of Sheba myth to have any basis. Great Zimbabwe, most likely ruled by the Karanga, amassed great

wealth from cattle and gold and so became a major regional power between 1200 and 1450.

Other precolonial cities were located near raw materials, such as Kano in northern Nigeria, which was positioned near deposits of iron ore. Still others were well situated along trade routes. Many of these commercial towns were Islamic cities located at the southern end of trans-Saharan trade routes. Timbuktu, Gao, and Jenne in contemporary Mali are examples of these types of cities, which were dominated by artisans and traders. These cities and others were at the center of a series of Islamic empires—Ghana, Mali, Songhai, Kanem-Borno, Hausa State, and Sokoto—that rose to prominence in the savannas of West Africa between the ninth and nineteenth centuries. All had agricultural economies, but control of major trade routes across the Sahara was their main source of wealth. The trans-Saharan trade route included Muslim spice traders in North Africa trading via caravans of camels across the Sahara to connect to the Senegal River and Niger River and the cities and civilizations around the headwaters of these west African rivers.

These cities were also centers of learning. Timbuktu, Mali, for example, is home to the oldest university in the world, which had 25,000 students by the twelfth century. Based on Islamic scholarship and designed around three

(a)

Courtesy of William Moseley.

(b)

Courtesy of William Moseley.

FIGURE 4.11   (a) **Great Zimbabwe, Zimbabwe** Located in the southeastern part of the country, the Great Zimbabwe was built between the eleventh and fifteenth centuries and served as the capital of the Kingdom of Zimbabwe. The walls of the great enclosure, shown in this photo, were built in the fourteenth century out of bricks made from sand and clay. (b) **Jenne, Mali** Located in present-day Mali, Jenne served as the end of a trans-Sahara trading route and dates to the third century BCE. The ancient city was built of clay and continued to serve trading routes when Islam diffused into North Africa more than a thousand years later. In the fourteenth and fifteenth centuries, an Islamic leader gained control of Jenne and brought an Egyptian architect to the city to build the Great Friday Mosque, which still stands and is used today.

mosques, the university offered four degrees. Jenne, Mali has the typical urban form of an Islamic city or town. That is, it is structured organically around a city center dominated by a large mosque and marketplace. Occupational groups traditionally are clustered in different wards of the city. Individual homes are found within walled courtyards along narrow winding streets (Figure 4.11b).

THE COLONIAL PERIOD

From 1500 to about 1850, the main focus of Europeans in Africa was the slave trade. The slave trade severely depop-

ulated coastal and West Africa, central West Africa, and Southeast Africa. Although the extent to which Africans participated in this process is hotly contested, it is indisputable that the high demand for labor on sugar, cotton, and tobacco plantations in the Americas was the fuel that fired this brutal trade in human lives.

By the mid-1800s, the morality of the slave trade was in question and demand for enslaved people had declined, setting the stage for a second colonial wave (see Chapter 2). Several European countries began to want to extract more from Africa than slaves. They commissioned explorations

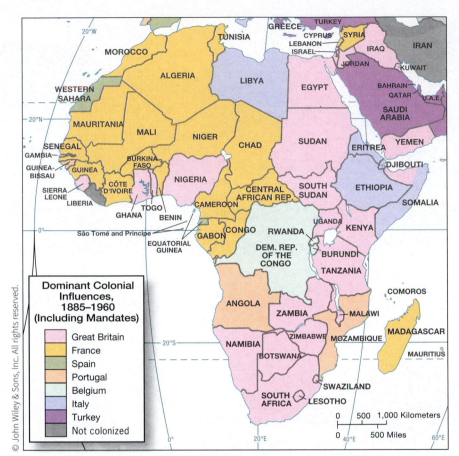

FIGURE 4.12 Colonial Africa in 1912. After the Berlin Conference in 1884, Europeans scrambled to assert control over their new colonies, as reflected in this map from 1912.

of the African interior. By the 1880s the European "Scramble for Africa" was well underway. At the 1884–1885 Berlin Conference, without a single African present, European colonial powers divided the continent of Africa among themselves and drew arbitrary boundaries among colonial territories (Figure 4.12). The Berlin Conference borders ignored the distribution of African ethnic groups, civilizations, resources, and history. The colonies included economically fragile, landlocked entities as well as large areas encompassing hundreds of ethnic groups.

Europeans took over or established cities as administrative centers containing military garrisons, the real source of power. In 1890, 200 British pioneers and about 400 soldiers raised the Union Jack on an empty plain in the middle of Mashonaland after a 425-mile (700 km) march north from South Africa, establishing the city of Salisbury (today's Harare, Zimbabwe) (Figure 4.13a). The land was under the influence of a prominent Shona chief named Harare. Nonetheless, the British built a fort and named it after the British prime minister at the time, Lord Salisbury.

Salisbury was designed to protect an area highly coveted by Europeans for its **site** characteristics, a cool upland

plateau, low levels of malaria infestation, and prime agricultural soils. The town would later be sustained by a railroad network connecting it to South Africa and European settlers farming tobacco with the use of African labor. As in many colonial cities, the British established separate neighborhoods for blacks and whites. In fact, even today, just outside Harare is the historically black township of Mbare with over a million people. The dense urban living of Mbare is quite different from the formerly white, elegant tree-lined neighborhoods to the north of the city center (Figure 4.13b).

Europeans also established colonial cities to facilitate resource extraction. Many were port cities, including Douala in Cameroon and Beira in Mozambique. In fact, if you examine a map of colonial rail and road networks, you will immediately notice that they do not form national networks, which would be helpful for building national economies (Figure 4.14). Rather, the rail lines extend from the port cities to the mines of the interior. With the exception of South Africa, most contemporary rail networks in Subsaharan Africa retain their colonial form.

Europeans sited other colonial cities in major mining areas, such as in the copper belts of Zambia and the Democratic Republic of Congo (DRC), or the gold and diamond mining areas of South Africa. The gold mines of "the Rand" area, or Gauteng Province, are only a short drive from

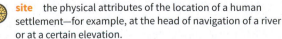

site the physical attributes of the location of a human settlement—for example, at the head of navigation of a river or at a certain elevation.

(a)

© Paul Almasy/Corbis.

(b)

ALEXANDER JOE/AFP/Getty Images/Newscom.

FIGURE 4.13   Harare and Mbare, Zimbabwe.   (a) Today the city of Harare still reflects its foundation as a home for Europeans during the period of British colonialism. The parliament building in this photo was built by the British before independence, when the city was known as Salisbury. **(b)** Children play football (soccer) in Mbare, a high-density, southern suburb of Harare that the British established as a township in 1907.

Johannesburg, a city built on mining. Even today, drivers can see old mine tailings as they approach the city on major highways. Mine workers from all over southern Africa, from Malawi to Lesotho, lived in the sprawling Johannesburg township of Soweto, which would become the intellectual incubator of leaders of the struggle against apartheid. Johannesburg is now a thriving, cosmopolitan city of 3.2 million people and is the commercial capital of South Africa.

Freed slaves established the site of Libreville, Gabon, in the mid-1800s. French colonizers favored Libreville's

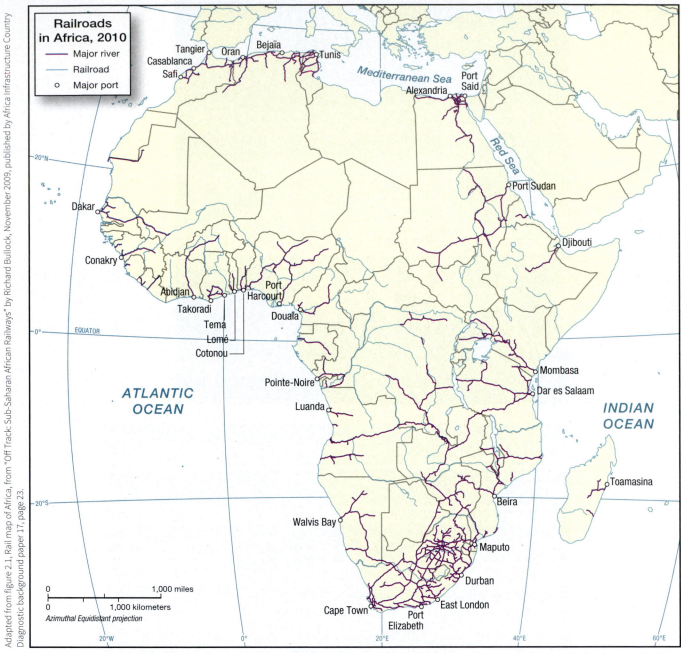

Adapted from figure 2.1, Rail map of Africa, from "Off Track: Sub-Saharan African Railways" by Richard Bullock, November 2009, published by Africa Infrastructure Country Diagnostic background paper 17, page 23.

FIGURE 4.14   Railroads in Africa.   The map of railroads in Africa today still reflects the central purpose of railroads during the colonial period, resource extraction. Railroads reach from port cities into the interior so that resources can be extracted and shipped out of Africa.

location as a port city on the Atlantic coast and its connection by railroad into the interior forest. During the colonial era, Libreville played a specific role in the larger context of France's colonies in West Africa. Today, Libreville is the capital city of Gabon, but Gabon's **situation** has changed since colonialism. Although Libreville's economy is still tied to France, the city now plays global and regional roles

as the capital of Gabon, one of the wealthiest countries in West Africa. The situation—that is, the position of a place relative to other places and the surrounding environment or context—changed for Africa's cities during and after the colonial period (Figure 4.15).

POSTCOLONIAL PERIOD

Colonies in Subsaharan Africa became independent countries after World War II. Africans called for the decolonization of the continent after India and Pakistan achieved independence in 1947. Some colonies were able to gain

 situation   The position of a city or place relative to its surrounding environmental or context.

GUEST *Field Note*   Pink House Hair Salon in a Cape Town Township

STEPHEN PEYTON
University of Minnesota

This makeshift beauty salon and the shack housing behind it were built on open land in a Cape Town **township**, one of thousands of underdeveloped areas on the outskirts of nearly every South African city originally designated as places for nonwhites to live during the apartheid era. Despite government efforts to build improved housing, it is not uncommon to find informal settlements or shack housing in and around townships. Although the quality of shack housing varies tremendously, its high density, pirated electricity (notice the multiple power lines running down from the pole on the right), and frequent use of paraffin stoves mean that shack fires are common. Furthermore, the location of this settlement on the low-lying, sandy Cape flats outside Cape Town makes it a highly undesirable place to live because of high winds, cold, and flooding during the winter months.

Although apartheid was the high point of brutality, state-sponsored racism was part of colonialism in South Africa from its start, beginning with the establishment of Cape Town by the Dutch in 1652. In addition to losing the right to own land in 1913, blacks received inferior education, were allowed to live only in designated neighborhoods, were barred from the best jobs, had their movements restricted, and, of course, could not vote.

Courtesy of William Moseley.

FIGURE 4.15   Cape Town, South Africa.   The Pink House Hair Salon is one of many informal businesses in the townships that encircle cities in South Africa.

independence through discussions and votes in a peaceful manner. Other colonies fought bitter wars against the colonizers to achieve independence. Whether independence came peacefully or by war depended on processes happening across scales. Regionally, Europeans opposed and fought against calls for independence in colonies with large European settler populations. In colonies such as Kenya where the British had taken the farmland and given it to European settlers, the British citizens in Kenya resisted calls for independence, afraid they would lose landholdings and become the minority in an independent Kenya. At the same time, in the global context, many western European colonizers suffered from the ravages of war on their own soil during World War II and in its aftermath could no longer afford to hold on to their African colonies.

In 1964, the newly independent African countries formed the Organization for African Unity and met at a summit in Cairo to determine whether the borders the Europeans had established at the Berlin Conference should stand. The newly independent countries agreed that keeping the arbitrary borders was preferable to warring over where the borders should be. Once an agreement was reached to maintain the colonial borders, new governments on the continent focused on growing their economies.

Most of the economic growth and financial investment in Subsaharan Africa has been concentrated in cities, which function as **islands of development** regionally (Figure 4.16). Around two-thirds of Subsaharan Africa is still rural, but the region has experienced high rates of urbanization since the 1960s as migrants are drawn to perceived greater opportunities, especially for employment, in cities. Migration to islands of development from rural areas and neighboring countries continues with or without the availability of jobs in a city's formal economy. As a result, significant levels of underemployment exist in most African cities. Many people are employed part-time in the **informal economy**. Street vendors and crafts people compose part of the informal sector in Subsaharan African cities.

The perception of a better life in the city is in part based on government policies that began after decolonization in the 1960s and lasted until the 1980s which favored urban residents in many African countries. For starters, the government provided employment for suitably educated people with proper connections. In addition, access to education and health care was typically easier in the cities. Finally, governments sometimes instituted policies that kept food prices artificially low, an approach that largely

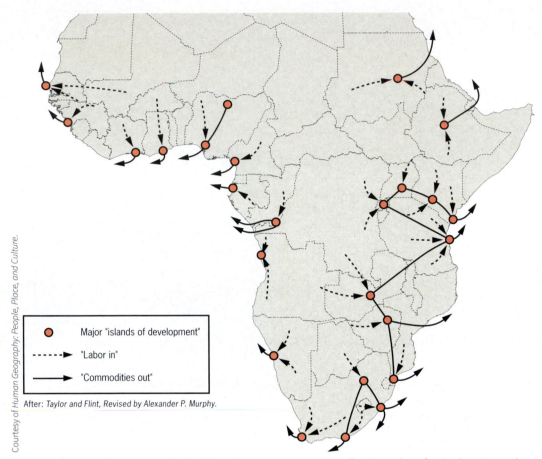

Courtesy of Human Geography: People, Place, and Culture.

○ Major "islands of development"

----→ "Labor in"

——→ "Commodities out"

After: *Taylor and Flint, Revised by Alexander P. Murphy.*

FIGURE 4.16 **Islands of development.** Islands of development are areas, usually cities, where foreign investment is concentrated in a developing country. Migrants move from rural areas and neighboring countries to work in the islands of development, and commodities are shipped out from the islands of development.

benefitted urban residents who had access to stores. This *urban bias* whereby policies are made to benefit urban areas over rural areas happens in part because urban voters in Africa are generally a more politically active constituency.

City-dwellers' favored status began to erode in the 1980s when the World Bank, which lends money to countries to aid their development, required governments to implement policies called structural adjustments to reduce the number of government employees, curtail subsidies, cut social services, and devalue currencies in an effort to reduce government debt to other countries and international institutions. These structural adjustments hit urban populations particularly hard because most of the jobs and social services were in the cities.

By the 1990s, rural incomes frequently exceeded those in urban areas. A key indicator of the changing fortunes of rural and urban populations has been a change in direction of **internal migration**, or migration within an individual country. Recent evidence suggests that movement from the country to the city has slowed dramatically, and in some cases has reversed. Whereas migration

accounted for a large proportion of urban growth in the 1960s and 1970s, the urban growth rate in Subsaharan Africa is now largely attributable to the natural growth of the urban population itself.

LANGUAGES AND RELIGIONS IN SUBSAHARAN AFRICA TODAY

Language and religion are two of the most important forms of cultural expression. The spatial distribution of language often tells us something about the history of a people and how they moved around in space over time, were conquered by some groups, and defeated others.

Take the case of the Khoisan language family, one of Africa's major language groups, yet relatively small in terms of its geographic extent (Figure 4.17). Today Khoisan has two remnant populations, KhoiKhoi herders in southern Namibia and San hunter gatherers in the Kalahari Desert. Related to this group are the pygmies of the Democratic Republic of Congo (DRC) who no longer speak their own language (they speak the language of the dominant group around them). Yet linguists know that many words used to

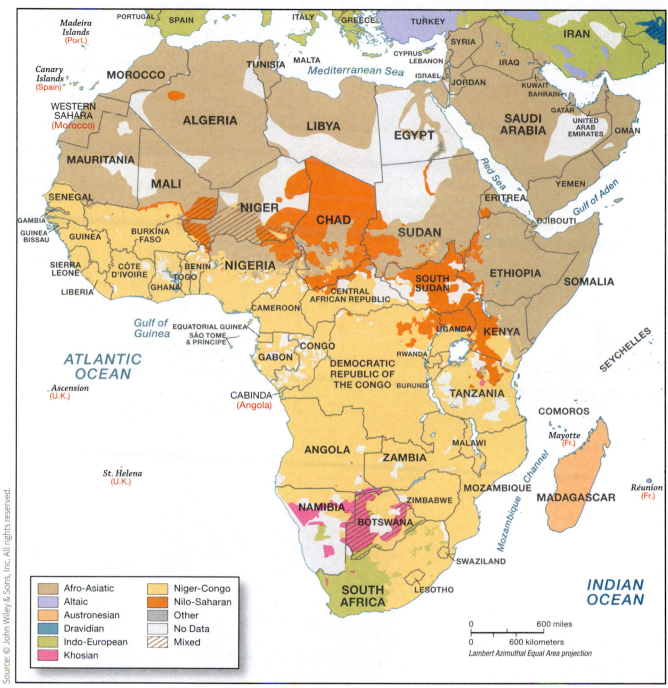

FIGURE 4.17 Languages of Subsaharan Africa. The prevalence of languages in the Niger-Congo family in Subsaharan Africa and of Afro-Asiatic languages of North Africa reflects the relative lack of spatial interaction, historically, between people in the two regions.

describe local flora and fauna all across southern Africa are Khoisan in origin, suggesting that Khoisan language speakers were once much more widely spread (Diamond 2004). Indeed, the Khoisan were once the dominant group throughout southern Africa, but they were gradually displaced into remote marginal areas, or subsumed, by Bantu speakers (who originated in Nigeria and spread south and east) between 5000 BCE and 1000 CE.

Over 1000 languages can be found in Africa, and they can be grouped into six large families. The Bantu languages referenced above are the largest branch in the Niger-Congo subfamily of the Niger-Kordofanian. This family covers most of West Africa and extends east and down to include most of southern Africa. The Nilo-Saharan family stretches from the Maasai in Kenya to the Teda in Chad. The Afro-Asiatic languages cover most of North Africa as well as the Horn of Africa and include Arabic, Tuareg, Somali, and Amharic. The remaining two language groups are Malay-Polynesian (Malagasay with origins in the Polynesians and Africans who colonized Madagascar some 2,000 years ago) and Indo-European languages (primarily English and Afrikaans) spoken by the white settler population in South Africa.

The religious map of Africa today reflects its triple heritage of indigenous, Muslim, and European influences (Mazuri 1986). Traditional, indigenous, or animist belief systems are the oldest and persist in a number of African countries (Figure 4.18) as majority (increasingly rare) or minority religions. Traditional belief systems have also deeply influenced the forms that Islam and Christianity have taken in Africa—just as these world religions have been influenced by local religions in other parts of the

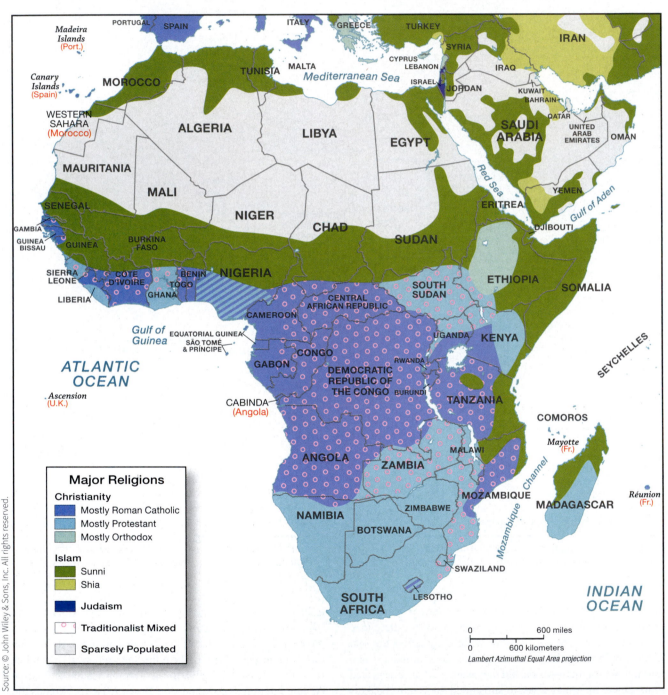

Major Religions

Christianity
- Mostly Roman Catholic
- Mostly Protestant
- Mostly Orthodox

Islam
- Sunni
- Shia

- Judaism
- Traditionalist Mixed
- Sparsely Populated

FIGURE 4.18   **Religions in Subsaharan Africa.**   Islam is the dominant religion in parts of east Africa and in North Africa. Christianity mainly diffused through European colonialism, and primarily to British colonies. Christian missionaries are still active in much of Subsaharan Africa today. Traditionalist religions are found in pockets throughout the region, but this map underestimates the influence of traditionalist religions because they have mixed syncretically with Islam and Christianity as the two universalizing religions diffused to the continent.

world. Traditional religions encompass all the belief systems that are specific to a culture. In Africa, these are often nature based (described as animist), believing that spirits dwell in certain flora or fauna. Worship of and communication with ancestors are also an important aspect of many African traditional religions. For example, the head of a family of Bamileke people of Cameroon is responsible for preserving the skulls of their ancestors. Various kinship and occupational groups may also depend on specific deities for protection. Practitioners of traditional religions often perform rituals to ensure good fortune.

Both Islam and Christianity are more common among urban and educated populations (and are often portrayed as a sign of modernity). In recent decades, radio and television have helped diffuse these beliefs into rural areas.

Islam dominates in northern Africa and the Horn (the point in the northeast where Somalia and Ethiopia are located) and dates back to the long history and influence of Arab traders in the region beginning around the twelfth century, as well as Arab traders along the east coast of Africa. In fact, the Arab trade led to the creation of the Swahili language, which is a mixture of Arabic and local languages. Islam was initially the religion of the ruling class in several Sahelian kingdoms, as well as among the traders in coastal East Africa. In the nineteenth century, there were a number of militant Islamic movements (or jihads) in, for example, Sudan, Nigeria, and Mali where large-scale forced conversions took place. West Africa was very chaotic during this period, and a number of fortified villages, or whole communities, moved into the hills or cliffs for defensive purposes (and to escape forced conversion). One of the most famous examples of this movement is that of the Dogan people in Mali. Today most African Muslims are Sunni (rather than Shia).

While much of Christianity in Africa dates to the European colonial period, one must not overlook the early Coptic Church in Ethiopia, dating to the fourth century CE. The area of Coptic Christianity is now surrounded by areas where Islam predominates. Christian missionary efforts were a major element of European colonization of Africa. The particular type of Christianity predominant in a given African country today typically has some relationship to the former colonial power. For example, Catholicism is dominant in Rwanda, Burundi, and the DRC, where the Belgians and French were active. The Anglican Church tends to be stronger in a number of former British colonies, although one sees sizable numbers of Presbyterians in Ghana, Malawi, Kenya, and Gabon.

More recently, a number of independent African churches have arisen because of the missionaries' intolerance of local customs, leading to the Africanization of a number of Christian faith groups in Subsaharan Africa. These independent churches combine Christian teachings with elements of African culture and religion. Examples include the Zionists of South Africa and the Cherubim and Seriphim, as well as the Kabunga, of Nigeria.

COLONIALISM AND RURAL LIVELIHOODS

The motivations for European colonialism in various parts of Subsaharan Africa were complex. Although notions of ethnic superiority, national rivalries, and religious fervor contributed to European colonialism in Africa, many scholars view expansion of the European capitalist economy as a main driver. According to this argument, the major purpose of colonialism after the Berlin Conference was to secure new, cheap sources of raw materials and to develop markets for European goods. Through colonialism and the establishment of capitalism and global trade, wealthy countries (many of which were colonizers) exploited poorer countries (most of which were first colonies), creating a relationship whereby the poor countries are economically dependent on the wealthy. Dependency theory (see Chapter 3), as originally envisioned by Frank, suggests that the Europeans needed to transform or underdevelop African economies so that they would operate to the benefit of European power. This often meant that the Europeans crippled or destroyed Africa's local, self-sufficient economies.

Regardless of the motivation for colonialism in Subsaharan Africa, the key problem for colonial authorities was getting Africans to produce commodities for export. How do you get someone to produce a crop for export and at a cheap price, if it is not really in that person's best interest to do so? Forced labor is one option, and some colonial powers such as Belgium made use of this labor much more than others. The Belgians in particular realized that the output of forced labor was low and that it required constant supervision. Another option was to pay people to grow certain crops. However, this approach also presented problems, especially if the prices offered were relatively low and local people did not need the money because they lived in an informal economy based on barter and trade.

Most colonizers solved the problem by creating a need for money. They imposed an annual head tax and fined or imprisoned those who did not pay the tax. The tax had the dual benefit of creating revenue for the colonial administration and forcing Africans to participate in a cash economy.

In West Africa, the European settler population was minuscule. Small family farms became the primary means by which local people earned income to pay the head tax. By their own labor, African farm families produced crops such as cacao in Ghana, coffee in Côte d'Ivoire, peanuts in Senegal, and cotton in the Sahelian countries. The same farmers often produced food for their families as well as crops for the market. Such **small-hold crop farming** could be contrasted with larger plantations, typically run by colonial powers or owned by settlers from the colonies, which employed labor and produced crops entirely for the market.

Even after imposition of a head tax, however, the colonial governments were sometimes thwarted in their efforts to meet production objectives. For example, colonial agricultural services often distributed higher yielding seeds as part of their effort to increase production. The assumption was that African farmers would seek to maximize output and therefore income. However, what if your goal was not to maximize output but rather to meet certain subsistence goals, that is, to pay a tax and then spend your time doing something else? Indeed, this is what many Africans did. They spent less time growing cash crops when they were given higher yielding seeds—a perfectly reasonable though decidedly uncapitalistic response, and one that provoked European colonial administrators. Colonizers soon saw the solution as either setting production quotas or raising the head tax even higher.

Unlike West Africa, where relatively few Europeans migrated, many more European farmers settled in East and southern Africa, especially in Kenya, Zimbabwe, and South Africa. Settlers in the south and east were looking to make permanent homes, own mine rights, or own land. West Africa, as noted, had fewer settlers and was largely colonized by the French. Large mines for gold and diamonds in South Africa and for copper in Zambia also attracted European settler communities. These farms and mines required a lot of labor, and, once again, the head tax was useful to the colonizers, forcing some family members, usually men, to leave their farms to work on European-owned farms or in mines. Colonizers soon realized that the head tax was not always sufficient to provide the levels of labor needed at the wages being offered. In South Africa, many white farmers found that they actually could not compete with small-hold African farmers, who were far more efficient (Bundy 1979). This led to passage of a law in South Africa in 1913 that prohibited blacks from owning land outside designated homelands. By destroying the livelihoods of black farmers, Europeans were able to capitalize on their labor.

NEOCOLONIALISM

Almost all Subsaharan countries acquired independence by 1970 (Figure 4.19), but many of the economic relationships developed under colonialism are still in place despite attempts by several African countries to break the chains of colonial dependency. One type of economic policy the newly independent countries frequently employed to increase self-sufficiency was *import substitution*—producing manufactured goods at home rather than importing them from other countries.

Given the limited amount of private capital in the majority of African countries, most import substitution industries were set up and run by African governments. These companies, known as *parastatals*, also benefited from tariff barriers (taxes on imported goods) and subsidies (wherein a portion of production costs was covered by government funding). Along with the development of import-substituting industries in the 1960s, several African countries nationalized (took over) large mining operations that had been owned by foreign corporations. For example, Zambia nationalized its copper mines in the early 1970s.

By the mid- to late 1970s, many of the parastatal industries were running into financial trouble. Three main factors contributed to the financial stress:

1. Many African countries were so small that their economies could not support certain types of industries.

2. While many of the tariff barriers and subsidies should have been temporary measures designed to help infant industries get off the ground, these costly measures continued and were a drain on state coffers.

3. Since these were state-run enterprises, management decisions were not solely based on financial realities.

In fact, politics often influenced how these businesses were run. For example, the government often hired more workers than needed because of political pressure or patronage demands. By the late 1970s, most countries in Subsaharan Africa were involved in the so-called Third World Debt Crisis, in which the loans taken out to build these new companies and run African governments were at massive risk for nonpayment or default.

Since most of the debt held by African countries was public debt owed to governments and international institutions rather than to private banks, by the 1980s the World Bank—a development organization based in Washington, D.C., to which the vast majority of countries in the world belong (described in Chapter 2)—entered into resolving the crisis. In order to reduce debt and balance the books, the World Bank used its considerable clout to force many African countries to carry out four broad *structural adjustments* in their economic policy:

1. Privatize or sell off inefficient parastatal companies.

2. Reduce the size of government bureaucracies. As a result, large numbers of civil servants were laid off, and many social services, including education and health care, were cut back.

3. Boost government revenues by returning to export orientation, often in the form of mining and commodity crops.

4. Devalue African currencies, making African exports on the international market more competitive and the price of imported goods for Africans relatively more expensive.

Governments that did not adopt these measures risked losing all access to credit, as well as development assistance. These adjustments led to a massive change in the way African governments do business.

FIGURE 4.19   Political geography of Subsaharan Africa.   The boundaries of African countries mostly follow those defined during the European colonial era.

Structural adjustment policies have been highly controversial. In most cases, the economic growth they were supposed to spawn did not materialize, and the cutbacks in education and health care were especially disastrous. Many economists would now argue that if a country is to develop, it is vital to invest in its human capital. Development specialists in Africa have strongly objected that these policies represent a new colonialism, or **neocolonialism**,

because African governments had to surrender control of their economies to outsiders. Also, as a result, the structure of many African national economies was returned to what it was during the colonial era—an economy producing primary goods for consumers outside of Africa.

Africa has given much to the world in terms of music, food, seed stock, agricultural practices, and its people, but the world has not always returned the favor.

CONCEPT CHECK

1. **Describe** some of the major differences between precolonial and colonial African cities.

2. **What** types of policies did the colonial powers implement that had a profound impact on rural livelihoods?

3. **Explain** why many analysts view structural adjustment policies as a form of neocolonialism.

IS AFRICA CAPABLE OF FEEDING ITSELF?

LEARNING OBJECTIVES

1. **Understand** how food is produced in different parts of Subsaharan Africa.

2. **Realize** the causes of famine in Africa.

3. **Analyze** whether Africa needs a New Green Revolution.

Subsaharan Africa is commonly portrayed as a region unable to feed itself. Movies show the most arid regions, landscapes filled with wild animals, and slums of cities. Search the Internet for any major news source and add the word "Africa" and you will find a plethora of stories about strife and starvation. Popular Western songs about Africa have lyrics that describe wild dogs and a lack of rain. Despite these perceptions, between 2000 and 2010 production of grain in Subsaharan Africa grew 4.1 percent annually, according to the U.S. Department of Agriculture, and approximately 75 percent of the female labor force and 62 percent of the male labor force are involved in agriculture and Subsaharan Africa.

AFRICAN RURAL LIVELIHOODS AND FOOD PRODUCTION

While we may typically think of farming as the main activity of rural households, most African rural **livelihoods**, activities pursued to secure food and shelter, encompass a broad set of activities, including crop farming, animal husbandry, fishing, hunting, gathering, petty commerce, paid labor, and seasonal and long-term migration to urban areas for additional employment. Many workers pursue more than one of these activities to secure food and cover expenses.

Pastoralists

Animal husbandry is a dominant component of pastoral livelihoods. Pure pastoralists migrate seasonally in search of water and good grazing lands for livestock, primarily in the semiarid grassland savannas of West, East, and north-central Africa (Figure 4.20a). Some are nomadic in that they move in irregular, yet cyclical ways. Pastoralists do not always follow one regular path from pasture to pasture, but their movements are purposeful and tied first to seasonal availability of fresh water. Many of the stops in their movement are reliable enough that they have established seasonal settlements at these locations. The majority are **transhumant pastoralists**, people who follow a regular movement pattern between dry-season and wet-season grazing lands. Pastoralists move livestock, whether zebu or cattle, to places where fresh water is abundant. During the rainy season, fresh water is plentiful. During the dry season in a savanna climate, transhumant pastoralists move their livestock to relatively permanent fresh water sources such as lakes or rivers. Examples of pure pastoralists are the Tuareg, who reside in northern Mali and Niger as well as southern Algeria, and also the Masai, who are largely found in Kenya and Tanzania.

Agropastoralists practice a mix of animal husbandry with farming. Livestock are usually as or more important than crop cultivation. Agropastoral households typically divide at some point in the year when boys and young men go off with the bulk of the herd to faraway pastures during the rainy season. Water is plentiful during the rainy season, so herders are able to move to pastures far from a home base knowing their livestock will not want for water. Women, children, and older men stay at the permanent encampment to farm and tend a milk herd. The Fulani, the most widely dispersed ethnic group in Africa, are agropastoralists who move their stock to find fresh water and who are also known for tracking the cycle and breeding grounds of the tsetse fly to keep their zebu away from disease. The Fulani live in the dry and wet savannas across West Africa, from Senegal and Mauritania to Cameroon.

Fisherfolk

Although fishing is a minor component in many African livelihoods, it is very important in some areas. Pure fishers or fisherfolk are rare. The Somono fisherfolk of the inland Niger Delta in central Mali live on large riverboats throughout the year (Figure 4.20b). This massive, low-lying inland delta expands with seasonal floods (supplied by rains in the Guinea highlands and southern Mali) from August to December, then recedes until the following June or July. Technically, these are transhumant fisherfolk because they have a regular migratory pattern that follows the ebb and flow of the river. They follow the fish northward during the flood season and southward when the river recedes. Major fishing communities also can be found along some stretches of the African coastline. The productivity of fisheries, including African fisheries, tends to relate strongly to the size of the continental shelf. Beyond the continental shelf, the ocean waters are too deep to create the conditions for sunlight and foodstocks eaten by fish. A broad continental shelf creates a natural fish habitat. While Africa is the least endowed continent in terms of the breadth of its continental shelf, the areas with the broadest continental shelf along the Namibian-Angolan coast, the Ghanaian and Guinean coasts, and the Mauritanian-Moroccan coasts are home to productive African fisheries.

(a)

Simon Rawles/Photolibrary/Getty Images.

ASSOCIATED PRESS.

(b)

(c)

HILIMON BULAWAYO/Reuters/Landov.

FIGURE 4.20. **The crop-farming systems of Subsaharan Africa.** (a) **Agropastoralists** A pastoralist stands in front of his cattle herd. (b) **Fisherfolk, Cotonou, Benin** Fishermen repair nets aboard their boats in between fishing trips, in the Akpakpa Dodomey neighborhood. (c) **Subsistence agriculture, Chivi, Zimbabwe** Martha Mafa, a subsistence farmer, stacks her crop of maize in southeastern Zimbabwe. Mafa faced a huge grain deficit during this harvest after a third of the maize crop was lost due to a prolonged dry spell.

As the term suggests, agrofishers are people who mix farming and fishing activities. These groups have permanent settlements where they farm, but they also fish for part or all of the year. The fishers are typically men. These groups inhabit the coastline of many of Africa's rivers, including the Nile, Niger, Congo, Zambezi, and Limpopo, and lakes, including Victoria, Malawi, Tanganyika, and Kariba. In the mid-1990s, agrofishers on the Malawi side of Lake Malawi earned about 30–40 percent of their income from fishing. Thus, while we consider the people fishermen, more than half of their income comes from farming and other nonfishing sources.

Farmers

Subsaharan Africa's crop-farming systems can be subdivided into three major types: slash-and-burn cultivation, rotational bush fallow, and permanent cultivation (Figure 4.20c). In the central African rainforest, a farmer will clear an area and burn cut vegetation, the stumps, and any remaining vegetation to release nutrients into the relatively impoverished soils. Rainforest soils are lacking in nutrients because the daily rain washes away nutrients, leaving behind heavy elements, including iron, which creates the reddish tint to the soils. After a few years of cultivation, the farmer abandons the plot to allow the soil to regenerate and the natural vegetation to recolonize the area. Traditionally, such a plot lies fallow (or rests) for about 30 years before the farmer returns to the area. To supplement the produce from slash-and-burn fields, farmers typically keep more permanent gardens closer to their communities in this region.

The process of clearing and burning, farming, and then moving on to leave fields fallow only to start again on another field is called **slash-and-burn agriculture**. Also called shifting cultivation or swidden, this form of agriculture dominates the tropical forest regions of Africa (Figure 4.20b).

The differences between commercial agriculture in North America and slash-and-burn agriculture in Africa may make slash and burn look wasteful and inefficient, but it is actually a highly rational and efficient approach in areas where land is plentiful and population is relatively sparse. If the population grows beyond the ability to allow fields to replenish by resting for 30 years, then soil erosion increases and sustainability becomes difficult. Common food crops for these systems include tubers such as yam and cassava, bananas, oil palm, and maize or corn.

In Africa's grassland and dry forest areas, farmers have created sustainable practices that avoid soil erosion and maintain soil fertility by rotating crops from year to year and allowing fields to go fallow for periods of time, a system called crop rotation. Farmers also mix complementary crops in the same field, such as nitrogen-fixing legumes and grain crops (a practice called **intercropping**), and apply fertilizer in the form of animal manure and household waste (compost) on fields at regular intervals. Grazing cattle on fields during the dry season in order to capture their manure is another efficient way farmers fertilize their fields. The mixture of these practices on Africa's grasslands and dry forest regions is called the **rotational bush fallow system**. Common food crops in these systems are sorghum, maize, millet, peanuts, and cowpeas. Ethiopia also has some unusual crops such as teff (a grain) and ensete (also known as the Ethiopian banana).

Areas where land is farmed every year, with only rare fallowing, are under **permanent cultivation**. Farmers practice permanent cultivation in relatively few areas in Africa on a broad scale and to a limited degree in a growing number of places. Typically, those areas closest to a village will be farmed more intensively than those farther out because of their accessibility—the ease of commute from the village to the field. Because they have the highest population densities, permanent cultivation occurs on a broad scale in the highlands of southern Uganda, Kenya, Rwanda, and Burundi, as well as parts of Nigeria.

With population densities around 250 to 500 persons per square kilometer, the densely settled area around Kano in northern Nigeria is a classic case of a permanently farmed area. Farmers cultivate all arable land every year with heavy applications of animal manure, household refuse, night soil, and (increasingly) commercial fertilizer. Hausa, an ethnic group found in West Africa, maximize yields from their small farms by intercropping beans and peanuts with millet and sorghum. Other areas of Subsaharan Africa that tend to have permanent agriculture are the floodplains. The main crop for floodplain agriculture in Subsaharan Africa is rice. If correctly managed, these systems are sustained by annual floods that bring a new layer of silt and nutrients to maintain the soils and make permanent agriculture production possible.

FAMINE

Westerners often perceive Subsaharan Africa as a region of famine. Great famines during the colonial era, including 1876–1879 in South Africa, 1889–1891 in Ethiopia and Sudan, 1897–1899 in Kenya, and 1900–1903 in the Sahel helped create this perception. In the postcolonial period, the semiarid grasslands of the Sahel (a region just south of the Sahara) have been particularly prone to drought and associated famines.

Major droughts in this region in the early 1970s and mid-1980s led to significant international famine relief efforts from Senegal to Ethiopia. The latter of these two droughts cemented an image of a drought- and famine-stricken Africa in the minds of many Westerners because television media showed starving children in Ethiopia on the nightly news. In response to images he saw on one television broadcast, music producer Bob Geldoff was inspired to raise funds for famine relief. In 1984, Geldoff co-authored "Do They Know It's Christmas?" about the famine in Ethiopia and released it right before the holidays. He assembled a group of famous musicians including members of the bands Duran Duran, U2, Bananarama, and Spandeau Ballet. The lyrics of the song, describing Africa by saying "Where nothing ever grows. No rain or rivers flow," demonstrate ignorance about agriculture and climates in Africa. Despite helping diffuse geographic ignorance, a follow-up concert called Live Aid held simultaneously at Wembley Stadium in England and JFK Stadium in Philadelphia in July 1985 raised over $200 million in revenue for famine relief in east Africa (Figure 4.21).

Subsequent droughts and related food shortages have occurred in southern Africa, especially Zimbabwe and Malawi in the early 1990s and in Sudan in the late 1990s and early 2000s because of civil strife.

It is commonly argued that famine results from an absolute shortage of food, either because the population is too large to feed or farming methods are insufficiently productive, or both. This argument is as old as that proposed by the British parson Thomas Malthus, who first suggested in 1798 that population numbers will inevitably outstrip food supplies. Population grows at a geometric rate, he argued, while agricultural production only grows at an arithmetic rate. Malthus was largely proved wrong because food production at the global scale has kept up with population growth.

A number of studies by geographers have also shown that population growth actually drives increasing food production in many cases (Turner et al., 1993). The primary reason for this increase is that more people means more labor to intensify agriculture. The classic case in this regard was a 1995 study in Machakos, Kenya, where geographers documented improved environmental stewardship and increasing agricultural productivity as population densities increased. In 1983, geographer Michael Watts demonstrated that while drought was a common occurrence in northern Nigeria, it hadn't always resulted in famine until the colonial era. In the precolonial period, Hausa chiefs often extracted wealth from local farmers for their own benefit. However, they also oversaw the collection and storage of surplus grain production. These stores served as a social safety net during drought years, a not uncommon occurrence in the Sahel, where rainfall is highly variable.

FIGURE 4.21   London, United Kingdom.   The 1985 Live Aid Concert was one of the first celebrity efforts to raise funds for African development on a large scale, in this case raising money to combat famine in Ethiopia. Performers at Live Aid included, from left to right, George Michael (Wham!), Bono (U2), Paul McCartney, and Freddie Mercury (Queen).

The British colonial administration also sought to extract wealth from the local population, but none of it was set aside for difficult years. Furthermore, the British required payment in cash rather than grains. Thus, the local population was forced to grow cash crops, cotton and peanuts in this instance, leading to a decrease in the area devoted to food crops. Whereas the Hausa chiefs understood the ebb and flow of Sahelian agriculture and were flexible about when and how much grain was collected, the British were unrelenting and rigid in their tax-collecting efforts. The result was four major famines in this area during the colonial period: in 1914, 1927, 1942, and 1951. Western scholars essentially wrote off these famines as the inevitable consequence of drought. However, Watts effectively showed that famine has as much or more to do with the organization of society as it does with natural factors.

Famines can occur in areas where poor households cannot afford the food that is available on the market. The Sahel famine of the early 1970s offers a classic case in which food resources were going to those most willing to pay, not to those most in need. Livestock and peanuts were exported to global markets as a massive famine struck the Sahel (Franke and Chasin 1980). In other cases, such as in Zimbabwe in the early 1990s, the region had sufficient food stocks, yet the poorest households could not afford the food.

In many cases, an absolute food shortage is not the problem, raising questions about solutions aimed primarily at increasing food production. For example, international policymakers frequently argue that to resolve their hunger problems Africans need only to adopt more productive farming methods. Since the 1960s, international agencies have encouraged the use of hybrid seeds, chemical fertilizers, and pesticides in order to boost production. Since 2008, large multinational companies and the U.S. government have been pushing the use of genetically modified seeds (Figure 4.22). Aside from a number of environmental concerns related to the use of such seed packages, one of the main problems it poses is that the poorest of the poor, the segment of the population most in need of help, typically cannot afford the improved seeds, machinery, and irrigation systems necessary to grow the crops. While these approaches may increase production, they do not really get food to those most in need.

In the 1940s, Norman Borlaug, the Nobel Peace Prize winning founder of the Green Revolution, created a series of agricultural improvements designed to improve the stability of the wheat supply in Mexico, a country that was not producing enough grains for its population. Through advances in hybrid seeds, fertilizers, and pesticides, grain production in Mexico stabilized. This set of advances in agriculture is known as the **Green Revolution** (see Chapter 6). In the 1960s, the Green Revolution expanded to rice production and diffused to Asia. While the new package of hybrid seeds, fertilizers, and pesticides did dramatically increase yields, the cost of such inputs was prohibitively expensive for the poorest farmers in Asia. The result was a silent reorganization of the Asian countryside as the poorest of the poor could not compete and then went to work for their wealthier neighbors or moved to the cities. Even wealthier farmers faced growing input costs as insects developed resistance to the most common pesticides, forcing them to apply more and more chemicals or switch to expensive alternatives.

• Reading the **CULTURAL** Landscape

A New Green Revolution for Africa

A family on the outskirts of Bamako, Mali, in the 1980s largely ate small grains (millet and sorghum) produced in the surrounding countryside. Today, they mostly purchase rice from Thailand or Vietnam, having developed a taste over several years for this import which, was relatively cheap until 2007 when food prices started to rise around the world. Research shows that this pattern has been repeated in cities across West Africa as structural adjustment policies encouraged the importation of cheaper Asian rice. As a result, markets for locally produced grains have shrunk, farmers have switched to other crops as a source of cash such as cotton, and rural people have abandoned farming altogether for life in the city.

The Green Revolution has focused primarily on Latin America and Asia; some development experts are now urging that it be applied to Subsaharan Africa. The Rockefeller Foundation, which funded the South American and Asian Green Revolution, and the Bill and Melinda Gates Foundation are teaming up to fund the New Green Revolution in Africa, in hopes of doubling agricultural output on the continent.

Certainly, curing hunger on any continent is an admirable goal. In the summer of 2008, as global food prices climbed dramatically, the "global food crisis" appeared particularly acute in urban West Africa where people protested rising food prices.

Courtesy of William Moseley.

FIGURE 4.22 **Bamako, Mali.** Corn growing in this field will be sold at market. Twenty years earlier, the family grew sorghum and millet for their own consumption on this field.

Many policymakers and commentators point to the apparent success of the last Green Revolution in the 1960s and 1970s, a concerted global effort focused on Asia and Latin America to disseminate a high-yield crop package of hybrid seeds, fertilizers, and pesticides. The problem, they argue, is that these innovations never reached Africa. The Green Revolution, however, did touch Africa in the form of cheap Asian rice, which began flooding African markets in the 1980s.

Political ecologists contend that the Green Revolution approach is flawed. For starters, many of the inputs required for higher yielding crops, especially fertilizers, are petroleum based. The cost of these inputs will only rise in step with the general upward trend in energy costs. Use of imported seeds including hybrid or genetically modified and other inputs also concentrates power in the boardrooms of global agrochemical firms rather than in the hands of the small farmers they are supposed to serve.

Many experts argue that an approach emphasizing local or national food provision and appropriate technology is more sustainable and empowering for small African farmers than the Green Revolution. Agricultural experiments comparing agricultural African methods that use manure and compost and rely on the intelligent mixing of multiple crops to conventional Western cropping strategies have repeatedly shown the African models to be more efficient in terms of energy consumed per unit of output. Nonetheless, African methods have been inhibited by cheap imports, especially of rice produced in Southeast or East Asia, and by agricultural agencies that emphasize industrial approaches to crop production.

It may also be necessary to protect national and regional food systems from unfair competition. Since the 1960s, the provision of cheap food has almost always trumped environmental or social costs in the countryside, reflecting an urban bias in global and national food policies. The results have been predictable: more underemployed urban residents hailing from rural areas and fewer small farmers. Africa has some of the best small farmers in the world. Perhaps we should find ways to support, not subvert, their genius.

CONCEPT **CHECK**

1. **Describe** different types of livelihood and agricultural systems in Subsaharan Africa.

2. **Understand** how the structure of economies and agricultural systems may drive food shortages.

3. **What** are the pros and cons of a hunger mitigation strategy focused on the dissemination of high-yield seed packages?

WHY ARE THE DYNAMICS OF HIV/AIDS IN SUBSAHARAN AFRICA DIFFERENT FROM THOSE IN OTHER WORLD REGIONS?

LEARNING OBJECTIVES

1. **Understand** how HIV/AIDS patterns in Africa compare to patterns in other world regions.

2. **Explain** how colonial patterns of development influenced the spatial pattern of HIV/AIDS in Africa today.

3. **Realize** how poverty and power influence HIV/AIDS disease dynamics.

FIGURE 4.23 **Winterton, South Africa.** People sit in the waiting room of an anti-retroviral clinic inside a rural hospital near the Drakensberg Mountains in the Kwazulu-Natal region of South Africa. The clinic is part of a plan to dispense anti-retrovirals to a larger part of the population.

Subsaharan Africa carries a heavy disease burden. The top two diseases on the continent are malaria and HIV/AIDS. Malaria is spread through a vector, in this case mosquitoes, making it a *vector-transmitted disease*. HIV/AIDS is spread directly from human to human as a *nonvector-transmitted disease*. In 1999, the disease surpassed malaria to become the number-one killer in the region.

HIV/AIDS IN SUBSAHARAN AFRICA

HIV/AIDS was first discovered in the late 1970s/early 1980s in East and Central Africa among men and women with multiple sexual partners. Around the same time, in 1981, doctors recognized it among homosexual men in the United States. Although scientists continue to debate the origins of HIV/AIDS, many researchers now assume the disease began with a nonhuman source. Both the green monkey in Central Africa and the mangabey (also a type of monkey) in West Africa carry Simian Immunodeficiency viruses, which are similar but not identical to HIV. It is possible that a mutation of these viruses created HIV. In some areas of Africa, people eat monkeys, so it is possible that blood from undercooked meat could have been directly absorbed and the virus transferred to humans. Although this theory is the most common one today, the origins of the disease are still murky.

Slowing the spread and impact of HIV/AIDS is among the greatest challenges facing contemporary Africa (see Chapter 1). Unlike some other diseases, HIV/AIDS is particularly devastating because it strikes the working-age population and thus has serious economic and social consequences (Figure 4.23). As of 2007, the Joint United Nations Programme on HIV/AIDS (UNAIDS) reported that 22.5 million adults and children in Subsaharan Africa were living with HIV/AIDS, a number that accounts for 68 percent of the infected population worldwide. Subsaharan Africa also accounted for more than three quarters of AIDS-related deaths in 2007, as well as for 68 percent of new infections (1.7 of 2.5 million). Unlike other world regions, the majority of people living with HIV in Subsaharan Africa are women (61 percent); the overall infection rate for women in Subsaharan Africa is estimated at 5 percent as compared to 0.8 percent worldwide.

Southern Africa is the most seriously affected region. Eight countries—Botswana, Lesotho, Mozambique, Namibia, South Africa, Swaziland, Zambia, and Zimbabwe—had a national adult prevalence rate that exceeded 15 percent in 2005. South Africa has the highest number of HIV/AIDS cases in the world.

The spatial pattern of HIV/AIDS has changed over time. The first AIDS belt was concentrated along major highways and in the urban centers of eastern Africa. In Africa, the disease is largely transmitted through unprotected heterosexual sex and unsafe medical practices. Truck drivers, sex workers, and military personnel all have above-average infection rates and are believed to play a significant role in the spread of the virus. In the 1990s, the epicenter of the virus moved to southern Africa, establishing the second HIV/AIDS belt in southern Africa.

Until the 2000s, much of the HIV/AIDS work in Africa focused on prevention rather than treatment. Policymakers touted Uganda as an example of a country that aggressively pursued prevention and was able to dramatically reduce its rate of new infections. The exorbitant costs of *antiretroviral drugs* also seemed to make such expenditures unrealistic. Increasingly, however, development specialists and health experts are framing access to such drugs at affordable prices as a human rights issue. Humanitarians also argue that African countries could not afford to lose so much human capital to early death.

COLONIAL PATTERNS OF DEVELOPMENT

Many scholars have sought to explain why HIV/AIDS prevalence rates are so high in Africa as compared to other world regions. Some have attributed it to a number of cultural practices, such as polygamy or sexual promiscuity among African cultures (Rushing 1995). Geographers Joseph Oppong and Ezekiel Kalipeni (1996) have been highly critical of such cultural explanations, which are often based, they argue, on inaccurate and gross generalizations. The polygamy argument breaks down fairly quickly when we realize that the most common areas for polygamy in Africa are Muslim, and that the Muslim regions of Africa have some of the lowest rates of HIV/AIDS on the continent. It is also inaccurate to equate polygamists with men and women who informally have multiple sexual partners because polygamous relationships are often quite stable. Oppong and Kalipeni also find it problematic to argue that Africans are more or less "sexually promiscuous" than any other people in the world. Africa has a huge diversity of cultures, with some being more permissive and others extremely restrictive with regard to sexual mores.

A more plausible explanation is the colonial powers' practice of identifying certain colonies, such as Malawi, as labor reserves. In these areas, authorities heavily recruited workers for jobs in other parts of the continent. Authorities contracted young men to work for long periods of time (one to three years) at great distances from their homes, particularly in East, Central, and southern Africa to support mining, plantation agriculture, and other economic activities. In many cases, families were essentially forced to send their young men away so that they would have the means to pay colonial head taxes. Oppong and Kalipeni call this practice and its role in the rapid spread of HIV/AIDS the *migrant labor thesis*.

One scholar estimated that in 1950, roughly one quarter of Malawi's male labor force was working outside the country. In most instances, employers highly discouraged families from accompanying a laborer. Thus, women were left at home to run the farm and raise the children. Male laborers living great distances from home often turned to alcohol and prostitution to deal with their loneliness and isolation. Persisting over time, this system has led to weakening familial bonds and a growing acceptance of multiple sexual partners. According to Oppong and Kalipeni, men rationalize having multiple sex partners because of their extended work away from home rather than looking to a cultural norm to rationalize behavior. This system was ripe for the spread of HIV/AIDS when it hit east and southern Africa in the 1990s. Men who contracted HIV/AIDS during their long stays away infected their wives when they came home to visit.

One cultural practice that has contributed to the rapid diffusion of HIV/AIDS in the region is the lack of male circumcision. Additionally, HIV/AIDS researchers confirmed in 2008 that malaria and the diffusion of HIV/AIDS are connected. Many programs that work to combat HIV/AIDS in Subsaharan Africa are now working to increase male circumcision rates and decrease malaria infections in hopes of decreasing the incidence of HIV/AIDS in the region (see chapter 1).

POWER, GENDER, AND HIV/AIDS

Women are often at a disadvantage with regard to their relative power within households, communities, and nations in most of Subsaharan Africa. Some of this power imbalance can be related to traditional structures, which often favor men, particularly older men, in terms of land tenure, roles, and responsibilities and decision making at the community level. However, women hold considerable power in many African societies, particularly matriarchal ones. The European colonial encounter also often worked to the disadvantage of women. Colonial authorities almost always provided new cash crops or labor opportunities to men, a situation that greatly undermined the economic power of women within households.

The limited power of women within their households and communities exacerbates their (already) higher vulnerability to HIV/AIDS. In heterosexual encounters, women are biologically at a disadvantage for contracting HIV if they engage in unprotected sex. That is, an uninfected female has a higher likelihood of contracting HIV from an infected male partner than does an uninfected male from an infected female partner. When women have less economic power, it is more difficult for them to deny the sexual requests of their husbands—even if they know he is infected. Female-headed households also have, on average, less earning power. This status may lead to riskier sexual behavior in order to earn money or stay in the good graces of male benefactors.

CONCEPT **CHECK**

1. **What** are the distinctive characteristics of HIV/AIDS patterns in Subsaharan Africa?

2. **Consider** the strengths and weaknesses of the migrant labor thesis for explaining HIV/AIDS patterns in southern Africa.

3. **How** does the limited economic power of women influence HIV/AIDS transmission in Subsaharan Africa?

SUMMARY

What Is Subsaharan Africa?

1. African climates are largely driven by four factors: latitude, pressure systems and wind movement, altitude, and ocean currents. The ITCZ migrates seasonally, creating distinct wet and dry seasons in the savanna climate zone.

2. In Subsaharan Africa, the way people perceive and are influenced by environmental change often differs because their interactions with environmental resources frequently vary based on their gender, ethnicity, age, and relative wealth.

3. Parks are controversial in many African contexts for several reasons. First, park developers evicted local people, often without compensation, from areas when national parks were established. Second, feelings of resentment and the inability of many national governments to effectively patrol park borders led local people to encroach on parks in search of sustenance. Third, big game animals rarely respect park boundaries and may wander onto the lands of communities abutting national parks. Large animals may pose safety risks and destroy field crops.

Who Are Subsaharan Africans?

1. Africa is the cradle of humanity. Africans also started complex civilizations, domesticated crops, built the world's first university, and developed a vibrant urban culture.

2. Roughly three-quarters of the largest, most important of today's cities were established prior to European conquest.

3. Colonialism devastated rural livelihoods across Subsaharan Africa. Many Africans lost their land to European settlers. Others were forced through taxation policies to produce cash crops.

4. The World Bank forced a series of economic reforms on African countries from the 1980s forward, which many would consider to be a form of **neocolonialism**. These reforms included the privatization of parastatals, a reduction in government spending on social services, a renewed emphasis on the export of raw materials and agricultural goods, and currency devaluations.

Is Africa Capable of Feeding Itself?

1. Major rural **livelihood** groups in Subsaharan Africa include **pastoralists**, fisherfolk, and farmers.

2. We often assume that famine results when there is an absolute shortage of food, either because there are too many mouths to food or farming methods are insufficiently productive, or both. A number of studies by geographers have shown that placing famine within the context of broader political economic structures is critical. Colonial histories and contemporary global economic forces may mean that many African farmers are producing products for the global market rather than food for the household, community, and nation.

3. A New Green Revolution for Africa may increase production among the wealthiest farmers but do little for the poorest of the poor who are the most food insecure.

Why Are the Dynamics of HIV/AIDS in Subsaharan Africa Different from Those in Other Regions of the World?

1. HIV/AIDS is a nonvector-transmitted disease and the top killer in Subsaharan Africa. Its transmission is primarily through heterosexual sex.

2. The migrant labor thesis asserts that the rapid spread of HIV/AIDS in Africa is related to the establishment of wage labor on the continent during the colonial era moving forward, particularly in East, Central and southern Africa, to support mining, plantation agriculture, and other economic activities. The more limited power of women within their households and communities exacerbates their (already) higher vulnerability to HIV/AIDS. When women have less economic power, it is more difficult for them to deny the sexual requests of a husband or benefactor—even if they know he is infected.

Geography in the Field   **YOUR**TURN

Courtesy of William Moseley.

Bamako, Mali.   A newly constructed neighborhood in the capital city.

This photo shows a relatively new neighborhood in the capital city of Mali, Bamako. Located on the Niger River, Bamako is the fastest growing city in Africa and the seventh fastest growing city in the world today. The growth in this neighborhood was not entirely unplanned—as streets were laid out and plots of land were sold—but investment in infrastructure by the city of Bamako was minimal. Do you notice that the nicer homes are built at higher elevations? We tend to see this pattern in many cities around the world as garbage and waste water roll downhill.

Thinking Critically

- Mali is a landlocked country. Would you expect to see this same pattern of wealth on hilltops and poverty at lower elevations in coastal cities of Subsaharan Africa (see Chapter 1)?

- Bamako is the fastest growing city in Africa. From where do you think all the people who are migrating to Bamako are coming?

KEY TERMS

insolation	ecotourism	informal economy	slash-and-burn agriculture
Intertropical Convergence Zone (ITCZ)	transboundary parks	internal migration	intercropping
subsolar point	hominids	small-hold crop farming	rotational bush fallow system
Harmattan winds	site	neocolonialism	permanent cultivation
cold water desert	situation	livelihoods	Green Revolution
marginality	townships	transhumant pastoralists	political ecologists
	islands of development	agropastoralists	

CREATIVE AND CRITICAL THINKING QUESTIONS

1. Why do so many Americans identify Africa as an arid **region**, when the continent has such large expanses of rainforest and savanna climates?

2. Why does the vegetation of the savanna climate zone consist of thick stands of grass with isolated trees? What is the **situation** of the savanna relative to the ITCZ, and how does that influence the amount of vegetation in the savanna?

3. Should the big game animals in national parks in southern Africa be seen as open-access resources or as common property? How does the **context** in which you view the animals affect the way you would manage the parks?

4. Read about Timbuktu, Mali (http://www.history.com/classroom/unesco/timbuktu.html) and explain how the **site** and **situation** of Timbuktu helped to make the city a cultural center for Islam in Africa.

5. Compare the maps of cities established before colonialism and cities of Africa today. What differences do you see in the **site** and **situation** of African cities between the two

maps? Which cities in Africa are most likely to be designated as **world cities**?

6. Dependency theory suggests that Europeans needed to transform, or underdevelop, African economies so they would operate to the benefit of European power. Do you agree with this theory of **development**? Why or why not?

7. How did getting African farmers to produce commodities for export help Europe and not Africa, and what role did it play in establishing the **networks** of **globalization** in place today?

8. Read about the structural adjustment policies of the World Bank in this chapter and in Chapter 2. How do you think the impact of the World Bank's structural adjustment policies on **development** vary by world region?

9. How do transhumant pastoralism and the savanna climate in Subsaharan Africa work together to sustain the agricultural practice? Through **globalization**, more grazing land has been transformed into agricultural land. How does this impact transhumant pastoralism and the savanna climate?

SELF-TEST

1. Virtually all of the continent of Africa lies within the:
   a. tropics
   b. equatorial region
   c. midlatitudes
   d. polar region

2. The Harmattan winds are caused by:
   a. wind coming off the Rift Valley mountains
   b. hurricanes on Africa's west coast
   c. the migration of the Intertropical Convergence Zone (ITCZ)
   d. global warming

3. On the map of population density in Africa, each of the following regions stands out as having dense concentrations of people, except the:
   a. Rift Valley

   b. Nile River
   c. Kalahari
   d. coast of West Africa

4. One of the major reasons that Europeans did not colonize the interior of Africa for several centuries is:
   a. a lack of guns
   b. the desert climate
   c. tropical diseases
   d. big game animals

5. Three-quarters of the cities in Africa today were founded during this period in the continent's history:
   a. precolonial
   b. colonial
   c. neocolonial
   d. none of the above

6. The oldest university in the world is found in:
   a. Lagos, Nigeria
   b. Johannesburg, South Africa
   c. Cairo, Egypt
   d. Timbuktu, Mali

7. Cities developed during the colonial period were established to function as administrative capitals for European colonizers or as sites for:
   a. plantation agriculture
   b. oceanic trade
   c. extracting resources
   d. railroad expansion

8. People who are employed as street vendors or crafts people are considered part of the _____ economy.
   a. neocolonial
   b. formal
   c. informal
   d. colonial

9. The continued growth of population in African cities today is largely a result of:
   a. rural to urban migration
   b. natural growth of the urban population
   c. the stronger impact of HIV/AIDS on rural areas than urban
   d. government acceptance of migrants from other African countries into major cities

10. A small-hold cash-crop farmer produces food for:
    a. export on the world market
    b. her family and a local market
    c. her family
    d. the European Union

11. During the colonial era, Europeans encouraged higher agricultural output among African farmers by:
    a. placing an annual head tax on each person
    b. funding a Green Revolution for Africa
    c. encouraging the operation of small-scale farms
    d. allowing large numbers of Europeans to migrate to West Africa

12. The main crop grown in floodplain agriculture in Africa is:
    a. cotton
    b. wheat
    c. rice
    d. millet

13. Geographer Michael Watts found in his study of northern Nigeria that famine was less common in precolonial Nigeria than it was in postcolonial Nigeria.
    a. true
    b. false

14. A good example of a vector-transmitted disease is:
    a. syphilis
    b. malaria
    c. H1N1
    d. HIV/AIDS

15. HIV/AIDS rates are highest in the regions of northern Africa where polygamy is commonly practiced.
    a. true
    b. false

ANSWERS FOR SELF-TEST QUESTIONS

1. a, **2.** c, **3.** c, **4.** c, **5.** b, **6.** d, **7.** b, **8.** c, **9.** a, **10.** b, **11.** a, **12.** c, **13.** True, **14.** b, **15.** False

5

SOUTHWEST ASIA AND NORTH AFRICA

When British artist John Frederick Lewis lived in Cairo, Egypt, from 1841 to 1851, he created hundreds of sketches of what he saw. He returned to England and spent the rest of his life painting, based on the sketches. In 1873, he made this painting, entitled "A Lady Receiving Visitors." He was part of a group of painters known as Orientalists, derived from the Latin *oriens,* which means "east." The Middle East fascinated this group of nineteenth-century artists, hailing from Britain, France, Italy, Germany, and Spain, who traveled east to paint local, archaeological and biblical scenes.

In 1978, about a century after Lewis had painted the "A Lady Receiving Visitors," Edward Said, a prominent intellectual of Palestinian heritage, published his influential text *Orientalism.* Said analyzed novels, travelogues, and academic books to argue that a dominant Western, imperialist lens underlay nearly all European and American representations of the so-called Orient, a concept originally applied to Southwest Asia and North Africa, but then extended to nearly all other developing regions of the world. Said described a process of "othering" in which Europeans and Americans not only viewed peoples of distant lands as exotic, but held them up for comparative purposes to make Europeans and North Americans look better.

Said calls on us to reconsider many issues, including Southwest Asia's common name: the Middle East, a term applied by western Europeans who saw the region as midway between Europe and East Asia (which they called the Far East). Since regional geographic names based on directions imply a perspective, they do not work for people with different points of view. In this case, "the East" meant areas east of Europe. The names Southwest Asia and North Africa are more accurate, since these terms use the center of continents as a reference point rather than a particular world region as the center from which other areas are measured.

Yale Center for British Art, Paul Mellon Collection.

Cairo, Egypt. British artist John Frederick Lewis painted "A Lady Receiving Visitors" in 1873. Paintings by European artists in the 19th century helped create a European perception of the region.

THRESHOLD CONCEPTS in this Chapter

Diffusion
Hearth

WHAT IS SOUTHWEST ASIA AND NORTH AFRICA?

LEARNING OBJECTIVES

1. **Describe** the physical geography of Southwest Asia and North Africa.

2. **Discuss** how the management of water resources has changed over time in the region.

3. **Explain** how climate has influenced agricultural practices in the region.

While the boundaries of any world region are somewhat arbitrary and porous, they typically have a rationale that is at least partially embedded in physical geography. Physical boundaries often interact with cultural geography when they contain the interaction among cultural groups in a region and slow down the transmission of ideas and practices from one area of the world to another. The region Southwest Asia and North Africa is bounded by seas and oceans, mountain ranges, and deserts (Figure 5.1). To the south, the Sahara Desert has in many ways acted like an ocean, creating a frontier realm between North and Subsaharan Africa. To the northwest, the Mediterranean Sea forms a boundary with Europe. To the north, mountain ranges running through Turkey and Iran form somewhat of a barrier. Finally, to the east, the Indian Ocean separates the region from South Asia, and to the west, the Atlantic Ocean is a natural boundary. These physical boundaries have helped create a cultural region distinguished by semiarid livelihood strategies, the widespread use of Arabic as a lingua franca, and the preponderance of Islam as a religious affiliation.

PHYSICAL GEOGRAPHY

More than any other world region, Southwest Asia and North Africa are characterized by dryland environments (Figure 5.2). The desert and semi-arid climate regions include some of the world's major deserts, most notably the Sahara Desert, the world's largest desert, which spans

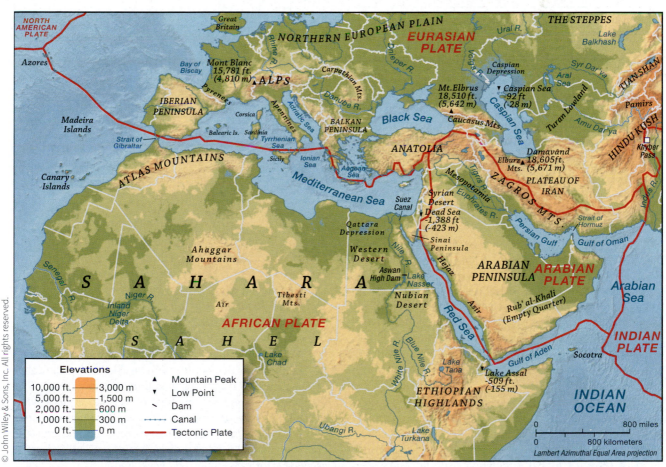

FIGURE 5.1 **Physical geography of Southwest Asia and North Africa.** The region has few mountain chains and vast areas of deserts. The plate boundaries in the region include a diverging or spreading boundary in the Red Sea, where the Arabian Peninsula is moving apart from the continent of Africa, and a converging boundary in the Mediterranean Sea, where Africa and the western end of the Eurasian plate are moving toward each other. The Atlas Mountains in North Africa run parallel to the Mediterranean plate boundary. Mountains in Southwest Asia run generally parallel to the plate boundaries and are created by compression along fault lines in the region from the divergence in the Red Sea, which is pushing the Arabian Plate northeast.

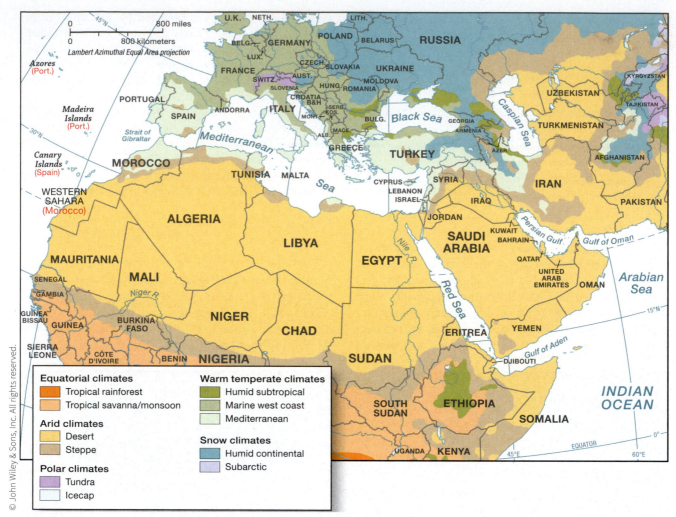

FIGURE 5.2   Climates of Southwest Asia and North Africa.   Arid climates cover the majority of the region. Farther north in the region, Iran and Turkey are located closer to the midlatitudes. Both countries have continental climates and experience snow at higher elevations and on windward sides of mountains.

North Africa, and the Arabian Desert, the third-largest desert, covering much of the Arabian Peninsula. Semiarid regions straddle the Sahara and Arabian Deserts, covering much of Iraq, Jordan, Syria, and Yemen (Figure 5.3). The Mediterranean climate weaves along the coast of the Mediterranean Sea, from Morocco to Libya and from Israel through Syria. Finally, Syria encompasses not only desert, semiarid desert, and Mediterranean but also a significant area of dry steppe grasslands due to higher elevations and latitudes.

Temperature and rainfall are two of the most important determining factors for a **biome**, or large ecological unit with similar vegetation and climate. The major deserts of North Africa and Southwest Asia are located where they are as a result of global circulation patterns. High average temperatures from year-round direct sunlight at the equator generate low-pressure systems, in which air rises, cools, and dumps moisture, leading to tropical rainforest vegetation around the equator. Air rises to the top of the troposphere, moves toward the poles, and then descends at

the subtropical high-pressure cells, north and south of the equator. When air descends at the subtropical high-pressure cells, the air warms. Descending air means little to no rainfall, creating the world's major warm deserts. In the Northern Hemisphere, the subtropics occur near the Tropic of Cancer (23.5°N), shown on Figure 5.2 as a dashed latitude line running straight through the region.

The difference between the true desert and the Steppe or semiarid areas that border them (also known as the Sahel in Subsaharan Africa) is a small amount of rainfall. The rain allows for greater amounts of vegetation, expanded opportunities for raising livestock, and sometimes crop farming. The Mediterranean's dry warm summer and wet cool winters make it the region's major agricultural zone, especially for fruits, nuts, and vegetables.

WATER MANAGEMENT

Water, a precious commodity in Southwest Asia and North Africa, is perceived both as a scarce resource

FIGURE 5.3 **Erg Mehejibad, Algeria.** The sand dunes in the Sahara Desert shift over time, as they are shaped by wind. An erg is a large expanse of sand also known as a sand sea.

and the root of ethnic tension. Major river systems in Southwest Asia and North Africa include, most notably, the Nile (the world's longest river) and the Tigris and Euphrates.[1] The upstream countries of the Nile include South Sudan, Uganda, and Ethiopia, and the lower reaches of the Nile flow north through Egypt (Figure 5.4a). The Tigris and Euphrates rivers largely run through Iraq, known historically as Mesopotamia (meaning between two rivers), with the upper reaches in Turkey and Syria (Figure 5.4b). The hearths of agriculture, the first urban revolution, and innovations in irrigation are all located along these river systems.

Traditional Irrigation

Cave paintings of lakes, grasslands and large game found in the midst of the Sahara Desert offer proof that North Africa and Southwest Asia were at one time much wetter, but most of the region has been arid for 9,000 years. Southwest Asians and North Africans have developed ingenious methods for managing water resources. Underground conduits, called **qanats**, transport water from the water table underground, using the slope of a hill and gravity, to agricultural areas or towns below (Figure 5.5). This sophisticated engineering approach, which requires no pumping, was first developed in Iran sometime in the first millennium BCE, predating the aqueduct system of the Roman Empire by several centuries. Qanats diffused to India, the Arabian Peninsula, North Africa, and Spain, and have taken on local names, such as *foggaras* in North Africa and *falaj* in the United Arab Emirates.

Despite being at least 2,000 years old, much of Iran's network of qanats is still functioning. Some 22,000 qanats make for more than 170,000 miles of underground channels. Until the 1960s, qanats supplied 75 percent of Iran's water for both agricultural and household purposes. Given limited water supplies in many areas in Southwest Asia and North Africa, this technology made it possible to sustain and develop crop agriculture. Qanats tap into aquifers that are replenished during the rainy season and rely on gravity, so the flow of water from a qanat varies based on the volume of water in the water table. Using gravity instead of pumping prevents the aquifer from being depleted and does not use energy. Modern irrigation systems often deplete aquifers, harm aquatic life in rivers, or use enormous amounts of energy to power motor pumps (Goldsmith and Hildyard 1984).

Contemporary Tensions over Water

Syria depends on the Euphrates River for over half of its water supply. Iraq depends more on the Euphrates for its water supply than it does on the Tigris River (Drake 1997). The Tigris and Euphrates Rivers start in the highlands of neighboring Turkey, which is the source area for more than 70 percent of the flow of both rivers. Turkey has taken advantage of its upstream position by constructing 9 dams (of a planned 22) to irrigate fields and generate power. These engineering plans, known as the Anatolia Development Project, raised tensions between Turkey and Syria and Iraq starting in the late 1980s. Syria and Iraq demanded that more water be released, a request that

---

[1] The River Jordan is sometimes discussed as a major river in the region, but it is much smaller.

Turkey refused in order for reservoirs to form behind the dams. Tensions among the three countries have waxed and waned since that time, erupting most recently in 2009 when Iraqi military police (MPs) complained about a water shortage created by Turkey. With spring water reserves down to 11 billion cubic meters (388 billion cubic feet) that year from nearly four times that amount three years earlier, the Iraqi MPs threatened to block all agreements with Turkey unless more water was released. Turkey eventually complied.

Secure access to water has also triggered arguments among some countries of the Nile Basin (Drake 1997). Egypt depends almost totally on the Nile as a source of fresh water, and its water resources are the most constrained among the ten countries that share its watershed. Over 85 percent of Egypt's Nile waters originate in Ethiopia, Sudan, and South Sudan. These countries want to claim a larger share of Nile waters in accordance with Helsinki and International Law Commission rules, but Egypt claims that its history of prior usage of a disproportionately larger share of Nile waters represents an entitlement. Egypt, with the largest army of the four countries, has warned its upstream neighbors that they risk war if they severely impede the flow of the Nile River (Drake 1997). Three of these countries are large by African standards: Egypt (75 million), Ethiopia (79 million), and Sudan (34 million), whereas South Sudan is smaller at 10 million (World Bank 2009 and 2011).

Some authors have argued that tension over water in the region is a classic **environmental security** problem. Environmental security is a body of political science scholarship broadly arguing that many political and ethnic conflicts stem from resource scarcity, and that such scarcity is driven by overpopulation (Homer-Dixon 1994). A

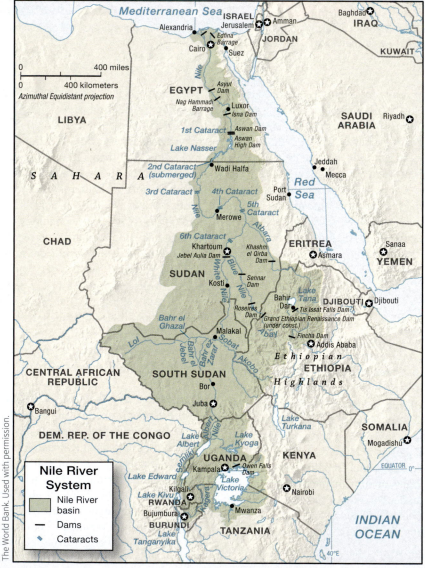

FIGURE 5.4a   Nile River system.   The Nile River system connects the highlands in Uganda, Ethiopia, South Sudan, and Sudan with Egypt, as each country relies at least in part on fresh water from the Nile River for irrigation and human water consumption.

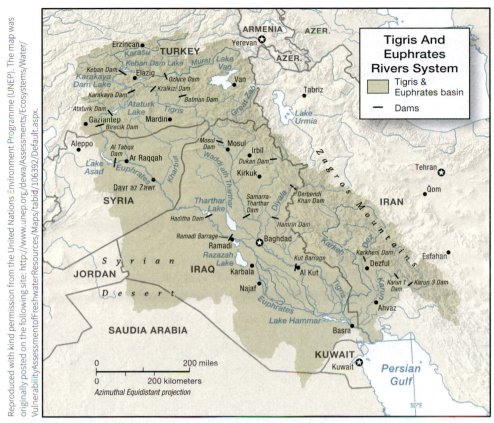

FIGURE 5.4b   **Tigris and Euphrates Rivers system.**   The Tigris and Euphrates Rivers dominate the physical landscape of Iraq, connecting the country with its neighbors to the north.

number of geographers have contested environmental security assertions.

In Southwest Asia and North Africa, water is a frequent focal point for environmental security debates. Experts intensely discuss both pieces of the environmental security argument. For starters, they debate what is driving the scarcity of water in the region. The political scientist Tad Homer-Dixon (1994) and the journalist Robert Kaplan (1994) have argued that population growth is leading to the scarcity of resources. Geographers (including Fairhead 2001 and Hartmann 2001) have suggested that resource scarcity often results from uneven distribution and extreme overuse by some groups. For example, both the Nile and the Tigris and Euphrates river systems have major dams. While these dams often provide hydroelectric power and municipal water, farmers are the major user group in almost all instances. Yet, it is a small group of select farmers who benefit from being inside irrigation schemes supplied by dams and water diversion. The vast majority of (often downstream) farmers, not to mention herders, fisherfolk, and smaller communities, actually lose access to water when such schemes are developed. In this case, resource scarcity is socially constructed (or the product of inequitable distribution) rather than a case of absolute scarcity.

The second part of the environmental security argument—whether or not resource scarcity (water in this case) leads to ethnic conflict—is another debate. A good example of such a discussion may be found in Sudan. In June 2007, the UN Secretary General Ban Ki Moon publicly argued that climate change was playing a role in the ethnic conflict in Darfur, Sudan (Moon 2007). Moon's comments were followed a month later by a report from a group of geologists at Boston University who had discovered a huge underground lake in Sudan. The scientists argued that drilling wells in Sudan would relieve tension in the region because "much of the unrest in Darfur and the misery is due to water shortages" (BBC 2007). While new sources of water were a welcome development, the belief that such discoveries would end the war in Darfur was misguided. An alternative (non-natural) explanation for the crisis may be found in the irrigation programs for commercial farming sponsored by colonial and postcolonial regimes that deprived the members of the Beja ethnic group of their best grazing land and fostered tensions between herders and farmers (Polgreen 2007). Geographers with experience in Sudan (Kevane and Gray 2009) have a role to play in questioning the environmental security argument and its related quick-fix, technical solution of sinking more wells.

Geographer Anthony Allan (2000) has argued that it is almost politically taboo to discuss water shortages in Southwest Asia and North Africa. In the 1950s, regional governments set a course for development based on modernization. This agenda includes a hydraulic mission, the belief that human ingenuity, large dams, and irrigation

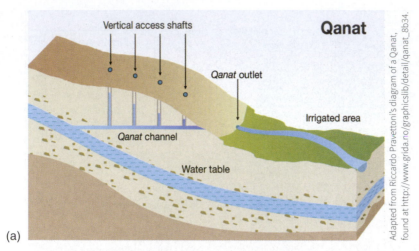

FIGURE 5.5   Diagram and photo of a qanat.   (a) A qanat taps into a deep-water table by digging a horizontal well and using gravity to create a dependable water flow. (b) **Adrar, Algeria** Aerial view of a qanat, an underground aqueduct system, with gardens protected by windbreaks in a Sahel village. The qanats here are very old but are still providing enough water for irrigation even though they have had no maintenance for 30 years.

projects could overcome natural constraints and turn marginal dryland areas into productive agricultural zones (Figure 5.6). To question this agenda, and to suggest that there are limits imposed by scarcity, has often been portrayed as unpatriotic. The problem is that the hydraulic mission has wasted water resources in a region that does not have them to spare. Vast amounts of water evaporate through dams and flood irrigation, which also destroy soil resources though salinization and seriously impair riparian (areas around rivers) ecosystems, problems that did not exist with the much more economical qanat technology. It may be that it just is not practical to try to produce so much food in dryland areas. By importing grain, countries are essentially importing the water, or **virtual water**, used to produce this grain. One strategy

for addressing water shortages in the region may be to simply import grain.

AGRICULTURE

People have not always farmed. Hunting and gathering are older forms of subsistence. In some parts of the world, the idea to farm crops and domesticate plants developed independently. In other areas, the idea to farm arrived from elsewhere. What is significant about Southwest Asia, and more specifically an area known as the **Fertile Crescent**, which lies between the Tigris and Euphrates rivers, and extends west to Syria is that this is the first place (according to most scientists) where crop agriculture started, about 10,000 BCE. The transition from hunting and gathering to farming

Tips Images/SuperStock.

FIGURE 5.6   Souss-Massa, Morocco.   The king of Morocco decided to build a dam in southwestern Morocco in 1968. Built in 1972, the Youssef Ben Tachfine Dam retains the Massa River. Water from the reservoir is used to support production of agriculture in the region. Today, this area of Morocco supports large-scale production of vegetables in greenhouses. The greenhouses use water from the river as well as groundwater for production.

is known as the **first agricultural revolution**, because it is the first of at least three important changes in agriculture over 12,000 years.[2] The first agricultural revolution occurred at different times in different parts of the world.

How and why crop farming actually began is uncertain. While we might assume that agriculture was a choice and an innovation representing a superior form of food procurement, geographer Jared Diamond (1987), among others, argued that scarcity probably forced us into it. Studies have demonstrated that many contemporary hunters and gatherers spend less time working to procure food than agriculturalists. Furthermore, hunting and gathering actually provides a much better diet. Archaeological evidence suggests that mortality and morbidity rates were higher for early agricultural communities than hunters and gatherers.

Crop farming probably developed slowly out of the gathering side of hunting and gathering. Early gatherers likely wished to propagate some of the more beneficial plants they were collecting and thus began by spreading around their seeds. These earliest farmers might have broadcast seeds in

an area and then returned after three to four months of itinerant hunting and gathering. Scarcity, as a result of growing and competing populations, would likely have forced people to become more reliant on these crops. Communities likely became more settled in order to protect the crops they planted. Crop domestication would have occurred over generations as people picked the seeds from the best plants and planted them.

The Fertile Crescent is not only the earliest known place of the first agricultural revolution, but is also an original **hearth** of domestication for important grain crops such as wheat and rye, and pulses such as peas. A hearth is the area from which an original idea develops and then diffuses to other areas. A hearth of domestication is the area where a plant or animal is first domesticated through the breeding of wild relatives in a process that can take hundreds of years. In some instances, certain plants or animals were likely independently domesticated at more than one location (Figure 5.7).

Theories differ as to what combination of circumstances led to the domestication of plants. One theory contends that agriculture began in an area with a diversity of plants, which enabled plant breeding to be successful, and that population grew after the food supply was stabilized. Another theory maintains that population pressure coupled with a variety of wild plants encouraged successful domestication (Diamond 2003).

---

[2] The other important junctures were the second and third agricultural revolutions. The second agricultural revolution refers to a system of more intensive and commercial agriculture that often accompanied industrialization. Agricultural output increased because of crop rotation and use of organic inputs. Like the first agricultural revolution, this occurred at different times in various areas of world. It is thought to have first taken place in Europe during the 1700s. The third agricultural revolution began in the late nineteenth century and continued through the twentieth. It was characterized by increasing use of mechanization, pesticides, and inorganic fertilizers in farming.

 hearth   an area or place where an idea, innovation, or technology originates.

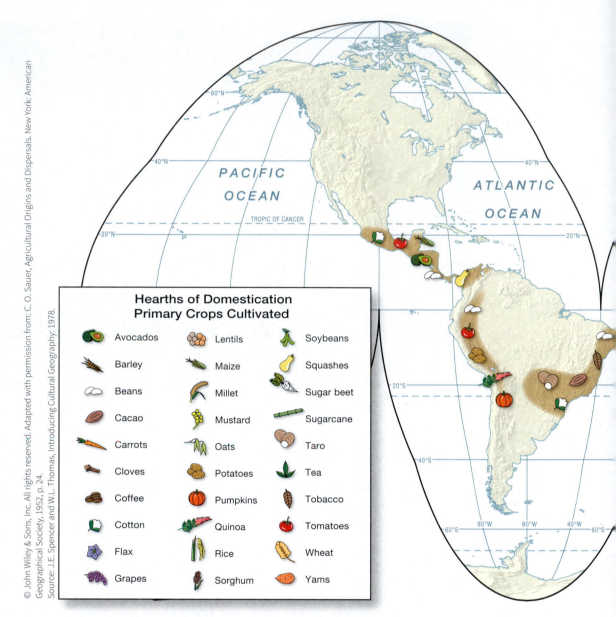

FIGURE 5.7   Hearths of early domestication.   The domestication of plants and animals occurred around the world. Each region of agriculture became the hearth of domestication of certain crops or animals. For example, the Fertile Crescent was the hearth for the domestication of goats and sheep as well as oats and lentils. Mesoamerica was the hearth of domestication for maizes (corn), tomatoes, avocados, and the turkey.

Today, some of the region's most vibrant and productive agricultural areas are found in the Mediterranean climate area, an environment with wet winters and dry summers. All along most of the Mediterranean coast, farmers cultivate a wide variety of fruits as well as winter wheat (Figure 5.8a).

Two unique forms of farming have developed according to the region's particular physical characteristics: floodplain and oasis agriculture. The Nile River Valley in Egypt is probably the best example of floodplain agriculture (Figure 5.8b). Farmers in ancient Egypt divided the year into three seasons: flood time, seeding, and harvest. Each October, the Nile River flooded for a few weeks, and then the water level would drop, leaving a layer of fertile, black mud. This mud fertilized the soil, and the floodwater was stored in a series of canals. A special government department was in charge of making sure the canals were kept in good repair. The farmers grew such crops as pomegranates, melons, figs, wheat, and barley. Today, Egypt's dams and irrigation canals have disturbed the Nile's natural flow, with advantages and disadvantages for farming systems in the country. On the positive side, dams allow for more even and predictable river flow. However, these dams have also reduced the flooding that brought nutrients to the soil. Now farmers must apply nutrients, often through the use of chemical fertilizers, in order to maintain productivity.

The other special form of crop production found in the region is oasis agriculture (Figure 5.8c). Oases are places of isolated vegetation in a desert, typically surrounding a natural spring or in an area where the water table is unusually high.

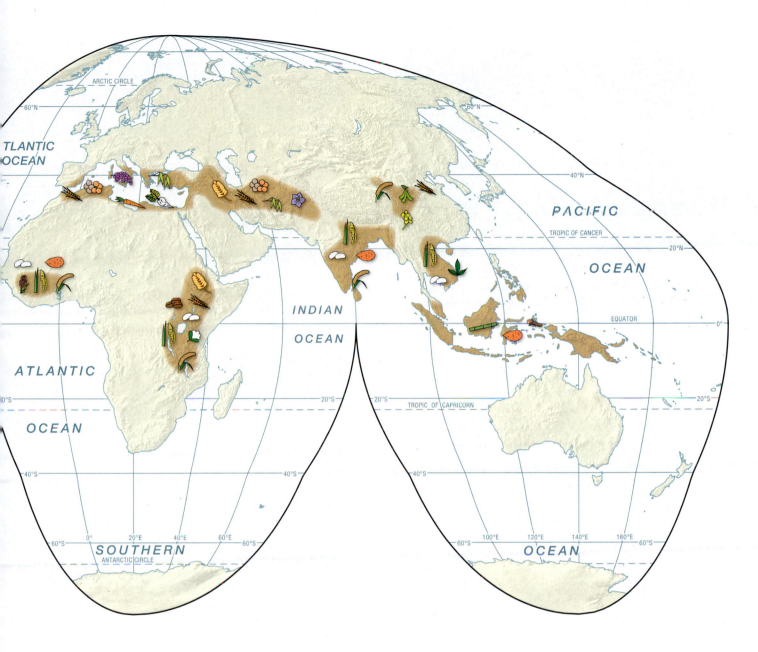

Oases dot both the Sahara and Arabian deserts. Oasis crops generally include dates, figs, olives, and apricots. A common, highly intensive strategy is to plant tall, heat-loving plants, such as date palms, as shade for smaller trees, like peach, which form a middle layer. Vegetables may then grow underneath. By growing plants in different layers, farmers make the best use of limited soil and water resources while protecting more fragile crops from too much sunlight.

In contrast to the intensive livestock operations found in North America or Europe, most animal husbandry in the region is quite extensive, with herders regularly moving their animals from place to place in relation to pasture conditions. This type of animal husbandry is called **pastoralism**, and it is by far the region's most common agricultural system. From an environmental standpoint, pastoralism is the most appropriate livelihood strategy in drylands with limited and variable rainfall. Pastoralism is widespread throughout all of the climate and vegeta-

tive regions except the driest and most isolated tracks of desert. Sheep, goat, cattle, and camels are the most common types of animals kept by herders in Southwest Asia and North Africa (Figure 5.8d).

In *Resurrecting the Granary of Rome* (2007), Diana Davis described how stories of deforestation and desertification in North Africa had been told from the Roman period to the present. Until recently, these stories of environmental decline in North Africa were still recounted by experts and were widely accepted without question. International organizations such as the United Nations frequently invoked these stories, also known as environmental narratives,[3]

---

[3] An environmental narrative is the dominant conception of an environmental problem in a region which may or may not be based on reliable empirical evidence. Many environmental narratives date back to the colonial era and have persisted over time in spite of evidence to the contrary.

FETHI BELAID/AFP/Getty Images.

Maisant Ludovic/ZUMAPRESS/Newscom.

(a)

(b)

Max Alexander/Dorling Kindersley/Getty Images.

Yadid Levy/Photo library/Getty Images.

(c)

(d)

FIGURE 5.8   Diversity of agriculture in North Africa and Southwest Asia.   (a) **Sabbalet Ammar, Tunisia** Workers pick olives, which are commonly grown in Mediterranean agriculture. Spain dominates the olive industry on the Mediterranean, followed by Italy and Greece. After investing in olive production, Tunisian olive growers have produced improved olive oil and now get Spanish prices, according to industry officials. (b) **Luxor, Egypt** Sediment from the Nile River brings nutrient-rich soil, and water from the river and is used to irrigate the fields. Just a few miles away from the river, the land does not support agricultural production. The UN Food and Agriculture Organization estimates that 80 percent of the renewable water in the Nile River region is used to support agricultural production. (c) **Al-Qasr, Egypt** The Dakhla Oasis, located about 350 miles west of the Nile River, has been occupied by people for thousands of years. Today, the oasis is home to about 75,000 people who focus agricultural production on growing and drying fruits. (d) **Atlas Mountains, Morocco** A shepherd stands with his goat herd in the Atlas Mountains of Morocco. Pastoralists rely on sparse grass and vegetation in this region to feed their herds, and they move the herds seasonally to find vegetation and water.

to justify environmental conservation and development projects in the arid and semiarid lands in North Africa and around the Mediterranean Basin. Davis's pioneering historical geographical analysis, including archival research in three different languages, revealed the critical influence of French scientists and administrators who

established much of the purported scientific basis of these stories during the colonial period in Algeria, Morocco, and Tunisia, illustrating the key role of environmental narratives in the colonial era. The processes set in place by the use of this narrative not only systematically disadvantaged the majority of North Africans but also led to profound

GUEST *Field Note* Using Historical Geography to Question Dominant Ideas About Agriculture and the Environment in North Africa

DR. DIANA DAVIS
University of California, Davis

Large trees like this one next to a religious shrine in southern Morocco nestled among apparently barren hills are some of the most striking scenes in North Africa. Considered deforested since the colonial period, reforestation efforts have been carried out over the decades in this region and in numerous other places in Morocco, many of which have failed. Current paleoecological data, however, suggest that many hills such as these were never heavily forested. Moreover, after a good rainfall, such areas that had looked barren other than for a few scrubby perennial plants will be covered with various grasses, flowers, and other plants adapted to the dry and highly variable climate. Listening to local peoples, many of whom do not describe their environment as deforested, desertified, or otherwise degraded, is an important part of conducting geographical research. Probing the official, "expert" stories of environmental degradation when the local inhabitants have a different story to tell can often result in some of the most productive research on environmental change.

Photo by James E. Housefield with permission.

FIGURE 5.9   Southern Morocco.   Relatively barren landscapes like this become lush after a good rainfall.

changes in the landscape, some of which produced the land degradation that continues to plague North Africa today (Figure 5.9).

CONCEPT CHECK

1. **How** does the physical geography of Southwest Asia and North Africa vary?

2. **What** are examples of traditional and modern modes of water management in the region, and what are the pros and cons of these approaches?

3. **Why** might agriculture have first developed in the Fertile Crescent, and how does farming vary across the region?

WHO ARE SOUTHWEST ASIANS AND NORTH AFRICANS?

LEARNING OBJECTIVES

1. **Discuss** the origins of urban living in Southwest Asia.

2. **Describe** the characteristics of a traditional Islamic city.

3. **Explain** contemporary patterns of urbanization.

Southwest Asia and North Africa have a storied history of cities and urban culture. Not only did the region give rise to the first cities in the world, but it is home to a unique urban spatial form known as the Islamic city. Today, the region is also one of the most urbanized in the world, after Europe and Latin America. As we ask who are Southwest Asians and North Africans, we focus on those living in cities and the unique Islamic city in this section of the chapter. We look at the origins and diffusion of religions and languages in the region in the last section of the chapter.

THE ORIGINS OF URBAN LIVING

The world's earliest cities developed independently in the various hearth areas of the first agricultural revolution. Most scholars agree that the production of agricultural surplus is a precondition for the development of cities, and that the first cities on the planet sprouted in the fertile valleys of Mesopotamia, around the Tigris and Euphrates (in present-day Iraq) between 4,000 and 3,500 BCE, which was between 4,500 and 7,000 years after the first agricultural revolution (Figure 5.10). These **city states** took on a particular form with a walled city surrounded by an agricultural **hinterland**. For the most part, they were largely self-sufficient. Since there was not much

long-distance trading, the size of a city state depended on the extent of its agricultural hinterland. The city states in the Fertile Crescent fought wars as they grew and needed more hinterland. The size of a city's hinterland was, however, limited by the distance food could be transported without spoiling. Archaeologists have unearthed Sumerian[4] city states at Eridu, Uruk, and Ur. The largest of these cities probably had about 10,000 people.

City states had defined social structures with priests at the top of the social hierarchy in theocratic city states, followed by craftsmen and farmers. The social structure was rigid, but the urban form was organic, taking shape as people saw fit to build, with winding, unplanned streets. The least noxious occupations were located closest to the city center, and the dirtiest trades such as tanneries were on the periphery. These cities did not have sanitation infrastructure. As the streets filled with garbage, the residents simply built over the debris. As a result, many of these cities and

[4] Sumer was a civilization and historical region in southern Mesopotamia (or modern-day Iraq) during the Chalcolithic and Early Bronze Age.

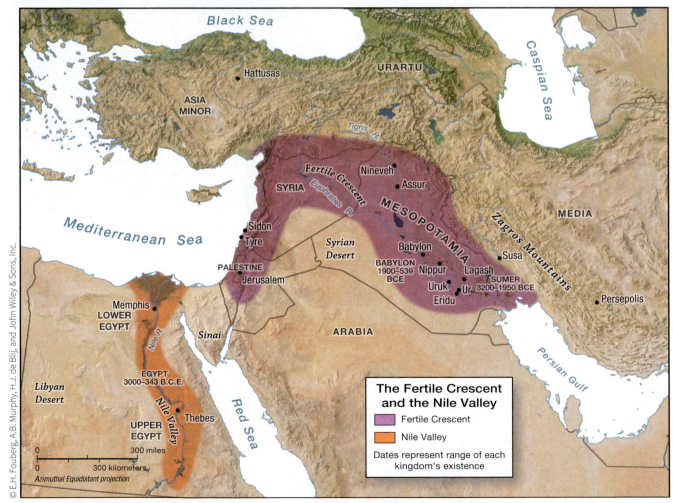

FIGURE 5.10 The Fertile Crescent and Nile Valley.   Civilizations including Sumer, Babylon, and Egypt occupied the Fertile Crescent and Nile Valley between 3,200 and 539 BCE. Several city states established walls and hinterlands in the Fertile Crescent region, including Ur, Uruk, and Babylon.

© Nik Wheeler/Corbis.

FIGURE 5.11   Uruk, Iraq.
Ruins of the city state of Uruk, founded around 4,500 BCE. As was common in most city states in the Fertile Crescent, Uruk was built on human-made hills, or tells.

their subsequent ruins are found on human-made mounds also called "tells" (Figure 5.11).

The beginnings of agriculture and the early rise of urban living in the region facilitated the building of civilizations. Sumerian city states were at first self-sufficient units, relying on agriculture produced in their hinterlands with little trading between cities. As trade among city states increased, cities and towns evolved into networks and, ultimately, empires. South-

west Asia and North Africa are home to the earliest great civilizations, including the Sumer, Akkadian, Babylonian, and Assyrian empires in Mesopotamia (3,200–539 BCE) and Ancient Egypt (3,000–343 BCE) (Figure 5.12). The Roman Empire (509 BCE–470 CE) depended on resources from the cities in its territories, which came to include substantial portions of North Africa and Southwest Asia after its defeat of Tunisia's Carthaginian Empire (Figure 5.13). The territorial

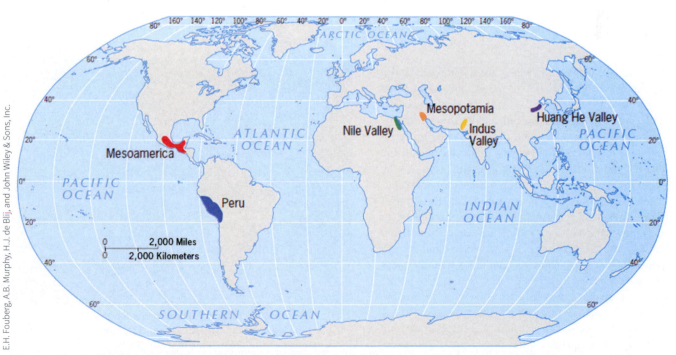

E.H. Fouberg, A.B. Murphy, H.J. de Blij, and John Wiley & Sons, Inc.

FIGURE 5.12   Hearths of ancient civilizations.   Based on archaeological evidence, academics have determined humans clustered in several early civilizations where they began agriculture, constructed the first cities, and developed systems of government.

Reading the PHYSICAL Landscape

Carthage Baths

In nearly all of their major cities, the Romans built sumptuous baths, with heated floors and pools of varying temperatures, a luxury that required significant labor and fuel resources to maintain. The baths were heated by fires fueled with wood from the city's hinterland. The Romans completely deforested several areas, including an island off the coast of Tunisia, in order to supply fuel for these baths in Carthage, Tunisia, which is now a part of the modern-day city of Tunis.

In addition to wood for fuel, the Carthage Baths needed water, and the Romans combined their aqueduct engineering with Carthage cisterns (designed originally to hold rain water) to provide a steady supply of water to the baths. The Roman aqueduct that fed fresh water to the baths of Carthage stretched a distance of more than 80 miles (132 km) from a spring in Zigus (now Zaghouan). To diffuse Roman bath culture to Carthage, the conquerors had to change the physical landscape, through deforestation and water supply engineering.

Courtesy of William Moseley.

FIGURE 5.13 **Carthage, Tunisia.** Ruins of the Carthage baths.

cities extracted resources from their rural hinterlands and reluctantly passed substantial portions on to the city of Rome as a form of tax or tribute.

THE ISLAMIC CITY

In the seventh century CE, Muhammad founded Islam on the Arabian Peninsula based on the teachings he received from God and recorded in the Qur'an. Over the next several centuries, the faith spread rapidly throughout the region, with its strongest and earliest foothold in urban settings. The faith was sufficiently influential in daily life that it came to influence urban form. The result was the **Islamic City**, considered a distinct urban form that originated in Southwest Asia and North Africa. Elements of the classic Islamic City include a central marketplace (or *suq*), mosques, a citadel, and public baths. The residential land use is often compact, with walled familial compounds and open courtyards. Buildings and alleyways are often designed to protect privacy. Entrances to homes may be L-shaped, walls are often windowless, and alleys are narrow and irregular to create blind spots.

Many contemporary cities in the region have long histories. As such, these cities often have an older section of town, frequently known as the Medina, which has the characteristics of an Islamic City (Figure 5.14).

Courtesy of William Moseley.

FIGURE 5.14 **Tunis, Tunisia.** The *Bab el Bahr* gate serves as the entry to the Medina or old city. The city dates to the fourth century BCE and was part of the Roman Empire. In the seventh century, Tunis became a Muslim city. Muslim architecture, including this gate, is evident in the Medina.

This older section of town is then surrounded by a more contemporary city with more of a Western urban form (Figure 5.15). However, even cities with shorter histories will typically have mosques and marketplaces that feature prominently on the urban landscape (Figure 5.16).

CONTEMPORARY PATTERNS OF URBANIZATION

The region of Southwest Asia and North Africa is one of the world's more urbanized areas, with relatively high proportions of the population living in cities in most countries (Figure 5.17). While 50 percent of the world's population lives in urban areas, the average for Southwest Asia and North Africa is 57.3 percent (World Bank 2009) (Table 5.1). Many, but not all, of these countries have oil resources or limited possibilities for crop agriculture. Some of the countries in the region with the lowest levels of urban living include Yemen and Egypt (World Bank 2009). These are countries that do not have large manufacturing or petroleum industries to support urbanization and thus are more dependent on agriculture.

The largest cities in the region include Cairo, Egypt (15.2 million); Tehran, Iran (12.8 million); Baghdad, Iraq (6.6 million); Riyadh, Saudi Arabia (5.1 million); Alexandria, Egypt (4.6 million); and Casablanca, Morocco (4 million).

Many of the region's cities are dynamic places of creativity and cultural innovation. For example, Cairo has long been known as an intellectual hub of the region. With several great universities, the city has a history of producing great thinkers and activists. Most recently, Cairo was a focal point of the **Arab Spring**, a democratic awakening in the Arab World that erupted during the spring of 2011 (Figure 5.18a). The movement began in Tunisia and then diffused to Egypt and eventually Libya, where the dictators in all three countries were toppled.

The outcomes of the Arab Spring are not confined to these three states. Demonstrations against the Assad regime in Syria led to a government crackdown, which in

FIGURE 5.15 Tunis, Tunisia. A grand avenue with wide boulevards, the Avenue Habib Bourguiba crosses through the contemporary city of Tunis from east to west. The avenue is lined with stores and cafes on the first floors of buildings and apartments above.

FIGURE 5.16 Tunis, Tunisia. La Souk de la Medina includes a mosque and a marketplace in central Tunis. The marketplace is usually crowded with tourists, but tourism declined after the fall of Dictator Ben Ali in 2011.

TABLE 5.1 Proportion of population living in urban areas, 2015

Bahrain	98.2%
Kuwait	98.5%
Qatar	96.2%
Israel	91.9%
Lebanon	87.9%
Libya	87.4%
Jordan	85.3%
Saudi Arabia	83.2%
Egypt	45.4%
Yemen	31.9%

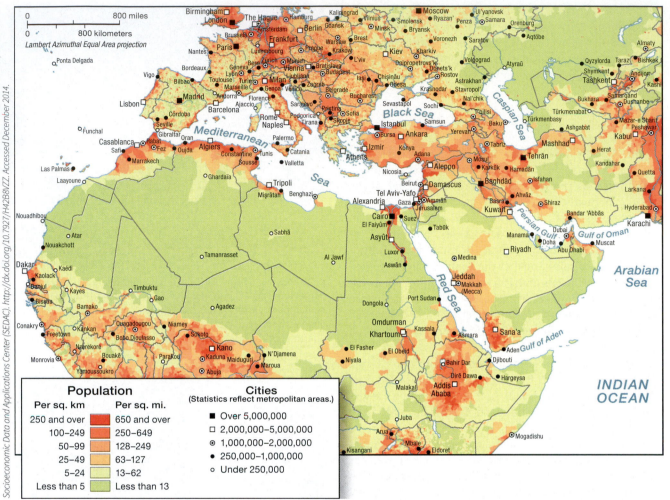

Data from CIESIN—Columbia University, United Nations Food and Agriculture Programme—FAO, and Centro Internacional de Agricultura Tropical—CIAT. 2005. Gridded Population of the World, Version 3 (GPWv3): Population Count Grid, Future Estimates. Palisades, NY: NASA Socioeconomic Data and Applications Center (SEDAC). http://dx.doi.org/10.7927/H42B8VZZ. Accessed December 2014.

FIGURE 5.17 **Population density of Southwest Asia and North Africa.** The highest population densities are along coastlines and throughout Syria, Turkey, Iraq, and western Iran. The arid regions of the Sahara and the Arabian peninsula are sparsely populated.

(a)   (b)

FIGURE 5.18 **Cairo, Egypt.** (a) Egyptian demonstrators gathered in Tahir Square on February 1, 2011, in a massive outpouring against the regime of President Hosni Mubarak. Protests began on January 25, 2011, and lasted 17 days. On February 11, 2011, Mubarak stepped down after 30 years of dictatorial rule. (b) Hosni Mubarak's suit was made of fabric that included his name woven into the pinstripes (zoom in on the photo to see the pinstripes). A newspaper in the United Kingdom researched custom cloth for suits and estimated that a two-piece suit made from personalized fabric like this would cost more than $16,000.

turn led to the outbreak of civil war, much of which could be followed through the social media, especially in the early days of the war (Figure 5.19). The states around the Persian Gulf that are controlled by monarchies, including Kuwait, Saudi Arabia, and Qatar, as well as the monarchical states of Jordan and Morocco, have felt relatively little upheaval from the Arab Spring.

One commonality among Arab countries in transition from authoritarian government to democracy is the issue that unemployment poses, especially among younger people in the countries. The United Nations Development Program reported that the unemployment rate for youth (ages 15 through 24) globally is 12.6 percent. The first two countries to experience the Arab Spring, Tunisia and Egypt, have youth unemployment rates of 30 percent and 30.7 percent, respectively. Syria and Iraq, which are in political upheaval, have youth unemployment rates of 18.3 percent and 43.5 percent, respectively. Several of the countries that have avoided upheaval have low unemployment rates, including Qatar at 1.2 percent and Kuwait at 7.7 percent.

While many of the countries share the issue of unemployment, especially for the youngest generation, the rights of women vary across countries in transition. On one end, Tunisia is seen as the forerunner for women's rights in the Arab world. In 1956, Tunisia became the first country in the Arab world to ban polygamy. Women in Tunisia were the first to have legal access to abortion (in 1973, the same year as women in the United States). In Tunisia, contraception is widely available, and women must consent to marriage and cannot marry before age 17. Tunisian women also have relatively high levels of education and are playing an active role in their country's political process. In 2012, a draft of the constitution labeled women as "complementary" to men, and in turn "Many women's rights activists bucked against the proposal and on the country's 2012 Women's Day, thousands marched in the streets in protest" (Hartmann and Keske 2013).

In Egypt, the hope was to topple a corrupt dictator who controlled the country from 1981 to 2011, taking a piece of major business deals while he ruled the country and amassing a family fortune valued at between $40 and $70 billion. While the rest of the country suffered high unemployment, Hosni Mubarak wore handmade custom suits with his name stitched into the pinstripe of the fabric (Figure 5.18b). In 2012, Mubarak and his two sons were tried and found guilty of the murder of protesters in Tahir Square. With regard to to women's rights, Egypt's women rank last among Arab countries: They are in 22nd place "for discrimination in law, sexual harassment and the paucity of female political representation," according to a report from a Reuters Foundation survey in 2013 (Aswany 2013). In the first democratic election in 2012, women won only 10 of Egypt's 508 seats in parliament. In spring 2014, Egypt underwent a military coup through which a longtime military general took control of the country, stomping out the progress the Egyptians hoped they had gained through protests in Tahir Square.

CONCEPT CHECK

1. Why did the first cities develop in the Fertile Crescent area?

2. How is the Islamic City a unique urban form?

3. What are the contemporary patterns of urbanization in the region?

 USING GEOGRAPHIC TOOLS

Crowdsource Mapping of Killings and Human Rights Violations in Syria

Syria Tracker is an online crowdsourcing site that has created points on a map for death and human rights violations that have taken place in the civil war in Syria since April 2011. Each report is made by someone on the ground in Syria, after which the incident is verified. Reports include killings, chemical poisoning, disease spreading, food tampering, revenge killing, and rape. International organizations report that the use of social media, including Twitter and blogs, in the Syrian civil war far exceeds the use in other wars since the invention of easily accessible social media. Some news agencies in the West have only analyzed and mapped Twitter reports in English. A report by the United States Institute for Peace found that the percentage of Tweets in English has declined over the course of the war, and the use of Arabic for messages in social media has increased precipitously. Syria Tracker includes reports in both English and Arabic, allowing a more complete picture of what is happening on the ground in Syria.

(Continued)

Thinking Geographically

1. Syria Tracker includes reports in both English and Arabic. How might the map differ if it only included reports in English?

2. Syria Tracker does not explain the conflict. Read the BBC website listed below and analyze the maps on BBC.

Then, go back to Syria Tracker see how your understanding of the map changes.

Sources: Syria Tracker https://syriatracker.crowdmap.com/bigmap BBC. Syria: Mapping the conflict. http://www.bbc.com/news/world-middle-east-22798391.

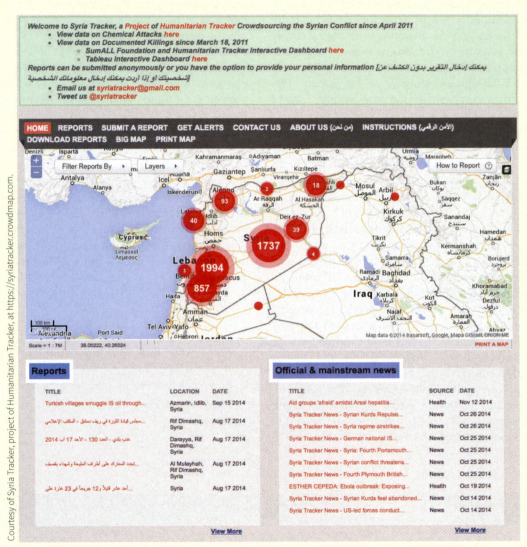

Courtesy of Syria Tracker, project of Humanitarian Tracker, at https://syriatracker.crowdmap.com.

FIGURE 5.19 Syria tracker. People on the ground in Syria's civil war have been able to report what they witness and experience through the crowdsourced map, Syria tracker.

WHAT IS THE RELATIONSHIP AMONG COLONIALISM, OIL, AND DEVELOPMENT IN THE REGION?

LEARNING OBJECTIVES

1. **Relate** the history of European colonialism in the region.

2. **Describe** the geography of oil and gas production in Southwest Asia and North Africa.

3. **Explain** the effects of oil extraction on governance and political-economic development in the region.

Southwest Asia and North Africa were colonized by Europeans, primarily the French and British, during the second wave of colonialism. The boundaries of countries in this region date back to the colonial era. In several countries, oil was discovered during colonialism and became a motivation for continued European involvement, by both governments and corporations, in the region. As the countries in the region gained independence, oil became an important part of the region's economy. Revenues from oil have been used both to provide income to the citizens of a country, as in Kuwait, and to prop up dictatorial regimes, as in Libya.

COLONIALISM

Beginning with the French conquest of Algeria in 1830, the British and the French became the major colonial powers in Southwest Asia and North Africa, an era that lasted until the liberation of most states in the late 1940s and 1950s. Italy and Spain played only minor roles. The French laid claim to much of North Africa, including Morocco, Algeria, and Tunisia, and to Lebanon and Syria during the colonial era, which lasted from the late nineteenth through mid-twentieth century. The British controlled the Arabian Peninsula, including modern-day Saudi Arabia, Oman, Yemen, Kuwait, Bahrain, Qatar, and the United Arab Emirates (UAE) as well as Iraq, Jordan, Israel, Palestine, and Egypt. Italy was the colonial power in Libya and Spain in the northern part of Morocco. Turkey and Iran (then Persia) were never occupied by European colonizers (Figure 5.20).

European colonialism came earlier in North Africa than in Southwest Asia. France began to conquer Algeria in 1830 and Tunisia in 1881. Morocco became a protectorate of France in 1912. All three of these areas were settler colonies where French families moved to live and work. The British invaded Egypt in 1882. Britain signed a series of treaties with leaders in the Arabian Peninsula from the mid-nineteenth through the early twentieth centuries, which were essentially protectorate arrangements requiring them to conduct foreign affairs through the British government (Figure 5.21). The Levant (Syria, Lebanon, Jordan, Israel, and Palestine) fell under the protection of Britain (Jordan and Palestine) and France (Lebanon and Syria) after these two countries defeated the Ottoman Empire in 1918.

Encouraged by Zionists[5], Jews who longed for a Jewish homeland in Palestine, Britain began to facilitate the creation of a "national home for the Jewish People" within Palestine under the Balfour Declaration of 1918. Between 1914 and 1946, the Jewish population of Palestine grew from about 60,000 to over 528,000, largely as the result of migrants from Europe fleeing the oppression in Europe before and during the Holocaust. Migration increased the proportion of Jews in Palestine from 13 to 31 percent of the total, and the Zionist dream of a State of Israel was realized with its formal creation in 1948, following a UN decision to partition Palestine in 1947 (Figure 5.22). Even before the creation of Israel, there was Arab opposition to Jewish immigration.

....

[5] Zionism, a term coined by Nathan Birnbaum in 1890, refers to an international movement for the return of the Jewish people to their homeland and the resumption of Jewish sovereignty in Israel.

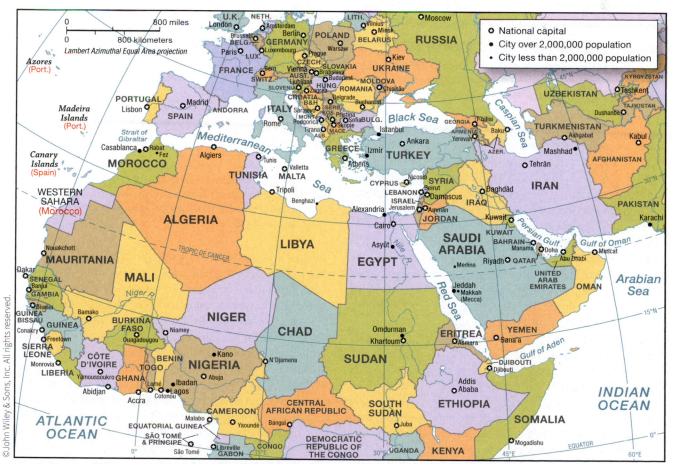

FIGURE 5.20 **Political geography of Southwest Asia and North Africa.** The Ottoman Empire, dating back to the 14th century, grew to dominate much of Southwest Asia until the Ottoman Empire fell during World War I. In 1919, France and Great Britain, under the purview of the League of Nations, divided up much of Southwest Asia into protectorates overseen by the two European powers. The borders of Iraq, Syria, Jordan, Turkey, Kuwait, Egypt, Lebanon, Qatar, U.A.E., Bahrain, and Israel/Palestine today basically follow those outlined during the British and French mandates. The borders of countries in North Africa were also generally drawn during the second wave of European colonialism.

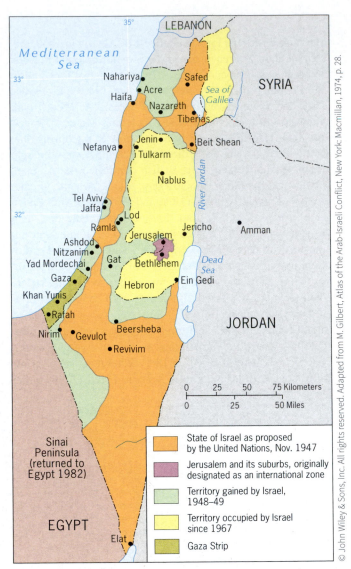

FIGURE 5.21   London, England.   This Egyptian artifact, the Younger Memnon, a colossal granite head of Pharaoh Ramesses II, stands in the British Museum in London. Giovanni Belzoni retrieved the sculpture in 1816 from the tomb of Ramesses in Thebes, Egypt. The British Museum also houses the Rosetta Stone, found in Egypt by Napoleon's troops and taken by the British upon the defeat of Napoleon in 1801.

This opposition was in part based on religion, but also on the perception that it was a continuation of European colonialism. While the Jewish state of Israel has flourished since 1948, its strained relationship with Palestine and neighboring states remains the most contentious issue in the region.

Tension between Israel and its neighbors, as well as with the Palestinian people, continued to simmer throughout the second half of the 20th century. Two decisive military victories allowed Israel to defeat its antagonists in the region and even expand its territory. The Six-Day War, from 5-10 June 1967, was fought between Israel and all of its neighboring countries: Egypt, Jordan, Syria and Lebanon. The war concluded with Israel expanding its territory significantly, to include the Gaza Strip and Sinai from Egypt, the West Bank and Jerusalem from Jordan, and the Golan Heights from Syria. A second pivotal fight was the Yom Kippur War, which took place between 6–24 October 1973. This war began with a surprise joint attack on two fronts by the armies of Syria (in the Golan Heights) and Egypt (in the Suez Canal). The Israelis eventually defeated Syria and Egypt but there were no significant territorial changes.

Despite victories on the battle field for Israel, tensions between Israel and the Palestinians have been a recurring flashpoint. From the 1967 war onward, Israel has occupied the territories of the West Bank, Golan Heights, and Gaza Strip. The Israeli government built housing developments

FIGURE 5.22   State of Israel in 1948 and today.   In 1948, Israel was the area in the map in orange. Today, Israel controls or occupies all of the land in orange, green, yellow, and purple on this map except the Gaza Strip.

Legend:
- State of Israel as proposed by the United Nations, Nov. 1947
- Jerusalem and its suburbs, originally designated as an international zone
- Territory gained by Israel, 1948–49
- Territory occupied by Israel since 1967
- Gaza Strip

referred to as **settlements** in the Occupied Territories, which has increased the tensions on the ground in the region. The United States has attempted, on many occasions, to broker peace accords between Israel and its neighbors, at times trying to address ownership of the Occupied Territories and more recently working to dissuade the building of more Israeli settlements. Most famous was a peace accord brokered by American President Jimmy Carter in 1979 between Israeli Prime Minister Menachem Begin and the Egyptian President Anwar Sadat in which Israel returned the Sinai Peninsula to Egypt and Egypt recognized Israel as an independent country.

In 1993, President Bill Clinton facilitated the first Oslo Accords between Yitzhak Rabin of Israel and Yasser Arafat of the Palestine Liberation Organization. The Oslo Accords created the Palestinian Authority, which was allowed to practice limited self-governance over the occupied Palestinian territories that comprise the West Bank (including East Jerusalem) and the Gaza Strip (Figure 5.23). Palestinian autonomy

The West Bank

- Israeli settlements inside Occupied Territory protected by Security Barrier
- Palestinian areas
- Israeli-controlled areas in the West Bank
- Syrian territory occupied by Israel
- Israeli settlements as of 2011
- Road
- Security Barrier, completed and planned

Source: Israeli Ministry of Defence

FIGURE 5.23 **The West Bank.** Israel limits the land on which Palestinians can build and also has built settlements for Jewish residents in the West Bank in order to exert further control of the lands.

expanded in 2005 when Israel withdrew from the Gaza Strip, giving authority to control the 223 square miles (360 square km) parcel of land to the Palestinians. Israel controls the borders around Gaza, and along with Egypt controls the trade in and out of the territory. The people of Gaza have little opportunity for employment and an unemployment rate of about 40 percent. The economically challenging situation for Palestine is exacerbated by ongoing ideological battles in the region. The majority of Israelis and Palestinians have at various times supported a two-state solution, but this plurality of opinion has eroded since fighting during the summer of 2014.

Of those states under European control, Egypt was the first state to gain independence in the region, with limited freedom in 1922 and full independence in 1936. Egypt was strategically important because it controlled the Suez Canal, an artificial sea-level waterway in Egypt connecting the Mediterranean Sea and the Red Sea. It opened in 1869 after ten years of construction work, which allowed transportation between Europe and Asia without traveling around Africa. It was an Egyptian nationalist movement and revolution that led the British to relinquish official control of the country, although their influence remained strong until the 1950s.

Most of the other states in the region gained independence in the 1940s and 1950s. The one exception was Algeria, with the largest French settler population in the region, Algeria fought a protracted guerrilla war against the French, eventually gaining independence in 1962 after over a million Algerians and 27,000 French had lost their lives.

A REGION OF FOSSIL FUEL PRODUCTION

In the first part of the twentieth century, the United States was the leading producer of oil in the world, supplying, for example, almost all of the allies with petroleum during World War II. In the 1940s, when it became clear that American supplies would soon taper off, a rush of oil exploration began in other parts of the world. Earlier discoveries in Southwest Asia—first in Persia (now Iran) in 1908 and then in Saudi Arabia in 1938—made this region a focus of ongoing exploration. Southwest Asia and North Africa possessed some of the most easily accessible reserves, which became central to industrialization in the twentieth century (Figure 5.24). British and American petroleum companies were particularly active in the region and owned wells until the 1970s when governments nationalized many drilling operations in the region. A key turning point for oil-producing states in the region was the formation of OPEC, the Organization of Petroleum Exporting Countries, which was established in 1965. OPEC burst onto the international scene in 1973 when it implemented an embargo because of the Yom Kippur War (which pitted Syria and Egypt against Israel). The international embargo created an oil crisis with high fuel prices and long gasoline lines around the world.

Today, Southwest Asia and North Africa are estimated to possess 50 percent of the world's crude oil reserves and over 40 percent of the known natural gas reserves (Figure 5.25). Saudi Arabia is the world's leading crude oil

Ray Ellis/Photo Researchers, Inc.

FIGURE 5.24   Jubail, Saudi Arabia.   The Jubail oil refinery located on the Red Sea in Saudi Arabia is jointly owned by the Saudi oil company, Saudi Aramco, and France's Total oil company. The facility processes 400,000 barrels of oil per day.

exporter. Based on 2013 figures, other significant oil producers in the region include the United Arab Emirates (#6), Iran (#7), Iraq (#8), Kuwait (#9), Qatar (#14), and Algeria (#18) (BP Statistical Review 2014).

The discovery of oil has arguably done tremendous good for the region's development and standard of living. Kuwait is a good example of a country whose fortunes were reversed by the discovery of oil. While Kuwait was never poor, it was not a center of power in the region. It was home to one of the best natural harbors in the Persian Gulf, and its population was mainly composed of pearl divers who harvested in the natural pearl beds located along the coast, traders, and Bedouin herders. The discovery of oil in Kuwait in 1938 revolutionized the sheikdom's economy and made it a valuable asset to Britain. After independence in 1961, Iraq tried to annex the country, claiming that it had been part of Iraq under the Ottoman Empire. The British, who were clearly interested in Kuwaiti oil, sent troops and made sure that such an annexation did not occur. Iraq attempted to annex Kuwait again in 1990, under the leadership of Saddam Hussein. Iraq invaded Kuwait and occupied it for some six months. This time the United States (also with an interest in Kuwaiti oil) drove the Iraqis out of Kuwait.

Oil not only insured Kuwait's survival as an independent country, but it also helped the country prosper. All Kuwaiti citizens (an important distinction, as two-thirds of the country's population is composed of immigrant workers) receive a generous subsidy from the government that makes poverty among nationals almost nonexistent. The government has invested heavily in health care and education, which are free to citizens. Kuwait was also the first country in the world to create a sovereign wealth fund, or public endowment (established in 1953, before the country was independent), to invest some of its oil profits in other businesses around the world. Since oil is a **nonrenewable resource** the fund will continue to generate revenue after the oil runs out.

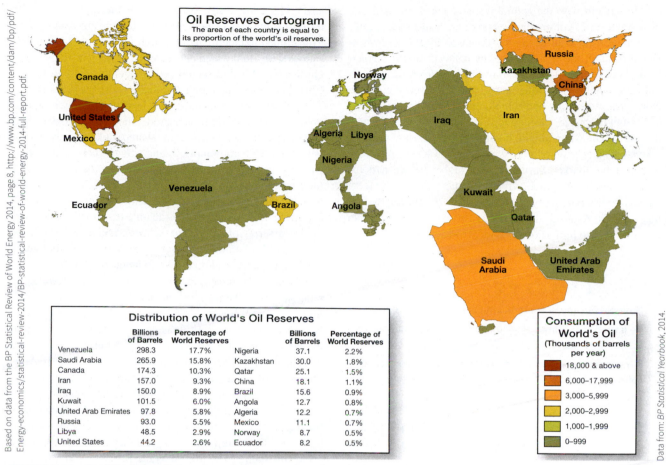

Oil Reserves Cartogram
The area of each country is equal to its proportion of the world's oil reserves.

Distribution of World's Oil Reserves					
	Billions of Barrels	Percentage of World Reserves		Billions of Barrels	Percentage of World Reserves
Venezuela	298.3	17.7%	Nigeria	37.1	2.2%
Saudi Arabia	265.9	15.8%	Kazakhstan	30.0	1.8%
Canada	174.3	10.3%	Qatar	25.1	1.5%
Iran	157.0	9.3%	China	18.1	1.1%
Iraq	150.0	8.9%	Brazil	15.6	0.9%
Kuwait	101.5	6.0%	Angola	12.7	0.8%
United Arab Emirates	97.8	5.8%	Algeria	12.2	0.7%
Russia	93.0	5.5%	Mexico	11.1	0.7%
Libya	48.5	2.9%	Norway	8.7	0.5%
United States	44.2	2.6%	Ecuador	8.2	0.5%

Consumption of World's Oil (Thousands of barrels per year)
- 18,000 & above
- 6,000–17,999
- 3,000–5,999
- 2,000–2,999
- 1,000–1,999
- 0–999

Based on data from the BP Statistical Review of World Energy 2014, page 8, http://www.bp.com/content/dam/bp/pdf/Energy-economics/statistical-review-2014/BP-statistical-review-of-world-energy-2014-full-report.pdf.

Data from: BP Statistical Yearbook, 2014.

FIGURE 5.25 World's major oil deposits. Southwest Asia and North Africa control approximately 50 percent of the world's known oil reserves. Saudi Arabia has more than 15 percent, Iran has 9.3 percent, and Iraq has 8.9 percent, as of 2014.

THE DOWNSIDE OF PETRO-BASED DEVELOPMENT

It might seem that an abundance of oil and natural gas could only bode well for a country, yet some scholars have argued that such resource wealth actually does more harm than good. Groups may squabble over control of locally abundant and lucrative resources such as oil or natural gas. The economist Paul Collier (2007) has asserted that this natural resource curse creates a greater risk of civil war in countries that depend on primary commodity exports. The resource curse may breed corruption among government officials and make them less responsive to public concerns, especially when the government becomes increasingly reliant on the revenues from resource extraction (and less dependent on tax revenues).

Analysts have suggested that Libya is a good example of a country that suffers from the natural resource curse. In October 2011, Libyan leader Moammar Gadhafi was killed clinging to power, trying to stave off rebels (inspired by the Arab Spring of 2011) who controlled large parts of his country. Many experts argue that the structure of Libya's economy is what allowed Gadhafi to remain in power as the world's longest serving dictator (41 years). Libya's economy is almost entirely based on resource exports, with 98 percent related to oil and gas (NPR 2011). Gadhafi's government made money by selling oil to the rest of the world, rather than collecting taxes. It used these funds to buy guns, mercenaries, and the loyalty of some citizens. Gadhafi also dissuaded Libyans from starting their own businesses, which would be a source of revenue he could not control and a potential rival source of power. But the resource curse is not inevitable, and Libya's next leader will not necessarily take over the oil wells and become another dictator. Libya's wealth from fossil fuels could also provide it with the opportunity to explore new forms of government.

The British-American geographer Michael Watts has been critical of the natural resource curse thesis. For Watts (2004), conflicts over resources are often a symptom rather than a cause of underlying tensions. Furthermore, to imbue a resource with the power to cause problems is to believe that the resource alone has a special transformative power—that is, to engage in "commodity determinism." Some commentators have argued that resource curses involving fossil fuels may deserve special consideration, given the dependence of the outside world on these energy resources (e.g., Moseley 2009). Energy-hungry countries such as the United States and China are often willing to support undemocratic and corrupt regimes if it means reliable access to oil or natural gas. In these instances, oil's use and meaning in the larger global economy help create the unhealthy dynamic inside exporting countries.

In tandem with the natural resource curse, overreliance on a small set of commodity exports may lead to the simplification of a national economy (Gylfason 2001). Known as Dutch disease, the problem gets its name from the Netherlands, because the Dutch economy suffered from such a simplification in the 1960s after natural gas was discovered in the North Sea. As exports of a commodity increase, so do inflows of foreign exchange, raising the value of the currency. Dutch disease occurs when a more highly valued currency makes a country's other exports more expensive and less competitive in the global market. Furthermore, a higher valued national currency also makes imports relatively inexpensive, and these cheap imports pose problems for domestic producers. Both of these trends, relatively more expensive exports and cheaper imports, tend to hurt a country's local industries.

CONCEPT CHECK

1. **How** did European colonialism vary across the region?

2. **Describe** the spatial distribution of fossil fuel reserves across the region.

3. **What** are the pros and cons of oil and gas reserves for a country's political and economic development?

HOW DO RELIGION AND GENDER ROLES VARY ACROSS THE REGION?

LEARNING OBJECTIVES

1. **Describe** the various forms of religious expression in Southwest Asia and North Africa.

2. **Compare** the rise and spread of major religions originating in the region.

3. **Explain** how gender roles vary across the region.

Religion looms large in Southwest Asia and North Africa. Not only is the region the cradle of three of the world's major faiths, but religion has exerted a considerable influence on regional politics and social norms. Religious differences have also been the source of conflict in the past and in the contemporary era.

SOUTHWEST ASIA AS THE CRADLE OF THREE MAJOR WORLD RELIGIONS

In the first and second millennia BCE, Judaism began in ancient Israel and Judah, and the Hebrew Bible, or Torah, was composed. Judaism began with the first covenant between Abraham and God which established the religion as monotheistic. Abraham agreed to worship only one God, and God agreed to protect Abraham and his people as the chosen people. Some 2,000 years later the Jewish followers of Jesus established Christianity, which is based on a second covenant. As taught in the books of the New Testament, God

sent his only son, Jesus, to earth to live, die, and rise, in order to forgive the sins of humanity. In the seventh century CE, Muhammad founded Islam through meditation and conversations with God, which are recorded in the teachings of the Qur'an. The Muslim sacred text frequently refers to Judeo-Christian figures including Abraham, Moses, and Jesus.

All three religions have their hearths in this region and hold sites in Jerusalem as sacred. They are also all called Abrahamic faiths (because of shared links to Abraham), which share a common origin and values, such as monotheism (Figure 5.26). Some scholars assert that there is a fourth (and much smaller) Abrahamic faith—Bahá'í. Bahá'u'lláh founded the Bahá'í Faith in nineteenth-century Persia (now Iran). The Bahá'í draw on Abrahamic traditions as they recognize a series of divine messengers, including Abraham, Jesus, and Muhammad, as well as Eastern religious figures (such as Buddha) and their own founder, Bahá'u'lláh.

Islam is the most recent of the trio of major Abrahamic religions and by far the most prevalent religious belief system in contemporary Southwest Asia and North Africa (only Israel has a non-Muslim majority). Muhammad was born in 570 in the city of Mecca (in present-day Saudi Arabia) and spent the latter half of his life in Medina, making these the two holiest cities in the Islamic faith. Muhammad died without having named a successor to be leader, or caliph, and a battle ensued over the succession. Some Muslims asserted that the successor ought to be one of Muhammad's closest associates, who had strong leadership skills and great piety. This group, which eventually came to be known as the **Sunnis**, won a decisive military battle in the 7th century and elected caliphs to follow Muhammad. Sunnis are the largest Muslim sect in the world today (accounting for more than 80 percent of all Muslims) and comprise the majority of Muslims in all countries in the region except Iran, Iraq, and Bahrain. Other seventh-century Muslims believed that a family member ought to succeed Muhammad. This group backed Ali ibn Abi Talib, a cousin and son-in-law of Muhammad, to be his successor. They eventually became the **Shi'a**, and they constitute about 15 percent of Muslims worldwide. Around the Persian Gulf, the Shi'a make up the majority of the Muslim population in Iran (90–95 percent), Iraq (65–70 percent), and Bahrain (70 percent).

Islam has five pillars or tenets that believers endeavor to fulfill: (1) professing the faith, that is, making a statement professing monotheism and accepting Muhammad as God's messenger; (2) praying five times a day at designated times; (3) giving alms to the poor; (4) fasting during daylight hours throughout the holy month of Ramadan, a 29-day lunar month, which shifts over time; and (5) making the Haj or pilgrimage to the holy city of Mecca, Saudi Arabia, at least once in a lifetime if one can afford to do so (approximately one in 50 Muslims make the Haj).

While Judaism, Christianity, and Islam have much in common, considerable tension also exists between them. Some of this tension arises from shared places of religious significance or **sacred space** (Figure 5.27). Jerusalem is sacred to all three religions. For Jews, Jerusalem is the center of the Holy Land, the land God promised Abraham at

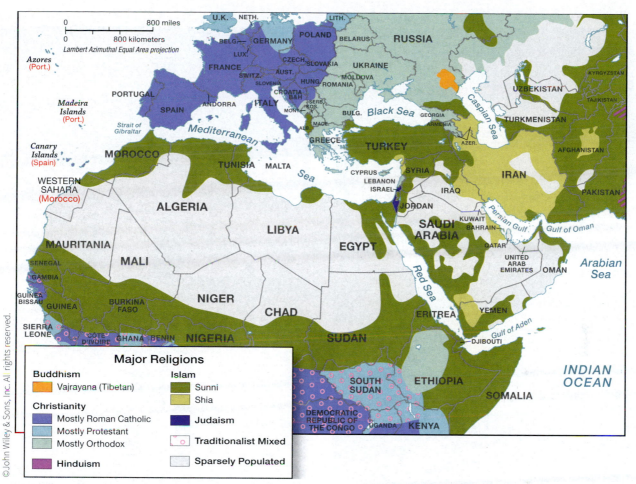

FIGURE 5.26   Religions in Southwest Asia and North Africa.   The hearths of three major world religions are in Southwest Asia and North Africa. Over 3,500 years ago, Judaism was founded by Abraham from Ur and the area of present-day Israel, including Jerusalem, was established as a sacred place. 2,000 years ago, Christianity was founded with the birth of Jesus in Bethlehem and his crucifixion and ascension into heaven in Jerusalem. Islam was founded by Muhammad about 1,400 years ago in Mecca and Medina. Jerusalem is also sacred to Muslims, who believe Muhammad ascended into heaven there.

the founding of the religion. For Christians, Jerusalem is sacred because it is the place where Jesus was crucified and then rose from the dead. For Muslims, Jerusalem is sacred because it is where Mohammad died.

PATTERNS AND DIFFUSION OF RELIGION

The rise and spread of Judaism differs markedly from the rise and spread of Christianity and Islam. Unlike Muslims and Christians, Jews do not actively proselytize. The diffusion of

Reading the **CULTURAL** Landscape

Sacred Spaces of Jerusalem

Over centuries, Jews, Christians, and Muslims have contested sacred spaces in the city of Jerusalem, which holds religious significance for all three. In the tenth century BCE, the Jewish King Solomon built the Temple of Jerusalem. The temple was destroyed, rebuilt, and destroyed again. Jews pray at the last standing wall that surrounded the temple, known as the Wailing Wall or Western Wall. Built in 326 CE, the Church of the Holy Sepulchre is believed to encompass both the sites where Jesus was crucified and the tomb where he was buried and then rose from the dead. Crucifixions took place outside the city walls; so the Church of the Holy Sepulchre is built away from the Temple Mount (where the Temple of Solomon and Al-Aqsa are located). Adjacent to the Wailing Wall, the Dome of the Rock mosque covers the site of the Al-Aqsa Mosque where Muslims believe Muhammad journeyed with the Angel Gabriel and ascended into heaven. Al-Aqsa was built in the eighth century CE.

(Continued)

FIGURE 5.27a **Western Wall.**

FIGURE 5.27b **Dome of the Rock.**

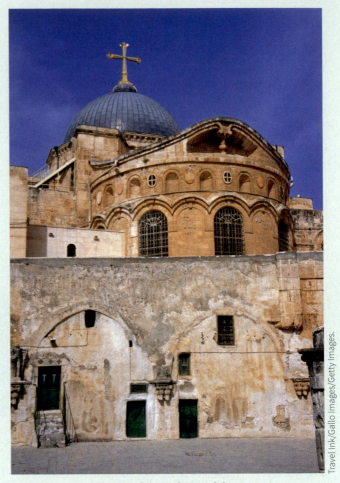

FIGURE 5.27c **Church of the Holy Sepulchre.**

Judaism, an ethnic religion, has most often occurred when Jews moved from one area to another, often not by choice. Judaism diffused to Europe when Jews lost their homelands in Southwest Asia, and then to North America from Europe when conditions became intolerable, most notably before, during, and after the Holocaust. The spread of a practice when a group moves or migrates from one area to another is called relocation **diffusion** (see Figure 5.28).

 diffusion the spread of an idea, innovation, or technology from its hearth to other people and places.

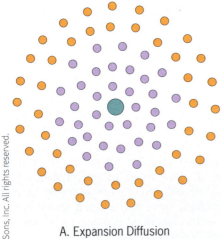

A. Expansion Diffusion

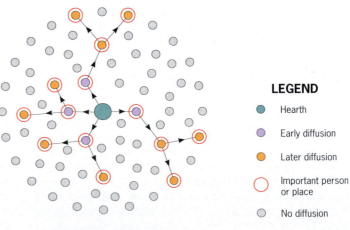

B. Hierarchical Diffusion

LEGEND

- 🟢 Hearth
- 🟣 Early diffusion
- 🟠 Later diffusion
- ⭕ Important person or place
- ⚪ No diffusion

FIGURE 5.28 **Expansion and hierarchical diffusion.** In expansion diffusion, the trait spreads from the hearth outward contiguously and evenly. In hierarchical diffusion, the trait spreads from the hearth to the most linked or most important people or places first.

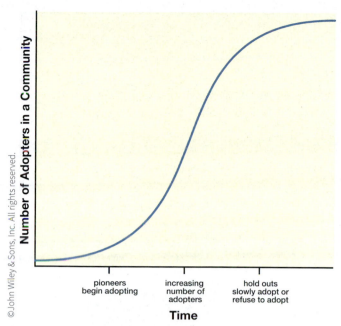

FIGURE 5.29 **The S-curve pattern of diffusion-adoption.** As time passes, the number of knowers or adopters increases, slowly at first, then quickly, and then plateaus.

Universal religions, such as Christianity and Islam, diffuse from neighbor to neighbor through proselytizing efforts or social pressures as the religion gains increasing popularity in a particular area. This simple spread of a concept or idea from one person to the next in close proximity to one another is called **expansion diffusion**. At the micro-level, the adoption of new ideas often follows an S-curve pattern (Figure 5.29). A few "pioneers" in the community are slow initially to adopt a new practice, followed by increasing rates of new adoption in the middle phase, and then tapering off with a slower rate of adoption as the few remaining "holdouts" in the community change, if ever. At broader scales, expansion diffusion appears as a practice gradually spreading across the landscape through time.

While each form of spatial diffusion is unique, they are not exclusive. For example, as ideas spread evenly across space through expansion diffusion, they may also spread from large city to large city, and then gradually down to medium-sized and then small cities, towns, and villages in a pattern of **hierarchical diffusion** (after the ranking of communities from small to large, known as an urban hierarchy) (Figure 5.30). Christianity spread through expansion

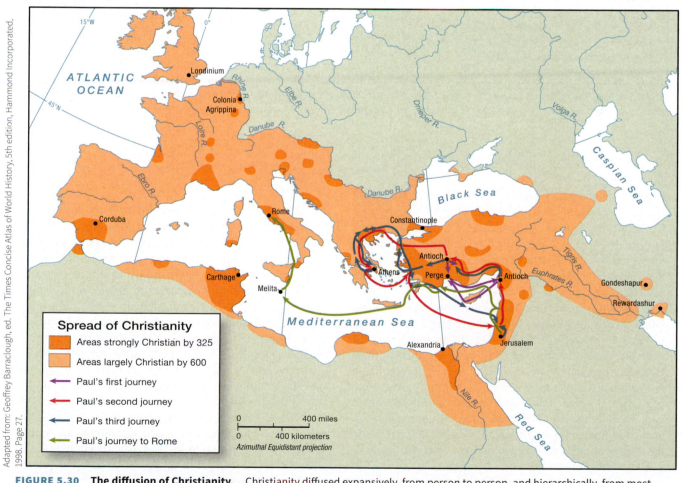

FIGURE 5.30 **The diffusion of Christianity.** Christianity diffused expansively, from person to person, and hierarchically, from most connected places to next most connected places. Through travels in the Mediterranean, Paul of Tarsus diffused Christianity hierarchically to key ports around the sea and cities on major trade routes.

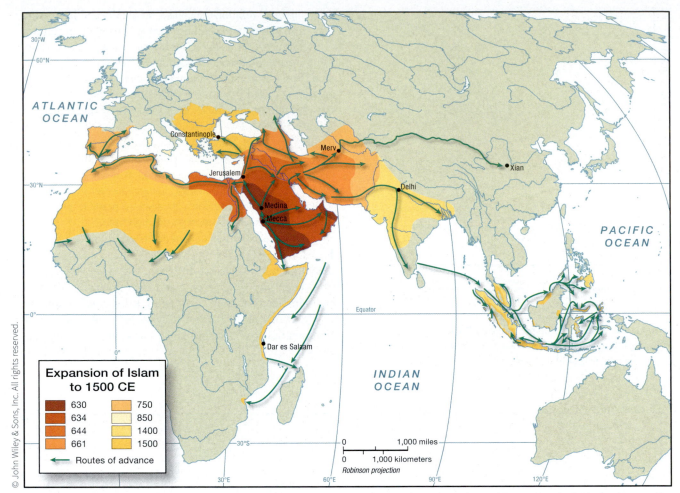

Expansion of Islam to 1500 CE

630	750
634	850
644	1400
661	1500

← Routes of advance

FIGURE 5.31   The diffusion of Islam.   Islam diffused from its hearth in Mecca throughout the Arabian Peninsula during Muhammad's lifetime. Islam then diffused across North Africa, into Spain, southeast Europe, central Asia, India, and Southeast Asia.

diffusion, contagiously from person to person as one taught the next the tenets of the faith. Christianity also diffused hierarchically, as Paul of Tarsus traveled to major trading ports on the Mediterranean Sea. Later, Christianity diffused through the political hierarchy of the Roman Empire when the Emperor Constantine converted to the religion and asked his subjects to do the same.

The diffusion of Islam throughout Southwest Asia and North Africa (and the world) is often thought to have followed a hierarchical diffusion pattern (Figure 5.31). For example, in North Africa, many cities were centers of trade that were connected to other cities in the region through networks of commerce. As such, Islam often spread from one city to the next via Muslim traders. Until the twentieth century and the advent of mass forms of communication such as radio that served to spread religious ideas, it would have been common to find practitioners of traditional religions in rural communities outside of these urban areas as they simply were not well connected to urban networks via circuits of exchange.

LANGUAGES

The diffusion of Arabic coincides with the diffusion of Islam today (Figure 5.32). Faithful Muslims should read the Qur'an in Arabic rather than in a vernacular language. The use of Arabic in religion and trade has helped the language maintain a stronghold in the region for centuries. Arabic is spoken throughout North Africa, with distinct dialects in the Maghreb, or Mediterranean, region and Egypt. The Arabic spoken in the Maghreb is influenced by the Romance languages found on the north side of the Mediterranean Sea, French and Spanish. Arabic is also the most common language spoken in the Persian Gulf region, aside from Iran where the common language is Farsi.

In the Levant region of the former Fertile Crescent, French and English (the languages of the former colonizers) are still commonly heard in addition to Arabic. Hebrew is also spoken in the Levant, especially in Israel where it is the official language. The diversity of the Levant is complete with the Aramaic language, which is commonly spoken among small clusters of people who

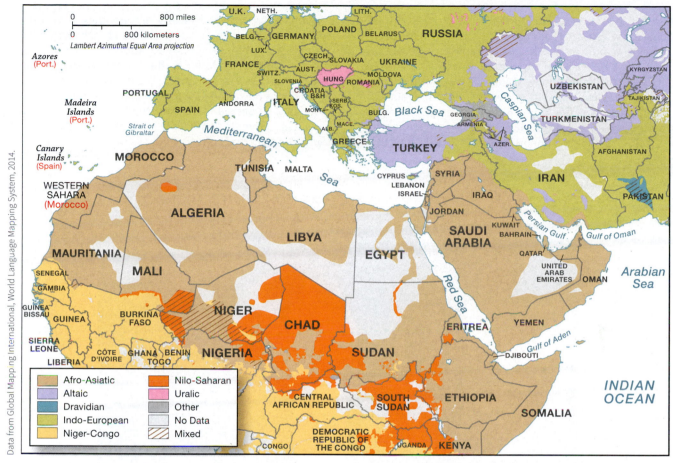

FIGURE 5.32 **Languages of Southwest Asia and North Africa.** Arabic, in the Afro-Asiatic language family, is the predominant language in the region and is the official state language of every country from Morocco east to Oman. Throughout the region, small pockets of vernacular languages belonging to other language families are still spoken.

form sects of ancient Christianity. Reaching from Syria across Iraq and Turkey, a large cluster of Kurds speak the Kurdish language.

GENDER ROLES

From an Orientalist perspective, the region of Southwest Asia and North Africa seems universally oppressive and restrictive vis-à-vis women. Images of exotic veiled women or the more restrictive tenets of the conservative or fundamentalist sects of Islam (such as the *Wahhabis* in Saudi Arabia) became particularly infamous after Islamic fundamentalist terrorists attacked the United States on September 11, 2001. In fact, while Islamic fundamentalism is on the rise, most countries in the region enjoy having a diversity of

Reading the **CULTURAL** Landscape

Faculty and Students in the Geography Department, University of Kuwait

This photo features geography faculty and students at the University of Kuwait. Their diverse attire and backgrounds tell us something about the region. The gentleman in the center was educated in the UK and is the chair or head of the department. He is wearing a dishdasha, the traditional long cotton gown worn by men in the Gulf region. Next to him on the left, wearing a Western tie, is a geography instructor from Egypt. Many university instructors throughout the Gulf area hail from Egypt, given this country's history of a strong university system. The three female students on the right nicely represent the diversity of attire worn by women in Kuwait's evolving social milieu.

(Continued)

Data from Global Mapping International, World Language Mapping System, 2014.

Closest to the center is a student wearing Western clothing, next is a student with her head covered, followed by a student wearing a hijab (or head scarf) and aba or abaya (black gown). While not common (and not pictured here), there are also some female students who cover their faces with a Niqab (which literally means "mask" in Arabic). On the opposite (or left) side are a female professor (educated in the United States) and a male student in Western attire, as well as two male students in dishdashas. Finally, these geography students and faculty are standing in a well-equipped university that is free to Kuwaiti nationals who qualify to study there. This educational infrastructure, as noted earlier, is paid for by the extraction and sale of Kuwait's lucrative oil resources. While instruction at the University of Kuwait has long been coeducational, a controversial law passed in 2009 by the Kuwaiti parliament stipulates that male and female students receive instruction in separate classrooms.

FIGURE 5.33   Kuwait City, Kuwait.   Faculty and students from the Geography Department pose for a group photo.

religious traditions (from conservative to liberal) and associated norms for women (Figure 5.33). For most countries in the region, it is acceptable for women to wear more conservative Western clothing (long pants or longer skirts), and the covering of the head or face is not obligatory.

Saudi Arabia is a notable exception because of the Wahhabist sect of Islam, followers of the eighteenth-century Islamic reformer Muhammad bin Abd al-Wahhab, places significant restrictions on women. Wahhabism dominates within Saudi Arabia because of the group's alliance with the founder of the Saudi dynasty. Wealthy Saudi patrons who support Koranic schools or Madrassas abroad have also promoted Wahhabism throughout the region. Wahhabism exhorts piety, rejection of non-Muslim beliefs and cultures, corporal and capital punishment, and complete segregation of men and women in public life. Women are to be covered in public; their participation in the workforce is limited (while 60 percent of university graduates are women, they make up only 5 percent of the workforce); and they are prohibited from driving. Saudi women are to travel with a male chaperone and are prohibited from voting, trying on clothes while shopping, and entering a cemetery (The Week 2014).

CONCEPT CHECK

1. **Which** major religions originated in Southwest Asia and North Africa?

2. **Why** have the major religions originating in the region spread at different rates throughout the world?

3. **How** do gender roles vary throughout the region?

What Is the Physical Geography of Southwest Asia and North Africa?

1. The region of North Africa and Southwest Asia is the driest world region. Dominated by subtropical high-pressure systems, the Sahara and Arabian deserts receive little to no rain over the course of the year.

2. In this largely semiarid region, water has long been a crucial resource, which has been a source of both ingenuity (leading to inventions like qanat) and conflict. The qanat uses gravity to move underground water downhill for human use. Egypt is a good example of the conflict and potential conflict over water in the region. Fully dependent on fresh water from the Nile River, Egypt is at the mercy of Ethiopia and the neighbors that control the upper part of the river, before it enters Egypt.

3. The region of Southwest Asia and North Africa is naturally bounded by deserts and oceans, with pastoralism being the dominant rural livelihood. Agriculture relies on water, and in southwestern Morocco 80 percent of the fresh water available is used to support agriculture. Pastoralists move their herds seasonally to access vegetation and fresh water.

Who Are Southwest Asia and North Africa's Urban Inhabitants?

1. The region also gave birth to the first cities and empires in the world in the hearth of the first urban revolution, between the Tigris and Euphrates Rivers in present-day Iraq. The first agricultural revolution preceded the urban revolution, as a larger, stable food supply was necessary for people to live in cities.

2. Cities in the region are known for a distinct morphology or shape called the Islamic City. Islam began in the seventh century CE and diffused primarily to cities during and after Muhammad's lifetime. Because Islam was such an integral part of everyday life, the cultural landscape of cities took on an Islamic character. Elements of the classic Islamic City include a central marketplace (or *suq*), mosques, a citadel, and public baths. Streets and alleyways were narrow and irregular, and homes were built with walls and interior courtyards to protect privacy.

3. Today, the region is one of the world's most urbanized, a phenomenon fueled by export revenues in many fossil fuel-rich states. Despite large amounts of oil wealth in some countries in the region, the high unemployment rates of youth ages 15 through 24 are a concern for several countries. The hearth of the Arab Spring was Tunisia, and the protest movement against authoritarian control diffused quickly to Egypt. Both countries have high youth unemployment rates.

What Is the Relationship Among Colonialism, Oil, and Development in the Region?

1. Iran and Turkey avoided being colonized by Europeans, but the rest of the region was controlled by Europe at some point between 1800 and 1950. North Africa experienced European colonialism first, as the Ottoman Empire controlled most of Southwest Asia until losing World War I in 1918. In North Africa, the French took control of Algeria in 1830, eventually possessing much of North Africa (including also Morocco and Tunisia), as well as Lebanon and Syria during the colonial era. The British controlled the Arabian Peninsula (which includes modern-day Saudi Arabia, Oman, Yemen, Kuwait, Bahrain, Qatar, and United Arab Emirates) as well as Iraq, Jordan, Israel, Palestine, and Egypt.

2. Oil was discovered in Iran in 1908 and in Saudi Arabia in 1938. British and American petroleum companies owned many of the wells until the 1970s when governments nationalized many drilling operations in the region. The region holds 70 percent of the world's crude oil reserves, which fueled industrialization worldwide during the twentieth century.

3. Oil has proven to be both a blessing and a curse, allowing for the financing of robust social services in some states (such as Kuwait) and retarding political development in others (such as Libya).

How Do Religion and Gender Roles Vary Across the Region?

1. The region gave birth to three major world religions: Judaism, Christianity, and Islam. All three religions are called Abrahamic faiths because of shared links to Abraham. Judaism is approximately 4,000 years old and was founded by a covenant between Abraham and God, which established the religion as monotheistic. Christianity is 2,000 years old and is built on a second covenant with God, where he sent his son, Jesus, to live on earth. Islam is approximately 1,400 years old, dating to Muhammad's conversations with God, which were written down in the Qur'an. All three have their hearths in this region, and all three religions have sacred sites in Jerusalem.

2. These religions have diffused differently throughout the world along expansion, hierarchical, or relocation patterns. Judaism, an ethnic religion, diffused primarily through relocation diffusion, when Jews moved to new areas of the world and brought their religion with them. Christianity and Islam spread through expansion diffusion, especially contagiously, from person to person. Islam also diffused hierarchically, from city to city, typically among Muslim traders. Likewise, Christianity diffused hierarchically, first through the travels of Paul of Tarsus through the Mediterranean and then through the political hierarchy of the Roman Empire.

3. Different religious traditions have also had an impact on gender norms. In much of the region, it is acceptable for women to wear conservative Western clothing. In Saudi Arabia and countries influenced by the Wahhabi sect of Islam, women and men are to be separate in public, and women are expected, at a minimum, to cover their hair. Tunisia, the hearth of the Arab Spring, is known as the most progressive country for women's rights and equality.

Geography in the Field   **YOUR**TURN

The interior of the Great Mosque is part of the Moorish landscape in Andalusia, Spain. At first the site of a Christian church, Spanish Muslims built the Great Mosque at Cordoba in the eighth century. The columns, archways, and geometric tiles reflect the aesthetic of Moorish architecture. In the thirteenth century, Ferdinand III the Saint recaptured Cordoba and

(Continued)

integrated the city into his Christian empire. As early as the fourteenth century Christian structures began to be built in and around the Great Mosque. In the sixteenth century, 300 years after Cordoba returned to the control of Christian monarchs, a cathedral was erected in the middle of the Great Mosque.

The wars between the ancient Mediterranean rivals Rome and Carthage are known as the Punic Wars. At the outbreak of the Second Punic War (218 BC), the Carthaginian military commander Hannibal marched an army, including war elephants, through Spain (then partly controlled by Carthage), over the Pyrenees and the Alps, and into northern Italy. In his first few years in Italy, Hannibal garnered three dramatic victories and won over several allies of Rome. He occupied much of Italy for 15 years, but a Roman counter-invasion of North Africa forced him to return to Carthage, where he was defeated by Scipio Africanus at the battle of Zama. North Africa's occupation of Italy is little remembered, perhaps because of the adage that it is the victors who often write history.

Ken Welsh/The Image Bank/Getty Images.

Cordoba Province, Andalusia, Spain.   Muslim design is apparent in the archways and geometric mosaics of tiles in the Great Mosque.

Thinking Geographically

- While the story of the Roman Empire and its control of areas in the Mediterranean Basin endures, our popular culture downplays the Southwest Asian and North African empires that controlled portions of Europe. Why do you think that is?

- What elements on the European landscape persist as signs of these occupations? Search the Internet for photographs of Tunis, Tunisia, Andalusia, Spain, and Belgrade, Serbia, and for evidence of similarity in the cultural landscape.

KEY TERMS

biome	first agricultural revolution	Islamic City	Shi'a
qanat	hearth	Arab Spring	sacred space
environmental security	pastoralism	settlement	diffusion
virtual water	city states	nonrenewable resource	expansion diffusion
Fertile Crescent	hinterland	Sunnis	hierarchical diffusion

CREATIVE AND CRITICAL THINKING QUESTIONS

1. Imagine the region of North Africa and Southwest Asia. What **identity** have you constructed for the region in your mind? Assess how Orientalism has influenced your perception of Southwest Asia and North Africa.

2. Dubai, United Arab Emirates, is a **world city**. Use the Internet to look at images of Dubai. Explain how **globalization** is represented in the **cultural landscape** of this world city.

3. Describe the water **networks** in the region historically and today. How is power over water distributed in the qanat system as compared to the modern water delivery and irrigation systems?

4. Describe pastoralism as an economic system. Sketch an image of the **mental map** of a pastoralist in North Africa 200 years ago and another image of the **mental map** of a pastoralist in North Africa today. How have the idea of **development** and the role of the state in creating development impacted the lives of pastoralists?

5. The **hearths** of what three major religions are found in Southwest Asia? Using the Internet, study the sacred sites of each of these religions in Jerusalem. Can you tell who has control over Jerusalem today based on the **cultural landscapes** you see?

6. At the **scale** of the individual, one way women represent their **identities** is through their dress. Explain how dress varies for women in Kuwait. Categorize the varying meanings of dress and identities in Kuwait.

7. What elements typically characterize the unique urban morphology known as the Islamic City? Describe the **site** of Tunis, Tunisia. Explain how the city's changing **situation** over time is reflected in the urban morphology of the city.

8. Explain the commodity chain for gasoline you buy at your local gas station. How much of the gasoline **commodity chain** has North Africa and Southwest Asia captured? What are the advantages and disadvantages of having an oil-based economy, in terms of **development**?

1. All of the following tend to be widespread characteristics of the Southwest Asia and North Africa region except:
   a. Semiarid livelihood strategies (most notably pastoralism)
   b. The widespread use of Arabic as a lingua franca
   c. The preponderance of the Islamic faith as a religious affiliation
   d. A fairly uniform conservative dress code for women
   e. Dryland environments

2. Prior to the introduction of modern Western technology, the people of this region largely lacked the ability to tap and move large quantities of water for agricultural purposes.
   a. true
   b. false

3. All of the following statements are true regarding environmental security, except:
   a. As a theory, environmental security originally emerged from the field of political science
   b. Environmental security is a body of scholarship which broadly argues that many political and ethnic conflicts stem from resource scarcity
   c. There is an active debate about what is driving the scarcity of the resources (population growth or uneven distribution) that may lead to conflict
   d. There is broad agreement about what is driving the scarcity of resources, which may lead to conflict

4. The Fertile Crescent, an area found between the Tigris and Euphrates rivers in Iraq, is significant because:
   a. this is the first place, according to most scientists, where crop agriculture started in about 10,000 BC
   b. most scholars agree that this was the first region of independent urbanization about 4,000–3,500 BC
   c. the first Islamic city was built in this region
   d. this is where the use of oil as a source of energy was first discovered
   e. Both a and b

5. All of the following crops were first domesticated in Southwest Asia except:
   a. tomatoes
   b. wheat
   c. peas
   d. rye

6. In addition to Judaism, Christianity, and Islam, another religion originating in Southwest Asia is:
   a. Hinduism
   b. Bahá'í
   c. Buddhism
   d. Confucianism

7. Which of the following types of diffusion best describes how Judaism has most commonly spread from one area to another?
   a. Expansion diffusion
   b. Hierarchical diffusion
   c. Relocation diffusion
   d. A-spatial diffusion

8. A major schism occurred within Islam after the death of Muhammad as he had not named a successor to be leader or caliph. As a result, there are two major branches or sects of Islam today, Sunni and Shi'a. All of the following statements about these two branches and their differences are accurate, except:
   a. The Sunnis asserted that the successor ought to be one of Muhammad's closest associates who had strong leadership skills and great piety, whereas the Shi'a believed that a family member ought to be the successor
   b. Both sects believe in the five pillars or tenants of Islam
   c. Sunnis are the most numerous Muslim sect in the world today (accounting for 70 percent of all Muslims) and comprise the majority of Muslims in all countries in the region of Southwest Asia and North Africa
   d. The Shi'a backed Ali ibn Abi Talib, a cousin and son-in-law of Muhammad, to be his successor

9. Places of religious or spiritual significance are sometimes referred to as sacred space. Scared spaces are unique to particular faith communities and never shared.
   a. true
   b. false

10. All of the following were characteristics of the first known example of urban living, the Sumerian city states, except:
    a. These city states took on a particular form with a walled city surrounded by an agricultural hinterland
    b. The size of these cities was very much dependent on the extent of their agricultural hinterlands
    c. These cities did not have modern sanitation infrastructure. As the streets filled with garbage, the residents would simply build over the accumulated debris
    d. The social structure of these cities was very egalitarian, with priests, craftsmen, and farmers all dealing with one another as equals
    e. The urban form was very organic, with winding, unplanned streets rather than a grid pattern

11. The Islamic City is an urban form characterized as follows:
    a. Elements of the classic Islamic City include a central marketplace (or *suq*), mosques, a citadel, and public baths
    b. The residential land use is often compact, with walled familial compounds and open courtyards
    c. Buildings and alleyways are often designed to protect privacy. Entrances to homes may be L-shaped, walls are often windowless, and alleys are narrow and irregular to create blind spots
    d. Many modern cities in the region have long histories and therefore have an older section of town, frequently known as the Medina, which has the characteristics of an Islamic city
    e. All of the above

12. All of the following were ancient civilizations which originated in Southwest Asia and North Africa except:
    a. Ancient Egyptian
    b. Ancient Roman
    c. Sumer
    d. Akkadian
    e. Babylonian

13. Which of the following was not a colonial power in Southwest Asia and North Africa?
    a. Spain
    b. France
    c. Italy
    d. England
    e. Netherlands

14. An internal characteristic of a religion that will not affect its rate of diffusion is whether it is a nonproselytizing or ethnic faith (such as Islam) or a proselytizing or universal faith (such as Judaism).
    a. true
    b. false

15. All of the following are characteristics of environmental narratives except:
    a. Refers to the dominant conception of an environmental problem in a region which may or may not be based on reliable empirical evidence

    b. Environmental narratives have been used to justify environmental conservation and development projects in the arid and semiarid lands in North Africa
    c. Long-standing understandings of deforestation and desertification in North Africa were based on solid science and are not examples of environmental narratives
    d. Diana Davis's research revealed that French scientists and administrators established much of the purported scientific basis for North African environmental narratives during the colonial period in Algeria, Morocco, and Tunisia

ANSWERS FOR SELF-TEST QUESTIONS

1. d, **2.** False, **3.** d, **4.** a & b, **5.** a, **6.** b, **7.** c, **8.** c, **9.** False, **10.** d, **11.** e, **12.** b, **13.** e, **14.** False, **15.** c

6

SOUTH ASIA

8:00 pm on a Friday evening. The streets are teeming with mopeds, motorcycles, and pedestrians. Hundreds of young, well-educated Indian men and women battle traffic to find their way to a restaurant, apartment house, or Café Coffee Day.

As they finish their work days, the 24-hour information technology (IT) economy in India is visible on the street. With the shift change, employees arrive for work at countless software engineering companies, consulting firms, and call centers. The 10.5-hour time difference between Bangalore, India, and the central time zone in the United States is perfectly suited for a night shift in India. Americans arrive at work to face computers and programs with technical difficulties, and they phone call centers in India for help. At software engineering companies, projects are advanced jointly by teams located in India and the United States.

The city's cultural landscape shows how much more than a call center India's economy has become. Hundreds of high-rise buildings and corporate campuses are clustered in the "science and technology capital of India." The buildings display signs of American corporations, including Microsoft and Target, as well as signs of Indian-owned companies, including Infosys, Tata and Wipro. Indian-owned companies are global leaders in informational technology, software engineering, and outsourcing, and in 2013 they generated revenues in India valued at over $100 billion (Mishra 2014).

The growth in IT and business process outsourcing (BPO) has brought higher wages and greater wealth to the 3 million Indians working in these sectors. At the same time India has more poor people than any other country in the world, with 42 percent of its 1.2 billion people living on less than $1.25 a day. Understanding the complexities of South Asia today requires looking beyond the call centers of India and seeing the vast diversity of experiences and everyday existences throughout the region.

© Stuart Forster India/Alamy

Bangalore, India. In a country where tea is the beverage of choice, coffee shops are a new feature in the cultural landscape. Café Coffee Day caters to the upwardly mobile young professionals who work in India's information technology industry and related consulting firms. With more than 1,500 locations in India, V.G. Siddhartha, founder of Café Coffee Day, also owns the plantations in India where the coffee beans are grown and produces the espresso machines used in the cafes. Entrepreneur Siddhartha also began a furniture company that manufactures the furniture used in all Café Coffee Day retail outlets. His hope is that his furniture company will become the IKEA of India.

CHAPTER OUTLINE

THRESHOLD CONCEPT in this Chapter

Green Revolution

WHAT IS SOUTH ASIA?

Physical geographers refer to South Asia as a **subcontinent** because in the history of plate tectonics, the region was once its own continent. After the supercontinent Pangaea broke apart 200 million years ago, what is now South Asia was part of Gondwana (the great southern continent), and came to rest at the South Pole. Approximately 120 million years ago, the subcontinent split from Gondwana and traveled northward, eventually colliding with Asia about 50 million years ago. This collision created the Himalaya Mountains and continues to build the mountain chain today (Figure 6.1). Mount Everest is growing in elevation at a rate of 2.4 inches a year.

The collision of the Indian subcontinent into Asia first created the Himalaya Mountains and Karakoram Range, and both mountain ranges continue to grow

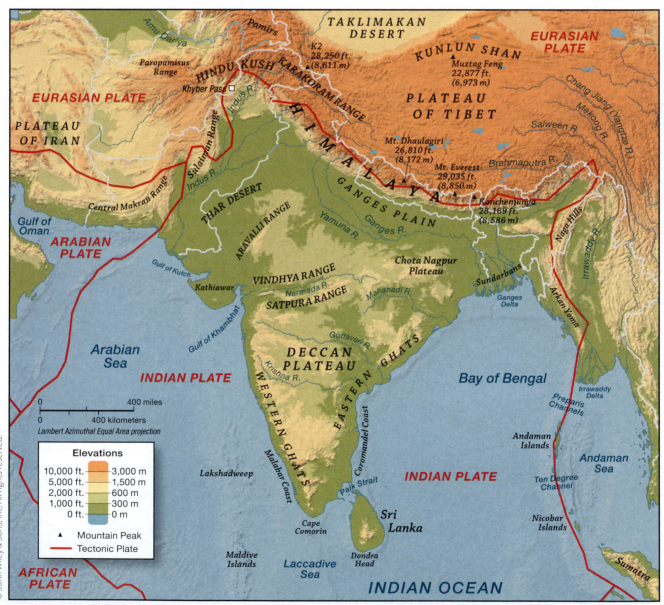

FIGURE 6.1 Physical geography of South Asia. The subcontinent of South Asia is still colliding with the Eurasian plate, building the Himalaya Mountains. Major rivers, including the Indus and Ganges, begin from meltwater in the Himalayas. The southern part of the region includes the Deccan Plateau, a high, flat plateau, surrounded by coastal plains.

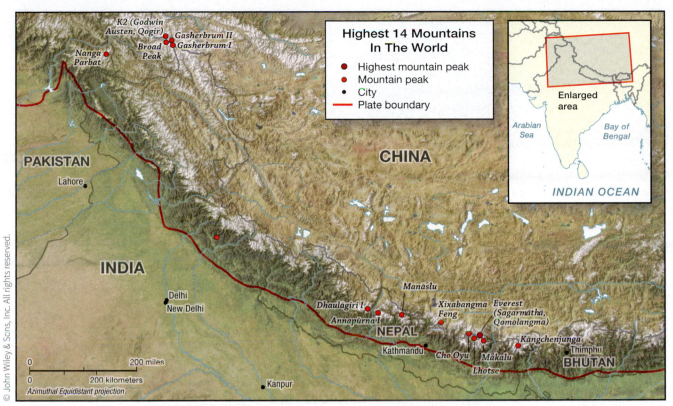

FIGURE 6.2 **14 tallest mountains in the world.** The 14 tallest mountains in the world, based upon height above sea level, are all found along the plate boundary where the subcontinent of South Asia is colliding with the Eurasian plate.

in elevation at the same time they are eroding, as the plates continue to collide. The two tallest mountains in the world are in this region: Mount Everest in the Himalayas and K2 in the Karakoram Range, which is a western extension of the Himalayas. South Asia has the 14 tallest mountains in the world (Figure 6.2).

MONSOON

As much as the Himalayas mark a physical geographic boundary of South Asia, the monsoon defines a common climate. Although only southern India, Bangladesh, Myanmar, and northern Sri Lanka are defined as having monsoon climates (Figure 6.3), the monsoon affects nearly everyone in the region.

The monsoon is not severe weather, like a tornado or a hurricane, which everyone hopes to avoid. Rather, the monsoon is a predictable climate pattern in South Asia that everyone needs to survive. Crops are planted to coincide with the predictable rains of the summer (wet) monsoon, festivals are scheduled to celebrate the monsoon, and people live their daily lives knowing the late afternoon or evening will bring rain during the wet monsoon.

In South Asia, during the summer monsoon, warm, moist winds blow from a high-pressure system in the Indian Ocean onto the Indian subcontinent toward a low-pressure system in southwest Asia (Figure 6.4). A

monsoon is a wind blowing from a predominant direction for a long period of time. The winds blow from the ocean north across India, forming the wet monsoon during the summer because a low-pressure belt called the Intertropical Convergence Zone (ITCZ) migrates to northern India in June.

Wind flows from high pressure to low pressure. The low pressure of the ITCZ draws moist air from the Indian Ocean north across India, creating the consistent pattern of the summer monsoon. The reason the ITCZ moves north in the summer is because the band of low pressure migrates over the course of the year with the **subsolar point**, the place on earth receiving direct rays of the sun. In June, the ITCZ is as far north of the equator as it reaches because the sun's rays are hitting far north. The summer monsoon blows from May to October, with rains peaking in June and July.

During the summer monsoon, the rain comes quickly in the afternoon. Pedestrians walking through streets teeming with carts, auto-rickshaws, and cars seek shelter under canopies. Life in the city responds quickly to the summer monsoon (Figure 6.5). Rain fills drainage grates in courtyards and irrigation ditches with water, slows motorcycles to a crawl, forces pedestrians to seek shelter, and brings traffic to a virtual standstill.

In December, the ITCZ is as far south of the equator as it reaches, along with the sun's rays. During

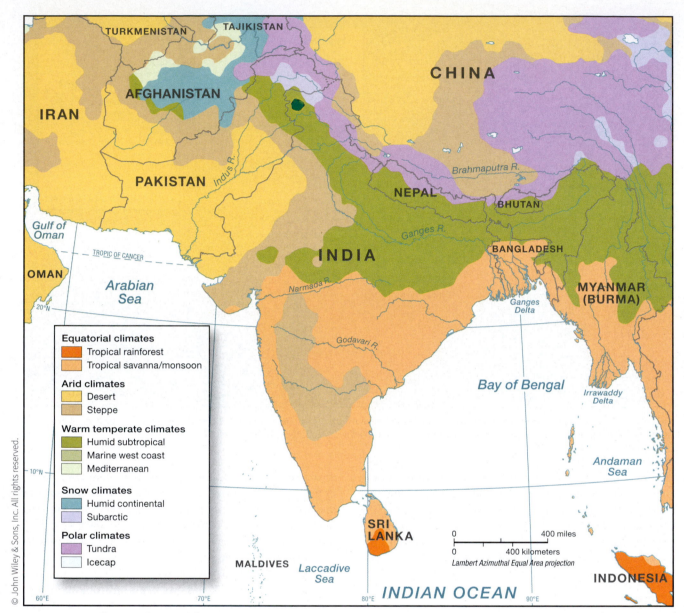

FIGURE 6.3 **Climates of South Asia.** The monsoon climate is found in southern India, Bangladesh, Myanmar, and northern Sri Lanka, but the monsoon rains affect the entire region of South Asia. In every climate zone in the region, precipitation peaks during the summer monsoon. Temperatures and sunlight peak during the summer monsoon, as the entire region is north of the equator.

the winter monsoon or dry monsoon, the ITCZ is far south and cool, dry winds blow from a high-pressure system in Siberia in East Asia to a low-pressure system in the Indian Ocean, bringing six months of little precipitation.

RIVERS

Rainfall from the summer monsoon and snowmelt from the Himalayas support life in South Asia. Two major water systems define the agricultural zones and are the source of water in the northern half of the subcontinent: the Indus in the west and the Ganges and Brahmaputra in the east. Each of these rivers originates in the Himalayas (Figure 6.6).

An estimated 1.3 billion live along rivers originating in the Himalayas. In Pakistan, agriculture and population are concentrated in the Indus River plain, and in India and Bangladesh, the Ganges and Brahmaputra river plains and delta are home to approximately 700 million people and much of the region's agriculture. The drainage basins of both river systems are much larger, and a total of 3 billion people (nearly half of the world's population) live within these drainage basins and depend in some way on the fresh water, either for drinking water or agricultural production.

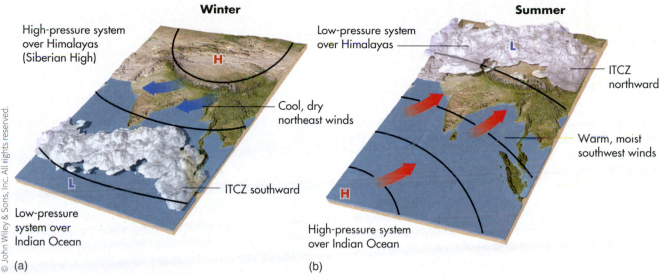

Winter

High-pressure system over Himalayas (Siberian High)

H

Cool, dry northeast winds

ITCZ southward

L

Low-pressure system over Indian Ocean

(a)

Summer

Low-pressure system over Himalayas

L

ITCZ northward

Warm, moist southwest winds

H

High-pressure system over Indian Ocean

(b)

FIGURE 6.4   The winter and summer monsoons.   Prevailing winds flow from Asia south to the Indian Ocean during the winter months, bringing dry air over South Asia and the dry monsoon. Prevailing winds flow from the Indian Ocean north to Asia during the summer months, bringing warm, moist air over South Asia and the wet monsoon.

AMIT DAVE/Reuters /Landov

FIGURE 6.5   Ahmedebad, India.   School children crowded into an autorickshaw after monsoon rains fell. Autorickshaws, also called tuk tuks, are designed to hold a driver plus three passengers. They run off compressed natural gas and are often used to transport up to ten school children when school bus transportation is not available or reliable.

The Indus River System

The Indus River originates on the Tibetan Plateau, north of the Himalayas. The Indus and its floodplain dominate the eastern half of Pakistan, and the headwaters of the Indus and several of its tributaries are located in India. The western portion of South Asia where the Indus River is located is a semiarid climate, not a monsoon climate (see Figure 6.3). Farmers in Pakistan and western India do not have predictable, plentiful rains of monsoon Asia, so they depend on the Indus River system for crop irrigation.

The tributaries of the Indus River originate in the border region between India and Pakistan. When the British partitioned the two countries in 1947, the border divided the Punjab region and left Jammu and Kashmir to decide whether to join India or Pakistan. Shortly after the August 1947 partition, Muslim Pakistanis in Jammu and Kashmir called for an alliance with Pakistan. The Hindu leader of Kashmir asked India for help, and in October 1947, India officially incorporated Kashmir. Pakistan resisted, and in 1949, India held 65 percent of Kashmir, with Pakistan holding the remaining 35 percent. India and Pakistan, both of which are nuclear powers, had several skirmishes and three brief wars, in 1947, 1965, and 1999, each time over Kashmir. The Kashmir question is still unsettled, and a line of control divides Pakistani-controlled land from Indian-controlled land in Kashmir (Figure 6.7).

Based on Kashmir's contentious history, it is somewhat surprising that India and Pakistan reached a treaty on the control and use of the Indus and its tributaries in 1960. Despite water scarcity, competition for use of the river, and their broader political disagreements, both India and Pakistan recognized they would need to agree on water sharing in order to secure their own food supplies.

To encourage India and Pakistan to sign the treaty, the World Bank, which negotiated the agreement, provided financial support to both countries to help build irrigation dams and canals. Under the 1960 Indus Waters Agreement, India has the right to divert water from the three eastern tributaries of the Indus that extend into the Punjab region of India, and Pakistan has the right to control the waters of much of the Indus itself and the western tributaries that flow through Kashmir. The countries mutually benefit from having secure and dependable water resources.

Reading the PHYSICAL Landscape

Himalayan Glaciers

When most people think about the effects of climate change, they usually consider the possibility of the Greenland and Antarctic ice caps melting and rising sea levels. Another, lesser known effect of climate change on the world's glaciers is occurring in the Himalaya Mountains in South Asia. The rugged Himalayas are the topographic backbone of South Asia (Figure 6.6a). Given their immense size and high altitude, the Himalayas contain a number of large alpine glaciers that are the source of water for major river systems, including the Ganges and Indus (Figure 6.6b).

Until the twentieth century, the Himalayan glaciers melted at a seasonal rate which, along with monsoon rains, steadily supplied the rivers with water (Figure 6.6c). In response to this rhythm, hundreds of millions of people living along these rivers in China and India, far downstream of their glacial source, are dependent upon the consistent supply of water (Figure 6.6d). As a result of ongoing warming in the Himalayas since the twentieth century, however, the mountain glaciers have been melting at alarming rates

Grant Dixon/Lonely Planet Images/Getty Images

FIGURE 6.6a   Mingbo La, Nepal.   Glaciers form where the climate is so cold that snow accumulates and compacts into glacial ice.

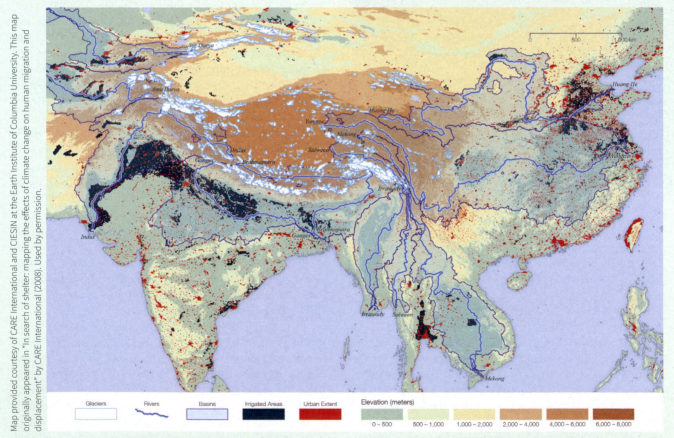

Map provided courtesy of CARE International and CIESIN at the Earth Institute of Columbia University. This map originally appeared in "In search of shelter: mapping the effects of climate change on human migration and displacement" by CARE International (2008). Used by permission.

Glaciers   Rivers   Basins   Irrigated Areas   Urban Extent   Elevation (meters)
0 – 500   500 – 1,000   1,000 – 2,000   2,000 – 4,000   4,000 – 6,000   6,000 – 8,000

FIGURE 6.6b   The highest elevations of the Himalaya Mountains are so cold that the region has approximately 15,000 glaciers (shown in white on this map). In the spring, when the leading edges of the glaciers melt, the meltwater feeds rivers that supply fresh water to about 3 billion people. The rivers fed by the Himalayas include the Ganges, Indus, Brahma-Putra, Yangtze, Huang He, and Mekong.

and thus have been retreating about 33–49 feet (10–15 m) each year. The Gangotri glacier in the Indian Himalayas is a perfect example, having retreated about 6 miles (1 km) since 1935 (Figure 6.6e).

The short-term impact of rapid glacial melting is that streams will receive more meltwater and are thus increasingly prone to flooding. A long-term concern is that as the glaciers shrink further in size, they will provide less meltwater annually, reducing the freshwater supply from the spring thaw.

Alvaro Leiva/age fotostock/Getty Images

FIGURE 6.6c   Uttaranchal, India.   A Hindu priest stands by the meltwater of Gangotri glacier, which is the source of the Ganges River.

Source: NASA image by Jesse Allen, Earth Observatory; based on data provided by the ASTER Science Team. Glacier retreat boundaries courtesy the Land Processes Distributed Active Archive Center.

FIGURE 6.6e   Satellite image of the Gangotri glacier and amount of retreat since 1780.   This glacier is the source of the Ganges River. Currently 30.2 km long and between 0.5 and 2.5 km wide, Gangotri glacier is one of the largest in the Himalayas. Gangotri has been receding since 1780, but studies show its retreat quickened after 1971.

Photo ©Tan Yilmaz/Flickr/Getty Images

FIGURE 6.6d   Varanasi, India.   The Ganges River is the most sacred river in Hinduism. The faithful perform ablutions in the river each morning, to purify themselves with the water.

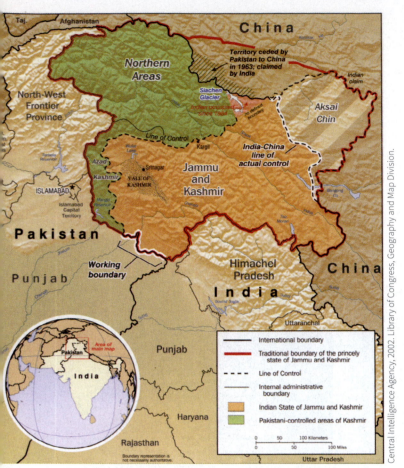

FIGURE 6.7   Line of control in Kashmir. Kashmir was a princely state during British colonialism, and since India and Pakistan became independent in 1947, the two countries have had three brief wars over Kashmir. A line of control divides the Pakistani-controlled northern areas from Indian-controlled Jammu and Kashmir. China claims part of the northern area of Kashmir.

Ganges–Brahmaputra Rivers

Bangladesh is marked by the world's largest delta: the Ganges-Brahmaputra Delta covers much of the country. The Ganges and Brahmaputra both begin in the Himalaya Mountains, fed by snowmelt in the spring. The Ganges begins on the southern side of the central Himalayas, west of Nepal, and flows through the northern portion of India and easterly into Bangladesh. The Brahmaputra begins in the Tibetan Plateau, north of the Himalayas. As the river flows east through the Tibetan Plateau, it is called the Yarlung Tsangpo (meaning water running from the highest peaks) River. The river then flows through the eastern Himalayas, curving to a westward flow across the eastern states of India and into Bangladesh, and then emptying in the Bay of Bengal, along with the Ganges (Figure 6.8).

The Ganges, Brahmaputra, and their tributaries carry sediment from the Himalayas and Tibetan Plateau and deposit it at the Bay of Bengal. The landscape of a river delta is constantly in flux. When the volume or speed of the river increases, the river **erodes** the landscape, and when the volume or speed of the river decreases, the river drops

silt in a process called **deposition**. The erosion and deposition processes in this region are amplified because flooding comes from three sources: snowmelt from the Himalayas (in the spring), monsoon rains (in the summer and fall), and cyclones (in the summer and fall). The relatively predictable combination of snowmelt and wet monsoon create a distinct seasonality to river levels and flows in the region.

In 1947, when Bangladesh was called East Pakistan and was part of a newly independent state of Pakistan, the region reported that less than 1 percent of its agricultural output depended on irrigation from the Ganges–Brahmaputra river system. Agriculture in Bangladesh was based almost solely on the monsoon rains. Both India and Bangladesh had their own plans for the river system, and neither had developed an extensive irrigation system (Brichieri-Colombi and Bradnock 2003).

Ironically, although the political relationship between India and Bangladesh is much friendlier than that between India and Pakistan, the treaty over the Indus between India and Pakistan was arrived at more easily and is on more solid footing than the agreement between India and Bangladesh over the Ganges–Brahmaputra. India and Bangladesh began negotiating a water treaty even before Bangladesh became independent in 1971, but did not agree to the Ganges Waters Treaty until 1996. The Ganges Water Treaty specifies water use for each country by season and is expected to last 30 years (until 2026).

The delta region has changed markedly since negotiations began, with the population of the world's largest delta doubling to over 200 million people and still growing (Brichieri-Colombi and Bradnock 2003). A growing population dependent on the waters from the Ganges could undermine the stability of the treaty. Concern is also mounting that climate change will alter current agricultural processes and settlement patterns in the region, which will also challenge the stability of the Ganges Waters Treaty.

TROPICAL CYCLONES

A tropical cyclone is a low-pressure system that grows over a warm ocean (at least 80 degrees Fahrenheit) between 5 and 15 degrees latitude north or south of the equator. Tropical cyclones are called hurricanes in the North Atlantic and cyclones or typhoons in the Pacific and Indian oceans. Bangladesh, one of the poorest countries in the world, is susceptible to Indian Ocean cyclones (Figure 6.9).

We expect flooding to have the strongest impact on the lowest lying, lowest elevation areas. However, the impact of a natural disaster varies by poverty levels and social/political norms that disadvantage certain populations in a country. The path of a natural disaster does not tell the entire story of its impact. Vulnerability to a disaster varies across populations within its path. Vulnerability is defined as "the characteristics of a person or group and their situation influencing their capacity to anticipate, cope with, resist and recover from the impact of a natural hazard" (Wisner et al. 2004, 11).

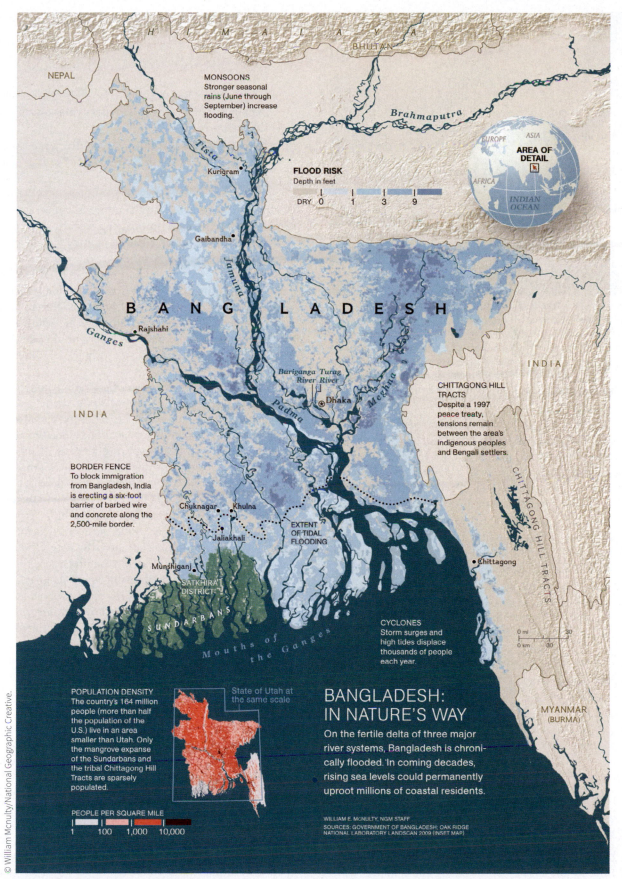

FLOOD RISK
Depth in feet

DRY 0   1   3   9

AREA OF DETAIL

MONSOONS
Stronger seasonal rains (June through September) increase flooding.

CHITTAGONG HILL TRACTS
Despite a 1997 peace treaty, tensions remain between the area's indigenous peoples and Bengali settlers.

BORDER FENCE
To block immigration from Bangladesh, India is erecting a six-foot barrier of barbed wire and concrete along the 2,500-mile border.

EXTENT OF TIDAL FLOODING

CYCLONES
Storm surges and high tides displace thousands of people each year.

POPULATION DENSITY
The country's 164 million people (more than half the population of the U.S.) live in an area smaller than Utah. Only the mangrove expanse of the Sundarbans and the tribal Chittagong Hill Tracts are sparsely populated.

State of Utah at the same scale

PEOPLE PER SQUARE MILE
1   100   1,000   10,000

BANGLADESH:
IN NATURE'S WAY

On the fertile delta of three major river systems, Bangladesh is chronically flooded. In coming decades, rising sea levels could permanently uproot millions of coastal residents.

WILLIAM E. MCNULTY, NGM STAFF
SOURCES: GOVERNMENT OF BANGLADESH; OAK RIDGE NATIONAL LABORATORY LANDSCAN 2009 (INSET MAP)

FIGURE 6.8   Bangladesh.   The majority of Bangladesh is prone to flooding, especially along the Ganges River coming in from the west and the Brahmaputra River flowing in from the northeast. The small map at the bottom shows that despite the risk of flooding, people live in dense settlements along the rivers where the soil is rich and rice and other crops can be grown.

Based on data from the National Oceanic and Atmospheric Administration, http://www.csc.noaa.gov/hurricanes/#.

FIGURE 6.9 **Tracks of cyclones over the last 30 years in South Asia.** Cyclones in the Indian Ocean are typically not as strong as in the Pacific or Atlantic because the low-pressure systems do not have as much space or time to cross the Indian Ocean, gather energy, and build in size. However, Bangladesh is particularly susceptible to cyclones because most of the country is low-lying delta lands of the Ganges and Brahmaputra Rivers.

In 1991, a cyclone hit Bangladesh and killed 140,000 people. The most vulnerable population, women and girls, were more adversely affected by the cyclone than men and boys. In the 1980s a famine had a deadly impact on women, who because of gendered eating practices, were left with little to eat (Neumayer and Plumper 2007). The gendered eating practices are a tradition found especially in rural areas: A mother feeds her father-in-law first, then her mother-in-law, husband, sons, daughters, and herself last. Bangladesh has the world's highest rate of **low-birth-weight** (LBW) babies, and half of its women are underweight (Shannon et al. 2008, 827), as a result of poverty and gendered eating practices.

In 2007, Cyclone Sidr flooded Bangladesh, killing 4,000 and displacing thousands of people. In 2009, the government of Bangladesh began a new cyclone warning system, whereby the government texts over 40 million mobile phone users (30 percent of the country's 130 million people) when a cyclone is headed to the delta region. Awareness of impending disasters will help millions of people in Bangladesh, but without structural changes in society, women and girls will remain most vulnerable.

CONCEPT CHECK

1. **Why** are the 14 tallest mountains in the world in South Asia?

2. **Why** does the monsoon have a wet and a dry season?

3. **How** did the 1991 cyclone in Bangladesh impact women and children more than men and boys?

WHO ARE SOUTH ASIANS?

LEARNING OBJECTIVES

1. **Discuss** the unique qualities of the Indus civilization.

2. **Explain** the historical and modern roles of caste in the region.

3. **Describe** where Hinduism, Buddhism, and Islam are found in South Asia.

Contrasting work and living conditions punctuate India, where Mumbai is home to both the world's most expensive house, a 27-story home valued at more than $1 billion, and the largest slum in Asia, Dharavi, where more than 1 million people live in squalor, paying rent as low as $4 a month. The growth of Mumbai's information technology sector has created pressure on Dharavi, which is located in the middle of the city. New buildings are being erected around the slum. Real estate developers would like to convert the land where Dharavi sits to high-rise offices and condominiums.

United Nations Habitat, an organization committed to reducing the number of slum dwellers globally, estimates that South Asia is home to 27 percent of the global total of people who live in slums, and that "India alone accounts for 17 percent of the world's slum dwellers" (United Nations Habitat, 2006). Although the economy of South Asia is growing at 6 percent a year, the region's rampant poverty stands testament to the fact that not everyone in the region is benefiting equally from economic growth.

The first civilization in South Asia formed on the Indus River in present-day Pakistan approximately 5,000 years ago. The Indus civilization left behind archaeological evidence that people had relatively equal wealth. Civilization in South Asia today stands in contrast to that arrangement with slums like Dharavi standing next to office parks that house multimillion-dollar corporations and the wealthy who run them. The wealthy and poor live in close proximity in one of the two most densely populated regions in the world (Figure 6.10).

INDUS CIVILIZATION

Archaeologists and historical geographers agree that the world's first cities had to have two characteristics to develop as cities and then to maintain order: an agricultural surplus and a leadership class. Theories vary on whether an agricultural surplus or a leadership class came first, but academics generally agree that the two elements were symbiotic because a leadership class can demand more work out of the "lower" class and force the generation of a surplus, and the creation of a surplus (perhaps from a new technology or seed or particularly good growing season) encourages a leadership class to come forward to determine how to distribute the surplus.

Data from CIESIN—Columbia University, United Nations Food and Agriculture Programme—FAO, and Centro Internacional de Agricultura Tropical—CIAT. 2005. Gridded Population of the World, Version 3 (GPWv3): Population Count Grid, Future Estimates. Palisades, NY: NASA Socioeconomic Data and Applications Center (SEDAC). http://dx.doi.org/10.7927/H42B8VZZ. Accessed December 2014.

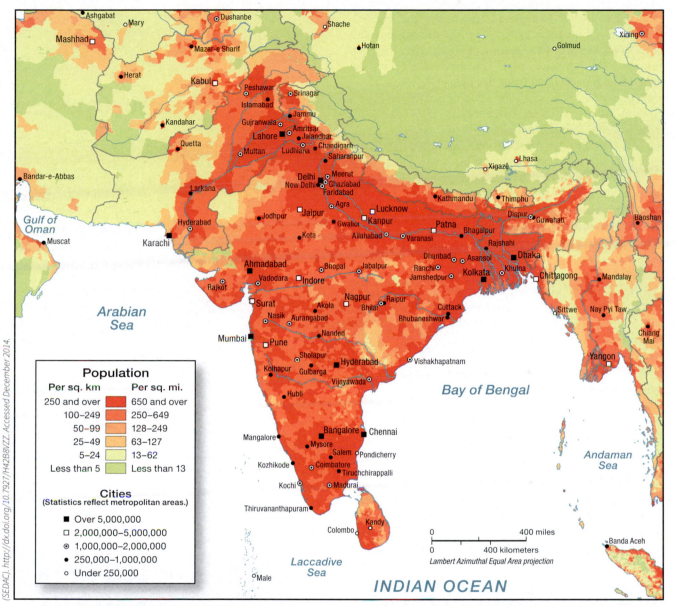

FIGURE 6.10   **Population density of South Asia.**   South Asia is densely populated, especially near major cities and along rivers and coasts.

The organization or arrangement of an ancient city reveals what the people and their culture valued and how their land use, art, engineering, and architecture represented their cultural values. In Mesopotamia and the Nile River Valley, researchers can clearly identify the leadership class by how the city was distinctly divided into zones that each had a certain purpose, known as the city's **functional zonation**.

In the Nile River Valley, elaborate structures such as pyramids where leaders were buried with riches beyond imagination point clearly to the leaders of society and also to the thousands of slaves who built the structures. In Mesopotamia, temples called ziggurats stood at the center of cities as a home for gods. Marking the connection between the leadership class (who took their domain from

gods) and the agricultural surplus, storehouses for the agricultural surplus were located adjacent to the ziggurats in ancient Mesopotamia.

The Indus civilization, which flourished from 2900 BCE to 1900 BCE, was a contemporary of the Nile civilization in Egypt and the Sumerians in Mesopotamia (Figure 6.11). The Indus civilization was different from civilizations in the Nile and Mesopotamia in two remarkable ways: First, the Indus cities have no palaces or temples, no obvious signs of who was in the leadership class (presuming there was a leadership class), and second, the Indus civilization was enormous. It included over 2,000 towns and up to 5 million people over a land area twice the size of civilizations in the Nile Valley or Mesopotamia.

Archaeologists struggle to understand the cultural foundations of the Indus civilization because despite hundreds of examples of writing from the Indus, they have not yet translated the Indus script. The archaeological record includes sets of weights from the Indus civilization that were used for trade or possibly to assess taxes. Evidence of trade with the Indus region extends to the Arabian Sea, to Mesopotamia by land and by sea, and into Afghanistan.

In the 1850s, the British removed millions of kiln-fired bricks from the ancient site of Harappa to build a foundation for railroad tracks. What remains leaves no clear evidence of who was in charge. The **urban morphology** of the city, that is, the layout of the houses and infrastructure, leads researchers to believe nearly everyone had access to the same housing and services (Figure 6.12a). Archaeologists hypothesize that perhaps the city had a large middle class or that the leadership class defined their power in a way that did not appear as a palace, monument, or temple (Wilford 1998). Artistically designed seals (Figure 6.12b) found throughout the Indus region may have distinguished the jobs one people had from another, including the leadership class (Edwards 2000).

A bath house measuring 39 feet by 23 feet (12 m by 7 m) with a watertight floor and two staircases leading into the tank stands at the highest point of ancient Mohenjo-Daro (Figure 6.13). This bath house may mark the cultural foundations of **ablution**, or ritual bathing, still

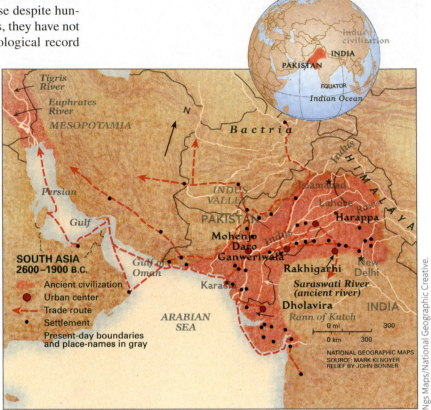

FIGURE 6.11 **Indus civilization.** The Indus civilization included cities located on the Indus River and its tributaries. Archaeologists believe the Indus civilization declined when the Indus River changed course.

Ngs Maps/National Geographic Creative.

Copyright J.M. Kenoyer/Harappa.com, Courtesy Dept. of Archaeology and Museums, Govt. of Pakistan

(a)

NYPL/Science Source/Photo Researchers/ Getty Images

(b)

FIGURE 6.12 **Mohenjo-Daro, Pakistan.** Indus housing appears to have been distributed equally along central streets. Archaeologists have excavated the housing on the left side of the street, but not on the right side (Figure 6.12a). Archaeologists believe the seals (Figure 6.12b) the Indus civilization left behind may explain how the civilization demonstrated wealth. This seal with a unicorn relief was discovered in Mohenjo-Daro. The city of Mohenjo-Daro was founded around 2600 BCE and abandoned around 1900 BCE. According to archaeologist Mark Kenoyer on Harappa.com, "The unicorn is the most common motif on Indus seals and appears to represent a mythical animal that Greek and Roman sources trace back to the Indian subcontinent."

carried out by Hindus and Muslims in South Asia today. Archaeologists also found brick platforms next to dwellings in Indus cities. Researchers believe the platforms were sites where residents of the dwellings could perform their ablutions.

Around 2000 BCE, Aryans entered the Indus region through the Khyber Pass. The Indus civilization peaked around 1900 BCE, and many scholars believe the course of the Indus River began to change around this time frame, marking the decline of its civilization. As the Aryans entered and the Indus River changed course, the pulse of life in South Asia shifted to the Ganges River.

LANGUAGE

The language families in South Asia reflect the historical spatial interaction in the region. Throughout northern South Asia, the languages belong to the Indo-European language family, which links Hindi and Sindhi with modern languages spoken in Europe today (Figure 6.14). In the southern part of South

FIGURE 6.13 Mohenjo-Daro, Pakistan. The Great Bath in the Indus city of Mohenjo-Daro was likely accessed by everyone and used for ablution, ritual bathing. The brickwork in the bath itself is watertight and designed at a slope with a drain so that it could be easily drained and refilled.

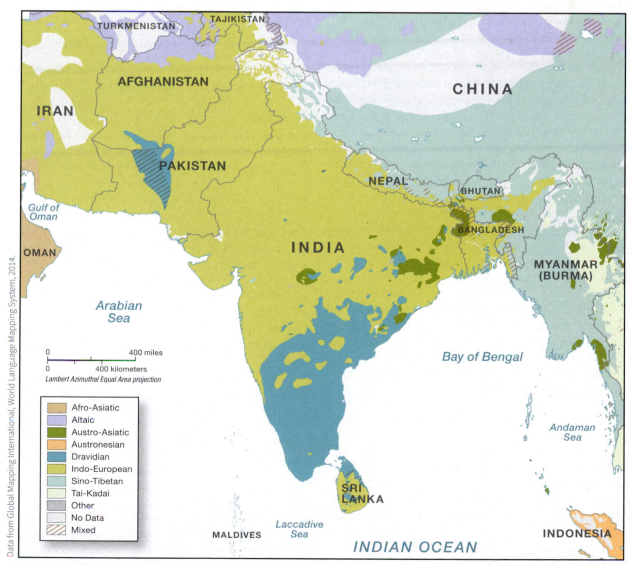

FIGURE 6.14 Languages of South Asia. Languages in the northern part of South Asia are part of the Indo-European language family, and languages in the southern region are in the Dravidian language family.

Asia, on the Deccan Plateau of India, the predominant languages are part of the Dravidian language family. The Indian constitution recognizes 22 official languages, and the 29 States of India are generally drawn around language regions.

English is prevalent among the educated class throughout South Asia. Learning English, earning a college degree, and moving to a city are helping lower-caste Indians break free from the limitations of their caste and earn a living. According to the Indian Census, English is the second most spoken language in India (125 million speakers), after Hindi (550 million speakers). About a quarter of the population of India speaks more than one language, and English is typically spoken as a second or third language.

CASTE

If, in fact, the artistic seals used among the people of the Indus civilization (see Figure 6.12) distinguished the kinds of work people performed, they may mark the early establishment of the **caste system**. Our inability to interpret the script of the Indus makes sorting out the origins of caste and even the origins of the Hindu religion quite difficult. We may never fully understand the origins or evolution of caste, but we can see the implications of caste for Indian society over the last 3,000 years.

The distinctly South Asian caste system divides people into an inherited social hierarchy based on a mutually sustaining order. Caste separates people by dictating a clear division of labor, who can dine together, who can marry, and where people can live. The four principles of caste are separation, division, hierarchy, and reciprocity, and these principles can be seen in the simplified list of castes any encyclopedia would give, dividing people among five groups, including four castes: the Brahmins (priestly class), the Kshatriyas (warriors), the Vaishyas (farmers, merchants, and artisans), the Shudras (laborers), and one group outside of the caste system, the Dalits (formerly called the "Untouchables"). Categorizing castes in this way is deceptively simple, however. Caste in South Asia is much more complex than it first appears.

Each caste can be divided into *jatis*, which are horizontal divisions within each caste along employment lines. Additionally, each jati has subcastes, which are established by place of birth and family lineage. For example, the Dom Raja Dalits have maintained the sacred fire for funeral pyres for more than 3,500 years along the Ganges River in Varanasi.

In 1950, the Indian constitution abolished and forbade the practice of "Untouchability" and prohibited discrimination based on caste, religion, race, sex, or place of birth. Caste is less important in India today than it has been historically; however, it remains a reality in the daily lives of hundreds of millions of Indians. Estimates are that India today has well over 3,000 jatis and 25,000 subcastes. The Indian government has established a list of scheduled[1] castes (Dalits), scheduled tribes (indigenous

peoples), and other backward castes (OBCs). In the 2011 census, India estimated that 16.2 percent of the country's population was in scheduled caste, which includes not only Hindus, but also Muslims, Sikhs, and Christians. The government categorizes another 8.2 percent of India's population as scheduled tribes, also called the Adivasis. The list of OBCs includes over 3,000 subcastes, primarily from the Shudra and Vaishya castes, and accounts for 37.1 percent of the country's population.

The government of India has practiced affirmative action for scheduled castes and scheduled tribes since 1950 and for OBCs since the 1980s, reserving a certain number of elected offices, places in universities, and positions in government or government-funded jobs for them.

RELIGION

The hearths of two world religions are found in South Asia: Hinduism, which began on the Indus River around 4000 BCE and relocated to the Ganges River after 1900 BCE, and Buddhism, which splintered from Hinduism and began on the Ganges River around 500 BCE (Figure 6.15). In addition to Hinduism and Buddhism, Christianity diffused to India as early as 52 CE, and Islam diffused into South Asia from Central Asia as early as 700 CE. Today, religion both unites and divides, as the region is divided into countries primarily along religious lines.

Hinduism

Although Hinduism has sacred scriptures, including the Rig Veda, Hindu beliefs are much more ritualistic than prescriptive. Of the approximately 1 billion Hindus in the world, over 90 percent live in India. Hindu temples punctuate cities and towns throughout India, and small altars or temples to particular gods are tucked into nearly every corner of daily life (Figure 6.16a and b). Different regions and peoples of India associate with different gods. Local cities, villages, and neighborhoods hold festivals for important or popular gods in their region or place. Some festivals are geared more toward women and others more toward men.

The Hindu calendar follows astrology rather than calendar dates. For example, the most sacred of Hindu holy days, the Kumbh Mela, happens four times every 12 years. On these days, Hindus make pilgrimages to rivers, including the Ganges, to perform ablutions (Figure 6.17). In 2013, over 100 million Hindus flocked to the confluence of the Ganges and Yamuna rivers in Allahabad, India during a 55-day festival called the Maha Kumbh Mela (great pot of nectar) because the planetary alignment made the holy days the most sacred in 144 years (Biswas 2013).

Each of the Kumbh Melas happens at a time when the sun, moon, and Jupiter are aligned in a conspicuous way, signaling that anyone who performs ablutions at the appointed places on these days can wash away all previous sins, even from previous lives, making *moksha*, or the breaking free from Karmic bondage, much more likely in one's lifetime.

---

[1] The term *scheduled* dates back to 1935 when the British colonizers made a list of 400 castes, all of which were Dalit, and gave them "special privileges in order to overcome deprivation and discrimination" (Deshpande 2010, 27).

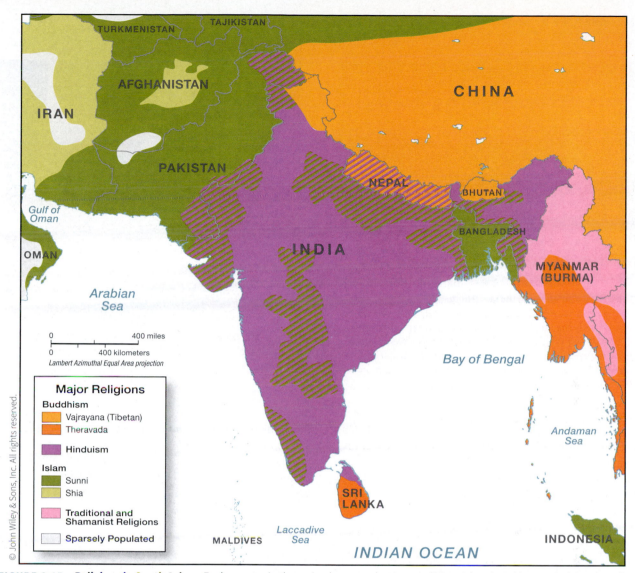

FIGURE 6.15 **Religions in South Asia.** Each country in the region has a predominant religion: Islam in Pakistan; Hinduism in India; Islam in Bangladesh; Buddhism in Sri Lanka; Buddhism in Bhutan; and Hinduism in Nepal. Relatively large minority religions are also found in the countries of the region. The Muslim population in India is so large that India has the third largest Muslim population of any country in the world, after Indonesia and Pakistan and ahead of Bangladesh, Nigeria, and Iran.

FIGURE 6.16a **New Delhi, India.** The Birla family built the three-story Laxmiinarayan Temple, also called the Birla Hindu Temple, out of red sandstone between 1933 and 1939. This Hindu temple, built for Lord Vishnu and Lakshmi, sits on over 7 acres and is open to Hindus and non-Hindus, regardless of caste.

FIGURE 6.16b **Rajasthan, India.** This home altar in the Mandawa region of Rajasthan is one of hundreds of thousands of home altars built on the outside of houses in India. Home altars are found inside houses throughout the country, and some families build altars to particular gods on the outside of their homes too.

Hindustan Times via Getty Images

FIGURE 6.17   Allahabad, India.   Hindu pilgrims take a holy dip in the most auspicious part of the Ganges River, known as the Sangam. This is the area where the Ganges and Yamuna rivers flow together with the mythical Saraswati River (a river referred to in the Rig Veda that no longer flows and whose former course is unknown). The Kumbh Mela occurs every 12 years. This photo was taken at the end of a 55-day Maha Kumbh Mela in 2013.

© INTERFOTO/Alamy

FIGURE 6.18   Nepal.   A Brahmin priest studies the Rig Veda. The Vedas date to 1500 BCE, and Hindus believe they were divinely revealed.

Although Hinduism is considered to be based more on orthopraxy (centered on prescribed practices or behaviors) than orthodoxy (centered on theological teachings), Hinduism does have a sacred text. The **Rig Veda** is a collection of over 1,000 hymns or poems that are the foundational scripture of the Hindu religion (Figure 6.18). While the Rig Veda is attributed to the Aryans, many scholars see influences of the Indus civilization in the text. By approximately 1000 BCE, the Rig Veda was passed orally among the most elite of the Aryan people, a group who became the priests of Ganges society, the highest caste, the Brahmins (Figure 6.19).

Reading the **CULTURAL** Landscape

Death Tradition in Hinduism

The city of Varanasi, India, is central to the death tradition in Hinduism. Hindus see Varanasi as the world of death and life, and some make pilgrimages to Varanasi to die. In Hindu tradition, if a person dies in the holy city of Varanasi on the Ganges River, he or she is freed from moksha, which is the cycle of death and rebirth. Pilgrims travel to Varanasi to cremate their deceased relatives on the ghats along the river. Faithful hire Brahmins to preside over the funerals of their loved ones.

Varanasi is also attractive to widows from India who may not easily remarry. They migrate to Varanasi, sometimes leading a life of prostitution in order to subsist in the holy city and wait years to die there. Not all Hindus travel to Varanasi to die. Each year, millions of Hindu pilgrims also travel to Varanasi to atone for their sins in order to live a better life.

LightRocket/Getty Images

FIGURE 6.19   Varanasi, India.   One part of the Hindu death tradition is that those who are praying over and saying farewell to a dead person will have their head and beard shaved. Along the ghats (steps) that enter into the Ganges River in Varanasi, faithful have their heads and beards shaved, purchase flowers for offerings, hire priests to preside, and arrange to have the departed cremated.

Jainism

Jainism is one of the first major reform movements of Hinduism, dating back to 550 BCE, approximately coinciding with the start of Buddhism. Jainism is based on a series of 24 tirthankara or prophets. The most recent one, Mahariva, born in 599 BCE, founded the modern Jain community. Jains follow a life of nonviolence and keep to strict vegan diets, as they believe animals and plants have living souls. Only about 4 million Jains remain in South Asia today, and they live primarily in a small area to the north of Mumbai.

Buddhism

Buddhism can be traced to a single founder: Siddhartha Gautama, the Buddha, who was born to a member of the upper caste in the 6th century BCE[2]. As an adult, the Buddha left the comforts of his father's home and sought to live an ascetic life, giving up worldly possessions and begging for food. In his ascetic life, the Buddha meditated and reached a point of enlightenment. He advocated the "middle path"—neither a life consumed with worldly possessions nor an ascetic life lacking all comforts. By following the middle path, Buddhists, too, can reach the point of enlightenment.

The Buddha had disciples and followers during his lifetime, and certain places where the Buddha spoke have become sacred sites for Buddhists. Near the Hindu city of Varanasi, the Buddhist city of Kushinagar marks where the Buddha died. Outside the city of Varanasi, in Sarnath, Deer Park marks the place where the Buddha gave his first sermon (Figure 6.20).

© Thierry Bresillon/Godong/Corbis

FIGURE 6.20 Sarnath, India. Located 15 kilometers from Varanasi, the Dhamek Stupa marks the site where the Buddha gave his first sermon to his five disciples after attaining enlightenment. The faithful make a pilgrimage to Sarnath and walk around the Dhamek Stupa.

[2] Based on results of recent research, some scholars contend Siddhartha Gautama was born in the 5th century BCE. http://www.bbc.co.uk/religion/religions/buddhism/history/history.shtml

Although Buddha did not see himself as a god and did not want to be worshipped, after his death many of his followers deified him or saw him as a god. Starting from a hearth in northern India, three primary forms of Buddhism were established and diffused in three different directions (see Figure 6.15). *Theravada Buddhism* translates as "the way of the elders" and was the first form of Buddhism established in India. Theravada Buddhists established a secondary hearth in Sri Lanka, and from there the religion diffused to Southeast Asia, including Myanmar (Burma), Thailand, Laos, and Cambodia. Theravada Buddhism emphasizes the teachings of the historical Buddha. Adherents have the goal of self-liberation, and many live monastic lives as monks or nuns to achieve enlightenment.

From the same hearth in northern India, *Mahayana Buddhism* was established later and diffused north into China, Japan, Korea, and Vietnam. Mahayana Buddhism is translated as "the greater vehicle." In Mahayana Buddhism, individuals achieve enlightenment, which helps the greater society of all living beings on earth. Mahayana Buddhists follow the teachings of the Buddha as well as the bodhisattvas, who are great spiritual leaders that have achieved enlightenment.

Vajrayana Buddhism also began in the hearth of northern India, later than Theravada and Mahayana. Vajrayana Buddhism, also called Tibetan Buddhism and Lamaism, translates as "the diamond vehicle." Diffusing primarily into Tibet and Mongolia, Vajrayana Buddhism stresses the importance of gurus (religious instructors) who teach mantras, tantras, and meditation so individuals can achieve enlightenment faster than in Mahayana Buddhism. Buddhists of all sects make pilgrimages to stupas to meditate on the teachings of Buddha and to reach enlightenment. **Stupas** are domes or rounded structures designed to house a relic of the Buddha or one of his saints, bodhisattvas.

Buddhism gained a large number of followers through hierarchical diffusion in approximately 250 BCE when the Maurya emperor Ashoka ruled much of the Indian subcontinent. Ashoka converted to Buddhism during his rule (268–232 BCE), and he oversaw the building of more than 80,000 stupas across his empire (Figure 6.21). Ashoka became a vegetarian, and he passed laws banning the unnecessary killing of animals. The majority of Indians under Ashoka's rule became vegetarians.

The wheel in the Indian flag is the symbol of Ashoka, and many of the stupas he built are still standing. However, the number of followers of Buddhism has declined precipitously in India since his rule. The map of religions shows few Buddhists in South Asia relative to Southeast and East Asia. Buddhism remains strong in Sri Lanka, Nepal, and Bhutan. However, in India, early followers of Buddhism typically fell back into the fold of Hinduism. Hindus began to incorporate the Buddha and his teachings into their religious beliefs. In fact, many Hindus see the Buddha as the ninth **avatar** or incarnation of the Hindu god Vishnu, who has ten avatars.

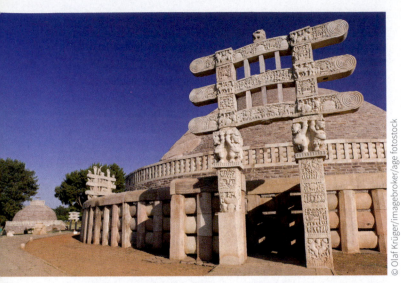

FIGURE 6.21   Sanchi, India. The Buddhist pillars and statues in Sanchi comprise the oldest Buddhist sanctuary, according to the United Nations Educational, Scientific and Cultural Organization (UNESCO). Ashoka, founder of the Mauryan Dynasty, converted to Buddhism around 250 BCE, and either built or added to the Buddhist sanctuary of Sanchi. Stupa 1 is a large hemispheric dome of sandstone surrounded by four gateways that are positioned along the cardinal directions.

Parsis

The Parsi religion in South Asia dates back to the tenth century CE. However, the religion began approximately 1500 BCE in Persia (present-day Iran) with the prophet Zoroaster. Zoroaster founded a monotheistic religion based on three precepts: "good thoughts, good words, good deeds."

Zoroastrianism is an **ethnic religion** that does not actively seek converts, whereas Christianity and Islam are **universalizing religions** that do actively seek converts. Nonetheless, Zoroastrianism had grown to millions of followers by the time Islam came into Persia, between 700 and 900. Persecuted by Muslims, a group of Persian Zoroastrians migrated to present-day India. According to a Zoroastrian oral tradition, the priests sent word to an Indian prince that they wanted to migrate to the coast of India, north of present-day Mumbai. The prince sent back a bowl of milk to tell them that his land was full and did not have room for them. The Zoroastrians returned the bowl with a gold ring gently dropped in the bottom—signifying that if they came to India, they would not disturb society; they would simply add wealth.

The Zoroastrians who migrated to India began to call themselves the Parsis and, over the last 1,000 years, have become among the wealthiest families in India. The Tata and Wadia families are Parsis who own automobile, mobile phone, satellite television, international consulting, tea, steel, ship building, and digital audio companies throughout the country (Figure 6.22).

Parsis control much of the Indian economy, but their numbers are dwindling. Today, most estimate the Parsi community globally to have dipped below 100,000, with fewer than 70,000 living in India. Parsis do not actively seek converts, and children of Parsi women and non-Parsi men are not considered Parsis. However, the declining Parsi population puts pressure on religious leaders to change this practice.

Islam

Islam first came to South Asia around 700. The diffusion of Islam into South Asia followed sea trade routes connecting the Arabian Peninsula, the Persian Gulf, and South Asia (see Figure 5.28). Islam also diffused to South Asia through the Khyber Pass from Central Asia to present-day Pakistan.

From 1206 to 1526 (with the decline beginning in 1398), the Delhi Sultanate ruled North India from Pakistan in the west to Bangladesh in the east. Muslim rulers sought to control Central Asia and used the riches of India, gained in part by sacking Hindu temples, to accomplish this goal.

The treatment of Hindus varied over a series of Islamic empires. The Delhi Sultanate recognized that the region simply had too many Hindus to convert, and so they instead employed the *jizya tax*, which made Hindus second-class citizens, and required non-Muslims to pay a tax. Although Hindus grew increasingly intolerant of the financial burden from the Delhi Sultanate over time, they were able to worship as they wished and prevent their temples from being sacked.

FIGURE 6.22   Chennai, India. The new campus of Tata Consultancy Services sits on a 70-acre campus and claims to be the largest campus in Asia. It was designed by Uruguayan architects to look like a Hindu temple and is combined with modern architecture. A conference room sits 100 meters high, under the 132-meter-high tower in the middle of this photo. The campus holds 24,000 professionals, has an underground garage that houses 2,000 cars, and looks like a butterfly from above.

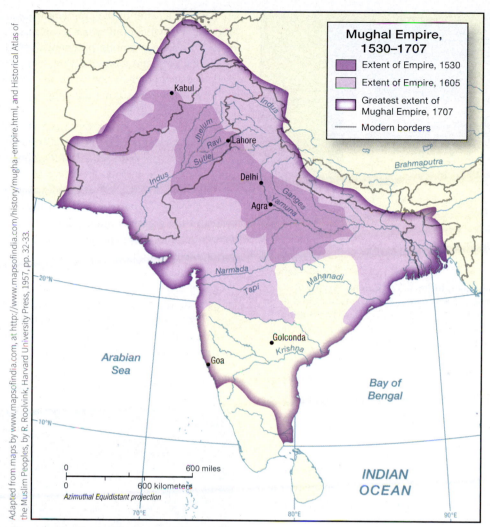

Adapted from maps by www.mapsofindia.com, at http://www.mapsofindia.com/history/mugha-empire.html, and Historical Atlas of the Muslim Peoples, by R. Roolvink, Harvard University Press, 1957, pp. 32-33.

FIGURE 6.23   India under the Mughal Empire.   The Mughal Empire was the last Muslim Empire to rule India. At its greatest extent, the empire incorporated almost the entire region. The Mughal Empire began declining in 1707, as the British East Indies Company gained economic and eventually political power in South Asia.

The Mughal Empire was the final Muslim empire in the Indian subcontinent from 1526 to 1857 (Figure 6.23). Akbar, who ruled from 1556 to 1605, was the greatest ruler of the Mughal Empire. He was an illiterate 13-year-old when he became the emperor. Akbar became known as one of the kindest leaders of India because he established works projects to improve the lot of people throughout the lands, removed the tax on Hindus, and encouraged all religions to interact peacefully.

After 1857, the British controlled the subcontinent, but hundreds of Muslim rulers remained in the princely states of the region. During the 1930s and 1940s and leading up to the end of British rule in 1947, Muslim leaders, including Muhammad Jinnah, lobbied the British and negotiated for the creation of the majority-Muslim country of Pakistan. At independence in 1947, Pakistan was one country physically divided in two parts, West Pakistan and East Pakistan, which were separated by 800 miles. In 1971, India helped East Pakistan gain its independence, ushering in the country of Bangladesh.

Today, Pakistan and Bangladesh have Muslim majorities. India is majority-Hindu as is Nepal, and Sri Lanka and Bhutan are majority Buddhist.

Sikhism

Sikhism was founded in the sixteenth century in the Punjab region, which since 1947 has been divided between India and Pakistan. Today, most of the world's 20 million Sikhs live in India, with the majority in the Punjab Province. The founder of Sikhism, Guru Nanak, was born in the fifteenth century near the city of Lahore, which is now in Pakistan. The most sacred city in Sikhism is Amritsar, where the Golden Temple is located at the site where the Guru Nanak once meditated (Figure 6.24).

Sikhism teaches that all people are equal, thereby denouncing the caste system. The Guru Singh established the 5 khalsa (5 Ks) of Sikhism: having uncut hair, which men typically wear up and under a turban, carrying a steel sword, and wearing a steel bracelet, a wooden comb, and cotton underwear. Practicing or following the

Malcolm P Chapman/Flickr Open/Getty Images

FIGURE 6.24   Amritsar, India.   The Golden Temple of Amritsar, also called the Harmandir Sahib, was built in the sixteenth century and is a sacred site for Sikhs. The Golden Temple complex includes a free kitchen that serves 100,000 people from all faiths and castes each day.

are not only a result of the pregnant woman eating last (see discussion of vulnerability earlier in this chapter). A study of food avoidance among pregnant women in Bangladesh found that pregnant women follow bad advice often delivered by the mother-in-law, such as avoiding protein and eating low calories so that the baby will be easy to deliver (Shannon et al. 2008).

In addition to who goes to school and who eats, the preference for boys over girls is also shown in a massive sex imbalance in South Asia. Demographers have found the ratio of males to females in nature should be 101 to 100, slightly more boys to "offset boys' greater susceptibility to infant disease"(*Economist* 2010). However, population pyramids of the region (Figure 6.26) demonstrate that female fetuses are aborted at an astounding rate in India, despite a government ban on selective abortion in 1994. A study on female feticide (abortion of girl fetuses) in 2010 found that "with up to 10 million female fetuses allegedly terminated in India in the last 20 years, there are now 36 million more men than women in India" (Ahmad 2010, 19).

India defies the global trend whereby more wealth and more education typically lead to an improved status

khalsa has established a shared identity among Sikhs since 1699.

GENDER

Whether a culture favors males or females and how a culture sees the place or roles of women and men are part of a culture's interpretation of gender. The region of South Asia has a historical preference for males over females, which can be found in who attends school, eats first, and is born. Boys are more likely to attend school than girls in much of the region. In Pakistan, for instance, girls are expected at a young age to take on much of the work of managing a home (Figure 6.25), while their brothers attend school or migrate to cities to earn wages for the family.

In northern Pakistan, a region with high incidence of diarrheal disease, the relationship between the mother-in-law and daughter-in-law helps explain which children are most susceptible to disease. When a woman marries a man, she moves in with her husband's family. The new wife is called the Bahu, and much of her life is determined by her mother-in-law, the Sass. A new mother may receive education regarding hygiene and water preparation that would help keep her children healthier, but whether or not that advice is followed is also up to her Sass. The new mom will work or perform chores at home while the Sass takes care of the children. Geographer Sarah Halvorson found that the children most vulnerable to diarrhea disease came from homes with low resources to address the problems and a lack of social support (especially from the Sass) for the mother who is trying to keep her children healthy (2003).

Bangladesh has very high rates of low-birth-weight babies and malnourished mothers. Low-birth-weight babies

Nadeem Khawar/Moment/Getty Images

FIGURE 6.25   Chitral district, Pakistan.   A young Kalash girl in the Chitral valley picks dried corn for her house's roof.

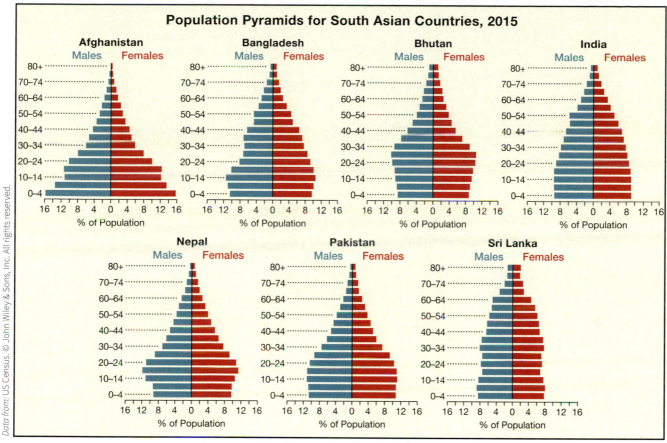

FIGURE 6.26   Population pyramids for each country in South Asia.   Countries with populations that are growing relatively quickly have pyramids with wide bases and narrow tops.

for women. Despite an annual economic growth rate of about 6 percent, the sex ratio in India has only worsened (Figure 6.27). In 1901, India had 972 females for every 1000 males, and in 1991, India had only 927 females for every 1000 males. In 2011, after more than two decades of positive annual economic growth, India had only 940 females for every 1000 males.

The cultural preference for sons remains strong in India. The custom in India is that when a daughter marries, her parents offer a dowry of gifts and cash to the husband and his family. The cost of providing a dowry leads many couples to believe that they cannot afford a daughter. Boys are still considered valuable to carry on the family name and to provide for parents in their old age.

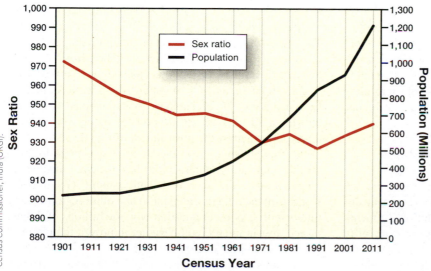

FIGURE 6.27   Sex ratio in India, 1901 to 2011.   Reflecting the cultural preference for boys, India has far more males than females. Although India's economy is growing, the sex imbalance is still far outside of normal. *Data from*: Census India, 2011.

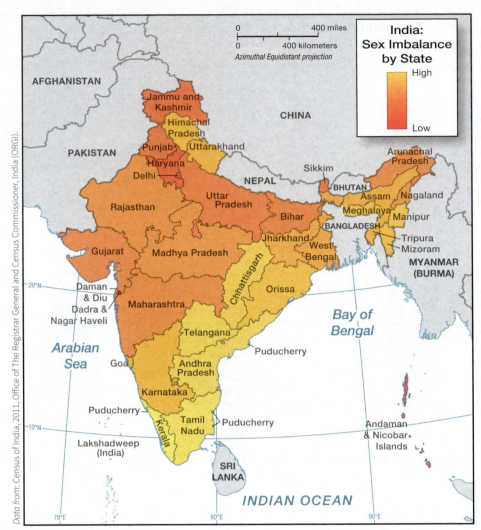

Data from: Census of India, 2011, Office of The Registrar General and Census Commissioner, India (ORGI).

FIGURE 6.28 **Sex imbalance by State in South Asia.** Northern India has the lowest number of females relative to males, while southern India and eastern India have sex ratios closer to normal, but still below the norm.

The selective abortion of upwards of one million female fetuses each year in India has dire consequences. The pervasive belief among many, especially in regions where sex imbalances are even worse than the national average (Figure 6.28), that a girl child is not as valuable as a boy child filters through nearly every aspect of society.

India is beginning to see longer-term, specific consequences of rampant female selective abortions. The demand for women for adult men to marry is so high now that families report daughters being kidnapped and sold into marriage. Rising sex crime levels in Delhi are also attributed, at least in part, to the sex imbalance.

Gender and Nongovernmental Organizations in South Asia

Nongovernmental organizations (NGOs) are a greater presence in South Asia than in any other region in the world. One of the most famous and revolutionary forms of nongovernmental organization activity, **microcredit lending**, was formulated in South Asia. Bangladeshi economist Muhammad Yunus responded to the plight of the poor in Bangladesh and founded the Grameen Bank when he issued a loan for $27 in the village of Jobra, Bangladesh, in 1976.

The fundamental principle of the Grameen Bank is that a small loan to a woman who has no collateral will be repaid if the woman is part of a loan community: a group of similar women who have similar loans (Figure 6.29). This system of lending is called microcredit, and since 1976, the Grameen Bank, which switched from an NGO to a traditional bank in 1983, has loaned $8.17 billion with $7.25 billion repaid. Of all microcredit loans issued by Grameen Bank, 97 percent have gone to women, and today 95 percent of the bank itself is owned by borrowers, with 5 percent owned by the government of Bangladesh.

Although Yunus won the Nobel Peace Prize in 2006 for his innovation, microcredit has received increasing criticism since the turn of the millennium. Geographers working in the field in Bangladesh and other regions where microcredit is popular question why the billions of dollars turned over in microcredit have not improved the plight of the poor to a larger degree.

Bangladeshi geographer Mokbul Morshed Ahmad (2003) interviewed non-governmental organization (NGO) field workers who were employed by microcredit programs to track loans and work with debtors. Ahmad did not study Grameen Bank because it is no longer an NGO. He did study several NGOs, including the Mennonite Central Committee and Save the Children United Kingdom. Approximately 1,000 NGOs operate in Bangladesh during any given year, the vast majority of whom are local NGOs that are typically funded by local donors or foreign governments. Most of the field workers who work for NGOs in Bangladesh are from the region and hired by NGOs.

Ahmad's fieldwork revealed that NGOs do not always target the populations they say they will (Figure 6.30). Many NGOs maintain that they target the poorest of the poor, the most vulnerable in society. The field workers employed by microcredit programs run by NGOs reported that although they were supposed to target the "poorest," they focused on the "less poor" with microcredit loans instead. They did so because their jobs were based, in part, on the borrowers paying back the loan, which was a safer bet with the "less poor" than with the "poorest" (Ahmad 2003, 69).

Whether efforts made by NGOs to improve the lives of women and help them out of poverty in South Asia

FIGURE 6.29 Bangladesh. Women who are part of a microcredit program meet in a rural village. The treasurer holds the accounting ledger, while the woman in the middle of the photo makes a payment to the fund and signs next to her contribution.

Burgler, Roel/Hollandse Hoogte/Redux

GUEST *Field Note*   NGOs in South Asia, RDRS Bangladesh

DR. MOKBUL MORSHED AHMAD
Asian Institute of Technology, Pathum Thani Thailand

The meeting of an NGO field worker (to the left, wearing pink) and her clients is taking place in a non-formal school after class hours. The other women in the photo are wearing sarees, the traditional South Asian dress, and are participating in the projects the NGO directs in the village. Several of the NGO clients brought their children to the meeting because there is nobody at home to take care of them. Some of the children in the photograph also attend the non-formal

school. The school is made of bamboo and built on land owned by a villager. Many children in Bangladesh attend non-formal schools during the day either because the public school is far away or the teachers do not regularly come to class in the public school.

When you look at the photo, you may assume the NGO worker comes to the meeting and then simply leaves. That is not the case, however. NGO field workers establish relationships with their clients and visit villages frequently, typically by foot or bicycle. This NGO field worker had to walk 4 kilometers (2.5 mi) from her home to reach the meeting venue. During the monsoon, the roads get muddy and travel becomes more difficult. NGO offices are typically quite modest and, like the non-formal school in this photograph, are not air conditioned.

In parts of South Asia, the literacy rate has increased, child mortality has decreased, and access to credit has increased mostly because of the hard work of NGO field workers and their partnerships with villagers in the region. NGOs mostly work on creating awareness of health, education, and women's rights. Often, NGOs provide services more effectively than the state through the field workers. Tens of thousands of children and also adults are literate because of the non-formal education provided by the NGOs. Many children do not die shortly after birth because the mothers and birth attendants are given basic health training. NGOs in the region also mobilize women to run small businesses and coordinate meetings weekly or monthly among the women entrepreneurs. The microcredit loans without collateral have helped many of the women and their families to come out of poverty.

FIGURE 6.30 Northern Bangladesh. An NGO field worker meets with women and children in front of a non-formal school.

Courtesy of Mokbul Morshed Ahmad, Associate Professor, Asian Institute of Technology

and the developing world are effective depends, in part on the situation of the women in the local culture. Societal and cultural structures differentiate the place of women in South Asia. Women who have no assets, are widowed, or divorced find themselves with a different realm of possibilities than other women.

EDUCATION

Most countries of South Asia adhere to the principle of universal public education for all children, whereby all children, through a designated grade level, have access to education.

The reality is quite different in Pakistan, whose government has failed to provide adequate education to its population. In 2013, only 59.8 percent of the population of Pakistan was literate. Literacy rates vary by gender and location (Figure 6.31). The male literacy rate is 71.1 percent, and the female is 48.1 percent. In the west in Balochistan, the literacy rate was 49.8 percent (only 26.0 percent for women). In the north, in the Khyber-Pakhtunkhwa Province, which used to be called the Northwest Frontier Province, the literacy rate is 54.2 percent. In the south in Sindh Province, the rate is 61.5 percent. On the whole, cities, at 75.0 percent, have much higher literacy rates than rural areas at 51.4.

Government corruption in Pakistan has led to **ghost schools** (Figure 6.32), which either exist only on paper or suffer from absent teachers whose students follow suit. Money flows to buildings (actual or pretend) and teachers, but children receive little, if any, education.

The failure of Pakistan's public education system has increasingly led to the popularity of **madrasas**, Islamic schools dating back to the eleventh century and funded by wealthy families from the Middle East. Pakistan's 13,000 madrasas educate 1.9 million children a year, and public schools educate 20 million kids a year. Students learn Arabic in order to read the Qur'an properly, and the vast majority of the students are boys. The curriculum is based on the Qur'an. However, the Musharraf government, which held power in Pakistan from 1999 to 2008, passed a rule that madrasas must register with the state and have a syllabus for scientific education. In many rural areas of the country, madrasas are the only functioning schools and thus the only choice for parents seeking education for their children.

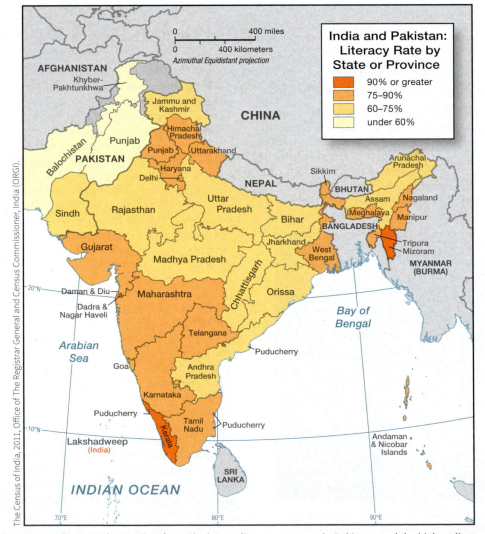

FIGURE 6.31 Literacy rate by State in South Asia. The lowest literacy rates are in Pakistan, and the highest literacy rates are generally in southern India.

FIGURE 6.32 **Basti Nawab Shahin, Pakistan.** A child runs through an empty school yard at a typical government school. Ghost schools provide few facilities or books, typically do not have running water or electricity, and lack furniture.

Some NGOs in Pakistan are working to provide alternative schooling in the country or are creating programs to encourage students to attend school. One Pakistani celebrity, a famous singer named Shehzad Roy, created an NGO called Zindagi Trust that encourages children in Karachi to attend school. Children often leave school before graduating to help earn cash income for families. To help counter this trend, Zindagi Trust pays children what they would earn for a day's work to attend school instead.

CORRUPTION IN PAKISTAN

The NGOs that function in Pakistan are mostly operated and funded by Pakistanis. The failure of the Pakistani state to provide education and necessities for its people encourages overseas Pakistanis to aid their country. In 2013, Pakistan received nearly $15 billion in remittances from Pakistanis working in the Middle East, Europe, and North America. Much of these funds go directly to NGOs.

Donations made directly to NGOs allow a system of services to be provided parallel to the state. The state of Pakistan is often referred to as a "failed state" because of the government's declining ability to provide education and basic services for its people and because violence often erupts within the country. As economist Feisal Khan states, "Frequent national power blackouts, the virtually complete lack of a municipal water supply to Karachi, or the all-important irrigation system are all symptomatic of declining state capabilities" (2007, 220). Khan contends that the failure of Pakistan is rooted in government corruption, and he accordingly refers to the Pakistani government as a **kleptocracy**—that is, a government where leaders steal from the state.

The National Anti-Corruption Report (2002) published by the Government of Pakistan and the National Accountability Bureau indicates that corruption is an everyday occurrence in all departments of Pakistan's government. Virtually every respondent to Khan's survey of 3,000 Pakistanis "reported having to pay a bribe or carry out some other corrupt action in order to complete their transaction" with the government (2007, 225) (Figure 6.33). The survey found that 92 percent of the respondents who had dealings

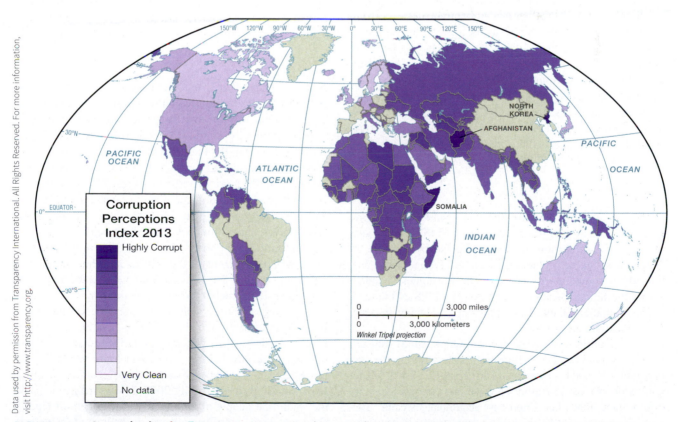

FIGURE 6.33 **Corruption level.** Transparency International, a non-political organization, calculates how corrupt a country's government is perceived to be. Afghanistan and Somalia have the most corrupt governments. Asia and Africa have relatively high levels of corruption compared to Europe and North America.

with the education department encountered corruption, and in the police and land administration departments, 100 percent of respondents encountered corruption.

CONCEPT CHECK

1. **How** was the Indus civilization different from the Nile and Mesopotamia civilizations?

2. **Why** have Dalits traditionally been discriminated against, and **what** benefits do Dalits and Other Backward Castes (OBCs) receive in India today?

3. **Where** is Buddhism located in South Asia and why?

WHAT ROLE DOES ETHNICITY PLAY IN THE POLITICS OF SOUTH ASIA?

LEARNING OBJECTIVES

1. **Describe** how British colonialism impacted South Asia.

2. **Discuss** the role of ethnic groups in politics in each country of South Asia.

The spice trade was one of the first forces of globalization: It linked China, Europe, South Asia, North Africa, and Southwest Asia together in a global trade network from about AD 500 to 1500. Arab traders from Southwest Asia traveled by ship to the Malabar Coast of India (the lush, monsoonal climate zone found on the west coast) and to the Spice Islands of Southeast Asia, carrying spices back to Europe. Travel accounts of Chinese explorers and traders demonstrate the presence of both Chinese and Arab traders on the Malabar Coast of India for over a thousand years before Europeans reached the area known for producing tea and spices, especially pepper.

EUROPEAN COLONIZATION

When the Portuguese sailed the west coast of Africa in 1419 and then around the tip of Africa in 1488, they began the Age of Exploration in Europe. Fueled by a desire to gain access to spices in India and Southeast Asia, Europeans looked for routes across oceans. The Portuguese headed across the Indian Ocean to South Asia in 1498 and landed in Kochi on the Malabar Coast of India, displacing an Arab trading settlement in 1503.

Kochi ultimately traded hands multiple times, reflecting the rise and decline of European powers during the age of exploration and **mercantilism**, an economic system where gold is the standard measure of value and countries seek to export more than they import to accumulate wealth. The Dutch took control of Kochi in 1663, and later the British controlled the city and coastline. British influence in South Asia began in the 1600s and reached its peak in the early 1900s.

For the people of the Indian subcontinent, after 600 years of Muslim empires, life under the British East India Company or even the British Raj was not that radically different. In fact, when the British colonized South Asia in the eighteenth and nineteenth centuries, they used the same regional divisions Akbar created for governance in the sixteenth century. From the 1700s to 1857, despite the presence and later control of the British East India Company, life in the villages of India continued to be dominated by religion, caste, and family.

The British East India Company continued to claim new lands for Britain, and each addition of land brought more potential for exports and for generating wealth for the company. In 1857, the British East India Company, which employed Indians—Hindu, Sikh, and Muslim alike—as soldiers called *sepoys*, introduced the Lee-Enfield rifle to the troops. The rifle required manual loading, and before placing the cartridge in the weapon, a sepoy had to bite the end of the cartridge, which was greased with pig fat and beef tallow. This cultural gaffe (Muslims do not eat pork; Hindus do not eat beef; and Hindus and Sikhs are often vegetarians) offended Indian sepoys, who rose up against the British in the city of Meerut, killing their officers and marching together to Delhi in protest.

The **Sepoy Rebellion** spread throughout northern India. After several months, the British reestablished control of north India, and from 1858 on, changed their posture and level of control in the South Asian colony. They began to rule much of South Asia as a colony of the British monarchy with direct control by the government of Great Britain rather than as a territory of the British East India Company.

PRINCELY STATES

At its height, the British colonial rule of South Asia, known as the British Raj, directly controlled about 60 percent of the subcontinent. Non-British princes in more than 600 states (Figure 6.34) continued to control more than 40 percent of the region. **Princely states** were a variety of sizes, distributed throughout present-day Pakistan, India, and Bangladesh. Each princely state was ruled by a prince who was the majority landowner. These states derived wealth not only from owning land in their states, but also from privy purses, sums of money the British paid for their loyalty and military support. All of the power in the princely states was in the hands of the princes and the landlords in this feudal system. The peasants only worked the land, paid rent, and satisfied debts through forced labor.

The largest princely state was an area around Hyderabad, India, approximately the size of Great Britain. A series of Muslim leaders called Nizams ruled Hyderabad beginning in the early 1700s. During World War I and World War II, the Nizam of Hyderabad loaned the government of Great Britain millions of dollars. In 2008, *Forbes* magazine listed the wealthiest people of all time, and the Nizam of Hyderabad was listed as fifth, with wealth estimated in modern dollars of $210 billion, ahead of Bill Gates, who placed twentieth at that time, with wealth estimated at $101 billion.

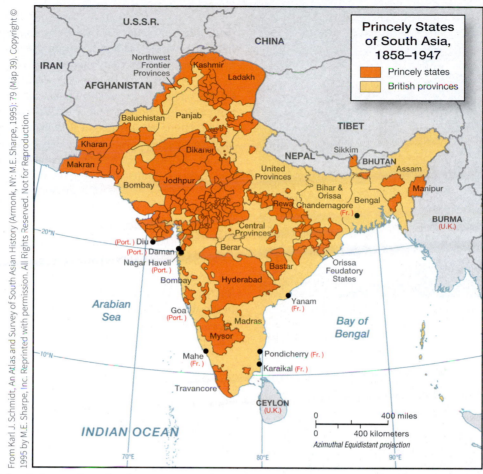

FIGURE 6.34   Princely States of South Asia.   During British colonial rule, large areas of India, Pakistan, and Bangladesh were not directly controlled by the British. The areas in darker orange on this map were princely states throughout the colonial era and had a large amount of autonomy in exchange for their allegiance to the British Empire.

INDEPENDENCE

The road to independence began when Indian leaders, educated primarily in Great Britain, established the Indian National Congress (INC) in 1885. Initially, they demanded more jobs in the Indian Civil Service and a greater say in the governance of India. In 1909, the British responded with the Government of India Act, which reserved a certain number of seats in civil service jobs for Indians.

In order to fulfill the Act, the British took a census of the population. The census made clear that Hindus outnumbered Muslims. The Act divided seats among Hindus, Muslims, and Sikhs, proportionate to the respective population. Muhammad Jinnah, who was born in Karachi and educated in London, established the Muslim League in 1909 and began to work directly with the British Raj to call attention to the need for greater Muslim representation in the subcontinent. Between 1909 and 1947, Jinnah and other Muslim leaders increasingly called for an independent Muslim state in South Asia. Indian National Congress leaders Mahatma Gandhi and Jawaharlal Nehru, both of whom were Hindu, wanted South Asia to become one country. Gandhi, in particular, believed strongly that Muslims, Hindus, Sikhs, and Christians should live together in one state.

Gandhi advocated nonviolence and a better life for the poor in South Asia. He led peaceful resistance movements against British rule, including a boycott of British clothes in 1921, a Salt March in 1930, and repeated hunger strikes (Figure 6.35). In the Salt March, Gandhi defied British control of the sale of salt, a staple in the Indian diet, by walking 240 miles from his religious retreat to the Arabian Sea to retrieve salt by hand from salt flats along the sea. During the march, which lasted from March 12 to April 5, Gandhi spoke to crowds along the route, spurring mass civil disobedience.

World War II exhausted Great Britain and the rest of Europe in both human and monetary costs, making colonizers more willing to grant independence to their colonies. When Pakistan and India became independent in 1947, a war immediately broke out in Kashmir, where the Hindu leader of a majority Muslim population in the princely state chose to integrate with India. From Kashmir to Punjab to Sindh, an estimated 14 million refugees fled across the hastily drawn border between Pakistan and India, with approximately 7 million people fleeing in each direction. In this chaotic, tragic migration, families lost homes and suffered much violence and malnutrition. When it was finally over, nearly one million people reportedly died. To this day, it remains a bitter, painful memory for many South Asians.

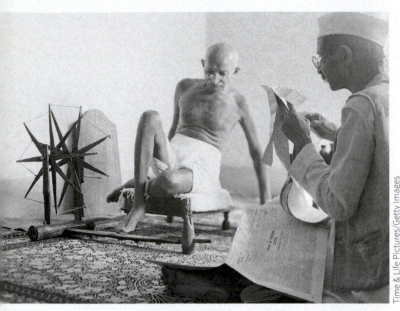

FIGURE 6.35 Mumbai, India. As one of his nonviolent movements, Gandhi encouraged Indians to spin their own thread, weave cloth, and sew their own clothing rather than buy clothing from British-owned companies.

ETHNIC GROUPS IN SOUTH ASIAN COUNTRIES

Governance in South Asia varies by country (Figure 6.36). India is the world's largest democracy; Pakistan is a kleptocracy; Nepal is an ancient kingdom that only recently ended its monarchy in 2008; Bhutan is an oasis whose government focuses on happiness; and Sri Lanka is attempting to recover from a bloody civil war that lasted 26 years. Throughout the region, ethnic groups play a role in politics both because they have a geography—a place they claim or identify with inside countries—and because they have persisted through a multitude of empires, including the British.

Bhutan

The kingdom of Bhutan protects its Mahayana Buddhist culture by limiting tourism and advocating growth in gross national happiness (GNH) over gross national product (GNP).

Bhutan has four major ethnic groups: the Ngalops, Sharchops, Drukpas, and Lhot Sampas. The Ngalops live in the western two-thirds of the country and tie their heritage to Tibet, which invaded Bhutan and introduced Tibetan Buddhism in 747 CE (Figure 6.37). The Sharchops live in

FIGURE 6.36 Political geography of South Asia. The Mughal Empire and the British Empire both colonized large expanses of the region. Most of the country boundaries were established by the British.

FIGURE 6.37   Paro, Bhutan. The Tiger's Nest Monastery sits on a cliff at an elevation of about 10,000 feet. The site first became a sacred Buddhist site during the eighth century when Buddhism was brought to Bhutan. The temple in this photo was built in 1692 and is accessible only by hiking trails, not by roads.

the remotest part of the country and see themselves as the first inhabitants of Bhutan.

The Drukpas are an indigenous population who raise livestock by moving them seasonally to snow-free regions. The Drukpas take the livestock to highlands during the summer months and lowlands during the winter months in a practice known globally as **transhumance**.

Together, the Sharchops, Ngalops, and Drukpas comprise 65 percent of the population. The remainder of Bhutan is populated by Hindu migrants from Nepal, called Lhot Sampas, who went to Bhutan for economic opportunities. Most are from lower castes and labor as agricultural workers in southern Bhutan. Many Lhot Sampas returned to Nepal during the twentieth century and now live in refugee camps in Nepal, even though their nationality is originally Nepaese.

India sees Bhutan as a buffer between itself and China. East of Bhutan is a territory claimed by both India and China. India places the border on the northern end of the region and refers to it as Arunchal Pradesh, whereas China places the border on the southern end of the region and labels it Southern Tibet (Figure 6.38).

Sri Lanka

The Tamils and Sinhalese on the island of Sri Lanka have fought one of the bloodiest civil wars in the post–World War II era. Tamils, Hindus who originally came from south India, live in northern Sri Lanka, in and around Jaffna. Sinhalese, some of whom claim to be descendants of Vijaya, a prince from north India who was banished to the island around 500 BCE, are Buddhists. Buddhism diffused to Sri Lanka from northern India around 300 BCE. The Sinhalese live primarily in the south, near Colombo, and also in the central, higher-elevation area of the country known as Kandy.

Portuguese and Dutch colonizers came to Sri Lanka for the cinnamon trade. The Portuguese traded with the Sinhalese in Colombo and attempted to gain control over the Tamils in Jaffna. The Tamils resisted twice, and in 1619, the Portuguese gained control of the Jaffna region and destroyed Hindu temples. The Sinhalese in the Kandy region also resisted

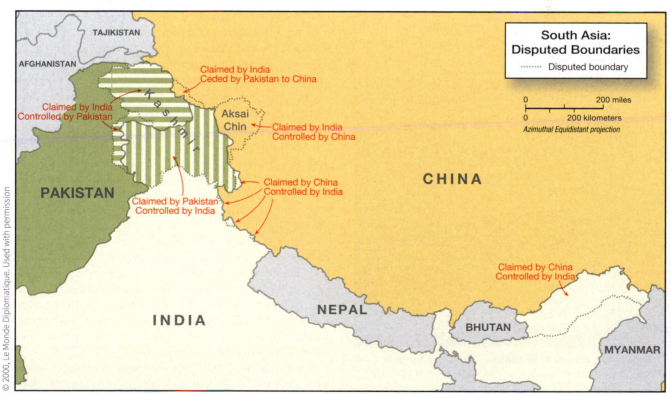

FIGURE 6.38   Disputed borders between India and China. India and China, the two most populated countries in the world, share several disputed boundaries.

colonization by the Portuguese and later the Dutch, as they saw themselves as responsible for preserving Buddhism.

The Dutch turned Sri Lanka over to the British in 1796. The British wanted the Sinhalese highlands for tea plantations, and by 1815, the British controlled the Kandy region. The British supplanted the coffee and rubber plantations in the highlands with tea plantations and then changed the demographic makeup of the island through the migration of British citizens and Tamils from India to Sri Lanka. The British migrants were looking for economic opportunity, and the Tamils from India were brought to work the newly developed tea plantations.

The British left in 1948 and the Sinhalese, who comprise 75 percent of the population, have dominated politics in the democratically elected government. The Sinhalese government denied the Tamil plantation workers (those who migrated to Sri Lanka from India during British colonial rule) citizenship in Sri Lanka. The government urged the Tamils to return to India. Those who did not became stateless. In addition, the Sinhalese government established Buddhism as the national religion and Sinhalese as the official language.

In 1983, a bloody civil war began. Led by Velupillai Prabhakaran, the Tamil rebels, who are called Tamil Tigers, were recognized as a terrorist group in 1997 by the United States. The Tamil Tigers consistently deployed suicide bombing attacks in Colombo, the home of the Sinhalese population. They recruited women as suicide bombers. And, remarkably, the Tamil Tigers raised enough money and received enough support that they had their own air force (Figure 6.39). Their goal was to split Sri Lanka into two countries.

In 2008, Tamil Tigers employed their air force to attack Colombo. They cut off electrical power to much of the city and destroyed the airport. In response to this major attack, the Sinhalese government killed Tamil leaders, including Prabhakaran.

FIGURE 6.39 Colombo, Sri Lanka. Sri Lankan Air Force Officers inspect the wreckage of a Tamil Tiger rebel aircraft after gunning it down. The Tamil Tiger rebels are one of few separatist groups that use aircraft in their campaign.

Eranga Jayawardena/ASSOCIATED PRESS

The massive offensive the Sinhalese government launched in the month leading up the end of the war in May 2009 is under scrutiny by the United Nations, human rights groups, and the global community. In one month, nearly 140,000 Tamil civilians went missing. The Tamil Tiger Rebels assert that the Sinhalese government pushed 40,000 Tiger rebels into a narrow strip of land in the north and then shelled them. The United Nations alleges that most civilian deaths caused by the government took place in no-fire zones or safe zones and hospitals. The United Nations estimates that 100,000 people died in the 26-year civil war and that a large proportion of the deaths occurred in the last month of fighting. On the other side, the Sri Lankan government argues that during the civil war the Tamil Tiger rebels abused human rights, accusing the Tamil Tiger rebels of "holding civilians as human shields, using child soldiers and killing people who tried to leave areas under their control" (BBC 2011).

Trying to rebuild the country of Sri Lanka after the end of the civil war has been an arduous process. Investigations into war crimes by both sides of the war continue, and the status of Tamils who migrated from India during British rule remains undetermined.

Nepal

Although Siddhartha Gautama, the Buddha, was born in Nepal, only 10 percent of the country's population is Buddhist. Almost all of the remaining 90 percent of Nepalese are Hindus.

During the Delhi Sultanate in India, between 1206 and 1526, groups of Hindus from India migrated to Nepal, in part to avoid paying the extra taxes the Muslim empire levied on non-Muslims. Among the Hindus who migrated to Nepal, most were Brahmins and upper to middle castes. In addition to avoiding taxes, Brahmins wanted to maintain the Hindu religion, a responsibility of their caste, and so they retreated to the Hindu kingdom of Nepal to do so. Nepalese see the kings of their country as incarnations of Lord Vishnu, and the cow is protected as it is seen as an incarnation of Laxmi, "the goddess of wealth and better half of Vishnu" (Sangraula 2006).

Over time, Hindu immigrants from India intermarried with the Nepali monarchy, which helped cement the tie between upper-caste Hindus in India and upper-caste Hindus in Nepal. Nepal was the world's only officially Hindu state until its parliament voted to end the official status of the religion in 2006.

In the 1800s, the British colonized the western third of Nepal. The British had an observer on the ground, but they did not control Nepal in terms of day-to-day activities or governance (as in India), and they did not control the eastern two-thirds of the country. During the 1857 Sepoy Rebellion, Nepal sided with the British. In return for their loyalty, the British handed some lands back to Nepal and allowed Nepal to function on its own.

Since the late 1990s, Nepal has been increasingly politically unstable. A bizarre incident in 2001 brought more instability to Nepal when the crown prince killed his father,

FIGURE 6.40 **Chitwan, Nepal.** Soldiers in the Maoist rebel group, the People's Liberation Army, assemble in their campaign against the government of Nepal. Many of the rebels are related to each other, as whole families have joined in the effort.

his mother, his siblings, several other family members, and himself. Since 2001, the country has gone through a painful period of intense political instability. A group of non-Hindu ethnic groups in Nepal has developed ties to China and now identifies as Maoists, demanding a say in government and representation (Figure 6.40). The Maoists fund their campaign by controlling the trade in *yarsagumba*, caterpillars with medicinal properties that enhance male sexual performance. Their campaign has reached about two-thirds of the country and has resulted in the deaths of over 10,000 people.

Parliament's decision in 2006 to remove the official status of Hinduism from the state was an attempt to assuage the Maoist rebels. The decision infuriated Hindu political parties in neighboring India, with the Bharatiya Janata Party (BJP), a political party controlled by Brahmins, officially denouncing the decision. The Maoist rebels continue to fight, and control of the government has changed hands and forms several times.

Pakistan

Although Pakistan has not suffered a civil war as have Sri Lanka and Nepal, the ethnic divisions

within Pakistan and the economic differences among ethnic groups in the country make governing the country a difficult task. The borders of Pakistan are arbitrary at best. The British established the border between Pakistan and Afghanistan, known as the Durand Line, in 1893 (Figure 6.41). Disagreement between Afghanistan and Pakistan on where the line should be located and the arbitrary division of ethnic groups by the line have resulted in constant instability along the Durand Line.

During the post-9/11 war in Afghanistan, the 1,500 mile border between Afghanistan and Pakistan has appeared in the news countless times. The border runs right through the homelands of ethnic groups and tribes, including the Pashtuns and the Baluchis.

In northwest Pakistan, the FATA (federally administered tribal areas) are not directly governed by Pakistan. The Swat River runs through the FATA Province. The Taliban has developed strongholds on both the Pakistani and Afghani sides of the Swat Valley. The post-9/11 War on Terror has pressured the Pakistani government to engage their military in the FATA Province, even though the Pakistanis, and before them the British, have controlled the FATA region in name only. The British allowed the tribes of the FATA to rule themselves in exchange for allegiance to Britain. The FATA then served as a buffer between British colonial possessions

FIGURE 6.41 **Pakistan ethnic groups and Pakistan borders.** The British drew Pakistan's borders with both Afghanistan and India. Neither the Durand, between Pakistan and Afghanistan, nor the Radcliffe Line, between Pakistan and India, corresponds with the locations of ethnic groups in the region.

in South Asia and Russian influence in Afghanistan, on the other side of the Durand Line. Likewise, the government of Pakistan has officially allowed the FATA to rule themselves and has historically stayed out of the region, which is a major reason why the Taliban was able to take root and continues to be able to hide in this area of Pakistan.

In the west, Baluchistan is the largest province in Pakistan but has the smallest population of any Pakistani province. The border also divided the Baluchis, and in 2005, Baluchi nationalists formed the Army of Baluchistan with the goal of seceding and establishing an independent country. The Army of Baluchistan operates as an insurgent group in Pakistan. In 2012, it claimed responsibility for a train bombing in the city of Lahore in Punjab Province.

Pakistan's eastern border, the Radcliffe Line, is as arbitrary as its western border. The British are also responsible for drawing the border between Pakistan and India. In 1947, the British Raj formally left its colonies in South Asia, but before they did, they carved out two states: one Muslim (Pakistan, which at the time included Bangladesh) and one Hindu (India). In 1945, needing to rebuild their own country after the devastation of World War II, Britain announced that it would withdraw from its colonies in South Asia.

Before withdrawing, Britain felt it needed to draw the border between the Muslim and Hindu states. British lawyer Cyril Radcliffe, who had who had not previously worked in South Asia, traveled to India, quarantined himself in a room, studied outdated maps of religion, and drew the borders between the countries. The Radcliffe Line between Pakistan and India divided Punjab and separated the Sikh cities of Lahore and Amritsar, placing Lahore in Pakistan and Amritsar in India. Radcliffe left the question of Kashmir unsettled, and in the end, left millions of Muslims in India and millions of Hindus in Pakistan, leading to the migration of 14 million South Asians.

Karachi, located in the Sindh Province of Pakistan, is the largest city in Pakistan, with a current population between 14 and 16 million people. At the time of partition, about half of the population of Karachi was Hindu and spoke Sindhi. After partition, the vast majority of the Hindus left Karachi for India. In their place, Muslims from India migrated to Karachi. The newer, Muslim residents of Karachi spoke Urdu and were called *Muhajirs*. They took over land, homes, and businesses that had been abandoned by Hindus who migrated to India.

The majority of Pakistan's exports flow through Karachi. The Sind Province has a large gap between rich and poor, with Muhajirs and Punjabis owning the land and businesses, and the Sindhi working the land. Tensions between peasants and landowners flare up frequently in Sind, while at the same time Karachi continues to attract job seekers from rural areas.

The city of Karachi is divided into zones by economic class (Figure 6.42). In the south, the most affluent neighborhoods are home to the city's wealthy elite. In the north and east,

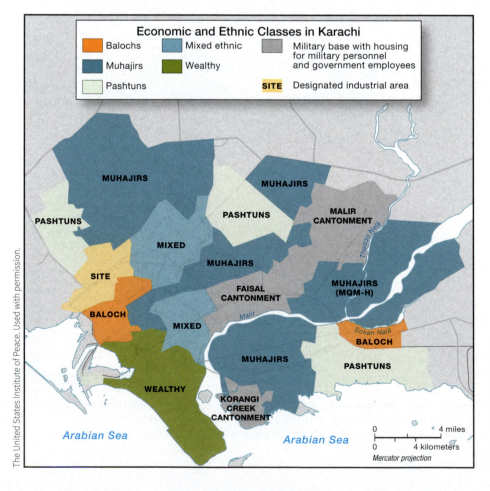

FIGURE 6.42   Karachi, Pakistan. Economic and ethnic classes in the city of Karachi, Pakistan, are largely segregated. Muhajirs comprise more than half the population of Karachi. In 2010, ethnic violence broke out between Pashtuns who migrated from Northwest Pakistan to Karachi and Muhajirs, some of whom are business owners in Karachi.

⊕ USING GEOGRAPHIC TOOLS

Combating Flooding in Bangladesh

The government of Bangladesh faces issues with severe flooding, as the country is home to 310 rivers, covering most of the country with floodplains. Bangladesh has built sea walls and coastal embankments, like this one, to combat flooding, and now the governments of the Netherlands and Bangladesh are now working together to create a comprehensive plan for Bangladesh. The plan, called Delta Plan 2100, will replace three separate water management plans now at work in Bangladesh.

Planning, including environmental, transportation, urban, and regional programs, is a part of many geography departments in the United States. Creating a comprehensive plan for Bangladesh that recognizes the interwoven nature of the 310 rivers and the Bay of Bengal as well as the ramifications of a solution in one part of the country for another part of the country will help Bangladesh as it stares down not just cyclones, as the people in this photograph are, but also rising sea levels from climate change.

ASSOCIATED PRESS

FIGURE 6.43   Chittagong, Bangladesh. A group of worried villagers wait for their family members who have gone fishing at a coastal embankment. A cyclone roaring in from the Bay of Bengal whipped up walls of water, flooding low-lying areas along the coast.

Thinking Geographically

1. Look at maps of polders in the Netherlands, the Venetian lagoon in Italy, and the coastline of Bangladesh. Compare and contrast the issues facing each coastline.

2. Bangladeshis are used the rhythm of flooding during the monsoon. How will climate change affect the ability of Bangladeshis to cope with flooding?

Karachi's middle class lives in a dozen city centers, similar to **edge cities** in North America, where businesses and shopping districts are clustered at major intersections. Slums are scattered between the middle-class zones, and around the city in the east, north, and northwest is a newer belt of squatter settlements, which are home to rural migrants.

Urdu, the language of the Muhajirs, became the official language of the newly formed Pakistan, although only 8 percent of the country speaks the language. Urdu is the language, in addition to English, commonly used among educated Muslims in South Asia since the Mughal period. The Muslim League used Urdu in their campaign. The vast majority of East Pakistanis (Bangladesh since 1971), however, spoke Bengala, and nearly half of West Pakistanis (Pakistan today) spoke Punjabi at partition.

Punjabi remains the most commonly used language in Pakistan. The Punjab Province is home to Pakistan's largest ethnic group and is also the most economically productive province of Pakistan. Most of Punjab's economic production comes from agriculture, as the tributaries of the Indus and the Indus itself deposit nutrient-rich soils. Farmers grow wheat, rice, cotton, and sugar on the *doabs*, the lands between the rivers.

The Punjab region also includes Pakistan's capital city of Islamabad, which was designated the capital more than a decade after Pakistan became independent. As a result,

the Punjab Province receives the economic flow that comes with government jobs found in a capital city.

Bangladesh

The Netherlands, which spent 1.07 percent of its gross national income on foreign aid in 2013 (the United States spent 0.19 percent and Canada spent 0.27 percent) has close ties with Bangladesh. The Netherlands contributed 58 million euros to Bangladesh in 2008 alone. According to the Dutch Minister of Foreign Affairs, the Netherlands government directly funds 15 NGOs that operate within Bangladesh, with a particular focus on alleviating poverty in both rural and urban areas. The Netherlands also uses its experience with land reclamation to help build infrastructure to protect Bangladesh from rising sea levels (Fig. 6.43).

CONCEPT CHECK

1. **How** did the British handle the princely states during the British colonial era?

2. **Why** are there Nepalese refugees in Bhutan, and what is their status in the country?

3. **How** do the high levels of corruption in Pakistan affect the government's ability to govern?

WHAT IS SOUTH ASIA'S ROLE IN THE WORLD ECONOMY?

LEARNING OBJECTIVES

1. **Explain** what outsourcing is and what role India plays in global outsourcing.

2. **Discuss** why uneven development is pervasive in South Asia and how it is reflected in cities.

3. **Describe** the Green Revolution and the controversy around it.

Globalization has helped bring NGOs to Bangladesh, an information technology revolution to India, and a new agricultural revolution to rural South Asia. The outcomes of economic development in South Asia are uneven. Software technology parks of India (STPIs) are located in cities throughout India, but the information technology sector employs only about 3 percent of the people. In rural India, the diffusion of the Green Revolution has intensified agriculture but has also made farmers more dependent upon seed and chemical companies. In order to bring development to more of its people, the government of Bhutan has developed a strategic plan based on improving gross national happiness rather than gross national product (GNP).

GROSS NATIONAL HAPPINESS

Unlike India, where Bangalore and Hyderabad are at the forefront of globalizing technologies, the monarchy of Bhutan banned Internet and television from the country until 1999. The monarchy of Bhutan works to maintain the Mahayana Buddhist culture in the country, requiring people to dress in traditional ways and maintaining tight controls over tourism. Tourists are required to use guides and visit prescribed places. At the same time, the government of Bhutan has asked the people to recognize that growing the country's gross national product (GNP; see Chapter 2) should not be a central goal of government. Rather, since 2009 the government of Bhutan has rated all government policies on whether the policy or action promotes growth in gross national happiness (GNH).

The monarchy argues that in order to keep the traditional Buddhist culture of Bhutan thriving, the country and its people must focus on growth in happiness rather than economic growth (Figure 6.44). In 2008, the monarchy rescinded executive control of the government and

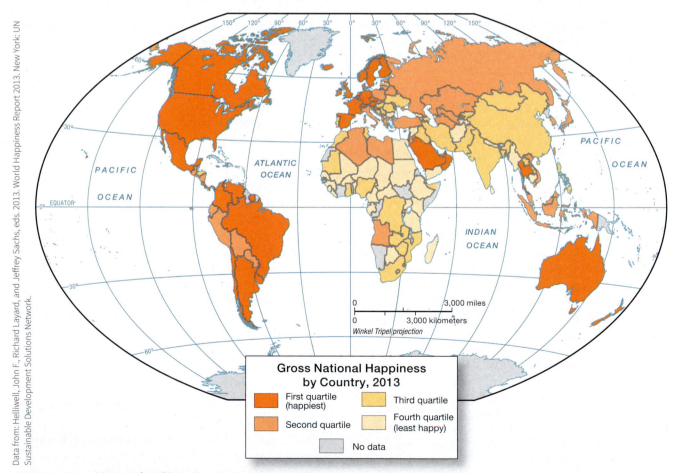

Data from: Helliwell, John F., Richard Layard, and Jeffrey Sachs, eds. 2013. World Happiness Report 2013. New York: UN Sustainable Development Solutions Network.

Gross National Happiness by Country, 2013

- First quartile (happiest)
- Second quartile
- Third quartile
- Fourth quartile (least happy)
- No data

FIGURE 6.44 **Gross national happiness by country.** Bhutan created an index of gross national happiness (GNH) using nine domains including psychological wellbeing, health, good governance, and ecological diversity. This map divides countries into quartiles, four equal size classes. Compare and contrast the map of GNH with a map of GNI.

established a parliamentary democracy. The monarchy is still in place, albeit in a different capacity.

At his coronation in 2008, the king of Bhutan explained that change is coming quickly to Bhutan and that happiness must continue to be the country's central goal, "Henceforth, as even more dramatic changes transform the world and our nation, as long as we continue to pursue the simple and timeless goal of being good human beings, and as long as we strive to build a nation that stands for everything that is good, we can ensure that our future generations for hundreds of years will live in happiness and peace."

OUTSOURCING

While Bhutan pursues growth in GNH, most of South Asia has become an integral part of the world economy, especially information technology, software engineering, and consulting. In the 1980s, the Indian government set the goal of establishing an export-oriented computer and software industry.

India's central government invested in the information technology (IT) industry in the 1990s. In 1991, India's central government provided economic incentives to draw export-oriented IT firms to newly established software technology parks of India (STPIs), located in 40 designated urban areas. Economic activity in India's STPIs grew quickly during the 1990s. In addition to the economic incentives provided by government, another major factor that attracted global IT firms to India was the labor supply. Thousands of Indians have college degrees, and they also have very strong English communication skills. University classes in India are offered in English, and thousands of other Indians have graduated from universities in the United Kingdom and the United States with degrees in computer software engineering.

The timing of the STPIs also helped spur their growth. Established in 1990, the STPIs were ready to help companies avert the looming Y2K disaster. In the 1990s, companies recognized that where their software had dates shortened to two digits (for example, 1978 was listed in the software as 78), it could malfunction when the date rolled to the year 2000. With the year 2000 looming, companies and governments feared widespread disaster (Y2K). Companies hired Indians in STPIs to patch software and thereby averted the crisis. The STPIs then created an agglomeration effect (see Chapter 2) whereby industries that serve IT firms clustered around STPIs to develop high-tech regions of cities (Figure 6.45).

The growth of STPIs in the 1990s marks the beginning of the **outsourcing** era. Under the original model of outsourcing, global companies outsourced lower-paying jobs to India, creating a number of decent-paying jobs for Indians who worked directly for Western companies, primarily in call centers. In this original model of outsourcing, a specific function or task of an international company was delegated to its employees in India.

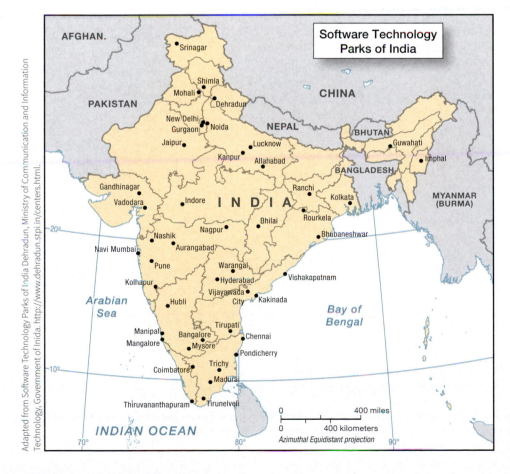

FIGURE 6.45 Sites of India's software technology parks (STPIs) funded by the government.

Outsourcing has markedly changed since the original model. India is no longer the world's call center. In fact, in 2011, the Philippines surpassed India with 350,000 call center employees compared to 330,000 in India. Instead of focusing on call centers, Indian entrepreneurs have established Indian companies with expertise in **global outsourcing**. Today, a global company can outsource entire segments of high-paid development jobs to Indian companies, which, in turn, outsource the lowest-paying, lowest-skilled jobs to the Philippines, Mexico, eastern Europe, and China.

For example, Infosys, an Indian company, has expertise in global outsourcing and draws from contacts around the world to find the right group to complete any task for which it is contracted. For instance, an American bank needed a website in Spanish. The bank hired Infosys in India, and Infosys went to Mexico and hired employees to create the website. As the *New York Times* reported, "Such is the new outsourcing: A company in the United States pays an Indian vendor 7,000 miles away to supply it with Mexican engineers working 150 miles south of the United States border" (Giridharadas 2007).

With global outsourcing, India is now home to a larger part of the commodity chain and increasingly to the portions of the commodity chain that generate the most wealth for people and places. Instead of having a few thousand Indians employed in call centers, entire Indian-owned companies are engaging with the global economy to create business solutions and to employ educated Indians. Infosys has over 75,000 Indian employees in India, and across its brands, Tata Corporation has more than 350,000 employees in India.

UNEVEN DEVELOPMENT

The impacts of globalization have not been even. Globalization has brought more of the world into the formal economy. Companies produce goods that are exported or counted and taxed by the government, and these goods are part of the **formal economy**. At the same time, other goods are not counted or taxed by government; they are part of the **informal economy**. The informal economy is what enables more than 2 billion people to live on less than $2 a day. In the informal economy, goods produced or services rendered are exchanged with others under the radar of the formal economy.

Although the informal economy is a thriving part of the global economy, it is not measured by GNP, GDP, or gross national income (GNI). The World Bank issued a report in 2010 authored by economists Friedrich Schneider, Andreas Buehin, and Claudio Montenegro that measured the informal economy of 162 countries (Figure 6.46).

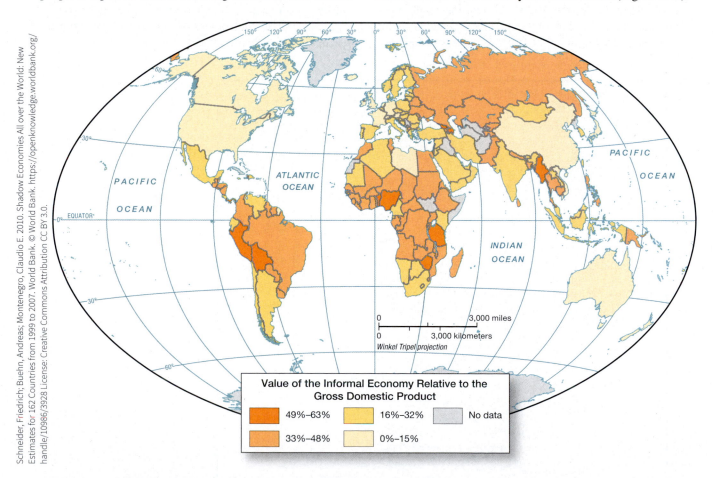

Schneider, Friedrich; Buehn, Andreas; Montenegro, Claudio E. 2010. Shadow Economies All over the World: New Estimates for 162 Countries from 1999 to 2007. World Bank. © World Bank. https://openknowledge.worldbank.org/ handle/10986/3928 License: Creative Commons Attribution CC BY 3.0.

Value of the Informal Economy Relative to the Gross Domestic Product

- 49%–63%
- 33%–48%
- 16%–32%
- 0%–15%
- No data

FIGURE 6.46 **Informal economy as a percentage of gross domestic product (GDP).** Based on econometric analysis, the World Bank estimates the value of the informal economy in 162 countries as a percentage of the country's official GDP.

In South Asia, the range as a proportion of the GDP is from 20.7 percent in India to 42.2 percent in Sri Lanka. In Pakistan, the informal economy is equivalent to 33.6 percent of the country's GDP.

Slums

Poverty in South Asia is most readily apparent in the teeming slums in its major cities, which are home to over 100 million in India, 30 million in Bangladesh, and 27 million in Pakistan (Figure 6.47). Sixty percent of the people living in Mumbai, India, 40 percent of the people in Dhaka, Bangladesh, and one in every two people in Colombo, Sri Lanka, live in slums. The World Bank estimates that in the entire region, "one in every four people is categorized as under 'informal population' or living in shanties or slums in the urban areas" of South Asia.

Hutments, the common name for slum dwellings in South Asia, are typically built in low-lying areas of cities and are therefore susceptible to flooding, especially during the monsoon. The dwellings and narrow alleys and walkways seem to be as densely populated as beehives. Mumbai's most famous slum, Dhavari, which was the setting of the film *Slumdog Millionaire*, has a population density of 1 million per square mile. Compare this to the population density of Mumbai as a whole, which is 12,698.84 people per square mile.

Hutments are either built on land to which the resident does not have title or legal right or in settlements that do not comply with planning and building regulations. If you visited a hutment in a slum like Dhavari, you would not only see people living there, but during the day you would likely see a small industry trying to make it big. Television commercials for Western charities show the poor standing around with empty bowls and distended stomachs,

© Andrew Parsons/I-Images/ZUMA Press/Corbis

FIGURE 6.48   Mumbai, India.   Two young male workers sleep after working on sewing machines making items of clothing in a hutment factory in the Dharavi slum in Mumbai, which is one of the largest slums in the world.

just waiting for help. In reality, Dhavari and other slums, whether in South Asia or South America, are teeming with people who work day and night to feed their families and improve their economic condition. Estimates are that Dhavari is home to 15,000 **hutment factories** consisting of "one or two jerry-built storeys [stories], stuffed with boys and men sewing cotton, melting plastic, hammering iron and moulding clay" (*The Economist*, December 19, 2007) (Figure 6.48).

GREEN REVOLUTION

Agricultural goods produced for self-consumption or sold in local markets are part of the informal economy because goods are exchanged for other goods or for cash and are not taxed. While much of the agricultural production in South Asia is still part of the informal economy, the Green Revolution, which seeks to increase yields and feed more people, has brought larger swaths of rural South Asia into the formal economy.

The hearth of the **Green Revolution** is in North America, where agricultural researchers, seed companies, and philanthropists sought a technological solution to feeding the world. In the 1940s, American scientist Norman Borlaug conducted experiments in Mexico and developed a disease-resistant strain of wheat. The Ford Foundation and the Rockefeller Foundation funded research in other crop varieties, including rice.

AFP/Getty Images

FIGURE 6.47   Colombo, Sri Lanka.   Residential slums in the foreground contrast with the housing of the wealthier and modern office buildings in the background.

 Green Revolution   intensified agriculture that uses engineered seeds, fertilizers, and irrigation to increase intensive agricultural practices.

In the 1960s, Borlaug's wheat and the Green Revolution diffused to India. Borlaug conducted research on rice in India and worked with researchers in the Philippines to develop a new strain of rice, IR8, which produced more rice grains per plant. To produce the high-yield varieties of crops, farmers needed access to the seeds and fertilizers, as each crop variety depended on specific fertilizers for growing. By engineering plant varieties that create more or larger grains, that respond well to fertilizers, and that grow on land previously considered marginal for that variety, Borlaug and the Green Revolution intensified agriculture in the developing world.

Farmers who use Green Revolution crops operate in the formal economy and have to part with a lot of cash each growing season. They need to purchase specific seeds, most of which are bioengineered by the seed companies so that farmers cannot preserve seeds for the next year's planting. Not only do farmers have to purchase new seeds each planting season, but also Green Revolution seeds require fertilizer. Farmers purchase and apply fertilizers and tap into water supplies for irrigation in order to grow their crops.

Because of the cash required, Green Revolution seeds and technology are more accessible to middle- and higher-income farmers than to low-income farmers in the developing world. Small-scale subsistence farmers who operate primarily in the informal economy cannot afford the seeds or the technologies to implement them. Middle- and higher-income farmers who can afford the technologies often become financially indebted to seed and chemical companies, trading in this year's crop for next year's seed and fertilizer.

Before the Green Revolution reached India, farmers collected seeds from the annual harvest and preserved them for the next year's crop. In adopting the Green Revolution, farmers and agricultural villages lost centuries-old **seed culture** whereby generations had hand selected the seeds best suited to grow in local conditions. Environmentalists argue that the lost seed culture was a better system than the Green Revolution because it prevented soil erosion and relied on organic fertilizers rather than chemical fertilizers.

Although rice yields have increased markedly since the Green Revolution began in India in the 1960s (Figure 6.49), rice yields in India have fallen since 2008. Scientists who are studying why India's rice yield has leveled off or declined are looking to the fertilizer *urea* and its widespread use in India as the cause. The Indian government has subsidized urea since the 1980s to make it affordable to farmers. Farmers in India have unwittingly followed a "more is better" approach, and now the overuse of fertilizer is depleting the soil. The *Wall Street Journal* reported: "Like humans, plants need balanced diets to thrive. Too much urea oversaturates plants with nitrogen without replenishing other nutrients that are vitally important, including phosphorus, potassium, sulfur, magnesium and calcium" (Anand 2010).

Weighing famine against feeding the world, the Green Revolution looks to be a great success. However, the Green Revolution is not without controversy. One of the world's most vocal opponents of the Green Revolution is Indian Vandana Shiva, who argues that the Green Revolution has benefited global agricultural companies, including agrochemical makers and seed companies, more than it has benefited farmers in India. She contends that the diffusion of the Green Revolution has displaced small farmers, created water shortages, amplified soil erosion, led to reduced genetic diversity, and made crops increasingly vulnerable to pests.

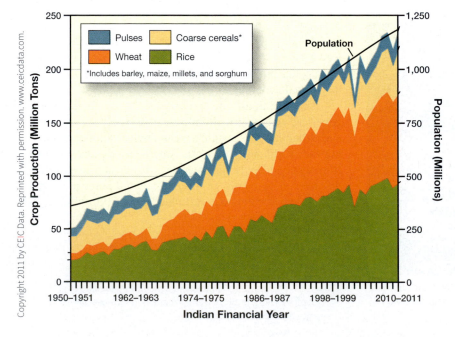

FIGURE 6.49 Crop yields and population growth in India, 1950 to 2011. Crop yields have risen over sixty years, along with India's population. Advocates of the Green Revolution credit the crops and farming methods from that revolution with producing enough to feed the growing population. Critics of the Green Revolution contend that Green Revolution is unsustainable and yields are dropping with overuse of soil and costly investments to maintain returns.

Despite these social, economic, and environmental impacts, high-yielding varieties of crops coupled with irrigation and fertilizers doubled cereal grain production between 1970 and 1995 in Asia. Proponents of the Green Revolution point to increased output and contend that feeding more people outweighs any costs or consequences.

The Green Revolution has increased agricultural output, but the long term consequences of agricultural intensification for people and the environment remain to be seen.

CONCEPT CHECK

1. **What** is outsourcing, and what role does India play in global outsourcing?

2. **How** do slum dwellers in South Asia fit into the formal and informal economies?

3. **How** did the Green Revolution increase rice yields in India, and what is controversial about the Green Revolution?

SUMMARY

What Is South Asia?

1. South Asia is a subcontinent of the Eurasian landmass. The Himalaya Mountains stretch along the northern edge of the subcontinent, and Tibet is to the north of the Himalayas. The subcontinent is still colliding with the Eurasian landmass, which is causing the mountains to continue to rise. As a result, among the Himalayas and the Karakoram Range, the 15 tallest mountains in the world are located.

2. A monsoon is a prevailing wind from a predominant direction over the course of six months. The wind direction shifts the other six months of the year, as the Intertropical Convergence Zone (ITCA) migrates with the direct rays of sun from south to north during the course of the year. The monsoon climate is located in coastal India and Bangladesh (Figure 6.3), but the entire region of South Asia is impacted by the monsoon. Most of the water flowing through the rivers of South Asia starts as rainfall during the wet monsoon.

3. Despite political differences and brief wars over the Kashmir region, both India and Pakistan continue to honor a treaty on the Indus River system that the World Bank helped negotiate in the 1960s. Although India helped Bangladesh in its independence movement against Pakistan in 1971, the two countries have not reached an agreement on the use of the Ganges–Brahmaputra River system.

4. Tropical cyclones (hurricanes) are low-pressure systems that form over warm oceans. The tropical cyclones that impact South Asia form in the Indian Ocean and typically track toward Bangladesh. Even without cyclones, about one-third of Bangladesh floods each year because of the country's low elevation and the dominance of river floodplains in the landscape. The physical geography of Bangladesh makes the country particularly vulnerable to tropical cyclones. In addition, women are especially vulnerable as a result of gendered eating practices.

Who Are South Asians?

1. The Indus civilization was the first civilization in South Asia. Archaeologists have found evidence that everyone had access to similar housing and likely to common bath houses in the cities of the Indus. The social structure of the Indus civilization was most likely demonstrated through clay circles that carry seals or insignias.

2. The caste system likely had its origins in the seemingly egalitarian society of the Indus civilization. The caste system uses four broad castes in addition to scheduled castes (Dalits) to define the place in society to which South Asians are born. Caste dictates who people can marry, what their livelihood can be, and where people can live. Caste is far more complicated than these categorizations and dictates, however. Including subcastes, India alone has more than 25,000 groups or castes into which a person can be born.

3. South Asia is the hearth of two major world religions: Hinduism and Buddhism. Additionally, Islam took root in South Asia as early as AD 700, and from South Asia, Islam diffused into Southeast Asia. South Asia is also the hearth of the Sikh religion, which has a relatively strong presence in India and in the United Kingdom.

4. The preference for boys in South Asia is profoundly evident in India, especially when examining the second and third births in a family. The 2011 Indian Census found that the sex ratio for the country is only 914 females for every 1000 males.

What Role Does Ethnicity Play in the Politics of South Asia Today?

1. The British directly controlled only about 60 percent of the subcontinent during British colonialism. The remaining 40 percent of the region was divided among princely states, each of which derived its wealth through landholdings and payments called privy purses from the colonizers.

2. Although it is the hearth of Buddhism, Nepal is a majority Hindu country and was the world's only officially Hindu country until 2006. Nepal sided with the British during the 1857 Sepoy Rebellion, which the British rewarded by allowing Nepal to remain relatively independent during the colonial era.

3. Pakistan is hampered by two arbitrary borders, neither of which is well defined or stable. The border between Pakistan and Afghanistan divides the Pashtun people and includes the FATA region of Pakistan. The country does not directly govern the Baluchistan region of Pakistan, which has a growing separatist movement.

What Is the Role of South Asia in the World Economy?

1. Bhutan chooses to pursue growth strategies focused on improving the gross national happiness of the country rather than its gross national income (GNI).

2. Although South Asia, specifically India, is often perceived as the call center of the world, outsourcing in India has moved far beyond being the world's call center. Indian-owned companies are now global leaders in outsourcing, and companies worldwide rely on and contract with Indian companies to meet business goals.

3. Economic development is uneven in South Asia. Slums, teeming with people, are the most visible manifestation of uneven development in the cultural landscape. People living in slums build hutment factories to earn money producing goods.

4. The hearth of the Green Revolution was in North America, but the impact of its diffusion in South Asia has been vast. The Green Revolution brings farmers who purchase seeds and fertilizers from large agrichemical companies into the formal economy and has replaced the traditional seed culture in many places.

YOUR**TURN**   Fieldwork in Geography

This photo from Maloya Village, India, shows women in a field with mounds of cow dung. They sell the cow dung to be used as cooking fuel or as fertilizer instead of urea, a nitrogen-based fertilizer that is heavily subsidized by the Indian government and is being overused on farms throughout the country. The recommended balance between nitrogen and potassium is 4 to 1, but in the neighboring State of Haryana, where urea is more heavily used, farmers are using nitrogen at a ratio of 32 to 1. The overuse of nitrogen depletes the soil, but urea is so inexpensive that farmers continue to use more.

Thinking Geographically

• Based on the women's work with cow manure, where do you think they fit into the caste system in India?

• Between 2004 and 2013, food prices rose 157 percent in India, outpacing inflation of other items. How will rising food prices and increasingly depleted soil affect farmers and consumers in India?

Read:
Anand, Geeta. 2010. "Green Revolution in India Wilts as Subsidies Backfire." *Wall Street Journal*. February 22.

© AJAY VERMA/Reuters/Corbis

Maloya Village, India.   Women dry paddies of cow dung.

Varma, Subodh. 2013. "Food Prices Rose 157% Between 2004 and 2013." *Times of India,* September 20. http://timesofin-dia.indiatimes.com/india/Food-prices-rose-157-between-2004-and-2013/articleshow/22777817.cms

KEY TERMS

subcontinent	ablutions	ghost schools	outsourcing
monsoon	caste system	madrasas	global outsourcing
subsolar point	Rig Veda	kleptocracy	formal economy
erosion	stupas	mercantilism	informal economy
deposition	avatar	Sepoy Rebellion	hutment factory
low birth weight	ethnic religion	princely states	Green Revolution
functional zonation	universalizing religion	transhumance	seed culture
urban morphology	microcredit lending	edge cities	

CREATIVE AND CRITICAL THINKING QUESTIONS

1. If you visited South Asia during the wet monsoon, what could you look for in the **cultural landscape** that would tell you the people welcome the monsoon rains?

2. Climate scientists contend that as temperatures rise during the **Anthropocene**, earth's atmosphere will have more water vapor. Consider how this may affect the monsoon.

3. The vulnerability of places and people to tropical cyclones (hurricanes) depends on the **scale** of analysis. Determine who, at the global, regional, and local scales, is most vulnerable to tropical cyclones.

4. This chapter describes the **cultural landscape** of the Indus city of Mohenjo-Daro. Look at the images of this Indus city and others on www.harappa.com. Then, using Google Earth, study the present urban morophology of Mumbai, India. Compare and contrast the cities in the ancient and modern civilizations.

5. Consider whether the caste system is a model of **unequal exchange**. Defend your answer.

6. South Asia is the **hearth** of both Hinduism and Buddhism. Choose two of the following countries: Bhutan, Nepal, India, or Sri Lanka. Discuss whether the country is predominantly Hindu or Buddhist today, describe the routes of **diffusion** of Hinduism and Buddhism into this country, and predict whether the religion will remain the same in 100 years.

7. Preference for male children is strong in much of India. Demographers are surprised because historically as **development** builds in a country, the preference for boys declines. However, India's preference for males actually rose in the 2011 census, especially for second- and third-born children. Explain why this **gender** preference remains so strong in India.

8. Describe the geography of **identities** in Pakistan, drawing a map and labeling each major ethnic group. Then, add to your map the prevailing ethnic **identities** found across Pakistan's borders in India and Afghanistan. Determine whether and how Pakistan can build an **identity** for Pakistanis that coincides with Pakistan's borders. If you governed Pakistan, how would you attempt to build a Pakistani **identity**?

9. India is the **hearth** of the modern outsourcing model. What **networks** have Indian companies created to capture more of the wealth generated in the **commodity chain** of outsourcing?

10. Assess whether the hutment factories in India's slums help alleviate or contribute to the problem of **unequal exchange**. Determine whether more successful hutment factories or more global corporations providing jobs in Mumbai would better advance the **development** of the poor in India.

SELF-TEST

1. South Asia is called a subcontinent because it:
   a. was once separate from the Eurasian plate and collided with it 60 million years ago
   b. is such a large peninsula that it is considered a subcontinent
   c. is at the southern edge of the Eurasian plate
   d. is south of the Tropic of Cancer

2. If you traveled in South Asia in the summer months, you could see evidence of the summer monsoon in all of the following **except**:
   a. crops planted to take advantage of the rainfall
   b. drainage grates in courtyards filled with water
   c. days of mourning or sorrow marking the first rains of the monsoon season
   d. pedestrians who seek shelter under the canopies in shopping districts

3. In 1960, India and Pakistan signed the Indus Waters Agreement. All of the following were motivations for both countries to sign **except**:
   a. Farmers in the Indus region in both India and Pakistan rely on water from the Indus and its tributaries for irrigation
   b. Pakistan wanted to build an enormous dam, and India did not want the dam built
   c. The World Bank promised to fund building dams and irrigation systems for both countries if they agreed
   d. The region has a semiarid climate, and neither country can afford to have the water depleted by the other

4. When India and Pakistan gained independence in 1947, Bangladesh was:
   a. an independent country called Bangladesh
   b. part of Pakistan called East Pakistan
   c. part of India called East Indies
   d. an independent country called Burma

5. Bangladesh is susceptible to flooding from the wet monsoon and cyclones. Aside from the low elevation and high poverty, researchers found high death rates because of:
   a. discriminatory eating practices that result in women being undernourished
   b. building practices whereby people build homes on islands in rivers
   c. the irrigation system of dams and dikes on the Brahmaputra River
   d. a cyclone warning system that is antiquated and typically fails

6. Which of the following pairs of predominant religion and country is correct?
   a. Sri Lanka and Sikhism
   b. Pakistan and Islam
   c. Bhutan and Hinduism
   d. Nepal and Buddhism

7. The Sepoy Rebellion took place in 1857 outside of Delhi, India. After the rebellion ended, the situation changed in South Asia because:
   a. the British East India Company began to colonize South Asia
   b. the ruler Akbar pushed the British out of South Asia
   c. the famous Indian leader Gandhi began a campaign to carve out a Muslim state, Pakistan, in South Asia
   d. the British government began to directly colonize South Asia instead of using the British East India company as its proxy

8. True or False
   The Pakistani government is described as a kleptocracy. One way this is evident is in the ghost schools where money flows to the school, but the school does not operate.

9. The British did not physically colonize the princely states of India. They gained the loyalty of the princes through:
   a. military intimidation
   b. assessing them taxes and giving them a cut of the tax
   c. paying them with annual privy purses
   d. converting them to Christianity

10. Which of the following rebel or separatist groups is paired with the correct state in which they are seeking a say in government?
    a. Maoist rebels in Sri Lanka
    b. Tamil Tiger rebels in Nepal
    c. Baluchi separatists in Pakistan
    d. Sindhi separatists in Bangladesh

11. When Radcliffe drew the border between India and Pakistan, which of the following did he do to determine where the border should go?
    a. He walked the border region
    b. He stayed in a room and studied outdated maps
    c. He surveyed people in the border region
    d. He chose features in the physical geography, including rivers

12. True or False
    India is the leading call center in the world.

13. The cultural landscape of the software technology parks of India (STPIs) reflect India's role in the current model of outsourcing in:
    a. the numerous schools that teach call center employees how to speak English with an American accent
    b. the headquarters and offices of Indian corporations, including Tata and Infosys
    c. the presence of American corporations, including Apple and Microsoft, who lead the world in outsourcing expertise

14. The 2 billion people who comprise the global poor and live on less than $2 a day conduct most of their economic activity in the:
    a. informal economy
    b. formal economy

15. Which of the following is **not** an impact of the Green Revolution in South Asia?
    a. The seed culture has been lost in many rural areas
    b. Fertilizers, including urea, are being overused in much of India
    c. More people are being fed by higher yielding crops
    d. Farmers who have adopted Green Revolution practices continue to farm in the informal economy

ANSWERS FOR SELF-TEST QUESTIONS

1. a, **2.** c, **3.** b, **4.** b, **5.** a, **6.** b, **7.** d, **8.** True, **9.** c, **10.** c, **11.** b, **12.** False, **13.** b, **14.** a, **15.** d

SOUTHEAST ASIA

The sweeping views of the Indian Ocean out the window of the suite invite you to walk the groomed sand beach of the InterContinental Bali Resort, Indonesia. Crystal clear water, lush vegetation, and four- and five-star resorts attract tourists and provide employment for locals. However, many economists contend that little of the wealth deposited by tourists goes to the locals. Tourists spend most of their money on flights, hotels, and meals, which tend to be paid to large companies, such as hotel chains InterContinental, Hilton, and Sheraton. Locals work in the hotels, restaurants, and tourist destinations, but wages paid in the tourist industry are among the lowest in the formal economy.

Tourism also brings an onslaught of Western culture and ideals. Bali is a small island in the midst of the Indonesian archipelago. It is part of Indonesia, but its Hindu culture, dating back to the eleventh century, sets it apart from the rest of the predominantly Muslim country. The distinct culture and beauty of Bali first attracted European tourists in the early twentieth century. The Russian-born German artist Walter Spies helped create the Western perception of Bali through his paintings. The first tourist hotel in Bali dates to 1926, and Spies took up residence in Bali in 1927.

Tourism increased dramatically in Bali in the 1970s, and its people have worked to preserve and protect their culture despite the onslaught of tourists. In turn, Bali's traditional culture has remained relatively intact. The Western-style night life is confined to the Kuta tourist district, where pubs, clubs, and budget accommodations draw surfers and those looking for fun (Figure 7.1).

The region of Southeast Asia has been balancing the protection of local culture and the engagement with the broader world for more than 1,500 years since the region's spices first drew traders from East Asia, South Asia, Southwest Asia and North Africa, and just 500 years ago, Europe.

Courtesy of INTERCONTINENTAL BALI RESORT

Bali, Indonesia. The InterContinental Bali Resort is a five-star hotel built in 1993 and renovated in 2011. The InterContinental Bali Resort group includes 183 hotels. The photos on the InterContinental Bali Resort website feature the pristine beaches, the pools in this photograph, and the unique Hindu local culture of Bali. On the hotel's website, one photo of the beach, pools, and hotels is overlaid by the text "suspended between heaven and earth" and a photo of local women features the message "honoring local cultural traditions."

THRESHOLD CONCEPTS in this Chapter

Tourism
Authenticity

FIGURE 7.1   Bali, Indonesia.   The Kuta district in Bali is the destination of many Western tourists who visit the nightclubs and pubs on Legian Street. The street is lined with shops, restaurants, and nightclubs featuring DJs and dancing.

WHAT IS SOUTHEAST ASIA?

LEARNING OBJECTIVES

1. **Explain** how mountains and rivers are distributed in Southeast Asia.

2. **Understand** the importance of the Mekong River system to the region.

3. **Determine** why natural disasters occur in the region and **understand** what places are most susceptible to natural disasters.

Aromatic clove trees line the coast of the Maluku Islands in eastern Indonesia. The grouping of more than 1,000 islands was once known as the Spice Islands and was at the center of a war among the British, Dutch, and indigenous Bandanese in the seventeenth and eighteenth centuries. Cloves, nutmeg, and other spices are native to the rich volcanic soils of Southeast Asia and were valued more than gold at the time. The region of Southeast Asia and its famed spice islands have attracted traders from around the world since the time of the Han Dynasty in China and the Roman Empire in Europe.

In this section of the chapter, we will gain an understanding of the physical geography of Southeast Asia, including how the thousands of islands and belts of mountains formed; the shifting monsoon winds that moved sailing ships for centuries; and the plate boundaries that resemble a jigsaw puzzle and pose a hazard in the region.

PHYSICAL GEOGRAPHY

Southeast Asia has two distinct physical geographies: one comprising thousands of islands in the Pacific Ocean and the eastern Indian Ocean and the other on the mainland of the Asian continent (Figure 7.2).

Mainland Southeast Asia is an extension of the Eurasian continent that includes the long, narrow Malay Peninsula to its south, including all of Laos, Cambodia, Vietnam, and Myanmar (Figure 7.3). Thailand includes part of the Malay Peninsula. Malaysia straddles the mainland and islands, as it incorporates the southern end of the Malay and a portion of the island of Borneo to the east. Singapore is at the tip of the Malay Peninsula, on the Strait of Malacca, a strategic waterway that separates the peninsula from the island of Sumatra (Indonesia).

The islands of Southeast Asia include the countries of Brunei, a tiny but incredibly wealthy country that shares Borneo with Malaysia and Indonesia; Indonesia, which is composed of more than 13,000 islands; the Philippines, which is made up of more than 5,000 islands; and the newest country in the region, East Timor, which gained its independence in 2002.

Mainland Southeast Asia is a series of cordilleras, plateaus, and rivers. Tectonic activity uplifted the region 50 million years ago, following the collision of the Indian subcontinent with the Eurasian plate. *Cordilleras* are parallel mountain ranges that run the same direction as one of the plate boundaries in the region. The boundary between the Indian plate and the Eurasian plate runs north–south between Myanmar and Thailand, forming the Tannen Range and the Bilauktaung Range. The Luang Prabang Range runs

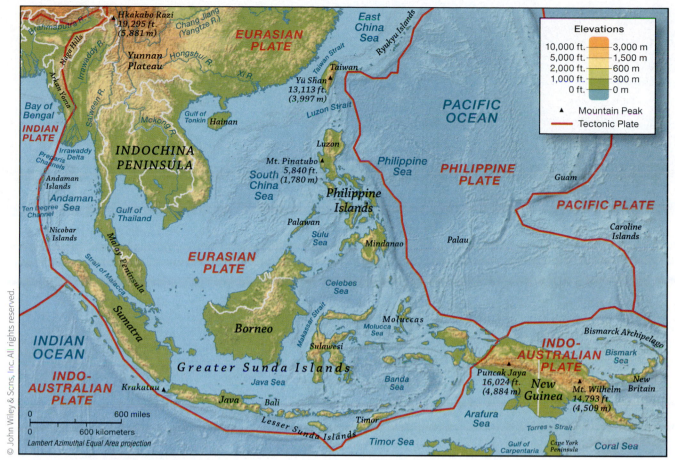

FIGURE 7.2 **Physical geography of Southeast Asia.** Water is a dominant theme in the physical geography of Southeast Asia. Mainland Southeast Asia has hundreds of miles of coastland, and even in the interior massive river systems, including the Mekong, direct life in Southeast Asia toward water. The thousands of islands of Southeast Asia are volcanic arcs formed along plate boundaries with highlands in the interior surrounded by miles of coastlines.

north–south, dividing Thailand and Laos. The Annam Cordillera is inland from the coast but parallels the coastline of Vietnam, dividing Vietnam from Laos and Cambodia.

Several major rivers cut between the mountain ranges in mainland Southeast Asia, forming plateaus, river basins, and deltas. From west to east across the mainland, the Irrawaddy flows through Myanmar, forming a wide plain and delta in the south. The Salween cuts deep through the mountains of China and eastern Myanmar and then forms a delta in Myanmar where it reaches the sea. The Ping and Yom flow through Thailand, creating marshy plains in the interior of the country and flowing together to form the Chao Phraya. The Mekong flows through and provides fresh water to China, Laos, Thailand, Cambodia, and Vietnam. The Red flows through China and supplies fresh water to northern Vietnam, including the capital city of Hanoi, and the rice fields of northern Vietnam.

MEKONG RIVER SYSTEM

The Mekong originates in the Tibetan Plateau, much like the Indus and Brahmaputra in South Asia. From the Tibetan Plateau, the Mekong River runs through China, forming the border between Myanmar and Laos; goes through part of northern Laos; then forms the boundary between Laos and Thailand; and widens as it runs through southwestern Laos. In Cambodia and Vietnam, the bedrock is typically sedimentary, and the lower Mekong cuts wider swaths of land, shifting course over time and creating a massive plain in Cambodia and a broad, fertile delta in Vietnam. The fertile plains of the Mekong Delta, coupled with predictable monsoon rains in summer, create the perfect environment for wet rice cultivation (Figure 7.4). Vietnam is the second largest exporter of rice globally, after Thailand.

The Mekong is a source of drinking water, irrigation, and fish for 65 million people. In the upper Mekong, Southeast Asians have grown rice in terraces that follow the contour lines of hillsides for centuries (Figure 7.5). The Mekong is the second most biologically diverse river system in the world, after the Amazon in South America. In order to protect the river, the four countries of Laos, Cambodia, Vietnam, and Thailand formed the Mekong River Commission in 1995.

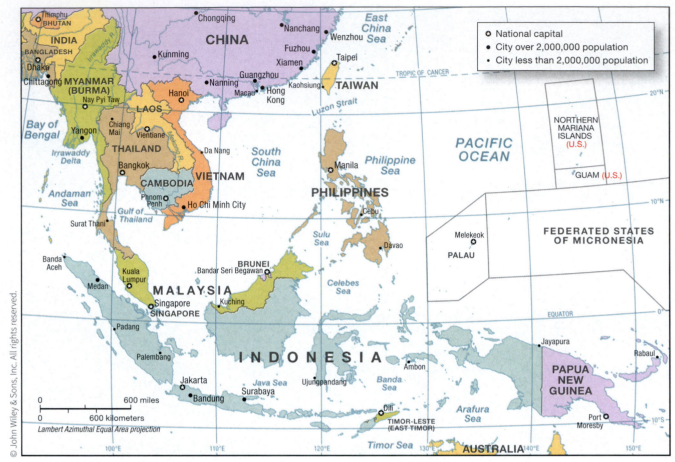

FIGURE 7.3 **Political geography of Southeast Asia.** Laos is the only landlocked country in Southeast Asia, as every other country has a coastline. Countries in Southeast Asia largely have the same borders established through European colonialism, including Indonesia (colonized by the Netherlands), Myanmar (colonized by Great Britain), and the Philippines (colonized first by Spain and then by the United States). Thailand was never a European colony. Laos, Cambodia, and Vietnam were colonized by France as one large colony.

Reading the **PHYSICAL** Landscape

Mekong Delta

River deltas mark the place where large streams meet an ocean or sea. From a physical geography standpoint, river deltas are interesting because they are places of abundant sediment deposition as stream velocity slows when it flows into the larger sea or ocean. These sediments are usually laden with nutrients and organic remains that make deltas extremely fertile places. Because they are also well watered, river deltas are often preferred places for agriculture and thus are usually densely populated.

One of the most important river deltas in the world is the Mekong Delta in Southeast Asia. This delta is at the terminus of the Mekong River, which originates in the Tibetan Plateau in China. From there, the river flows in a generally southerly direction along the border between Laos and Thailand, through Cambodia, and into southern Vietnam where it splits into several streams before it empties into the South China Sea.

Like the better known Amazon, Mississippi, and Nile deltas, the topography of the Mekong Delta is virtually level (or flat) because stream sediments build up horizontally over time. The fertile and frequently flooded landscape of the delta is covered with rice paddies (Figure 7.4a), contributing to Vietnam's role in the world as a major rice exporter. Another major industry in the delta region is fishing (Figure 7.4b) for species such as basa, which is a type of catfish. To support these activities, numerous towns and villages also dot the landscape.

FIGURE 7.4a **Can Tho, Vietnam.** Workers plant rice in the Mekong Delta region.

FIGURE 7.4b **Mekong Delta, Vietnam.** A fishing village abuts the river, along with dozens of fishing boats and other vessels.

In 2010, Laos proposed building the Xayaburi Dam in the upper Mekong, and began building the dam in 2012. Laos plans to sell power generated by the dam to neighboring Thailand, and the dam is being built with financing from banks in Thailand. In the lower Mekong, Vietnam and Cambodia fear the dam will damage the fertile plain and delta where they fish more than 800 species of freshwater fish and grow rice for global consumption.

Vietnam and Cambodia are concerned that the dam will retain silt and sediment, preventing the nutrient-rich material from flowing to the lower Mekong. The constant supply of silt and sediment from the upper Mekong are what rejuvenate and bring nutrients to the plain and delta in the lower Mekong. Building the Xayaburi Dam was suspended and several environmental groups, including the World Wildlife Fund, are working to stop the construction all together. If the flow of silt and freshwater slows, salt water from the South China Sea will encroach the delta region, and the government of Vietnam contends that salinization of the rice fields is a real threat if the dam is built. Laos, supported by China, contends that weirs (fish passes) can be built with the dam that will enable fish to safely find their way around the dam. Whether and how the dam is constructed is the first major test for the Mekong River Commission.

ISLANDS OF SOUTHEAST ASIA

The islands of Southeast Asia were formed primarily through volcanic activity, and active volcanoes, earthquakes, and tsunamis are serious concerns in the region.

FIGURE 7.5 **Yuanyang, China.** A farmer works rice terraces along the upper Mekong River. For centuries, farmers have developed and maintained terraces that follow the contours of the hills in order to create fields for growing rice, prevent erosion, and capture rainfall.

Plate boundaries along the islands of Southeast Asia include subduction zones, where denser oceanic plates subduct under less dense continental plates. The islands of Indonesia are arcs of islands parallel to the subduction boundary between the Australian and Eurasian plates. The Philippines, which run north–south, are parallel to the subduction boundary between the Philippine and Eurasian plates. Where subduction zones occur, a trench, or area of extremely low elevation, commonly forms along the plate boundary.

The ocean in Southeast Asia is divided into relatively shallow seas. Figure 7.6 shows the topography of the ocean floor in the region, and most of the seas around Malaysia and Indonesia are less than 500 feet deep. The shallowest

portions of the sea are the places where parts of the Eurasian plate are just below sea level.

In addition to volcanic rock, the islands of Southeast Asia are composed of limestone, which forms when calcium-rich sea life decomposes in shallow waters. The shallow seas of Southeast Asia are well suited for limestone formation. Limestone can be easily eroded by rainfall that reacts chemically with carbon in the atmosphere or soil, creating mildly acidic solution called carbonic acid that erodes the calcium carbonate in limestone into **Karst topography**.

CLIMATE

To understand the climate of Southeast Asia, begin by looking at Figure 7.7 and notice the location of the equator, 0°

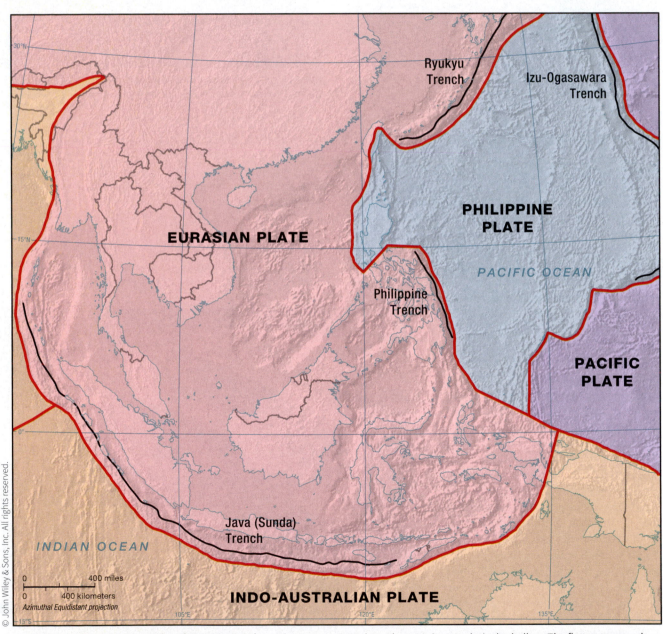

FIGURE 7.6   Ocean topography of Southeast Asia.   The oceans around Southeast Asia are relatively shallow. The flat area around the Malay peninsula, the Mekong peninsula, and western Indonesia is the continental shelf, which forms a shallow sea. The deeper areas surrounding the shallow seas are located where trenches form on subduction zones.

latitude. The equator cuts through the country of Indonesia, which is made up of over 17,000 islands, 6,000 of which are inhabited. Over the course of the year, the equator receives more incoming solar radiation (insolation) than any other latitude on earth. The rainforest climate extends along the entire equator, including the equatorial region of Southeast Asia.

The islands of Southeast Asia are dominated by wet equatorial (rainforest) and monsoon climates. The equatorial regions of Southeast Asia absorb consistent heating from the sun and then release relatively warm air, which rises, cools, condenses, and releases the daily rains that create the rainforest climate.

North and south of the rainforest, Southeast Asia is marked by the monsoon climate. As the **subsolar point** migrates north and south of the equator over the course of the year (Figure 7.8), the wet monsoon affects different parts of the region. The Intertropical Convergence Zone (ITCZ) is in the Northern Hemisphere between March and September with the direct rays of the sun. In June and July, the ITCZ is located closer to 23.5°N, which draws moist air from the Indian Ocean on to South Asia and mainland Southeast Asia, creating the wet monsoon. The subsolar point is in the Southern Hemisphere between September and March, closer

to 23.5°S near the end of December. Moisture is then drawn on to the southern islands of Southeast Asia and the northern coast of Australia, creating a wet monsoon in this area south of the equator. Because Southeast Asia straddles the equator and the ITCZ migrates across the equator twice over the course of a year, the northern part of the region has a wet season from June to October and the southern part of the region has a wet season from December to March.

The monsoon climate extends to mainland Southeast Asia. Cambodia and parts of Thailand, Vietnam, Laos, and Myanmar have monsoon climates, which are well suited for rice cultivation. Proximity to the coast and summer precipitation create a **humid subtropical** climate in the north.

HURRICANES

Southeast Asia is susceptible to hurricanes, which are low-pressure systems that form and build strength from warm ocean water with temperatures at least 80°F. Consistent heating of the ocean in the tropics (between 23.5°N and 23.5°S of the equator) creates ocean temperatures warm enough in the respective late summer and early fall in each hemisphere for hurricanes to form.

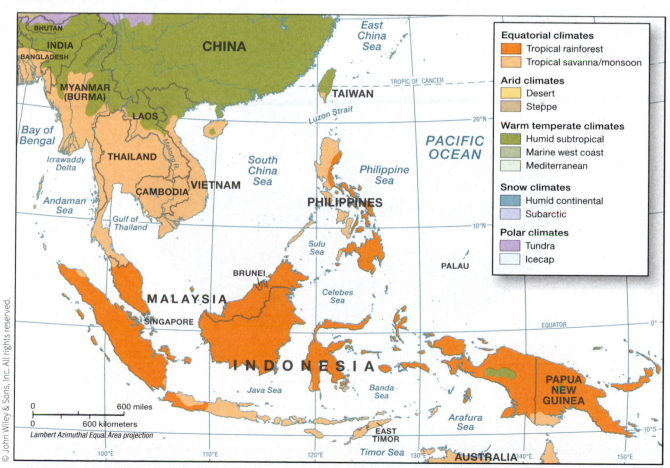

FIGURE 7.7   Climates of Southeast Asia.   The rainforest climate straddles the equator, which receives the greatest amount of incoming solar radiation over the course of the year. Monsoon climates are found on either side of the rainforest.

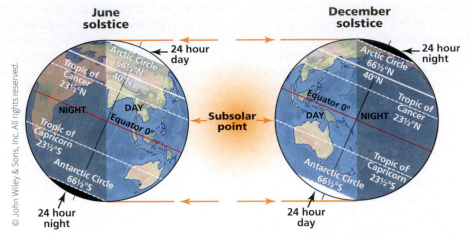

FIGURE 7.8 **Subsolar point in Southeast Asia.** The subsolar point is the latitude where the direct rays of the sun hit Earth at a 90° angle. The farthest north the subsolar point reaches is 23.5°N on the June solstice, and the farthest south the subsolar point reaches is 23.5°S on the December solstice. In this annual cycle, the subsolar point is at the equator twice during the year, on the spring (March) and fall (September) equinoxes. Thus, the equator receives the most consistent heating of any place on earth over the entire year.

In Southeast Asia, hurricanes are called typhoons in the Pacific Ocean and cyclones in the Indian Ocean. The typhoon and tropical cyclone season in Southeast Asia is somewhat complex and is directly tied to the migration of the subsolar point and thus to when the oceans in the region reach their highest temperatures of the year. North of the equator, the typhoon season stretches from May to November (summer and early fall in the Northern Hemisphere).

Typhoons do not occur within about 5 degrees north or south of the equator (Figure 7.9). Ocean waters along the equator are certainly heated enough for cyclones to form. However, a cyclone starts as a tropical depression, a low pressure that begins to rotate (counterclockwise in the Northern Hemisphere or clockwise in the Southern Hemisphere). Along the equator, where the earth rotates its fastest, the winds do not deflect or rotate. In fact, the winds of the equatorial region are called the doldrums

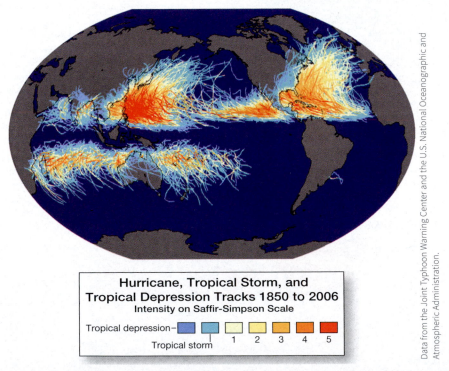

Hurricane, Tropical Storm, and Tropical Depression Tracks 1850 to 2006
Intensity on Saffir-Simpson Scale

Tropical depression—
Tropical storm— 1 2 3 4 5

FIGURE 7.9 **Typhoon and cyclone tracks.** The path of typhoons and cyclones throughout the world between 1850 and 2006. Cyclones do not from around the equator, as there is not enough rotation in the winds for a tropical depression to form that close to the equator. Cyclones do not form off the west coast of Africa or South America because cold water currents keep the ocean too cold for cyclones to form. The east coast of South America is relatively free of cyclones because there is not enough room for a cyclone to organize in the areas of the ocean that are warm enough for one to form.

because of the lack of movement; so, hurricanes do not form.

In the Indian Ocean, tropical cyclones occur between April and June and again between September and November. The western portion of Southeast Asia, including Myanmar, is most susceptible to tropical cyclones in these seasons. The break in the Indian Ocean tropical cyclone season occurs because the sun's rays hit directly so far north, on the Asian landmass, during the June solstice and into early August that the sun is heating the land more than the Indian Ocean, and cyclones do not form on land.

The enormity of the Pacific Ocean gives typhoons plenty of opportunity to build strength. Thus, the typhoons that hit the western Pacific, including the Philippines, are typically stronger than the hurricanes hitting the western Atlantic, including the southeast coast of the United States. The smaller size of the Indian Ocean generally allows fewer tropical storms to form, and the storms are typically of less intensity. The year 2008 was an exception because Tropical Cyclone Nargis hit Myanmar with great strength.

When tropical cyclones do hit Myanmar, they can cause a great deal of destruction to the lowlands along the coast, the Irrawaddy River Delta, and the floodplain of the Irrawaddy River. On May 2, 2008, Cyclone Nargis hit the coast, tracked northeast through the low-lying Irrawaddy Delta, and devastated the country's largest city of Yangon, before dissipating inland (Figure 7.10a). Cyclone Nargis caused widespread death and destruction (Figure 7.10b)

(a)
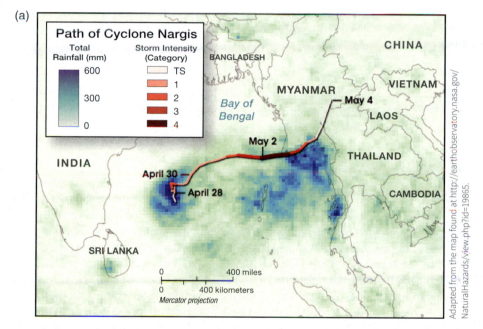

(b)

FIGURE 7.10a and b Cyclone Nargis. (a) Path of Cyclone Nargis through Myanmar. (b) **Laputta, Myanmar.** Cyclone Nargis shredded the buildings of Laputta, the largest town in the Irrawaddy Delta. Some 80,000 people in this district alone perished in the disaster, and estimates place the death toll at 134,000. Nearly 2.5 million people were left in severe need of humanitarian aid.

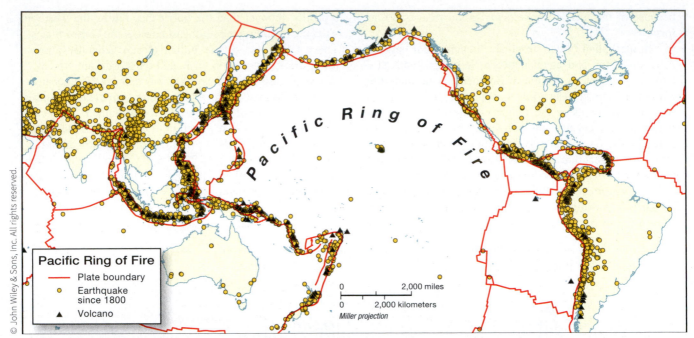

FIGURE 7.11 **Pacific Ring of Fire.** The concentration of volcanoes and earthquakes along the plate boundaries of the Pacific Plate form the Ring of Fire.

and revealed to the world the total control of the military junta that ran Myanmar and their lack of concern for human rights. The military junta refused international assistance from the global community in response to Cyclone Nargis and for a time denied the scope of the impact of the cyclone.

With the low elevation of the Irrawaddy River Delta and its floodplain, climate change will only make Myanmar more susceptible to flooding from typhoons and storms. Climate change is already bringing more frequent and extreme storms to Southeast Asia. In 2010–2011, "42 million people in Asia were displaced by 'extreme' weather" (*Economist* 2012, 2).

TSUNAMIS

Southeast Asia is located on the Pacific Ring of Fire (Figure 7.11) where plate boundaries meet and where 81 percent of the world's earthquakes take place. Southeast Asia is home to several plate boundaries, including boundaries along the Java Trench, the Mariana Trench, and the Philippines plate (see Figures 7.2 and 7.6).

In a subduction zone, earthquakes occur after tension that has built up along the subduction zone is released with movement of a plate. Earthquakes can also occur along transform boundaries, where two plates of similar density slide past one another, and along faults within plates. The strongest earthquakes occur along subduction zones when the overlying plate uplifts. In 2004, an earthquake registering 9.15 on the Richter scale took place just off the west coast of the island of Sumatra, Indonesia, under the Indian Ocean. The epicenter was on the boundary between

two plates, where the Indian plate is subducting under the Eurasian plate. The earthquake lasted 10 minutes when the Indian plate dropped approximately 50 feet along the 750-mile plate boundary. The high-magnitude earthquake generated a massive seismic sea wave, **tsunami**, that devastated much of the western coasts of Thailand and Indonesia.

The Indian Ocean tsunami had waves up to 100 feet high. Vertical movement of a plate "allows a quick and efficient transfer of energy from the solid earth to the ocean" and this energy is released in an upward thrust of water and the generation of massive waves (NOAA). Tsunami waves grow in height as they reach the shore because the seafloor becomes shallower along the coastline. The first sign of a tsunami is often the ocean receding from the beach, a phenomenon known as **drawdown**. Witnesses of the Indian Ocean tsunami reported that, unfortunately, thousands of tsunami victims did not recognize the drawdown as a sign that a tsunami was imminent, and instead they walked into the ocean to examine and photograph the newly exposed ocean floor.

During the Indian Ocean tsunami, locals and tourists alike were unaware of the earthquake and unprepared. The first place hit, within about 20 minutes of the earthquake, was the Indonesian island of Sumatra and the beach town of Banda Aceh (Figure 7.12). The waves hit Banda Aceh in some places as far as 5 miles inland. Within 2 hours, the tsunami hit the coast of Thailand (including Phuket) to the east of the earthquake site, and Sri Lanka and India on the west side of the earthquake. Within ten hours of the initial earthquake, the tsunami had killed more than 275,000 people, including approximately 300 victims in the East African country of Somalia.

USING GEOGRAPHIC TOOLS

Using Satellite Images to Assess Damage from a Tsunami

The impact of the Indian Ocean tsunami on Banda Aceh was disastrous, as the before and after satellite images of the tourism hotspot attest. Few photographs of the Indian Ocean tsunami exist because most people had no warning a tsunami was arriving, and when it did arrive people fled for their lives. Geographers used satellite images taken before, during, and after the tsunami to assess the damage, estimate the size of the waves, and help guide humanitarian and rebuilding efforts.

Thinking Geographically

1. What other layers of information could you add to satellite photographs in order to help rebuilding efforts in Banda Aceh?

2. Search the Internet for updated photos of Banda Aceh. Which parts of the beach were rebuilt first and why?

FIGURE 7.12a **Banda Aceh, Indonesia.** True color satellite image of Banda Aceh, Indonesia, collected on June 23, 2004, months before the tsunami.

FIGURE 7.12b **Banda Aceh, Indonesia.** True color satellite image collected on December 28, 2004, of the northern part of the island, along the shoreline where the tsunami struck.

DigitalGlobe/Getty Images.

CONCEPT CHECK

1. **Why** are farmers in the Mekong Delta concerned about the proposed dam on the upper Mekong?

2. **How** does the migration of the ITCZ affect climates in Southeast Asia?

3. **What** is the warning sign of a tsunami, and **why** does it take place before a tsunami hits?

2. **Describe** how language policies in Malaysia, Indonesia, and the Philippines differ.

3. **Understand** how the diffusion of Hinduism and Buddhism into the region can be seen in the cultural landscape.

WHO ARE SOUTHEAST ASIANS?

LO

LEARNING OBJECTIVES

1. **Identify** the history of spatial interaction in the region by understanding the distribution of languages.

The physical geography of Southeast Asia varies between the mainland and the thousands of islands of the region. The population distribution and corresponding **spatial interaction** in the region are closely tied to the physical geography (Figure 7.13). In the island region of Southeast Asia, people are concentrated in cities on coastal areas. Major coastal cities include Manila, the Philippines, the fifteenth most densely populated city in the world, and Jakarta, Indonesia, the seventeenth most densely populated city in the world with 10,500 people per square kilometer (Figure 7.14).

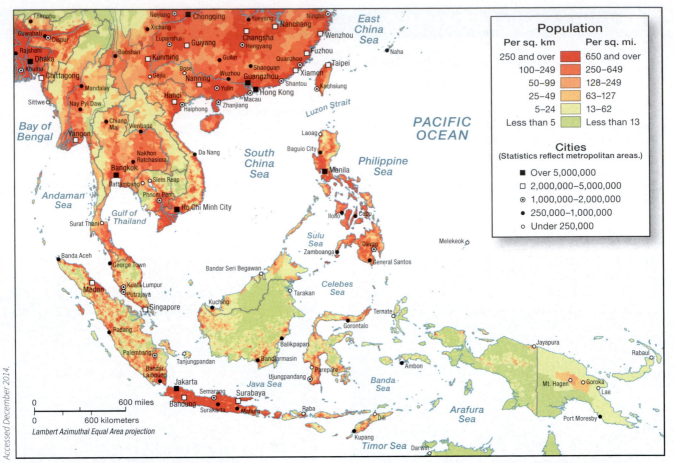

Data from CIESIN—Columbia University, United Nations Food and Agriculture Programme—FAO, and Centro Internacional de Agricultura Tropical—CIAT. 2005. Gridded Population of the World, Version 3 (GPW3): Population Count Grid, Future Estimates. Palisades, NY: NASA Socioeconomic Data and Applications Center (SEDAC). http://dx.doi.org/10.7927/H42B8VZZ. Accessed December 2014.

FIGURE 7.13   Population density of Southeast Asia.   People are concentrated mainly in coastal regions of Southeast Asia and along major rivers. The interior of islands and the areas between rivers are higher elevations and are generally less densely populated.

THE FIRST SOUTHEAST ASIANS

Modern humans arrived in Southeast Asia approximately 50,000 years ago, when humans migrated from Central Africa, across Southwest Asia and South Asia into Southeast Asia. Two different hypotheses have developed in the literature about the migration of humans after this point. One theory focuses on DNA evidence and the other examines linguistic evidence.

The theory focusing on DNA analysis contends that peopling the region came from this one wave of migration out of Africa into Southeast Asia. Then, from Southeast Asia, humans migrated north into East Asia. According to *Scientific American*, "East Asians, the analysis suggests, share a large degree of common genetic background with southeast Asians but very little with central Asians" (Cyranoski 2009). By discerning the genetic relations among Central Asians, East Asians, and Southeast Asians, DNA evidence suggests that "diverse peoples

FIGURE 7.14   Manila, the Philippines.   Manila is the fifteenth most densely populated city in the world. Makati City, part of Manila shown here, is the financial district for the Philippines.

Stuart Dee/Photographer's ChoiceRF/Getty Images.

FIGURE 7.15 Migration of humans based on genetic tracking. This map of human migration is based on DNA data gathered through the National Geographic's human genome project, and it supports one wave of migration into South Asia, Southeast Asia, and East Asia.

living in southeast Asia migrated northwards" to inhabit east Asia (Cyranoski 2009). The recent DNA evidence that supports this theory shows a decline in human genetic diversity as one goes from south to north, supporting the idea that Southeast Asia was settled first and then East Asia later (Figure 7.15).

The second theory is based on linguistic evidence and contends that long after the first wave from Africa, a second wave occurred when people migrated from Central Asia through present-day China, then into present-day Taiwan, and finally into Southeast Asia about 5,000 years ago.

LANGUAGE

For at least the last 2,000 years, many different peoples have traveled through Southeast Asia for trade, diffusing world religions, languages, and economic and political systems. Spatial interaction through trade in Southeast Asia is reflected in the complexity of languages and cultures in the region (Figure 7.16). The islanders of Southeast Asia speak

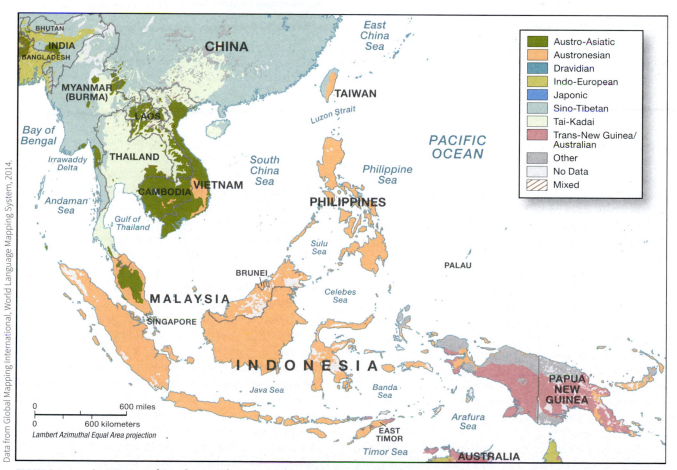

FIGURE 7.16 Languages of Southeast Asia. Languages of Southeast Asia reveal a difference in spatial interaction between the continental and island regions of Southeast Asia. Languages on the islands, including Malaysia, Indonesia, and the Philippines, are part of the Austronesian language family. Languages in continental Southeast Asia belong to the Sino-Tibetan language family.

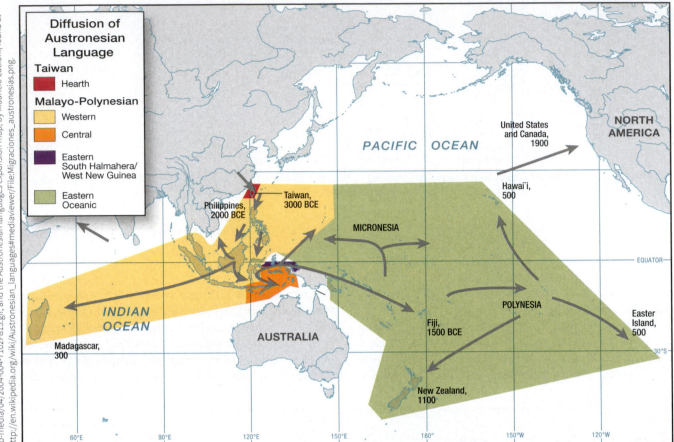

FIGURE 7.17 **Austronesian language family.** The hearth of the Austronesian language family is the island of Taiwan. From there, the language diffused beyond Southeast Asia as far west as Madagascar and as far east as Polynesia.

languages in the Austronesian language family. The map of Austronesian languages (Figure 7.17) reveals historical connectedness among New Zealand, Hawaii, Madagascar, and the islands of Southeast Asia, both through peopling of the places and through trade. Remarkably, the language and people of Madagascar (located 250 miles off the coast of Africa) are more closely related to Southeast Asia than to Africa.

Indigenous populations in Southeast Asia have created **local cultures** found today mainly in the interior or mountainous country of both the mainland and islands. Many indigenous peoples continue to practice traditional or **animist** religions. The great diversity of indigenous populations and the lack of spatial interaction among indigenous peoples are evident in the linguistic diversity of the region. Indonesia is composed of over 17,000 islands that stretch over 3,000 miles (4,800 km) from west to east and a population that speaks over 600 **vernacular** or local languages.

To communicate across a multitude of islands and peoples, Indonesians have adopted a common language for trade. The official language in Indonesia today is Bahasa Indonesian (commonly called Indonesian), which is part of

the Austronesian language family. Indonesian is a form of the Malay language that has been used as a **lingua franca** in Indonesia for over 1,000 years. The Indonesian language likely began on the island of Sumatra in present-day Indonesia. The language took hold on both sides of the Strait of Malacca: the Malay Peninsula and Sumatra. Traders used Indonesian in order to communicate across the hundreds of vernacular languages in the region. Missionaries also used Indonesian to diffuse Islam.

The Netherlands, through the Dutch East Indies Company, began a 350-year colonial presence in Indonesia in the sixteenth century. The Indonesian language persisted over more than three centuries of Dutch colonization of Indonesia, in large part because the Dutch East Indies Company and the Dutch colonial government, like the British in South Asia, did not actively control every part of Indonesia. Sultans in Aceh and Mataram fought Dutch control in the seventeenth century. The Dutch colonial presence focused on trade, and the colonizers adopted Indonesian in order to communicate with the diverse peoples of Indonesia. Christian missionaries from the Netherlands also used Indonesian to communicate and diffuse Christianity in parts of Indonesia.

The Indonesian language was an important component of creating an Indonesian nationalism to help the country gain independence from the Netherlands. During the colonial era, the Netherlands used Dutch as the language of government. In 1928, the Second Congress of Indonesian Youth chose Bahasa Indonesian as the official language of the country they hoped to free from Dutch colonialism. In response, the Dutch saw the Indonesian language as a threat and banned its use in education in 1932. When the Japanese occupied Indonesia in 1942, they immediately banned the use of the Dutch language in addition to the Indonesian language. At the end of World War II, Japan lost its colonies and occupied territories, including Indonesia, and the people of Indonesia claimed their independence from Japanese occupation and Dutch colonialism in 1945.

The newly independent Indonesia chose Indonesian as the country's official language instead of Dutch, which only the most educated Indonesians spoke, or Javanese, the language spoken by the country's largest ethnic group (47.8 percent at the time). Upon independence, Indonesian was the native language, or first language, of less than 5 percent of the population (Paauw 2009). One advantage of choosing Indonesian was that the few native speakers did not have any political privilege in the country, which meant the government was not reinforcing a political or social stratification, established by the colonizers, as the choice of Dutch or Javanese as an official language would have done.

The Indonesian government sees having a common language as a major force of national cohesion. The constitution recognizes Indonesian as the national language and also recognizes vernacular languages. Today, over 87 percent of the people in Indonesia are literate in the Indonesian language. "Indonesian has a dual function in Indonesian society, as it is the language of national identity, and also the language of education, literacy, modernization, and social mobility" (Paauw 2009, 5).

In contrast to the success of Indonesia's language policy, the national language policies in Malaysia and the Philippines have caused problems (Paauw 2009). Since 1957, Malaysia has recognized Malay (the language from which Indonesian is derived) as the country's first official language and English as the second official language. Malay is the native language of 46 percent of the country's population and the Malay ethnic group is politically privileged in the country. **Overseas Chinese**, who have great economic clout, make up 35 percent of the population and control a majority of the wealth. The Malay government gives Malay people access to university seats and government jobs in order to preserve their political privilege over the well-to-do Chinese.

Overseas Chinese in Malaysia speak one of several dialects of Chinese and are usually educated in Chinese or English schools. Despite the fact that more than 30 percent of the population speaks Chinese, the Malaysian government does not use Chinese as a language for public schools or government. The Malaysian government does use English, the language of their former colonizer, as a lingua franca, particularly in law, and since 1993 in university instruction. Through affirmative action for ethnic Malay, the Malaysian government is trying to overcome the economic strength of the ethnic Chinese population.

PHILIPPINES

Language policies in the Philippines are also influenced by colonialism, as local ethnic languages vie for use with Spanish and English. The Spanish colonized the Philippines beginning in 1565. In the last decades of the 1800s, the Philippines began an independence movement against the Spanish, culminating in the 1896 Philippine Revolution. The United States sided with the revolution in the Philippines. In April 1898, the United States was fighting Spain in the Spanish-American War. After the war ended in December 1898, the United States "won" the colony from the Spanish and became the colonizers of the Philippines. The Philippines fought back, battling with the United States for independence from 1899 until acquiescing to American colonialization in 1902. In 1937, in a movement toward full independence, the Commonwealth government of the Philippines chose neither Spanish nor English as the national language (Paauw 2009). The Commonwealth government chose Tagalog as the national language, which is the native language of 21 percent of the population, who live primarily in and around the city of Manila.

In 1946, the Philippines reached full independence. Other ethnic groups resented the ascendancy of Tagalog because wealth and power were concentrated in the primate (largest, leading) city of Manila, which is also a primarily Christian region of the country. Members of smaller language groups (many of which are not Christian) outside of the Manila region had little economic or political power in the new country, and many felt their language was being lost under the ascendancy of Tagalog.

In 1971, the Philippines constitution removed Tagalog as the national language. Since the 1980s, the Philippines have used Filipino as their national language, which is derived from Tagalog and incorporates Spanish and English vocabulary words. Both Filipino and English are taught in the country's public schools (Figure 7.18).

One advantage the language policies of Malaysia and the Philippines have over those of Indonesia is that English, a major language of secondary school and university instruction in both countries, is a **global language** in academia, diplomacy, and trade. Indonesian, though it ranks fifth on the list of languages with the most native speakers in the world, is a regional, but not a global, language.

The importance of learning English to competing in the global economy has led thousands of students from Asia and Russia to travel to the Philippines to take English classes. The Philippines are a draw both because the people speak English with an American accent and because the classes are affordable. The Philippines is home to approximately 500 schools of English, and a fifth of the schools are located on the

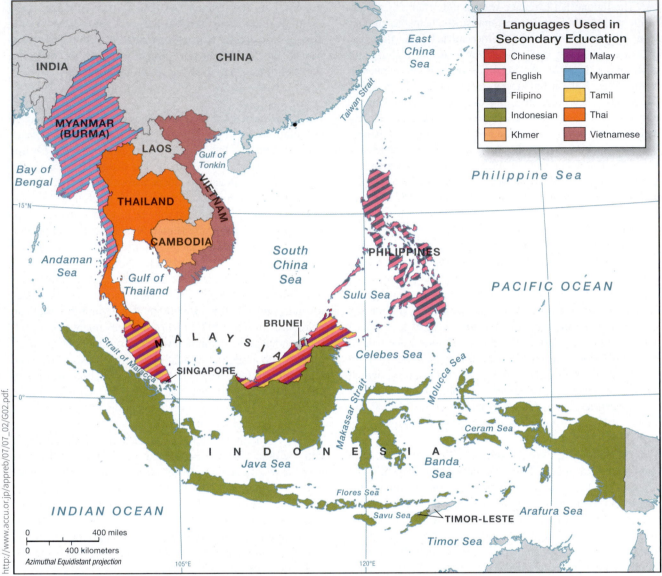

FIGURE 7.18 **Languages used in secondary education in Southeast Asia.** While local languages are typically used in primary school, English is used in secondary schools in Myanmar, Malaysia, and the Philippines.

island of Cebu. The island's white sand beaches draw students from Europe, Asia, and Russia who attend schools with beaches, yoga classes, and hotel accommodations (Figure 7.19).

RELIGIONS

The spices native to the islands of Southeast Asia were the focal point of the global economy in its infancy. Located 3°S, the Maluku Islands have a rainforest climate, which allows valuable spices, including cloves and nutmeg, to grow in abundance in the forested highlands of the islands (Figure 7.20).

The Chinese and South Asians traded with the Maluku Islanders for spices for centuries before Arab traders reached the islands in the 1300s and the Portuguese traders arrived in the 1500s. Europeans had access to spices

from China via the Silk Road. Europeans also accessed Southeast Asian spices from the Mediterranean, where Italian traders had a network with Arab traders who in turn traded with South Asians who regularly traded with the Maluku Islands. In the 1300s, Arab traders who sought to expand their direct trade network and to diffuse Islam established trading posts on the Spice Islands of Southeast Asia. In the 1400s and the 1500s, one of the main reasons Europeans set out to explore the world by ship was to cut out the "middleman" of the Chinese or Arab traders in the spice trade.

When the Portuguese first arrived in the Maluku Islands on a voyage that lasted from 1512 to 1513, they found that the Arab traders had already peacefully diffused Islam into the islands of the region. At the time European explorers and colonizers reached Southeast Asia, the region was dominated

© ERIK DE CASTRO/Reuters Corbis.

FIGURE 7.19 **Cebu, the Philippines.** South Korean students pose for a picture at the campus of Cebu Pacific International Language School. English is widely spoken in the former American colony. Approximately 500 English-language proficiency schools are found in Cebu, attracting Asian and European students looking to combine English learning with tropical tourism.

by independent kingdoms called temple states (Figure 7.21) and by Muslim sultanates. Each state was centered on Hindu and Buddhist temples established by political powers. The sultanates were Muslim political territories controlled by a sultan, who gained his wealth and power from trade in spices. When Europeans arrived in the 1500s, they brought Christianity and added to the complex mosaic of religions in Southeast Asia (Figure 7.22).

Hinduism

The first world religions to diffuse into Southeast Asia were Hinduism and Buddhism. Hinduism predates Buddhism, beginning sometime around 2000 BCE in South Asia (Figure 7.23). Today, Hindus do not actively seek converts; however, nearly two thousand years ago, Hinduism had a universalizing period in which the religion diffused from South Asia to Southeast Asia.

Hinduism diffused from South Asia to Southeast Asia along trade routes. The Maluku Islands in Southeast Asia, like the west coast of India (called the Malabar Coast), had an abundance of native spices. Before Arab traders established direct trade with Southeast Asia, Indian traders from the Malabar Coast functioned as middlemen between the Arabian Peninsula and Southeast Asia. Indians sailed from the west coast of India to the islands of Southeast Asia to trade and returned to India where Arab traders purchased spices and returned to the Arabian Peninsula.

Through trade, Indians brought Hinduism to the region nearly 2,000 years ago. Brahmin priests (see Chapter 6) resettled in Southeast Asia and taught Hindu beliefs

Ludovic GALKO-RUNDGREN/Moment Open/Getty Images.

FIGURE 7.20 **Banda Island, Maluku, Indonesia.** The Banda Islands, also known as the Spice Islands, were at the center of global trade in the sixteenth and seventeenth centuries. First claimed by the Portuguese and later the Dutch, the islands were the only place where nutmeg grew, which was in great demand in Europe because of its medicinal properties.

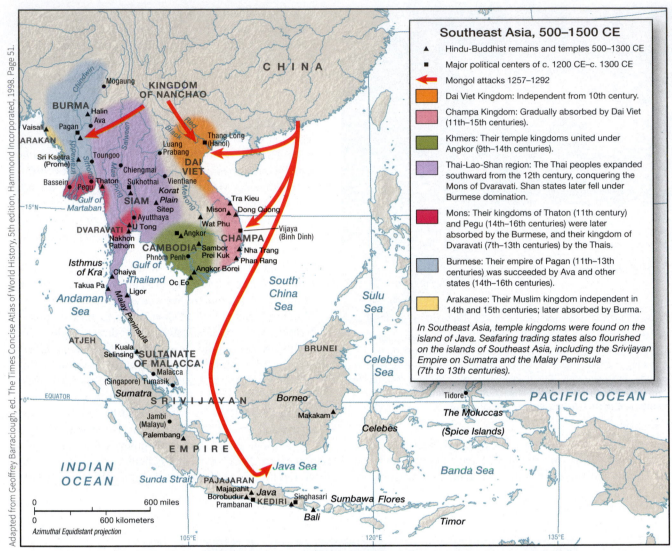

Adapted from Geoffrey Barraclough, ed. The Times Concise Atlas of World History, 5th edition, Hammond Incorporated, 1998. Page 51.

FIGURE 7.21   Temple states of Southeast Asia.   Prior to European colonialism, continental Southeast Asia was politically divided into temple states between 500 and 1500. Many of the world heritage sites in the region today were founded by the temple states during this time, including Angkor Wat in Cambodia.

to the people. Evidence of Sanskrit, the writing used for ancient Hindu texts, is found on the island of Borneo and dates to the fourth century CE. The Sanskrit writing is evidence that Brahmins were in Borneo and diffusing Hinduism during this time period.

Evidence of Hindu temple building attests to the far reach of Hindu teachings before Islam supplanted Hinduism in much of the region. Between 800 and 1400 CE, the Khmer Empire was perhaps the biggest influence for the growth of Hinduism in mainland Southeast Asia. Centralized in the city of Angkor in northern Cambodia, the Khmer Empire fluctuated in size and also in religious persuasion over time. The empire was first Hindu and later became Buddhist.

In northern Cambodia in the twelfth century CE, the Khmer Empire built the Hindu temple complex Angkor

Wat to honor Lord Vishnu. Approximately 100 temples made of stone comprise Angkor Wat (Figure 7.24). A subsequent emperor built an enormous city called Angkor Thom near the temple site. This emperor was a convert to Buddhism and adopted Angkor Wat as a Buddhist temple. He also built hundreds of other Buddhist temples throughout the region during his reign. Later, during the 1500s, Thai armies invaded northern Cambodia, and the Khmer Empire moved its capital to the south of Cambodia. Hinduism had fallen from favor in Cambodia, the government had moved from Angkor Wat, and in turn forests grew in, around, and covered the great temple. The temple was hidden but not forgotten. Cambodians knew its location and Spanish, Dutch, and later French visitors sent reports home of the grand Angkor Wat. Angkor Wat is considered the largest religious structure in the

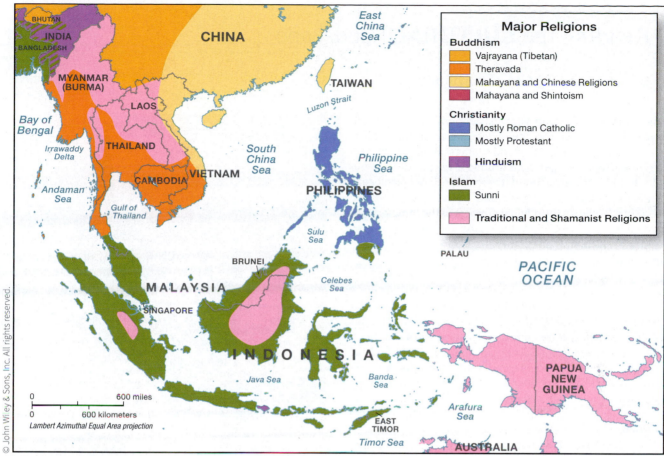

FIGURE 7.22 **Religions of Southeast Asia.** Buddhism and traditional religions are found primarily in continental Southeast Asia. Religions on the islands of Southeast Asia reflect the paths of traders and colonizers through the region between 500 and 1900. Hinduism came first and is found in small pockets, including Bali. Islam diffused through the region through relocation diffusion by Arab traders between the thirteenth and sixteenth centuries. Spanish colonizers brought Roman Catholicism to the Philippines after 1500.

world today and draws approximately 1.5 million visitors a year.

Although Hinduism is not widely practiced in Cambodia today, it is still practiced in pockets of Southeast Asia (see Figure 7.21). Hinduism's diffusion to Southeast Asia is evidenced by Angkor Wat and other smaller temples in the cultural landscape of the region.

Buddhism

Buddhism is divided into three major sects or practices (see Chapter 6). **Theravada Buddhism** originated in northern India, established a secondary hearth in southern India and Sri Lanka, and then diffused to Myanmar, Thailand, Malaysia, and Laos, and Cambodia. Theravada Buddhism holds that enlightenment is based on the individual and is achieved through good acts, religious practice, and service as a monk or nun. **Mahayana Buddhism** originated in northern India and diffused to China, Japan, and Korea. Mahayana Buddhism is tied closely to Daoism (Chapter 8) as practiced in China. **Vajrayana**

Buddhism is located in Tibet and is seen by some as the third school of Buddhism and by others as part of Mahayana Buddhism.

Evidence suggests the Buddhism practiced in mainland Southeast Asia was at first Mahayana and later, through influence and interrelations with Sri Lanka, became Theravada Buddhism. The Buddhism practiced in mainland Southeast Asia grew during the thirteenth century through hierarchical diffusion when the Thai king established close ties between the Thai monarchy and Theravada Buddhist teachers in Sri Lanka.

The cultural landscape of Buddhism in Southeast Asia is marked by stupas, which dot the landscape. Stupas are temples usually built around a relic of Lord Buddha, whether a bone, a hair, a footprint, or some other attachment to Lord Buddha, making each stupa a sacred site. The diffusion of stupa building into Southeast Asia marks the foundation of Buddhism in present-day Thailand and Laos (Figure 7.25). In Thailand and Laos, stupas were typically built at higher elevations, adopting sacred sites

Reading the **CULTURAL** Landscape

Buddhist Sacred Site of Borobudur

The massive stone temple called Borobudur, sited between two volcanoes near the city of Yogyakarta on the Indonesian island of Java, is the world's largest Buddhist stupa—a dome-shaped temple built around a relic of or to honor the Buddha. The structure dates to 775 CE. Archaeological evidence points to the Sailendra Dynasty, a Mahayana Buddhist family, as the original builders of the temple. Once completed, the Borobudur temple's top tier housed one large central stupa that stands empty and 72 smaller stupas, each of which has a statue of Buddha.

At some point between 1000 CE and the early 1800s, volcanic ash buried Borobudur. Around 1400, the Javanese shifted religions again, this time to Islam, and Borobudur was largely forgotten. When British delegate Sir Thomas Stamford Raffles took control of Java in 1814, he heard about the existence of Borobudur and worked to find the temple. Raffles's followers cleared the vegetation and volcanic ash to reveal the temple to the world. Unfortunately, this also revealed the temple to the elements, and the temple began to deteriorate from the elements and from thieves.

In the 1970s, the Indonesian government dedicated funds to restore Borobudur with the support of the United Nations Educational, Scientific, and Cultural Organization (UNESCO).

© Tuul/Robert Harding World Imagery/Corbis.

FIGURE 7.23   Java, Indonesia.   The Buddhist archaeological site Borobudur built during the eighth and ninth centuries CE.

from indigenous or animist religions. Political leaders who prescribed their citizens to make pilgrimages to stupas helped advance the diffusion of Buddhism in the region.

Islam

The map of the diffusion of Islam in Southeast Asia between the thirteenth and sixteenth centuries (Figure 7.26) shows the extent of Arab trade networks in the islands of Southeast Asia between 1200 and 1500. The most important travel routes, including straits, and

important ports for trade were the focus of the earliest diffusion of Islam.

The diffusion of Islam into the region was both hierarchical, focusing on the most important trading ports, and contagious from the people living in ports to people in the surrounding regions. The hierarchical diffusion amplified when Islamic governors established **sultanates**. A sultanate is a territorial area ruled by a sultan, who functions like a governor, a political leader of the sultanate. Sultans are the political leaders, not the religious leaders, of Sunni Islam. Each sultan centralizes power over the people in his region.

FIGURE 7.24 **Siem Reap, Cambodia.** Angkor Wat began as a Hindu temple and later was developed into a Buddhist temple, when the religion in northern Cambodia changed.

Although the sultan is not a religious leader, he leads with the blessing of religious leaders, and thus the power of the sultan is supported by the persistence of Islam in the place. Through a pyramid, centralized, or top-down structure of power, the sultan can help diffuse Islam by demanding that his people follow the religion. The sultan also gained wealth by controlling whatever resources were in the sultanate and demanding the payment of taxes from the people in the sultanate.

Sultans in Southeast Asia, including the Sultan of Brunei who today represents the wealthiest royal family in the world (Figure 7.27), established their power during the Islamic diffusion period in the region. When the Portuguese arrived in Southeast Asia in the 1500s, sultans ruled key economic parts of the region. The Portuguese pragmatically worked with sultans to efficiently establish ports and to tap into the spice trade. Today, Indonesia is the largest Muslim country in the world, in terms of total adherents, with a population of over 170 million Muslims.

CONCEPT CHECK

1. **Where** are the major religions and languages of Southeast Asia found and **why**?

2. **Why** is the language policy more unifying in Indonesia than in Malaysia?

3. **How** do Borobudur and Angkor Wat temples reflect the diffusion of Hinduism and Buddhism into the region?

FIGURE 7.25 **Vientiane, Laos.** Gilded perimeter and spire of Pha That Luang (Great Stupa). A study of stupas in Southeast Asia described how pilgrimage advances the diffusion of a religion in three steps: First, the pilgrim separates from his or her home and ordinary life in traveling to the sacred site; next, the pilgrim interacts with the sacred space, creating a stronger sense of his or her religion through the pilgrimage; and, finally, the pilgrim returns home with a stronger attachment to the religion (Wichasin 2009).

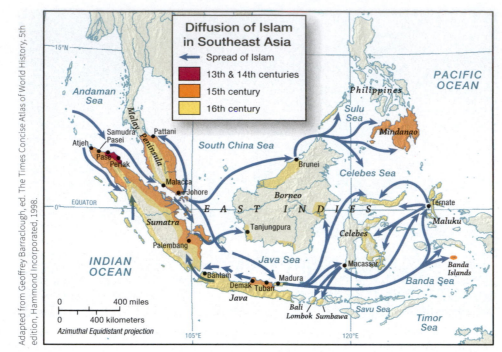

Adapted from Geoffrey Barraclough, ed. *The Times Concise Atlas of World History*, 5th edition, Hammond Incorporated, 1998.

FIGURE 7.26 **Diffusion of Islam in Southeast Asia.** The diffusion of Islam followed the routes of Arab traders into the region with important focal points in western Sumatra, where ships entered the Strait of Malacca, and in the east in the Maluku Islands.

FIGURE 7.27 **Bandar Seri Begawan, Brunei.** Stilt houses in Kampong Ayer stand in front of the opulent Sultan Omar Ali Saifuddin Mosque. Named for the 28th Sultan of Brunei, the mosque was built in 1958. The architecture of the mosque reflects the extreme wealth of the sultan of Brunei, as the minarets are made of marble, the mosque is located on a human-made lagoon, and the dome of the mosque is covered in pure gold. The mosque features a total of 28 golden domes because it was built for the 28th sultan.

WHAT ROLE DOES SOUTHEAST ASIA PLAY IN THE GLOBAL ECONOMY?

LEARNING OBJECTIVES

1. **Determine** how ASEAN and global shipping secure a role for Southeast Asia in the global economy.

2. **Understand** how the informal and formal economies affect land use in the region.

3. **Explain** how global tourists impact local cultures in Southeast Asia.

The streets bustle with thousands of people moving in all directions. Jakarta, Indonesia has a Central Business District (CBD) with teeming skyscrapers housing insurance companies, banks, two stock exchanges, and multinational companies. Southeast Asia, the centerpiece of global trade for more than 1,500 years, is capitalizing on its ethnic and religious diversity to corner a growing market in world trade (Figure 7.28). Through trade, Islam diffused to Indonesia (the largest Muslim country in the world today) and Malaysia. Today, Indonesian financial companies are vying with nearby Malaysian companies to serve the growing market for Islamic banking and insurance.

The teachings of Islam instruct (much as did Christianity until the 1500s) that interest should not be charged for lending money. Indonesian and Malaysian banks and insurance companies operate not on interest but on sharing profits. When an entrepreneur comes to an Islamic bank with a business proposition, the bank will lend money by matching the entrepreneur up with investors who then share the business risk. If the entrepreneur makes money, he shares his profit with his investors through the bank. If the entrepreneur loses money, the investors likewise lose.

Jakarta is only one of many places in Southeast Asia that works to draw economic strength from its cultural complexity. Singapore capitalizes on its connections to China to serve as a financial center investing and advising in China's growing economy. The hill tribes, or indigenous peoples, in northern Thailand and Laos draw tourists interested in ethnic tours. Cambodia and Bali draw from their Hindu history (although Cambodia is predominantly Buddhist today) to attract tourists to famed ancient temples and spaces of solitude. The Philippines and Vietnam are known for their well-manicured and maintained hillside rice terraces, many of which have been growing rice for more than a thousand years.

CRONY CAPITALISM

During the 1980s and 1990s, economists described the rapidly growing economies of Southeast Asia as "tigers." In 1997, a financial crisis hit Southeast Asia, followed by slow to negative growth rates in the region. Since 1997, fewer studies have focused on Southeast Asia as a region of "tigers" or "miracles." In turn, economists and political scientists have looked behind the curtain and found Southeast Asia's economic growth was no "miracle." It was a result of **crony capitalism** that left millions of urban and rural poor behind.

The tiger states of Southeast Asia included Singapore, Taiwan, and Hong Kong (Figure 7.29), followed by Thailand, Malaysia, Indonesia, and the Philippines (Boyle 2002). Under crony capitalism, the government of the state and those in its favor propped up industries by delivering state contracts to businesses with links to the government. Many of the industrial owners who led the Southeast Asian tigers were overseas Chinese entrepreneurs. The businesses artificially propped up by crony capitalism failed to gain global market share. Today, very few global brands come from Southeast Asia.

ASEAN AND GLOBAL INFLUENCE

The Association of Southeast Asian Nations (ASEAN) is a regional trade organization with a focus on political stability and peace that has operated in the region since 1967. The founding members, Indonesia, Thailand, Malaysia, the

FIGURE 7.28 Kuala Lumpur, Malaysia. The Petronas Towers stand among skyscrapers in the financial center of Kuala Lumpur.

FIGURE 7.29   Victoria Harbor, Hong Kong. Hong Kong straddles Southeast Asia and East Asia, and was a British colony, much like Singapore. While Singapore gained its independence after World War II, the British handed Hong Kong over to China in 1997 in order to fulfill a 99-year-old agreement. Hong Kong is on the South China Sea, which is strategically protected by the Chinese.

Philippines, and Singapore, are considered the ASEAN-5, and since 1984, five more states, Brunei (1984), Vietnam (1995), Laos (1997), Myanmar (1997), and Cambodia (1999) have joined to make an association of ten states. The ASEAN-5 were the economic tigers of the 1980s and 1990s. In 1997, the economies in the region (aside from Singapore) went into a financial crisis when currencies were devalued and companies had to default on loans (Bloomberg 2012).

Companies that made it through the 1997 financial crisis and the many global companies with subsidiaries in Southeast Asia generally have experienced growth since 2009. ASEAN countries have focused on growing domestic consumption of their products instead of relying on export income. In 2000, the ten countries of ASEAN shipped 72.4 percent of their exports to the United States and European Union, and in 2010, that proportion was down to 32.9 percent. An increasing share of exports went to other Asian countries, especially China, with whom ASEAN negotiated a trade agreement in 2010.

By not relying only on sales to the United States and the EU and by increasing domestic consumption, the ASEAN countries have reinvigorated their economies. Between 2008 and 2011, Thailand's economy experienced an average quarterly growth rate of 1.8 percent and Singapore 6 percent as compared to the United States with 0.3 percent and the euro zone of the EU with –0.3 percent (Bloomberg 2012, 2). In 2012, the International Monetary Fund (IMF) predicted that the economies of Indonesia, Thailand, the Philippines, Malaysia, and Vietnam (along with India and China) would "outpace the rest of the world" in the short term (Bloomberg 2012, 2).

Among the states of Southeast Asia, Singapore, which is one of the smallest, has one of the largest global economic impacts. Singapore, made up of one main island and about 60 smaller islands, is located at the tip of the Malay Peninsula,

on Singapore Strait, which is at the southern end of the Strait of Malacca. British colonizers established this former fishing village as a port in 1819. Settled by overseas Chinese who make up more than 75 percent of the population, Singapore is ethnically distinct from its neighbors. It was colonized separately from Malaysia, and when each gained independence, Singapore remained separate from Malaysia. With about 4 million people, Singapore is by far the wealthiest country in Southeast Asia and is the second wealthiest country in the world as measured by GNI per capita ($76,850 in 2013).

Singapore has used its strategic location to become a global leader in container shipping and the **entrepôt** for Southeast Asia. Its port system rivals the Hong Kong and the Pearl River Delta port system in China. The port of Singapore is the largest **bunkering port** in the world (Figure 7.30), and it fuels many of the 182,000 vessels, including 19,000 container ships and 21,000 tankers, that travel through the port each year.

Singapore is also a regional and global center in electronics manufacturing and in financial services. Singapore business leaders have played an integral role in the capitalist developments in China since the reforms of 1979 (see Chapter 10).

INFORMAL ECONOMY

With much of the formal economy dominated by crony capitalism, the informal economy plays an enormous role in providing the livelihoods of Southeast Asians. The **informal**

FIGURE 7.30   Singapore. In addition to being the largest bunkering port in the world, Singapore is the second busiest port in the world overall.

economy is the portion of the economy that is not taxed or regulated by government, where goods and services are exchanged in barter or cash systems that are not reported to the government. In the 1950s, economists began measuring the informal economy and predicted that developing countries would become more industrial over time and their informal economies would shrink. Since that prediction in the 1950s, economists have found the exact opposite to be happening! In 2004, a World Bank report found that the informal economy has actually grown in poorer countries in the world, across all developing regions.

The same World Bank study found that of the working population around the world, women (especially in countries that prohibit women from owning land), young people, internal migrants (within a country, usually from rural to urban areas), and international immigrants are most likely to be working in the informal economy. Subsaharan Africa and South Asia have larger portions of the population in the informal economy than Southeast Asia. However, the 1997 Asian financial crisis hit women in Southeast Asia harder than men, increasing unemployment rates for women (over 30 percent of women in Indonesia were unemployed in 2008) more drastically than men and pushing a growing number of unemployed women into the informal economy.

Globally, food peddling, domestic help, and trash picking are among the most common jobs in the informal economy. In mainland Southeast Asia, a major part of the informal economy is trade in opium and heroin. The Golden Triangle is one of the two largest heroin-producing regions in the world (the other being Afghanistan), stretching over northern Thailand, Myanmar, and Laos. The Shan State in northern Myanmar is the main production area for opium in the Golden Triangle. Historically, opium was grown in the Shan State, transported to Thailand, then to Hong Kong, and finally to the global market. Opium is still produced in the Shan State, but it now follows a different path in response to crackdowns on the opium trade in Thailand. From production in the Shan State, opium now flows into Yunnan, China, then to the Chinese provinces of Guizhou, Guangxi, and Guangdong, then Hong Kong, and onto the global market (Figure 7.31). The Golden Triangle dominated the opium trade from the early 1900s until the 1990s, when Afghanistan's opium production surpassed it.

TOURISM

Tourism is short-term travel to a destination away from home with the central purpose of recreation and relaxation. Thousands of miles of coastline in Southeast Asia are home to pristine tourist destinations, including Bali

---

tourism short-term travel for the purpose of recreation and relaxation

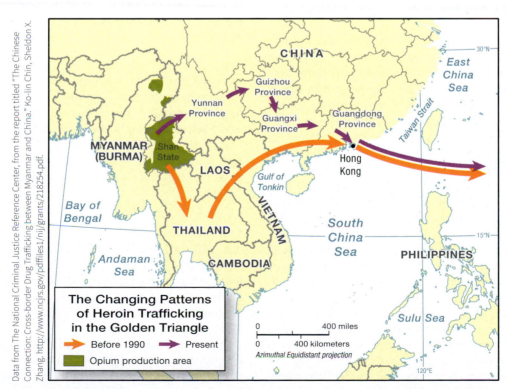

Data from The National Criminal Justice Reference Center, from the report titled "The Chinese Connection: Cross-border Drug Trafficking between Myanmar and China." Ko-lin Chin, Sheldon X. Zhang. http://www.ncjrs.gov/pdffiles1/nij/grants/218254.pdf.

The Changing Patterns of Heroin Trafficking in the Golden Triangle
→ Before 1990 → Present
■ Opium production area

FIGURE 7.31 **Opium and heroin trade.** The world's largest opium production region is the golden triangle, which includes Myanmar, Laos, and northern Thailand. The export path for opium produced in the region has changed, both in response to crackdowns in Thailand and the growth in opium demand in China. East Asia now accounts for 25 percent of global heroin use.

(Indonesia) and Phuket (Thailand). In 2008, over 65 million international visitors arrived in Southeast Asia, up from 20 million in 1991, according to the ASEAN Tourism Association.

The earliest tourists throughout the world were pilgrims who traveled to sacred sites. In Southeast Asia, Hindu and Buddhist temples, such as Angkor Wat and Borobudur, drew pilgrims for centuries before tourism for purely recreational purposes began in the region in the 1800s. The Industrial Revolution in the 1700s and 1800s in Europe helped generate wealth for factory owners, managers, merchants, and large-scale farmers in Europe. Wealth, specifically disposable income, is required for tourism to occur. In addition, tourists need time off from work and access to transportation. In the 1800s, improvements to the steam engine and its use in powering steam ships and trains made travel easier both by sea and land.

The first tourists in Europe headed to two major destinations still common among tourists today: the mountains and the beach. Originally, tourists traveled to the mountains during the summer, but by the late 1800s, the ski industry and winter tourism were well established in the European Alps. The first beach destination in Europe was the French Riviera, which is on the Mediterranean Coast. In the 1800s, tourism to the beach was mostly in the winter months, but by the 1920s the beach tourist season shifted to the summer months when workers had vacation days. In 1883, the French government required 12 days of paid vacation for employees. Today, full-time employees in France receive a minimum of 30 days of paid annual leave, by law (Figure 7.32).

The paid time off employees in wealthy countries receive is central to understanding tourism in Southeast Asia. The vast majority of tourists in Southeast Asia come from Europe and Australia (Figure 7.33). Tourism in Southeast Asia is year round, with peak seasons in dry months (opposite the wet monsoon) and around the Christmas holiday. On mainland Southeast Asia, the tourism season stretches from December to March, during the area's winter or dry season.

On the mainland, the dictatorial and communist regimes of Cambodia, Laos, and Vietnam deterred most Western tourists in the 20th century. Tourism was unheard of in Cambodia (the home of Angkor Wat) until the 1980s because the Khmer Rouge, a communist party led by Pol Pot, waged a genocidal campaign against Cambodians from 1975 to 1979. The Khmer Rouge tortured, executed, and starved 1.7 million people on Cambodia's killing fields. Although Cambodia is still extremely poor and the government is corrupt, Siem Reap is the closest city to Angkor Wat, and the city has "recently been transformed into a city of art galleries, gelaterias, and chic, architect-designed hotels such as the Hotel de la Paix" (Cummings 2008, 21).

On the islands of Southeast Asia, the tourism season includes the weeks before and after Christmas and also the months of July through September, which is the dry season for areas south of the equator in Southeast Asia. After

Paul Beinssen/Lonely Planet Images/Getty Images.

FIGURE 7.32   Bali, Indonesia. European tourists on holiday on Kuta Beach.

the India Ocean tsunami destroyed resorts and beaches in many popular tourist destinations, governments and nongovernmental organizations (NGOs) from around the world helped to rebuild the local tourist economies. The upscale resorts and hotels rebuilt first in Phuket, Thailand, with the first rebuilt less than 6 months after the tsunami devastated the region's people and economy.

COMMODIFICATION IN TOURISM

Although tourism is considered a form of economic development (see Chapter 3), it does not necessarily benefit local communities. A World Bank study estimated that people living in tourist areas receive only about 10 percent of what tourists spend. Much of the tourism revenue flows out of the country to developers, hotel and restaurant chains, and tour operators.

Businesses that stand to gain wealth from tourism have become increasingly **vertically integrated** as businesses serving different parts of the commodity chain have merged into one larger company or set of companies that work with each other to corner a market. Developers build global hotel chains that have contracts with global

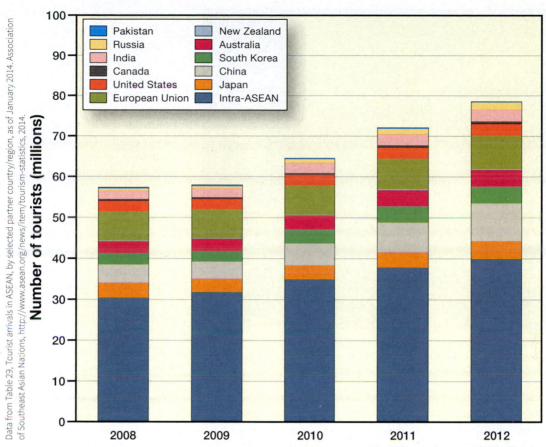

Data from Table 29, Tourist arrivals in ASEAN, by selected partner country/region, as of January 2014. Association of Southeast Asian Nations, http://www.asean.org/news/item/tourism-statistics, 2014.

FIGURE 7.33 **Tourists to Southeast Asia.**   The annual number of tourists in Southeast Asia increased between 2008 and 2012. The majority of tourists to Southeast Asian destinations originate within the ASEAN region. Tourists from European Union countries and from China account for the greatest proportion of tourists from outside the region. The number of Chinese tourists increased over the five years charted.

restaurant chains and give contracts to tour operators that send tourists to their properties (Hemingway 2004). Global hotel and restaurant chains employ locals, but tourism employment typically offers low wages and few benefits, and is often seasonal.

In addition to tourist dollars flowing out of the countries, tourists impact local communities by consuming scarce commodities and resources including food, fresh water, and electricity. Tourists also require the building of large-scale infrastructure rarely used by locals, including international airports and miles of freeways. Governments receive income from tourism in the form of taxes, but much of the tax revenue is reinvested in airports, cruise ship ports, and other infrastructure to support tourism.

One of the biggest effects of tourism on local cultures is the commodification of cultural festivals and practices. **Authenticity** is the idea that one place or experience is the true, actual one. In tourism, cultural brokers and their

brochures offer one image or experience with a local culture that is typecast as the authentic image or experience of that culture; it is that image or experience the tourist or buyer desires. The commodification of local cultures is reinforced by **cultural brokers** who are tour guides, operators, and resort companies that sell an "authentic" image of a local culture, thereby commodifying a people or place.

Through commodification and authenticity, cultural brokers work from a stereotype or perception of a local culture to create a desired cultural experience for tourists. Cultural brokers freeze the culture or place in one time. The irony is that local cultures change over time and place; rarely do members of a local culture agree to one right way of practicing their culture. Through tourism, the local culture and place become commodified in order to mesh with tourists' expectations. Commodified local cultures are often sold to tourists as **ethnic tourism**, and most tourists believe they are experiencing an authentic local culture.

Australian geographer Sallie Yea (2002) studied the experiences of European tourists who visited longhouses

---

authenticity   the idea that one place or experience is the true, actual one.

FIGURE 7.34   Sarawak, Malaysia. A traditional Iban longhouse has a veranda stretching the length of the structure. Much of the community life takes place on the veranda, from drying rice and laundering clothing to preparing food.

© Luca Invernizzi Tetto/AGE fotostock.

on the Malaysian portion of the island of Borneo. Yea asserts that minority ethnic groups in Southeast Asia have become popular for ethnic tourism because tourists want to visit people whose lifestyles are much different from their own. In Malaysia, tourists go on longhouse tours to see how people live in traditional longhouses along the rivers in Borneo.

The Iban traditionally built longhouses along rivers in the Sarawak area of Borneo. Longhouses are a series of apartments built on stilts, connected by a long veranda, typically holding between 3 and 40 families (Figure 7.34). The Iban practice **shifting cultivation** (swidden), an agricultural system whereby they build a longhouse, clear a section of forest, farm the land, and then, once the land loses its fertility after about two or three years, clear another nearby field. They continue to live in the same longhouse for about 15 years, until they move on to allow the forest in that area to regenerate. At that point, the Iban move to another section of the river, build a new longhouse from materials near the river and materials salvaged and carted from the previous house, and begin the shifting cultivation process again with a new area of the forest.

Tour operators in Sarawak, most of whom are Chinese and not Iban, act as cultural brokers, selling ethnic tours of Iban longhouses as "authentic" tours of a local culture. However, what tourists experience is anything but authentic. Tourists experience what tour operators construct for them, an experience that meshes with common perceptions or images of the Iban. Yea found that cultural brokers encouraged the Iban to build more

permanent homes closer to coastal cities so that tourists would not have to travel as far inland on the tours; asked the Iban to stay home, rather than farm, during tours; suggested the Iban dress traditionally; and told the Iban to make their homes look more "authentic" in order to construct the kinds of experiences tourists expect. One longhouse headman explained to Yea that the tour operators gave him "specific and detailed instructions for traditionalization of the longhouse, which included tacking bark to all of the interior walls, hanging skulls, beads, and tapestries, and other trinkets from the rafters and walls, and hiding any modern electrical appliances and other signifiers that this Iban community were living in the late twentieth century" (2002, 185) (Figure 7.35).

Geographer Yea recorded the comments tourists, most of whom were from Europe, made about their visits. A tourist from the Netherlands commented, "'Let the Iban stay the way they are'" (2002, 182). Another tourist from France wrote in a guest book: "'What a beautiful life here. Don't change a thing'" (2002, 182). In a content analysis of tourist comments, Yea found the most common remarks from tourists were "unique," "traditional culture," and "don't change."

The Iban in Malaysia are not the only local culture being commodified for tourism. Remote locations in Southeast Asia report a marked increase in the number of European, American, Chinese, and Japanese tourists seeking authentic, noncommodified experiences with local cultures. Tourists increasingly desire to observe or participate in long-standing rituals, practices, or festivals of local cultures. For example, in Luang Prabang, Laos, each morning monks from more than 80 temples in the city file into the streets in saffron-colored robes and participate in *tak bat*, the traditional Buddhist practice of asking for alms, or donations of food (Figure 7.36). The locals have, for centuries, awakened early to prepare sticky rice and then take it with bananas and other foods to perches along the streets to share their food with the monks. Both the locals and the monks perform the morning ritual in silence or a meditative state.

More than 300,000 tourists visit the town of Luang Prabang, Laos, each year to see the Tak Bat and observe the local culture. The increasing number of tourists is placing pressure on the ritual. A local prince explained, "'For many tourists, coming to Luang Prabang is like going on safari, but our monks are not monkeys or buffaloes'" (Gray 2008). Tourists who observe the Tak Bat as a show or demonstration often behave in ways that undermine the solemn, centuries-old practice.

GUEST *Field Note*   Iban Longhouse Tourism

DR. SALLIE YEA

National Institute of Education, Singapore

The impacts of tourism on indigenous peoples and ethnic minorities has been the subject of extensive discussion in tourism geography. Through my fieldwork with five different Iban communities in the Batang Ai region of the East Malaysian state of Sarawak, I found that tourism led to new spatialities, landscapes and social relations in the longhouses. Over time longhouses generally increased their economic dependency on tourism, but economic improvement came with social costs as new inequalities emerged within and between longhouses, and between longhouses and external stakeholders, such as tour operators. Communities were compelled to alter the physical landscapes of their communities to make them appear more traditional and "authentic" for the tourist gaze. Reflecting on these findings, I feel it would have been extremely difficult to gain the insights into the dynamics of tourism in the longhouses through rapid surveys. Staying with the communities, conducting in-depth interviews and focus groups, and engaging in participant observation all provided nuanced insights into the community dynamics as they were transformed through tourism. A return visit to some of the longhouses several years later has revealed that many of the longhouses are splitting into two semi-communities of traditional and modern longhouse. The traditional longhouse is retained for tourist visits, but most of the community lives in the modern longhouse, which is concealed from tourists. Further fieldwork would reveal how this remapping of the Iban longhouses is productive of further alterations in social relations, everyday routines, and identity.

© S K Chong/PhotoAsia/agefotostock.

FIGURE 7.35   Kapit, Sarawak, Malaysia. Iban longhouses.

© DEGAS Jean-Pierre/h/age fotostock.

FIGURE 7.36   Luang Prabang, Laos.   Monks participate in the procession of the Tak Bat ceremony. Each morning at sunrise, monks receive offerings from the community.

SEX TOURISM

Sex tourism impacts the entire region of Southeast Asia, and child trafficking is closely associated with sex tourism. Sex tourism took off in Thailand during the Vietnam War, when American troops were sent to Thailand for rest and relaxation. Since then, Thailand has been joined in competition by Cambodia, Laos, and other countries in the region. Child trafficking feeds the sex tourism industry by taking boys and girls from their homes and forcing them to work as sex workers in other parts of the region. Not all children trafficked end up in the sex industry. Some children are trafficked for illegal adoptions, to work as beggars, aid in criminal activities, or become domestic servants (Rafferty 2007).

Geographers can use scale to grasp the forces that conspire to make Southeast Asia ripe for child trafficking. At the regional scale, some countries in Southeast Asia have been known as economic tigers and other countries are quite impoverished in comparison. Uneven economic development within the region opens poor areas in poorer countries to exploitation and trafficking (Figure 7.37). Currently, girls in Laos and Cambodia are particularly susceptible to trafficking because of the lack of employment in the countries, especially for women and children. In Cambodia, between 15 and 33 percent of the prostitutes are children between the ages of 9 and 16, and many of the child prostitutes in Cambodia were trafficked from Vietnam and China.

© epa european pressphoto agency b.v./Alamy.

FIGURE 7.37 **Manila, Philippines.** Filipino demonstrators hold up signs against human trafficking during a protest in 2013. Human rights activists demanded the government intensify efforts against human trafficking, referring to alleged cases where the Malaysian state of Sabah has been used as a transit point for trafficking Filipinos into Malaysia.

In Southeast Asia, Thailand is a well-known destination for victims of sexual trafficking, in large part because of its vibrant economy and well-established sex industry. An estimated 40,000 children are enslaved in the sex industry in Thailand, and a growing number of the children are boys.

At the national scale, weak law enforcement, lack of gender empowerment, low social status, and lack of access to education play roles in predicting the likelihood of child trafficking. "The UN Special Rapporteur on the Sale of Children highlights limited access to education and perceived social value as some of the disadvantages faced by girls in certain Asian countries" (Hemmingway 2004, 279). In response to the lack of brides as a result of the one-child policy in China (see Chapter 8), a growing number of girls and women are trafficked from Thailand, Vietnam, and Laos through Laos into China.

At the scale of the family in Southeast Asia, females are vastly more likely to be trafficked as children than males. Upwards of 98 percent of children trafficked for the sex trade are female (Rafferty 2007). Other risk factors within families that increase the likelihood of a child being trafficked are being the oldest female, belonging to a dysfunctional family, and living in poverty.

Pedophiles are attracted to Southeast Asia's sex tourist industry because the region is known for "weak law enforcement, a large sex industry, and much poverty" (*Economist* 2008, 36). The United States and Germany are working to stop pedophiles from leaving their home countries to travel to Southeast Asia, but other sending countries have done little to stem the tide. In addition, pedophiles from countries in Southeast Asia help perpetuate child sex tourism.

CONCEPT **CHECK**

1. **Why** are the Straits of Malacca so important in global shipping?

2. **Where** is opium produced in Southeast Asia, and **how** is it distributed?

3. **How** has the growth of tourism in Borneo impacted the livelihoods and lifestyles of the Iban?

HOW HAS MIGRATION SHAPED SOUTHEAST ASIA?

LEARNING OBJECTIVES

1. **Explain** the draw of primate cities for migrants and **understand** why primate cities in Southeast Asia are susceptible to flooding.

2. **Describe** when and to where overseas Chinese migrated to Southeast Asia and **understand** the role of overseas Chinese in the region's economy.

3. **Understand** why women in Indonesia migrate to Southeast Asia as guest workers.

Wealth in the city of Jakarta, Indonesia is not distributed equally. The outskirts of the city are teeming with slums (called kampung, which translates as village), and within the city temporary shacks are squeezed into unused spaces, such as along railroad tracks. An estimated 26 percent of Indonesians live in slums. The country of Indonesia has a Gini index of 0.41 in 2011, which is up from 0.31 in 1999. The Indonesian Central Statistics Agency reported that in 2011 the top 20 percent of income earners in the country controlled 48.42 percent of household income, while the bottom 40 percent controlled 16.85 percent of household income (Sinaga 2012).

At the national scale, **primate cities**, which are lead a country in size and influence, including Jakarta, have attracted migrants who are drawn by employment opportunities. At the regional scale, overseas Chinese who migrated to trade centers in Southeast Asia over centuries are influential in the regional and global economies. Globally, Indonesian guest workers and refugees from hill tribes migrate to find a better life, whether economic or with respect to human rights. In the process, migrants build transnational identities and connect places.

MIGRATION THEORY

Migration theory considers **push and pull factors**, through which migrants weigh their reasons for leaving home and for going to a particular place (see Chapter 11). The decisions and lives of migrants are not as simple as weighing a balance sheet of push and pull factors in a sterile environment and then turning in one identity or home country for another.

The decision to migrate is made in a context of global, national, local, familial, and individual concerns. When migrants leave home for a new place, they take part of their home culture with them through relocation diffusion and bring it to the new destination. Migrants create **transnational** identities, negotiating how they make sense of themselves by balancing their home identity with their identity in their country of destination.

NATIONAL SCALE: RURAL TO URBAN

Approximately 1 billion people who make up the urban poor in Southeast Asia, South Asia, Subsaharan Africa, and Latin America live in slums skirting major cities (UN Habitat 2003). United Nations Habitat, an organization whose mission is to "promote socially and environmentally sustainable cities and towns with the goal of providing adequate shelter for all," estimates that 28 percent of Southeast Asians live in urban slums (Figure 7.38a and b). Slums

(a)

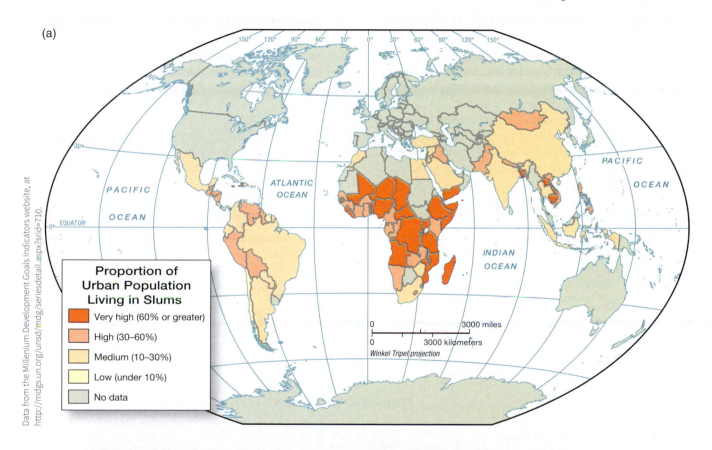

Proportion of Urban Population Living in Slums

- Very high (60% or greater)
- High (30–60%)
- Medium (10–30%)
- Low (under 10%)
- No data

(b)

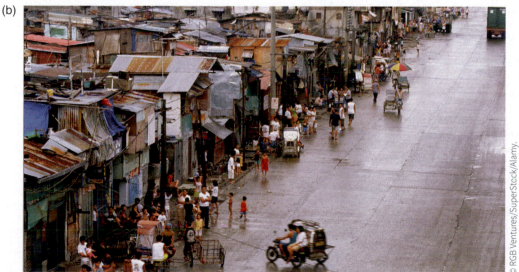

© RGB Ventures/SuperStock/Alamy.

FIGURE 7.38a, b   **Urban slums.**   **(a) Urban population living in slums.**   Subsaharan Africa and Southeast Asia have the highest proportion of city-dwellers living in slums. **(b) Manila, the Philippines.**   Some of the largest slums in Manila are close to the city center, and newer slums form a ring around the outside of the city.

grow fastest where massive numbers of rural poor migrate to the cities. In Southeast Asia, the largest slums are in the Philippines, Malaysia, and Indonesia, each of which has a primate city to which internal migrants are drawn: Manila, Kuala Lumpur, and Jakarta, respectively.

The growth in number of urban poor in primate cities has resulted in an increasing number of people susceptible to urban flooding. In Southeast Asia primate cities are located on rivers and are often built on floodplains. Slums and industrial areas, which are often located in low-lying areas, are susceptible to flooding. Jakarta is sited near 13 rivers, which are inundated with silt and waste that has not been dredged "for centuries" (*Economist* 2012, 2). Like Venice, Italy (see Chapter 9), Jakarta is sinking under the weight of the city and as a result of groundwater drawdown bringing fresh water to the burgeoning population (*Economist* 2012).

Even cities that do not have large slum populations are susceptible to flooding. Thailand's primate city is Bangkok, but not as many people live in its slums because the government has a proactive program to provide housing improvements for the city's slums. Nonetheless, in 2011, Bangkok suffered $46 billion in damages when the Chao Phraya River flooded after a heavy rainfall (*Economist*

2012). Much of the flooding damage occurred in an industrial center that was built on former rice paddies. The problem with paving over rice paddies is twofold: (1) water runs off pavements instead of being absorbed by soil and (2) rice paddies are sited where they are "precisely because the land floods regularly" (*Economist* 2012, 2).

The majority of people in Cambodia, Laos, and Vietnam live in rural areas. About 70 percent of Cambodians are subsistence farmers, in large part because the Khmer Rouge "abolished money and private property and ordered city dwellers into the countryside to cultivate fields" (BBC News, July 3, 2012). More than 75 percent of the people in Laos live in rural areas of the country. The rural population of Laos continues to grow despite rampant poverty in a country where, according to the World Bank, "two-thirds of households have no access to electricity, half have no safe water supply and half of all villages are unreachable by all-weather roads during the rainy season." Vietnam, a communist state with an increasingly open economy, still has a rural population of over 70 percent. As the economies of Cambodia, Laos, and Vietnam increase trade and connectedness with the global economy, economic growth in their major cities will lead to greater rural-to-urban migration.

Data from: Gerard Chaliand and Jean-Pierre Rageau. *The Penguin Atlas of Diasporas.* Penguin Books, 1997 and http://www.wcec-secretariat.org/en/pdf/06/sycip_w_11.pdf

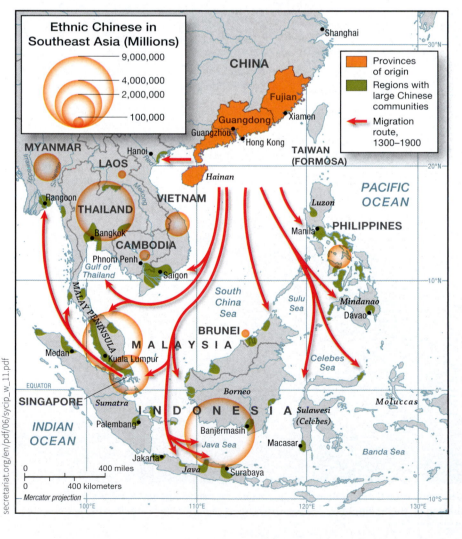

FIGURE 7.39 Overseas Chinese in Southeast Asia. Chinese migrated into Southeast Asia, settling primarily in port cities and along coastlines.

REGIONAL SCALE: OVERSEAS CHINESE IN SOUTHEAST ASIA

Of the 35 million overseas Chinese, 30 million live in Southeast Asia (Figure 7.39). Between the seventh and sixteenth centuries, Chinese migrated to Southeast Asia during transitions among the Tang, Song, and Yuan dynasties and also in response to trade relationships between each dynasty and centers in Southeast Asia (Lai 2004). Between the seventeenth and twentieth centuries, millions of Chinese migrated to Southeast Asia for employment opportunities in European colonies in Southeast Asia, and a smaller proportion migrated to the Americas.

In the second and larger wave of migration from China to Southeast Asia, Chinese migrants sought to tap into growing economic opportunities in Southeast Asia. When Chinese migrants reached Southeast Asia in the twentieth century, they arrived in **trade centers** where earlier Chinese migrants had already set up businesses.

During the twentieth century, a series of events from a civil war in China to Japanese invasion during World War II and the establishment of communism led to thousands of Chinese weighing push factors and making the decision to leave China. In 1882 the United States banned Chinese immigration, which intensified the pull of Southeast Asia for Chinese migrants. Although the Chinese who migrated to Southeast Asia came from diverse regions of China, they were generally poor peasants (Thomson 1993, 2). Once in Southeast Asia, Chinese settled in trade centers, created ethnic neighborhoods, perpetuated Chinese identities, and maintained cultural and economic links by keeping contacts with China (Lia 2004).

During both waves of the Chinese diaspora to Southeast Asia, the relationship between the Chinese **enclaves**, the home base (often an ethnic neighborhood) of migrants in their new country, and people in their home provinces in China remained tight. Migrants in Southeast Asia sent remittances home to China, much as Mexican and Latin American migrants in the United States and Canada do today (see Chapter 12). The first generations of overseas Chinese in Southeast Asia focused on sending remittances home and worked diligently to maintain Chinese customs and culture along with their ethnic identity, often by settling in Chinatowns.

The tight relationship between overseas Chinese and their families in their home provinces helped establish **chain migration**, where migrants find economic success in their new location and encourage others back home to follow the same path (Figure 7.40). Until

the 1800s, the majority of overseas Chinese hailed from coastal China's Fujian Province and after the 1800s, Guangdong Province.

Efforts by governments in South Asia to protect the production of their staple crop, rice, helped pull Chinese to urban centers. In Thailand, for example, the government prohibited overseas Chinese from engaging in rice cultivation (Thomson 1993). Unable to work in rice production, overseas Chinese became unskilled workers, miners, fruit and vegetable growers, merchants, and traders. Chinese tended to live in cities and became the "commercial class, or laborers, with most of their economic activity" in the service economy in Thailand (Thomson 1993, 3). With the growth of trade and industrialization, overseas Chinese have been well positioned to play the role of entrepreneurs and to benefit financially.

After overseas Chinese found economic success, they became settlers and leaders in Chinatown who saw themselves as responsible for maintaining Chinese language and culture, especially for their children (Lia 2004). Eventually, economic success encouraged overseas Chinese families to move out of Chinatown and focus more on assimilation than on preservation of Chinese culture.

Ethnic Southeast Asians often resent the economic successes that overseas Chinese have realized. Since the 1950s, most governments in Southeast Asia have closed their countries to immigration from China. Between 1950 and 1976, as a result of immigration policies in Southeast Asia and the Maoist government in China barring communication with overseas Chinese, a gap between overseas Chinese and mainland Chinese developed.

Bruno Morandi/The Image Bank/Getty Images.

FIGURE 7.40   Kuala Lumpur, Malaysia.   The petaling market in the Chinatown district of Kuala Lumpur attracts tourists and locals.

Governments in Southeast Asia seek to protect the indigenous population from the economic power of overseas Chinese. Ethnic Chinese control the majority of companies in the stock markets of Thailand, Singapore, Malaysia, the Philippines, and Indonesia. The Malaysian government uses a racial quota system to reserve spots at universities for ethnic Malays. Additionally, Malays receive a larger proportion of jobs in the government as well as greater access to business loans and property than overseas Chinese. In Indonesia, non-Chinese are afforded access to land ownership and businesses. Thailand's King Rama VI once called Chinese in Southeast Asia the "Jews of the East," implying that their distinct ethnicity and commanding role in the businesses and economy of Southeast Asia made them targets of resentment and discrimination by Southeast Asians.

GLOBAL SCALE: INDONESIAN GUEST WORKERS IN SAUDI ARABIA

In the late 1800s and early 1900s, Indonesian women migrated internally from rural areas to urban areas of Indonesia to work as domestic servants in the hopes of earning enough money to support their families back home. Since the 1970s, Indonesian women have increasingly migrated internationally to work primarily as domestic servants. The Indonesian government encourages workers to go abroad, legally allowing over 2 million citizens to emigrate between 1999 and 2004. More than 80 percent of Indonesians working abroad are women, and more than 90 percent of the women abroad are working as domestic help.

Between 1999 and 2004, Indonesians sent more than $2.5 billion in remittances home, a statistic not lost on the government. The largest flow of Indonesian women migratory workers is from the island of Java to Saudi Arabia in Southwest Asia. Geographer Rachel Silvey traced the cycle of migration from a worker's conversations with Saudi Arabian labor recruiters in Indonesia to her work in Saudi Arabia to her return to life in Indonesia.

Saudi Arabian recruiters abound in Indonesia, especially in Islamic boarding schools for girls, where they expect to find devout followers of Islam who desire to visit Mecca. The number of migrant workers in Saudi Arabia today is more than 780,000, making Saudi Arabia the country in the world with the most legal immigrant guest workers. To recruit Indonesian women, recruiters contend that Indonesian women should work in Saudi Arabia because the country is pious and the Muslim middle and upper classes will not force Indonesian women to cook or eat pork or to take care of dogs, cultural practices forbidden for Muslims (Silvey 2007).

Indonesian women working in Saudi Arabia live as transnationals. Although women who were recruited from Islamic boarding schools know some Arabic, they do not know enough or have a position to truly integrate into Saudi Arabian society. Upon the women's return to Indonesia, many of their husbands separate from them, and they are accused of being prostitutes because of the money they

accumulated or sent home, or become victims of corruption or harassment.

Silvey found that the two major reasons for the massive migration flow of women immigrants from Indonesia to Saudi Arabia are remittances and cultural ties. First, women migrate to find work and send remittances home as part of their motherly duty. Second, Indonesian women see Saudi Arabia as an attractive destination because it is home to Mecca. One of the five pillars of Islam instructs followers to take the hajj, pilgrimage to Mecca, at least once in their lifetimes if financially possible. Many of the Indonesian women Silvey interviewed hoped to participate in the hajj during their work time abroad or to visit Mecca during another time of the year.

ACROSS SCALES: KAREN REFUGEES, MYANMAR, THAILAND, AND NORTH AMERICA

Until 2011, Myanmar had the longest running military dictatorship in the world. The military junta, comprised of several prominent families, came to control the country in 1948 when Myanmar (then Burma) gained independence from its colonizer, Great Britain. The Karen, a major ethnic group who live along the border with Thailand and who often converted to Christianity during the British colonial period, did not recognize the newly independent government in Burma, which began what is now the world's longest running civil war.

In 1988, Than Shwe became the leader of the military junta and took control of the country. Under Shwe's leadership, the military junta changed the toponym of the country from Burma to Myanmar in 1989 (Burma and Myanmar are one and the same, but Myanmar is the more formal name of the country) and then moved the capital city from Yangon, which is in the south, approximately 200 miles (321 km) north to the middle of a malaria-infested jungle. The new capital city, named Nay Pyi Taw (meaning Royal City), is in the middle of the country, far removed from everything.

Some political observers saw the move of the capital as the government's transition deeper into isolation. The government isolated itself from much of the outside world, purportedly fearing an attack by the United States. Others in Myanmar saw the move of the capital as a reflection of the military junta's faith in soothsayers or astrologers. Officials inside Myanmar reported that the government moved the capital because General Shwe's astrologer said the capital needed to be moved in order to avert an impending disaster (MacIntyre 2007). The caravan of trucks moving officials and government furnishings to Nay Pyi Taw left Yangon on November 6, 2005, at precisely 6:37 A.M. because General Shwe's astrologer declared that day and time as the most fortunate moment to move the capital.

In 1990, an opposition party won legislative elections, but the military junta refused to recognize the elections. Shwe continued to rule the country without regard for human rights or freedoms. The government confiscated land, forced people into forced labor, raped, and forcibly attacked members of minority groups. The Christian Karen who actively rebelled against the government were targeted. At the time, people in Myanmar had little access to the Internet, no freedom of the press, and suffered under the stiff control of the military.

Approximately 350,000 Karen fled from Myanmar to neighboring Thailand as refugees from the civil war and growing political persecution by the Shwe government. An organization of 11 international NGOs still oversees and provides services to the Karen refugee camps in Thailand. In 2005, the United Nations High Commissioner on Refugees began a program to resettle Karen refugees in Western countries, including Canada and the United States.

In 2007, the International Committee of the Red Cross (ICRC) publicly condemned the government's violation of its people's human rights. In 2007, Buddhist monks, who are revered by the people of Myanmar, led protests against the oppressive Shwe government after the government raised the prices of fuel. Aung San Suu Kyi, the chief opposition, pro-democracy leader in Myanmar, has been in and out of house arrest in the country since 1989. In 2007, the military junta allowed Aung San Suu Kyi to leave her home for the first time since 2003 to greet the protesting Buddhist monks.

Under pressure from international sanctions, human rights organizations, other countries, and its own people, the government of Myanmar held elections in 2010. Although Aung San Suu Kyi's party boycotted the elections, a new leader named Thein Sein, handpicked by Shwe, took control of the country in 2011. Sein has reached a ceasefire with the Karen. In response to Sein's reforms, then U.S. Secretary of State Hillary Clinton met with him, the European Union withdrew its economic sanctions, and the prime minister of India visited the country. By 2014, the new government was embroiled in battle again, this time with rebel groups in the Kachin and Shan States in the north. Civilians in the region remain caught in the crossfires, and human rights groups report that the government military is committing war crimes, including using civilians as shields in battle, raping women, and destroying villages.

CONCEPT CHECK

1. **Why** are primate cities in Southeast Asia susceptible to flooding?

2. **How** can you see the presence of overseas Chinese in the cultural landscape of Southeast Asia?

3. **Where** do Indonesian women migrate as guest workers and why?

SUMMARY

What Is Southeast Asia?

1. Southeast Asia has two physical regions: mainland Southeast Asia, which is the southeastern edge of the Eurasian plate, and insular Southeast Asia, which is the thousands of islands that extend from western Indonesia to eastern Indonesia and south to north from Australia to the Philippines.

2. The mountain chains in mainland Southeast Asia run parallel to the plate boundary between the Indian plate and the Eurasian plate. Major rivers run from higher elevations in the Tibetan Plateau and Yunnan Province of China to the Indian and Pacific oceans between mountain ranges in Southeast Asia.

3. The continental shelf of Eurasia extends beyond the Malaysian Peninsula and creates relatively shallow seas in the western portion of Southeast Asia. The Eurasian plate meets the Indian Ocean plate, the Philippine plate, and the Australian plate in this region. The islands of Southeast Asia parallel the plate boundaries.

4. The plate boundaries are areas of active tectonic activity, making the region susceptible to earthquakes, tsunamis, and volcanoes.

Who Are Southeast Asians?

1. Indonesia is composed of more than 13,000 islands, and the Philippines comprise more than 5,000 islands. Both countries have enacted language policies to try to connect the people and inhabited islands. Indonesia has successfully used the Indonesian language as a lingua franca for over 1,000 years. The language functions as a unifying force in the country. The Philippines have become a center for English education in the region, drawing students from Europe, Asia, and Russia who hope to learn the global language of English with an American accent.

2. Hinduism, Buddhism, Islam, and Christianity have diffused into Southeast Asia over the last 2,000 years. The diversity of religions can be seen in the cultural landscape of temples. In Indonesia, Prambanan is a Hindu temple built to honor Lord Shiva, and Borobudur is a Buddhist temple that was buried under volcanic ash for centuries. Angkor Wat in Cambodia was built as a Hindu temple, coopted into a Buddhist temple, and then overgrown by rainforest. Today, the site of Angkor Wat is the largest religious structure in the world and draws 1.5 million visitors a year.

3. Islam diffused to Southeast Asia by trade. The religion took root hierarchically in important trading ports first and then contagiously from the ports to people in surrounding areas. Sultans, Islamic leaders who controlled trade and spice production in certain areas of the region, became powerful, wealthy leaders through the spice trade. The Sultan of Brunei, a small country on the island of Borneo, whose wealth is derived largely from oil resources today, is worth $20 billion.

What Role Does Southeast Asia Play in the Global Economy?

1. The rapid growth of Southeast Asian economies in the 1990s and early 2000s was built on crony capitalism, where the government propped up industries owned by those most connected to the government. ASEAN, a regional trade organization, is working to encourage the growth of companies that produce for domestic and regional consumption.

2. Singapore is a port city built by overseas Chinese. The city-state is the wealthiest country in the region. The island state is also an entrepôt, or gateway for trade, in Southeast Asia and the largest bunkering port, where container ships refuel, in the world.

3. Tourism is a growing industry in Southeast Asia. Tourists flock to sacred and historic temple sites, including Angkor Wat in Cambodia. Ethnic tourism is a growing industry on the islands in the region. Tourists seek an authentic experience with or observing a local culture. Cultural brokers commodify local cultures to present a prescribed experience to tourists.

How Has Migration Shaped Southeast Asia?

1. At the national scale, rural residents are migrating to primate cities in Indonesia, Thailand, Malaysia, and the Philippines. Located on floodplains of rivers, the cities are increasingly susceptible to flooding.

2. Overseas Chinese live in trade centers in Southeast Asia that were built over time through chain migration. Overseas Chinese are often entrepreneurs and business leaders in the region of Southeast Asia.

3. Globally, women from Indonesia have migrated to Saudi Arabia as guest workers. Muslim Indonesians are drawn to Saudi Arabia both for employment and for the opportunity to complete the hajj while they are in the country. Indonesian guest workers typically send remittances back to their home country.

4. Myanmar (Burma) has been independent since 1948 and has been in a state of civil war ever since. The Karen, who live in southeastern Myanmar, near the border with Thailand, have rebelled against the increasingly oppressive government. Since the 1990s, the government has taken land and homes from the Karen and forced them into labor for the military. Karen refugees fled first to Thailand, and many are now being resettled by NGOs in the United States and Canada.

YOUR**TURN** ▶ Geography in the Field

This photo from Singapore Harbor shows cranes lining the waterfront and hundreds of containers waiting for shipment. Each crane is ten stories high. When a container ship reaches its berth in the harbor, each crane moves the containers at a rate of about 35 containers an hour. The port of Singapore is a break-of-bulk port, which means container ships move into the port; containers are removed; and then the containers are placed on trucks or rail cars or assembled onto smaller ships or barges. The Strait of Malacca is tight, extending only 1.5 nautical miles (2.8 km) wide at the Phillips Channel, near Singapore. Like air traffic control at a busy airport, traffic control in the Phillips Channel guides enormous container ships through the narrow passage.

© HOW HWEE YOUNG/epa/Corbis Images.

Singapore. Cranes and containers line the harbor.

Thinking Geographically

- Examine a map of possible shipping routes between the Indian Ocean and the Pacific Ocean. List three possible paths for ships to travel. Determine whether the Strait of Malacca is the best possible route to move goods between East Asia and Europe.

- Examine the map of shipping lanes and roads at: http://news.nationalgeographic.com/news/2007/06/070628-human-footprint.html. Explain how this map demonstrates why the geologic era we live in can be described as the Anthropocene.

KEY TERMS

Karst topography	vernacular	crony capitalism	ethnic tourism
subsolar point	lingua franca	entrepôt	shifting cultivation
humid subtropical	overseas Chinese	bunkering port	primate cities
tsunami	global language	informal economy	push and pull factors
drawdown	Theravada Buddhism	tourism	transnational
spatial interaction	Mahayana Buddhism	vertical integration	trade center
local cultures	Vajrayana Buddhism	authenticity	enclaves
animist	sultanates	cultural brokers	chain migration

CREATIVE AND CRITICAL THINKING QUESTIONS

1. Imagine you are a **tourist** in Bali, and describe the **commodity chain** for the hotel you stay at or the meal you eat at a restaurant. At which step(s) in the commodity chain is most of the wealth accumulated and by whom?

2. Choose one of these three temples: Angkor Wat, Borobudur, or Prambanan. Research the temple at: UNESCO.org. Describe the **cultural landscape** of the temple site. Examine what the predominant religion is in the country where this temple is located today. Justify why the temple still stands despite the **diffusion** of a different religion to this site.

3. Describe the **network** of countries connected through the Mekong River. Determine whether you think a new dam should be built in the upper Mekong.

4. Explain how humans have changed the coastal environment of Southeast Asia in the **Anthropocene**. Using the Indian Ocean tsunami as an example, describe how human-induced changes to the environment amplified the impact of the tsunami in some places.

5. The Indonesian government believes the Indonesian language helps create a national **identity** in the country. Compare and contrast Indonesia and Malaysia's language policies to determine whether and when a common language builds or undermines a national identity.

6. Through ASEAN, corporations in Southeast Asia are working to produce more goods to be consumed within the region. If goods are produced and consumed at the same **scale** (in the same region as opposed to those in one region and consumed globally), how does that impact **unequal exchange**?

7. Describe how cultural brokers construct an "**authentic**" tourist experience for visitors to Iban longhouses in Malaysia. Determine whether the commodification of local culture can be achieved for tourist consumption without undermining the **authenticity** of experience.

8. Draw your **mental map** of the Iban region of shifting cultivation before **tourism**. Then, draw your **mental map** of the Iban region after globalized tourism. How does the increasing dependence on the formal economy change the land use in the place?

9. Look up Singapore on a map, and describe its **site.** Explain the **situation** of Singapore in the world economy today.

10. Describe the waves of **migration** of overseas Chinese to Southeast Asia. Using the **context** of agriculture and rice production in Southeast Asia, explain why overseas Chinese ended up living in cities and trade centers in the **region**.

SELF-TEST

1. The mountains of mainland Southeast Asia:
   a. run parallel to the plate boundary between South Asia and Eurasia
   b. run perpendicular to the plate boundary between South Asia and Eurasia
   c. are a volcanic arc on the north side of a subduction zone
   d. are a volcanic arc on the south side of a subduction zone

2. The ocean around the Malay Peninsula and in western Indonesia is generally _____ because the _____ of Eurasia extends into that region.
   a. deep/continental shelf
   b. deep/continental slope
   c. shallow/continental shelf
   d. shallow/continental slope

3. The Indian Ocean tsunami happened along _____, with the greatest amount of damage to the _____ of the plate boundary.
   a. a subduction zone/north and south
   b. a subduction zone/east and west
   c. folded mountain belt/north and south
   d. folded mountain belt/east and west

4. Indonesia's official language is Indonesian, which:
   a. has been used on the islands for more than 1,000 years
   b. is a lingua franca invented by the Dutch
   c. is an indigenous language used in East Timor that was adopted by the country at independence
   d. has no relation to the Malaysian language

5. Students from Russia, South Korea, France, and Germany attend school in the Philippines to learn this global language:
   a. Arabic   b. Chinese
   c. English   d. Spanish

6. Angkor Wat is the largest religious structure in the world and was originally built as a _____ temple and then coopted into a _____ temple.
   a. Buddhist/Muslim
   b. Muslim/Christian
   c. Christian/Hindu
   d. Hindu/Buddhist

7. Islam diffused into Southeast Asia _____ by first taking root in important trading ports and then _____ by spreading to people in surrounding areas.
   a. hierarchically/contagiously
   b. contagiously/hierarchically
   c. relocationally/stimulusly
   d. stimulusly/relocationally

8. The Sultan of Brunei belongs to one of the wealthiest royal families in the world. The family's wealth began in the spice trade and is now derived largely from:
   a. rare earth minerals
   b. coffee plantations
   c. ethnic tourism
   d. oil resources

9. Under crony capitalism, government gives contracts and tax incentives to industries that are owned by:
   a. multinational corporations with headquarters in the West
   b. nongovernmental organizations that have a charity presence in the region
   c. people within the country who are most connected to the government
   d. the government

10. The Iban traditionally live in longhouses as part of their economy built on _____. Today, some Iban have become increasingly dependent on a cash-based economy (the formal economy) as a result of a rise in _____.
    a. shifting cultivation / ethnic tourism
    b. ethnic tourism / shifting cultivation
    c. ethnic tourism / palm oil plantations
    d. palm oil plantations / ethnic tourism

11. Singapore is an entrepôt in Southeast Asia, which means it is a:
    a. bunkering port
    b. gateway for trade
    c. sacred site
    d. all of the above

12. Overseas Chinese:
    a. are often entrepreneurs and business leaders in Southeast Asia
    b. migrated to trade centers and not the rural, agricultural areas of Southeast Asia
    c. migrated through chain migration, connecting places in China with places in Southeast Asia
    d. all of the above

13. Primate cities in Southeast Asia include all of the following except:
    a. Jakarta, Indonesia
    b. Bangkok, Thailand
    c. Managua, the Philippines
    d. Kuala Lumpur, Malaysia

14. Women who migrate from Indonesia to Saudi Arabia as guest workers hope to:
    a. earn money and send remittances back to Indonesia
    b. have the opportunity to complete the hajj while in Saudi Arabia
    c. form transnational identities, bridging Indonesia and Saudi Arabia
    d. all of the above

15. Karen refugees from Myanmar have mainly settled in refugee camps in this nearby Southeast Asian country:
    a. Indonesia
    b. Thailand
    c. India
    d. Malaysia

ANSWERS FOR SELF-TEST QUESTIONS

1. a, **2.** c, **3.** b, **4.** a, **5.** c, **6.** d, **7.** a, **8.** d, **9.** c, **10.** a, **11.** b, **12.** d, **13.** c, **14.** d, **15.** b

8

EAST ASIA

Seoul, South Korea. Gwanghwamun Gate, at the entrance to the Gyeongbukgung Palace complex, was first built in 1395, but it does not look 600 years old. The gate, like South Korea, has changed in a multitude of ways since the establishment of the Choson Dynasty. The current gate was opened in 2010.

Erin Fouberg.

Gwanghwamun Gate in Seoul stands as a microcosm of South Korea's history. The gate marks the entrance to Gyeongbokgung Palace, a collection of buildings that originally numbered 500 when the first king of the Choson (Joseon) Dynasty built it in 1394. Choson architects designed the palace buildings and its main gate, Gwanghwamun, to align with the Bukaksan Mountains.

Japan invaded Korea in 1592 and burned down Gwanghwamun Gate. In 1867, in a surge of Korean nationalism, the king restored the gate at its original site, on the south end of the palace. In 1910, Japan invaded again and colonized Korea. The Japanese moved the gate to the east end of the palace in 1926 and built an enormous Japanese Government General Building on the site of the Gyeongbokgung Palace (Figure 8.1).

During the Korean War, from 1950 to 1953, bombing campaigns destroyed the Gwanghwamun Gate completely. At the end of the war, President Park Chung-Lee, the first leader of South Korea, rebuilt the gate on the south side, but the new gate was 6 degrees off of its original axis because the architects designed it to align with the Government General Building instead of the original palace.

In the 1980s, South Korea's economy began growing, and in 1989, the country became a democracy. In 1995, the government decided to destroy the Government General Building and to rebuild the Gyeongbokgung Palace. As part of that restoration, in 2010, the government restored Gwanghwamun Gate to its original location, on the south side and aligned it with the palace and mountains.

The gate now stands sentry at the palace of the Choson Dynasty, rebuilt by the democratically elected government of an economically thriving South Korea. Reflecting centuries of change and external influences, the Gwanghwamun gate tells us the arc of the story of South Korea in the context of East Asia.

THRESHOLD CONCEPT in this Chapter

Commodity Chain

FIGURE 8.1   Seoul, South Korea.   When Japan colonized South Korea, they tore down the palace of the Choson Dynasty, Gyeong-bukgung Palace. In its place, the Japanese built this Government General Building in 1926. After the Korean War, South Korea became independent. In 1995 the government of South Korea tore down this neoclassical building and reconstructed Gyeongbukgung Palace.

WHAT IS EAST ASIA?

LEARNING OBJECTIVES

1. **Describe** the physical geography of East Asia.

2. **Understand** what the Three Gorges Dam is and why the Chinese built it.

3. **Explain** how the islands of Japan were formed and **understand** what potential natural disasters Japan faces because of its location on two subduction zones.

Mount Fuji stands 62 miles (100 km) southwest of Tokyo, Japan, a city of over 13 million people (37 million in the metropolitan area) with global influence (Figure 8.2). Last erupting in 1707–1708, Mount Fuji is classified as an active volcano. In addition to Mount Fuji, Japan has more than 100 other active volcanoes. Japan is located on the western edge of the Pacific Ring of Fire. The plate boundaries around Japan are complicated, with two subduction zones and multiple connecting plates (see Figure 10.4). Japan has not only suffered from earthquakes and tsunamis, such as the one that caused the Fukushima Daiichi nuclear disaster in March 2011, but at some point, one or more of Japan's active volcanoes will invariably impact the country.

East Asia is defined by the Pacific Ocean as its eastern border. In the south and west, mountain chains divide East Asia from South and Central Asia. East Asia is not as easy to distinguish from Russia in the northeast or from Southeast Asia in the southeast. The boundaries of existing countries help define East Asia today, although historically peoples across the regions have had cultural ties.

FIGURE 8.2   Mount Fuji, Japan.   Mount Fuji is an active stratovolcano that last erupted in the 18th century. The volcano is located in an active tectonic region of Japan.

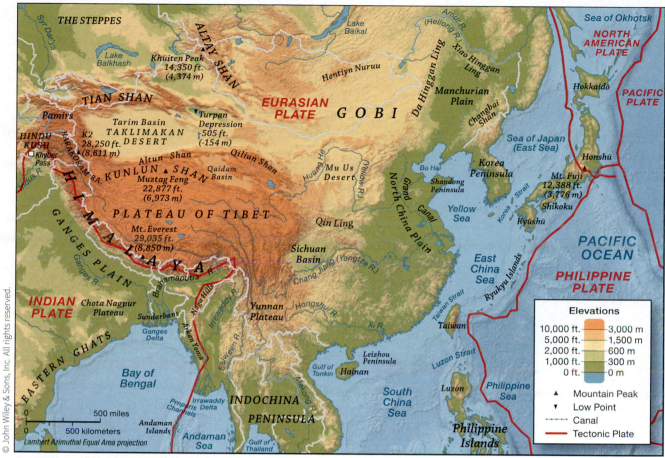

FIGURE 8.3 **Physical geography of East Asia.** The physical geography of East Asia includes the high plateau of Tibet in the west with average elevations of over 10,000 feet, the major rivers of the Huang He and Yangtze that flow from west to east, the Korean Peninsula, and the archipelago of Japan.

HARNESSING RIVERS

Civilizations in East Asia began along the Yangtze and Huang He valleys (Figure 8.3). The Great Wall stands as a reminder of the longevity of civilization in East Asia. The Zhou Dynasty began building the wall in 770 BCE. Linked together during the Qin Dynasty around 200 BCE, the sections of wall grew to over 5,500 miles (8,841 kilometers) long over the course of centuries in attempts to keep the Mongols out of China and to define the space of subsequent empires. While primary and secondary school textbooks focus on the Great Wall as a marker for the power of a series of Chinese empires, understanding how the Chinese have controlled the Huang He over time actually better illumines the might of Chinese empires.

Managing the Huang He

Windblown sediment, called **loess**, was blown southward onto the North China Plain from Siberia 2.5 billion years ago. Winds deposited the heaviest particles, the sand, in the Gobi Desert and carried the lighter particles, the silt, to form the Loess plains around the Huang He (Figure 8.4). The Huang He, which takes its name from the yellow

FIGURE 8.4 **Shanxi Province, China.** The Loess plains of China stretch over an area of about 250,000 square miles (approximately 650,000 sq km) along the western end of the Huang He. Loess is windblown sediment that is easily eroded.

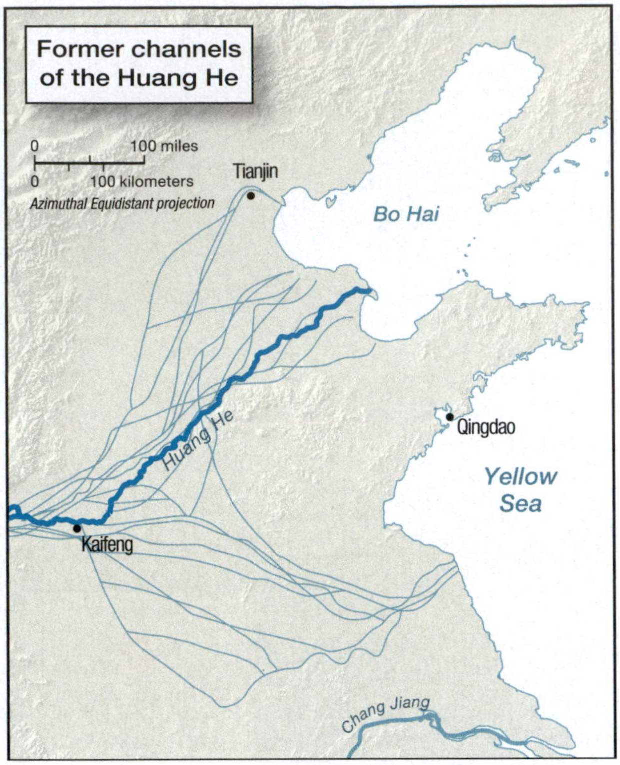

Former channels of the Huang He
0 100 miles
0 100 kilometers
Azimuthal Equidistant projection
Tianjin
Bo Hai
Huang He
Qingdao
Yellow Sea
Kaifeng
Chang Jiang

Adapted from map of the Huang He, from the website, Asia for Educators, at http://afe.easia.columbia.edu/china/geog/HuangHeMap.gif.

FIGURE 8.5 Channels carved by the Huang He. The Huang He, which flows through the easily eroded loess plateau, has carved 26 different channels over the last 2,000 years.

Erin Fouberg.

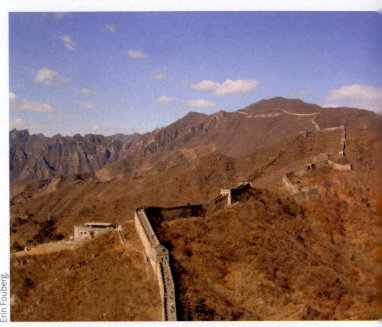

FIGURE 8.6 Mutianyu, China. The Great Wall of China was built in segments from the seventh century BCE to the time of the Ming Dynasty in the seventeenth century. The Great Wall includes watch towers and served primarily as a defensive structure.

color of the loess, originates in the Kunlun Mountains on the northern end of the Tibetan Plateau and flows east to the Bo Hai.

Windblown sediment erodes easily, and as a result, the Huang He easily cuts through the loess, frequently flooding and often changing course. The Huang He has changed course 26 times in the last 2,000 years (Figure 8.5). Flooding coincided with each major change in course.

Agriculture in East Asia dates to 6000 BCE with the cultivation of rice in the Huang He Valley. Almost 6,000 years later, the Qin emperor began the first large-scale irrigation system, overseeing the construction of a canal from the Huang He to the Wei Valley. With a stabilized food supply from production in the irrigated Wei Valley, the emperor extended the size of his empire, united China, and grew the population. He forced thousands of laborers to create the Great Wall by uniting walls that earlier states had constructed to defend their agricultural livelihoods from nomadic raiders on the steppe (Figure 8.6). He also commanded the creation of well over 8,000 terracotta warriors to guard his tomb at Xi'an (Figure 8.7). Not surprisingly, his subjects rebelled against the forced labor, making China's first dynasty also its shortest.

Each subsequent dynasty sought to control the Huang He. Over the last 2,000 years, the major approaches to controlling the Huang He have included channelizing the river, building high levees along the river, and building lower levees on floodplains. The goal of **channelization**

© Gavin Hellier/Robert Harding World Imagery/Corbis.

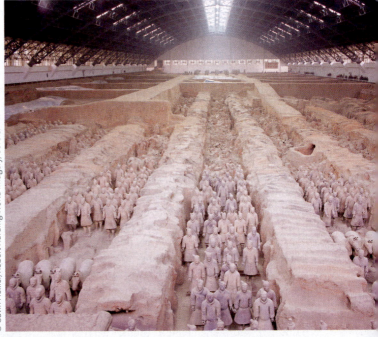

FIGURE 8.7 Xi'an, China. The Emperor Qin Shi Huang designed his mausoleum in the third century BCE with over 8,000 warriors, horses, chariots, and other battle scene pieces made from terracotta. The Terracotta Warriors, as they are now known, were unearthed by archaeologists after peasants digging a well accessed Pit 1 in 1974. Today, the museum housing the mausoleum and its three pits is one of the most visited sites in China.

is to speed the flow of the river so that it can more easily carry silt, and the goal of building high levees is to keep the river from flooding over its banks. When the Huang He flows, it carries sediment in suspension, and when the river slows down or the sediment load is too heavy, it deposits silt. As a result, in the region of the lower Huang He, near Kaifeng, so much silt has been deposited in the river bed that the river itself, blocked in by high levees, is at an elevation higher than the surrounding towns and floodplain (Figure 8.8).

Three Gorges Dam

In southern China, the Yangtze is the lifeblood of fresh water, and more recently the source of power for millions of Chinese. Beginning in the Tibetan Plateau and fueled by snowmelt and rainfall from the Himalayas and western highlands of China, the Yangtze flows west to east, reaching the sea at Shanghai. Concerned with bringing electricity to a growing population and controlling flooding downstream, the Chinese government decided in 1992, despite protests from global and local environmentalists and geologists, to build a massive dam, known as the Three Gorges Dam, across the Yangtze at a cost of $37 billion (Figure 8.9).

Construction of the Three Gorges Dam began in 1994 and has displaced more than 1.3 million Chinese. The dam crosses geologic fault lines, and scientists are now concerned

FIGURE 8.8   Kaifeng, China.   Located on the floodplain of the Huang He, Kaifeng has suffered flooding, caused both by nature and by humans. The Ming army flooded the city of Kaifeng in the seventeenth century to quell a peasant rebellion.

that the dam is "triggering landslides, altering entire ecosystems," and "endangering the millions who live in its shadow" (*Scientific American* 2008, 31). Among the environmental impacts, climatologists and meteorologists contend that the dam is at the root of a large-scale drought and that residents of Shanghai, which is at the mouth of the river, "are experiencing water shortages" (*Scientific American* 2008, 33).

Reading the **PHYSICAL** Landscape

The Three Gorges Dam

Humans have long used rivers and streams as transportation arteries and sources of power. One way that people modify streams is by building dams, which block the downstream flow of water, thus creating a large reservoir of water upstream. The first successful dams may have been built in Mesopotamia in about 2000 BCE. These first dams were watertight earthen structures that allowed farmers upstream to divert water for agriculture. Since that time, literally thousands of dams have been built along streams around the world. Many of these structures are now used to generate hydroelectricity, which is created when water rushing through intakes in the dam spins giant turbines.

When geographers think of dams, they often see the environmental concerns associated with them. Among other things, dams are controversial because they can block migrating aquatic species, trap sediment, and change water temperatures. In this context, one of the most controversial dams ever built is the Three Gorges Dam along the Yangtze River in east-central China (Figures 8.9a and b). Construction was completed in 2012, and Chinese officials claim the dam provides flood control along the Yangtze River and contributes significantly to the growing energy demand through generation of hydroelectric power. Energy generated by the dam helps fuel Shanghai as well as surrounding provinces.

(Continued)

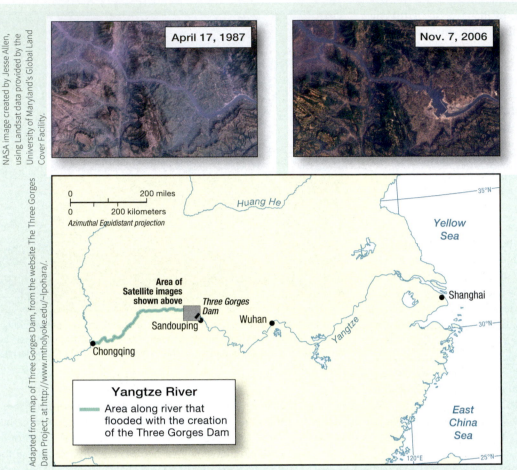

April 17, 1987

Nov. 7, 2006

Area of Satellite images shown above

Three Gorges Dam

Sandouping

Chongqing

Huang He

35°N

Yellow Sea

Shanghai

Wuhan

Yangtze

30°N

East China Sea

120°E   25°N

0   200 miles
0   200 kilometers
Azimuthal Equidistant projection

Yangtze River
— Area along river that flooded with the creation of the Three Gorges Dam

FIGURE 8.9a   Three Gorges Dam on the Yangtze River.   The satellite image on the top left shows the Yangtze River in 1987, before dam construction began in 1994. The dam is built and shown as a beige line crossing the river in the 2006 image. To the west of the dam, 'behind' it, the Yangtze is backing up, forming a reservoir. The area of these photos is shown by the gray box on the map. The map shows the flooding extended far west of the area covered by the satellite images.

FIGURE 8.9b   Yichang, China.   The sluice gates of the Three Gorges Dam are opened as flood water hurtles down the swollen Yangtze River in 2013.

FIGURE 8.9c   Yangtze River.   Much of this landscape is being inundated by the reservoir created by the dam. The river is dark brown because of a high amount of sediment and pollution brought to the reservoir by the tributaries of the Yangtze.

At the same time the dam provides these benefits outlined by the Chinese government, it also has a variety of negative environmental impacts. One obvious impact is that the reservoir has flooded 395 square miles (632 square kilometers) of the spectacular Yangtze gorges (Figure 8.9c), thus forcing approximately 1.3 million people to resettle. In addition, wetlands used in winter by the endangered Siberian Crane have been inundated. The dam has likely contributed to the functional extinction of the Yangtze River dolphin (Figure 8.9d) through habitat destruction and will also have negative impacts on the Yangtze sturgeon.

© Alex Hofford/epa/Corbis.

FIGURE 8.9d   The dam has likely contributed to the functional extinction of the Yangtze River dolphin. In 2010, the National Geographic reported the bajj, the Yangtze River dolphin, as extinct.

CLIMATE

The climate of East Asia is greatly influenced by the Pacific Ocean in the east and by the Himalayas in the west. The relatively consistent temperature of the Pacific Ocean over the course of the year creates a maritime effect on the surrounding lands in Asia, keeping the temperature range (difference between high and low temperature) in coastal areas smaller than in interior locations.

Climate in East Asia ranges from humid subtropical and humid continental in the east to steppe and desert in the west (Figure 8.10). If you look at the world map of climates, the climates of East Asia are similar to the climates of eastern North America (see Figure 1.10). On the world map, you should note that the regions extend across the same range of latitudes, from the Arctic Circle south to the Tropic of Cancer.

Within East Asia, the **humid subtropical climate** is found in the southeast, which brings summer rains as the Pacific High migrates north in the winter months. North of the humid subtropical is the **humid continental climate**, which also receives summer precipitation when the Pacific High migrates north with the subsolar point in the summer months.

North of the humid continental climate zone is the subarctic climate, which is bitterly cold in the winter months when East Asia is dominated by the **Siberian High**, a high-pressure cell that develops where a lack of sunlight creates extreme cold in Siberia. During the winter months, the region around Lake Baikal where the Siberian High is centered receives about 7.3 hours of sunlight each day, for weeks (see Figure 10.6). In a high-pressure system, winds descend from aloft to the surface. At the surface, air flows out from the high, creating a dry, cool effect on East Asia during the winter months.

In the summer, the same latitude in Siberia receives much more solar radiation, with 16.7 hours of sunlight on the June solstice. More sun creates warmer conditions in Siberia, and the high-pressure cell dissipates in the summer months.

Moving from east to west, East Asia becomes increasingly dry, and the climate is influenced by mountains and high plateaus. The Himalaya Mountains create a **rainshadow effect** for the Tibetan Plateau. **Monsoon** winds bring moist air from the Indian Ocean on to the Indian subcontinent during the summer monsoon. As the warm, moist air lifts over the Himalayas, it continues to cool, reach its dew point, and condenses into **orographic precipitation** on the windward side. On its trip over the Himalayas and onto the Tibetan Plateau, the air descends and warms, taking away the conditions necessary for precipitation. The dry, warm air creates an arid environment in the Tibetan Plateau known as a rainshadow.

The steppe and desert climates in the west are located where they are because the region is in the rainshadow of a mountain range. Others are dry because the midlatitude jet stream travels from west to east, which brings dry air masses from Eurasia into the region.

PHYSICAL GEOGRAPHY

The physical geography of East Asia ranges from volcanic, mountainous islands in the Pacific Ocean in the east to the vast Tibetan Plateau in the west. The Himalaya Mountains were formed 50 million years ago when the Indian subcontinent collided with the Asian plate. On the north side of the Himalayas, the Tibetan Plateau stands at an average elevation of over 16,000 feet. The Tibetan Plateau was pushed up when the leading edge of the Indian

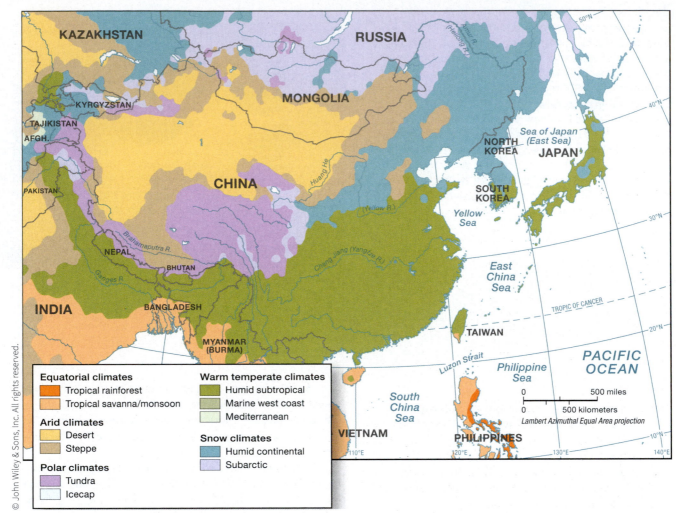

FIGURE 8.10 **Climates of East Asia.** The climates of East Asia range from dry, elevated desert in the west to humid subtropical in the southeast. The climate zones on the map generally follow lines of latitude, with the greatest temperature ranges farthest north and the smallest annual temperature ranges in the south.

subcontinent began colliding with Eurasia up to 120 million years ago. The collision pushed the Asian plate upward, lifting the Tibetan Plateau, and as a result the continental crust of the Tibetan Plateau is some of the thickest crust on earth. Once the subcontinent of South Asia collided with the Eurasian plate 50 million years ago, a continental–continental collision commenced and the two plates both thrust upward, building the Himalaya Mountain chain. The Himalayas divide East and South Asia, with the Tibetan Plateau in the north and the Deccan Plateau in the south. The two plates continue to collide today, and the Himalaya Mountains continue to grow.

The Tarim Basin, an area of low elevation dominated by the Taklimakan Desert, stands in the rainshadow of the Himalayas. The Tien Shan (Heavenly Mountains), which folded into mountains when the Indian subcontinent and Asia collided, border the Tarim Basin on the north (Figure 8.11). The Himalayas border the Tarim Basin on the South.

Several plate boundaries along the Pacific Rim of East Asia create active tectonic zones in the region (Figure 8.12). Where oceanic plates and continental plates collide, the denser oceanic plate subducts underneath the less dense continental plate. In a **subduction** zone, the plate boundary creates a trench, and the magma generated by the grinding of the oceanic plate under the continental plate forms an arc of islands or mountains further inland along the continental plate (Figure 8.13). In the case of East Asia, the islands of Japan are an arc of volcanic islands formed parallel to the plate boundaries and trenches that mark them.

The line of subduction between the Pacific plate and the North American plate is marked by the Japan Trench and the Kuril Trench. On March 11, 2011, one plate moved 80 feet along a 280-mile fault line. The magnitude 9.0 earthquake lasted five minutes and generated a massive tsunami. A **tsunami**, as noted in Chapter 7, is a seismic sea wave created by an earthquake, volcano,

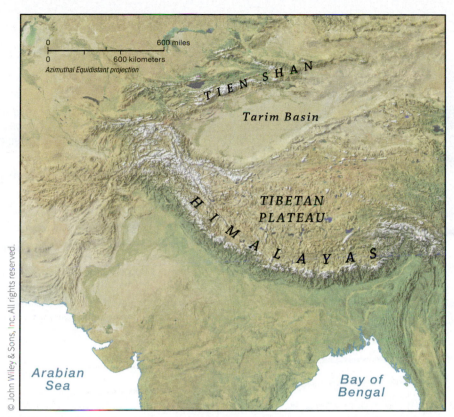

FIGURE 8.11   **Tibetan Plateau and Tien Shan Mountains.**   The Tien Shan Mountains are located to the northwest of the Tibetan Plateau. Both the Tien Shan and the Himalayas were formed when the Indian subcontinent collided with Asia.

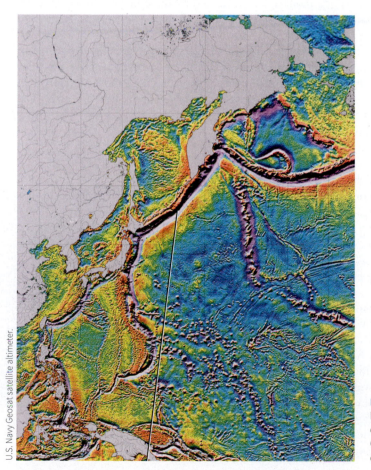

landslide, or other rock movement under water. In the 2011 Japan tsunami, the movement of plates generated an earthquake and displaced a massive amount of water. The water moved into the space created by the displaced plates, and then the potential energy in the water changed to kinetic energy, creating tsunami waves.

When a tsunami occurs, the size of the waves and the scale of the destruction depend on several factors, including the configuration of the coastline, the amount of plate displacement, and the distance of the earthquake epicenter from the coastline. As the wave travels toward the coastline, the **continental slope** and **shelf** becomes increasingly shallow, and the wave height or amplitude increases. If you are standing on shore and a tsunami is coming landward, you will first notice that the ocean recedes, pulling water away from the coastline. Sand and shells are exposed as water near the shore is pulled deeper into the ocean and into the wave. The crest of the wave then inundates the coast. Not all tsunami

FIGURE 8.12   **Plate boundaries along the Pacific Rim.**   Trenches (the dark lines on the image that show greater depth) form along subduction zones in the western Pacific Ocean where oceanic crust is descending under continental crust.

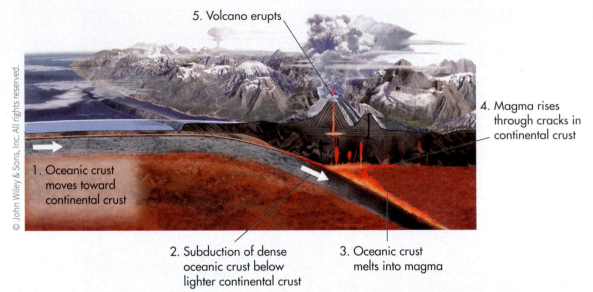

5. Volcano erupts

4. Magma rises through cracks in continental crust

1. Oceanic crust moves toward continental crust

2. Subduction of dense oceanic crust below lighter continental crust

3. Oceanic crust melts into magma

FIGURE 8.13 **Subduction Zone.** A subduction zone occurs where a denser oceanic plate descends under a less dense continental plate. The leading edge of the oceanic plate melts under heat and pressure, generating magma. The magma pools and eventually erupts as an arc of volcanoes.

waves are massive 32 foot (10-meter) waves like the one that hit the east coast of Japan. The Japanese tsunami was massive because of the magnitude of the earthquake and the crescent-shaped harbors of the Japanese coastline. As NASA's Earth Observatory explained, the "Crescent shaped coasts and harbors, such as those near Sendai, can play a role in focusing the waves as they approach the shore. Also, since land elevation is low and flat along much of the Japanese coast, many areas are particularly susceptible to tsunamis" (Figure 8.14).

The earthquake in Japan occurred 15.2 miles (24.5 kilometers) below the seafloor, and the seismic waves from the earthquake flowed upward along the plate boundary between the North American plate, on which the islands of northern Japan are, and the Pacific plate, which is subducting. *Scientific American* reported that based on seismography readings from around the world, the movement occurred along a 186–248 mile (300–400 kilometer) plate boundary, and the vertical shift between the plates was 32 to 65 feet (10 to 20 meters). The quake "released the energy equivalent of 8,000 Hiroshima bombs" (Folger 2012, 58).

The Japanese tsunami killed between 16,000 and 20,000 people, mostly along the coasts. Those who fled inland were not necessarily safe, as they moved closer to the Fukushima Daiichi nuclear power plant. The earthquake cut the power supply to the nuclear reactors, and the first sign of trouble was when the Unit 2 reactor lost power to its cooling system. In a nuclear reactor, fission of atoms occurs, which generates massive amounts of heat energy. A functioning nuclear reactor is always pumping water into the reactor to cool the core. Reactors have backup

generators to continue the cooling system if energy is lost, but the earthquake also disabled the plant's diesel generators. Three of the six reactors were operating when the earthquake occurred, and all three melted down when the cooling systems shut down as a result of the quake and tsunami.

On the day of the earthquake, the Japanese government ordered the evacuation of everyone within a 1.8 miles (3 kilometers) radius of the reactor. By the next day, March 12, 2011, an explosion occurred at the Unit 1 reactor. The

FIGURE 8.14 **Kesennuma, Japan.** Houses, cars, and other debris were washed away by tsunami waves, caused by an offshore earthquake in March 2011.

Japanese used seawater to attempt to cool the reactors, but the reactors melted despite their efforts. The government ordered the evacuation of 170,000 people within a 12.4 mile (20 kilometer) radius of Fukushima Daiichi. The same day, the government reported radioactivity and confirmed that Japanese citizens had been exposed to radiation.

Still another consequence of the massive tsunami in Japan is the debris it created. The Japanese government estimates that the tsunami launched 5 million tons of debris into the ocean and that 70 percent of that debris was quickly deposited just off the coast of Japan. The other 30 percent of the debris traveled across the Pacific Ocean, following ocean currents. At first, a majority of the debris floating across the ocean traveled as a mass that was visible from space (Figure 8.15). How far debris travels depends on the buoyancy of the debris. As less buoyant objects sank and more buoyant objects continued to float, the debris was no longer discernible from space. The most buoyant particles traveled the farthest the fastest. In winter 2012, reports confirmed that debris from the Japanese tsunami had already washed ashore in coastal Alaska. Debris will continue to surface for years to come in and around the northern Pacific Ocean.

In addition to the debris that polluted the ocean, the tsunami and the Fukushima Daiichi nuclear plant failure generated radioactive debris on land in Japan. The Japanese government estimates that the earthquake and tsunami created more than 20 million tons of rubble. The government planned "to destroy 4 million tons of potentially radioactive earthquake debris in garbage incinerators around the country," but many Japanese people have protested this plan with concerns over radioactive waste being incinerated in their backyards (McAteer 2012). The tsunami and nuclear reactor meltdown most impacted northern Japan, where the disasters took place. But the Japanese government is transporting nuclear waste to other, far-flung parts of the country to be incinerated. In June 2012, the government carted radioactive debris to the southwestern island of Kyushu so that waste from Ishinomaki City, which is 70 miles (112 kilometers) from the Fukushima Daiichi reactor, could be incinerated more than 600 miles (965 kilometers) from its place of origin as part of a plan called "wide area incineration." People in the southwest pushed back and protestors "blocked the road for 8 hours over fears that incinerating the debris would spread radiation to areas that have not yet been contaminated by nuclear disaster" (McAteer 2012). Advocates of wide area incineration argue that it helps erase the line between "contaminated" and "clean" to forge a stronger national Japanese identity.

Japan's location at the intersection of four plates and two major subduction zones means that earthquakes and

© HO/Reuters/Corbis.

FIGURE 8.15 **Pacific Ocean, off the coast of Japan.** Two days after a devastating tsunami hit Japan, debris drifted through the Pacific Ocean.

tsunamis will continue to impact the region. How Japan recovers from the March 2011 earthquake and tsunami, the largest and most destructive in its history, will likely guide its decision making in events to come.

CONCEPT CHECK

1. **Why** has China tried to control the Huang He over time?
2. **What** is the Three Gorges Dam, and **what** are some of the environmental impacts of the dam?
3. **How** were the islands of Japan formed, and **what** are the potential natural disasters Japan faces because of its location on two subduction zones?

WHO ARE EAST ASIANS?

LEARNING OBJECTIVES

1. **Describe** how religions, languages, and diseases diffuse on the Silk Road.
2. **Understand** Japan's role as a colonizer in East Asia and its lasting effects in South Korea.
3. **Explain** why Mahayana Buddhism is practiced with Daoism and Shintoism in East Asia.

Although the flow of commodities along sea trade routes was actually more valuable than that along the Silk Road, the diffusion of culture, especially religions, and Christian practices. Religious ideas and artistic ways of expressing them traveled back and forth along the Silk Road between Central and South Asia and East Asia. (Figure 8.17).

LANGUAGES IN EAST ASIA

The pattern of world language families in East Asia follows the trade routes and connections formed over thousands of years of spatial interaction through trade and cultural links (Figure 8.18). The Altaic language family extends from northeastern Russia through Mongolia, western China, and into Central Asia. The ties among languages in Central Asia and western China are a record of the spatial interaction among the people in that region. The Sino-Tibetan language family dominates much of China and into Myanmar and Thailand. The Japanese and Korean language families are distinct from Sino-Tibetan, but Korean extends across the border from North Korea into China where the Chaoxian minority population, which numbers nearly 2 million, lives.

The Sino-Tibetan language family includes seven dialects of Chinese. The dialects are

FIGURE 8.16 **Dunhuang, China.** A frescoed wall of one of the hundreds of temples dug inside Mogao caves that depicts Buddha meditating, surrounded by two Bodhisattvas carrying lotus flowers and other goods. Mogao caves are one of the most ancient places of worship for Chinese Buddhist culture, as they are sited near an important traffic junction along the Silk Road.

along the Silk Road along with travelers' accounts of arduous journeys across desert and mountains give the trade route a well-noted place in weaving together the economies of Asia and Europe. The Silk Road connected Xi'an, China, with points to the east, including Beijing, China, Seoul, Korea, and Kyoto, Japan. The Roman Empire on the west end and the Han Empire in China on the east end were both "self-sufficient in all essential commodities, and foreign trade was essentially a luxury trade" of goods, including Chinese silk, porcelain, jade, and opium (Hammond 1995, 24).

Archaeologists have found evidence of the diffusion of religions along the Silk Road in paintings in the loess caves of northern China on the eastern side of the Taklimakan Desert (Figure 8.16). The caves are filled with "500,000 square feet of murals and 2,000 statues of various sizes" (Weightman 2006, 223). Stories of men and women living ascetic lives, as monks and nuns with no personal wealth and in cloistered monasteries, diffused from Buddhist regions of South and East Asia to Europe, inspiring Christians to establish religious orders. Other religious practices, such as tying pieces of cloth to trees as offerings or signs of devotion are found throughout Eurasia in Hindu, Buddhist,

FIGURE 8.17 **Bamiyan, Afghanistan.** Built in the fourth and fifth centuries, the 175 foot tall standing Buddha helped diffuse Buddhism from Central and South Asia to China. In 2001, the Taliban took control of the region and destroyed this massive statue. This alcove stood empty from 2001 to 2011, when UNESCO began the painstaking process of reconstructing the statue.

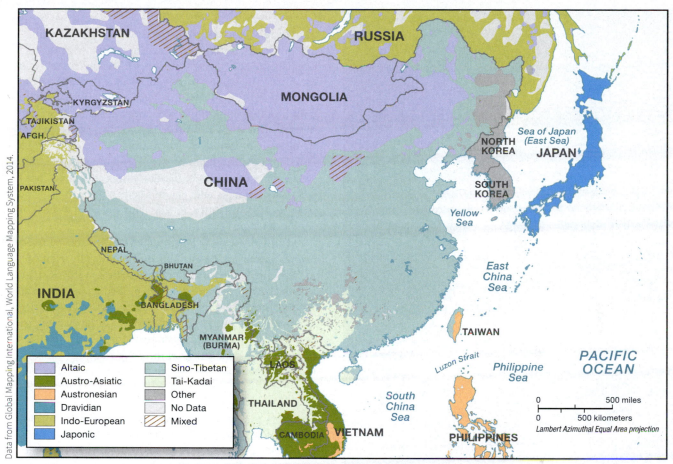

FIGURE 8.18 **Languages of East Asia.** Chinese is in the Sino-Tibetan language family, which extends to parts of Southeast Asia. Japan has its own language family, as does Korea.

regionally based in China and when spoken are not mutually intelligible. Chinese dialects have shared a common script of characters for more than 2,000 years. Sharing the same characters enables Chinese to read across dialects even when they cannot speak across dialects. The Han Chinese speak the Mandarin dialect, which has become the modern standard and national language in the country since the 1900s.

The languages in Korea and Japan are closely linked, and for centuries, Koreans used Chinese characters to write their language. With the rise of Korean nationalism in the fifteenth and again in the twentieth centuries, Koreans set aside the Chinese characters and adopted a new lettering system. The lettering above Gwanghwamun Gate in Seoul, South Korea (see chapter opener) is written in Chinese characters, which Koreans used for their language when the gate was originally built in 1394. All of the tourist information signs in the Gwanghwamun plaza and inside the palace are written in *hangul*, which is a Korean character system created to replace the Chinese characters used for the Korean language in the fifteenth century and widely adopted in the twentieth century.

COLONIAL HISTORY

If we were alive in 1400 and aware of all of the civilizations of the world, we would have been impressed by the efficient, centralized government of the Ming Empire in China; the enormous cities and grand boulevards of the Incan and Mayan Empires in South and Central America; the ornate temples of India; and the trading empire of the West African kingdom. When world history is taught today, students sometimes leave after a semester with the false impression that Europe was destined to be a colonial empire. Empires in Asia, Africa, and the Americas were stronger than Europe before 1500. European colonization changed the flow of goods, intertwined the world economy in a cash economy, and overthrew the political organization of peoples around the world, but in East Asia, colonization by Japan impacted a larger area of the region than did colonization by Europe.

European colonization can be seen in the cultural landscape of East Asia in the Bund region of Shanghai, China (Figure 8.19) and in the southern island of Dejima, Japan (Figure 8.20). Although European colonization left an imprint on East Asia, the greater colonial impact in East Asia came from East Asian colonizers who asserted territorial control from within the region. Both China and Japan were major colonial powers in the region, and the lasting impacts of those colonial relations are at the root of some political disputes today.

Andrew Rowat/Stone/Getty Images.

FIGURE 8.19   Shanghai, China.   The European style of buildings in the middle of Shanghai does not seem to fit. Europeans built the Bund trading district, sited on the Huangpu River, during the nineteenth and twentieth centuries. Shanghai has more than doubled in population since 1990, but the city has maintained the Bund, improved traffic patterns around the district, and encouraged tourism in the area.

CHINESE EXPANSION

The Tang Dynasty lasted from 618 to 907 in China, and much like Asoka and later Akbar did in South Asia (see Chapter 6), the administrative structure the Tang established became so influential that its impacts are still visible today. The Tang "was a centralized empire with a uniform administrative organization of prefectures, in which the old ruling aristocracy was replaced by officials recruited by an examination system which lasted into the twentieth century" (Hammond 1995, 50). The Tang expanded the landholdings of China, built roads and canals, and encouraged internal migration from northern China to the rice-growing areas of southern China (Figure 8.21).

The Chinese directly controlled the Korean Peninsula dating back from 109 to 106 BCE and again from 668 to 676. The Chinese withdrew from Korea in 676 after a Korean resistance movement pushed them out. The Korean Empire established a centralized state modeled after the Tang Empire in China. In 645 CE Japan's empire established both its administrative structure and capitals of Kyoto and Nara based on the Tang Empire's administrative structure and its capital in Chang-an (Xi'an).

MODERN COLONIALISM IN EAST ASIA

European traders and colonial empires were motivated first by trade in spices and then by the trading profits generated by early capitalism to establish ports and colonies in Southeast Asia (see Chapter 7). First the Portuguese and Dutch and then the British and French sought to move beyond Southeast Asia and establish port colonies in Japan and China. During the first wave of European colonialism from 1500 to about 1850, the Qing (also called Manchu) Dynasty in China, which lasted from 1644 to 1911, and the Tokuguwa Shogunate in Japan, which lasted from 1603 to 1867, both resisted European colonization.

Reading the **CULTURAL** Landscape

Dejima (Deshima), Japan

The Portuguese were the first Europeans to reach Japan and establish trade in Nagasaki. In the 1600s, after a Christian uprising, the Japanese sought to contain and stop the diffusion of Christianity. Japanese merchants built the island of Dejima off the coast of Nagasaki by digging a canal to separate the island from the mainland and then moved the Portuguese to the island.

During the 1600s, a number of trading ships from the Netherlands reached Japan. The Japanese gave the Dutch merchants the island of Dejima as their port (Figure 8.20a). Between 1603 and 1867, Japan was in an isolationist period, and Dejima served as Japan's point of contact with European traders. In 1859, Japan opened other ports to the outside, and Dejima lost its significance. Eventually, the island was enveloped by Nagasaki, and the fan shape was no longer distinguishable.

In 1951, the city of Nagasaki decided to restore the island of Dejima and the buildings that stood when the

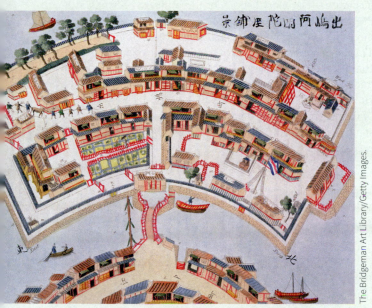

FIGURE 8.20a **Dejima, Japan.** Platte of Dejima based on a woodblock print by Toshimaya Bunjiemon in 1780,

island was a Dutch port. In 1996, Nagasaki began rebuilding the island, with much of it complete today (Figure 8.20b). The cultural landscape of Dejima, as it stands today, is a recreation of an historical site. The Japanese government rebuilt the island in recognition of the central role the island played during this era of Japanese history.

Source: http://www1.city.nagasaki.nagasaki.jp/dejima/en/new_dejima/contents/index001.html

FIGURE 8.20b **Dejima, Japan.** The city of Nagasaki rebuilt the island of Dejima to look similar to the way it did during the Japanese isolationist period when the island was a Dutch trading port.

FIGURE 8.21 **Tang Empire.** The empire headquartered in Chang-an (Xi'an), covered a massive territory, extended into the Korean Peninsula and encompassed upwards of 50 million people.

The Portuguese made headway in China by occupying the port city of Macau starting in 1557. Aside from Macau, the Qing Dynasty kept European colonizers at bay until the nineteenth century when they attempted to put an end to illegal British opium trade. The Opium War ensued, and at its end in 1842, the Chinese conceded to European influence in the region. The British were granted, through treaty, the port of Hong Kong in 1842 and then by lease in 1898. The Chinese set up five additional treaty ports and opened them to foreigners, primarily Europeans and Japanese, to live and establish businesses. The most famous of these ports is Shanghai. In the 1800s, foreign merchants from Britain, France, Russia, the United States, Germany, and India gathered in Shanghai. Traders established a cluster of buildings in a region called the Bund along the Huangpu River (see Figure 8.17). Through trade and intermarriage among residents, Shanghai created a culture called "Shanghai style" that is distinct from the rest of China. By the end of the nineteenth century, from north to south the Russians, Japanese, Germans, British, and French each claimed spheres of influence in China (Figure 8.22).

Adapted from Geoffrey Barraclough, ed. The Times Concise Atlas of World History, 5th edition, Hammond Incorporated, 1998. Page 107.

FIGURE 8.22   Colonialism in East Asia.   European colonizers controlled ports in China and asserted control over spheres of influence.

JAPANESE COLONIALISM

During the Tokuguwa Shogunate, which lasted more than 250 years, samurais held positions of power in a feudal system in which they were lords (Figure 8.23). The Tokuguwa would only trade with the Netherlands during the first wave of colonialism, largely because the Dutch, unlike the British and French, were interested almost exclusively in trade and not in diffusing their religion and culture. Despite attempts to keep Japan sequestered from the wider world, their trade with the Netherlands and others in East Asia brought a cash-based economy to Japan during the shogunate period. The cash-based economy significantly undermined the position of power samurais had held under the feudal system.

In 1868, a revolutionary movement in Japan called the Meiji Restoration opened Japan, economically and politically. Japan defeated China in a war in 1895, gaining territory in northeastern and eastern China. Japan went to war with Russia in 1904, winning and gaining territory in southeast Russia. In 1910 Japan overtook the Korean Peninsula (Figure 8.24). The Japanese justified the colonization of Korea by arguing that in order for Japan's industrial sector

Independent Picture Service/UIG via Getty Images.

FIGURE 8.23   Kyoto, Japan.   One of Japan's most powerful shoguns, Tokugawa Ieyasu, oversaw the construction of the Nijo Castle, stone wall, and moat. The reign of Tokugawa Ieyasu marked the beginning of the Tokugawa Shogunate.

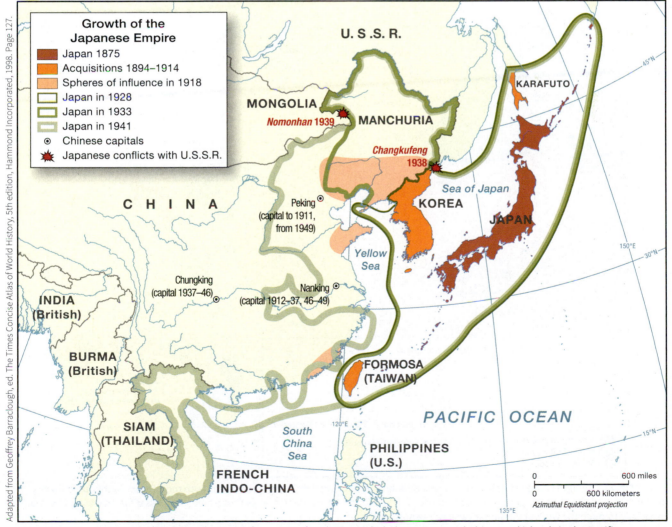

Adapted from Geoffrey Barraclough, ed. The Times Concise Atlas of World History, 5th edition, Hammond Incorporated, 1998. Page 127.

FIGURE 8.24   Japanese colonialism in East Asia.   Japan's colonies included Korea, coastal China, and islands in the Pacific.

to grow, the Japanese needed cash revenue from selling agricultural goods raised in Korea to fund increased manufacturing in Japan.

Expanding colonial influences in China signified a weakness in the Qing Empire and led to political unrest in China that lasted until the empire ended in 1911 when civil war broke out. Also in the nineteenth century, the isolationist Tokugawa Shogunate in Japan succumbed to external pressures with the Meiji Restoration in 1868. The new Japanese government not only opened itself to the world but also built its own colonial empire in East Asia and the Pacific. Japanese colonies stretched from southeastern Russia and northeastern China across the Korean Peninsula, the northern part of eastern China, including Beijing and Shanghai, into French colonies in Southeast Asia, and across southeast Asia, into the islands of the Pacific (see Figure 8.24).

The lasting effects of Japanese colonialism in East Asia are felt in Korea. The Japanese gained control of Korea in 1910. The Japanese colonizers worked to assimilate Koreans by repressing their language and forced thousands of Korean women into sex slavery as "comfort women" to Japanese soldiers.

Japan's colonization of East Asia and the Pacific drew the United States into World War II when the Japanese attacked Pearl Harbor in Oahu, Hawaii, on December 7, 1941. The Allied powers defeated the Japanese in the Pacific theater of the war, with the final blow being the Americans dropping nuclear bombs on Hiroshima and Nagasaki in August 1945. Japan's loss in World War II ended its colonial domination in East Asia and the Pacific.

RELIGION

The hearths of Daoism, Shintoism, and Confucianism are in East Asia, but the global religion that has had the most influence in the region, diffusing first to Tibet and then to China and the rest of East Asia, is Buddhism. In China, Buddhism mixed with indigenous religions, including Daoism, and in Japan, Buddhism combined with Shintoism (Figure 8.25).

Daoism

Daoism (also spelled Taoism) is a philosophy on the nature of reality and how to live and originated in the hearth of China. Daoists seek to live in accord with

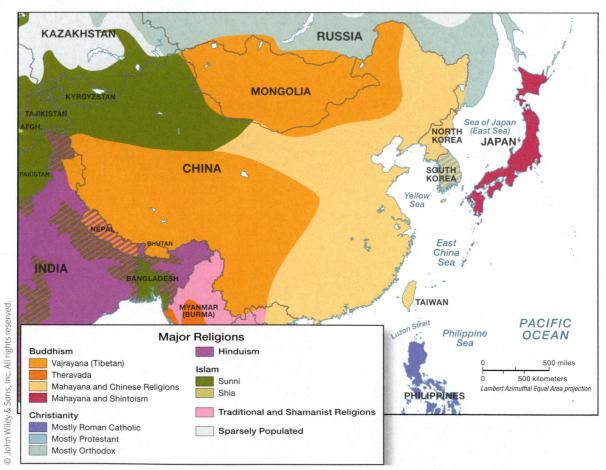

FIGURE 8.25   Religions in East Asia.   Mahayana Buddhism is influential throughout East Asia, with the largest number of followers in China. In Japan, Mahayana Buddhism is combined with Shintoism. Christianity (mainly Protestant sects) has diffused to South Korea, especially through the work of missionaries since the Korean War.

the *dao*, or the way, in order to live healthy, long lives. The *dao* is "the spontaneous process regulating all beings and manifested at all levels—the human body, in society, in nature, and in the universe as a whole" (Teiser 1996). Daoism focuses on how human beings fit into the cosmos as a whole.

During the Han Dynasty, scholars outlined the teachings of Daoism, and the teachings and practices mixed "with folk beliefs centered on the worship of nature and spirits" to evolve into a religion (Weightman 2006, 220). The Chinese philosopher Lao Zi taught the principles, often using nature as a guide. For example, the eighth chapter of the *Daode Jing* states: "The best of man is like water, Which benefits all things, and does not contend with them, Which flows in places that others disdain, Where it is in harmony with the Way." Daoism teaches that with disciplined meditation on their place in the cosmos, human beings can achieve higher and higher levels of deities. Li Yuan, founder of the Tang Dynasty, claimed to be a descendent of Lao Zi and helped to diffuse Daoism in China.

Shintoism

Shintoism dates back to a period between 8000 and 300 BCE in Japan. Like Hinduism and many traditional, indigenous religions, Shintoism does not have a single founder. Shintoism means the way of the gods, and its teachings are based on nature. Around 500 CE, in response to the diffusion of Buddhism into Japan, Shintoists began to define the beliefs and texts of the religion into writings.

Shinto shrines throughout Japan draw followers through a **torii**, a gateway that marks the sacred nature of the site (Figure 8.26). During the nineteenth century, the Japanese emperor elevated Shintoism to the state religion, and the emperor became the leader of the state and the state religion. After World War II, the Japanese government separated Shintoism from the state.

Buddhism

Buddhism began near the Ganges River in present-day India. As of 600 CE, China was primarily Daoist and Japan was primarily Shinto, but Buddhism diffused from its hearth in the Ganges River area of India to Lhasa (in Tibet) and Kandy (in Sri Lanka) and then throughout East Asia (Figure 8.27). Buddhists carved caves and adorned them with art on the western and eastern expanses of the Silk Road.

Buddhism diffused into western China through Silk Road trade in the second and third centuries. Just east of the Taklimakan Desert, near Dunhuang, China, travelers and traders rested in caves carved into the rock face of a cliff. Hundreds of the carved caves became grottoes dedicated to the Buddha (Figure 8.28a).

Daoism and Confucianism influenced how Buddhism diffused in China. Daoism focuses on meditation and on individuals finding their place in the cosmos. Its theology coincides closely with Mahayana Buddhism, whereby individuals seek enlightenment through meditation and right acts. The political structures established through Confucianism also aided or suppressed the diffusion of Buddhism, as deference and loyalty to authority gave political

Danita Delimont/Gallo Images/Getty Images.

FIGURE 8.26 Fushimi-ku, Kyoto, Japan.   A series of toriis, marking the transition from this world into a sacred space, leads to the Fushimi Inari Taisha shrine, which is part of a shine complex that dates to 711. In the Shinto religion, Inari represents rice, business, and prosperity, and is represented by the fox. The shrine draws several million worshipers over the Japanese New Year.

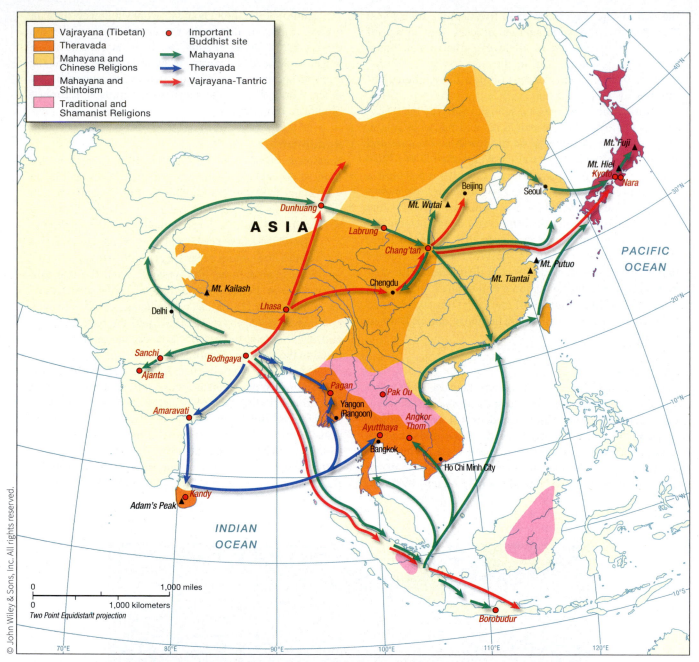

FIGURE 8.27 **The Diffusion of Buddhism in Asia.** The hearth of Buddhism is near the Ganges River in northern India. Although Buddhism is no longer predominant in India, the religion diffused broadly through East and Southeast Asia. Three major sects of Buddhism now exist.

leaders and emperors a direct say in the hierarchical diffusion of Buddhism. For example, in northern China, Buddhism suffered in the fifth century when the Emperor Taiwu of the Wei Dynasty banned Buddhism and destroyed Buddhist temples. He was followed by Emperor Wencheng who encouraged the diffusion of Buddhism by funding the building of temples (Figure 8.28b).

In the process of Buddhism diffusing from its hearth in India, the Buddha was thought of in two primary ways. First, Buddha is the name given to Siddhartha Guatama, the man who gave up his worldly wealth to meditate, become enlightened, and give sermons to followers during his lifetime. In Southeast Asia, followers of Buddhism build stupas or shrines on relics of the Buddha, and in India followers build stupas in places where Buddha preached. Second, **buddha** is a name used for any enlightened being, including Buddha. Related to the concept of buddha is **bodhisattva**, which refers to a person who is intent on becoming enlightened. Several bodhisattvas are "particularly popular in China" (Columbia University Asia for Educators). Some historians contend that dynasties were the most important element in the diffusion of Buddhism in China, and other evidence suggests that the diffusion of Buddhism in China was not centrally planned because

FIGURE 8.28b **Shanxi Province, China.** The Yungang Grottoes were built on the east end of the Silk Road in the 5th and 6th centuries, beginning during the rule of the Emperor Wencheng. The 252 caves include more than 51,000 statues.

FIGURE 8.28a **Dunhuang, China.** Mogao Grotto was built over a millennium, between 336 CE and 1400. On the western side of the Silk Road, 492 caves along the Dachuan River house more than 2,000 painted statues along with murals depicting trade and the diffusion of culture, including Buddhism, on the Silk Road.

there is no one set of common beliefs in Chinese Buddhism. Syncretism is evident, as Buddhists in China both interpreted Indian texts and created texts native to China.

CONCEPT CHECKS

1. **How** did religions, languages, and diseases diffuse on the Silk Road?

2. **Why** did Japan colonize parts of East Asia?

3. **What** are Daoism and Shintoism, and **what** similarities do they have to Buddhism?

HOW DO CONFUCIANISM AND POLITICAL POLICIES INFLUENCE POPULATION GROWTH IN EAST ASIA?

LEARNING OBJECTIVES

1. **Understand** what gender imbalance is.

2. **Explain** the role of Confucianism in East Asia and its role in creating gender imbalance.

3. **Compare** and **contrast** population policies in China, South Korea, and Japan.

More than half of the world lives in Asia, a continent home to 4.2 billion people. East Asia's population is closing in on 2 billion people (Figure 8.29). The people of East Asia are amalgamated in cities, with the majority of people in Japan and South Korea living in urban areas. China is home to more than 160 cities with populations over 1 million. Shanghai and Beijing, China, both have populations teetering around 25 million. Nonetheless, China is not majority urban (though by other countries' definitions, many villages in China would be considered urban) in large part because the Chinese government has policies that incentivize Chinese to stay in rural areas. The North Korean government also restricts the mobility of its citizens to maintain a population that is majority rural.

In China in 1953, the communist government conducted the first modern census and marked the population at 583 million. By 1979, the population of China grew to

Data from CIESIN—Columbia University, United Nations Food and Agriculture Programme—FAO, and Centro Internacional de Agricultura Tropical—CIAT. 2005. Gridded Population of the World, Version 3 (GPWv3): Population Count Grid, Future Estimates. Palisades, NY: NASA Socioeconomic Data and Applications Center (SEDAC). http://dx.doi.org/10.7927/H42B8VZZ. Accessed December 2014.

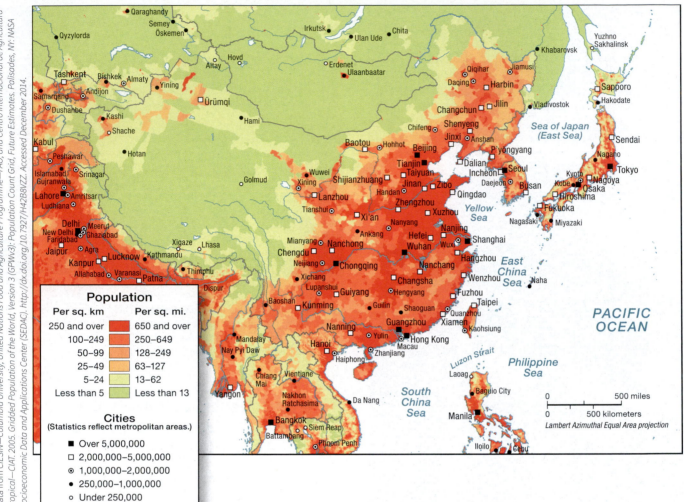

FIGURE 8.29 Population density of East Asia.   East Asia is densely populated except for arid regions in the west, including Tibet and Mongolia.

975 million people. Concerned about the rapidly growing population in East Asia, China instituted a one-child policy in 1979, and South Korea instituted a small family campaign in the 1960s and 1970s. As a result of restrictive population policies, both China and South Korea have experienced massive gender imbalances as sons have been favored over daughters. At the same time, neighboring Japan experienced the onset of population decline, where TFRs have fallen so far below the replacement level that the total population of the country is actually falling.

GENDER IMBALANCE

With a basic understanding of probability, one would expect a fairly even ratio of 100 boys born for every 100 girls. Demographers have found that over time, in the absence of human intervention, most countries average 103 boys for every 100 girls at birth, and researchers think it is because boys are more likely to die in infancy than girls. China, South Korea, and Hong Kong all have a sex ratio (the number of boys for every 100 girls) beyond, and in some cases far beyond, the state of nature. For every 100 girls

born in China in 2010, 123 boys were born. In South Korea, 111 boys were born for every 100 girls, and in Hong Kong, 108 boys were born for every 100 girls. The only explanation for the radical **gender imbalance** in parts of East Asia is human intervention.

Demographers have found that certain cultural norms or societal conditions, including low levels of education for women and patriarchal societies, lead to undervaluing girls. In East Asia, understanding the gender imbalance is not that simple. To understand this phenomenon, one needs to start with cultural or societal norms and also examine the differences within the region to better discern the number of factors affecting sex ratios in the region.

CONFUCIANISM

Preference for sons in East Asia correlates with the diffusion of **Confucianism**. In Confucianism, which is followed in East Asia as more of a cultural philosophy than a religion, son preference is tied to the hierarchical structure of society.

Although Confucius did not see himself as starting a new philosophy, he came to be one of the greatest intellectual and moral influences in East Asia. Confucian values include being kind toward others, knowing what is right, and serving one's superiors loyally (Teiser 1996). Confucian values applied not only in political contexts where subjects were loyal to rulers but also in familial contexts where families were loyal to the father.

Confucianism elevates loyalty across scales and uses a lineage system that focuses on the birth of sons to reinforce order at each scale. Traditionally, women have been marginalized to the home, and fathers and sons represent the household in the public sphere. When a woman marries, she leaves her family and becomes part of her husband's family and shifts her loyalty to them. Traditionally, sons care for their parents in old age, and thus it becomes a woman's primary duty to bear a son to maintain her husband's lineage. In the public sphere, fathers and sons work for and are expected to be loyal to the state or the government authority. Through this rigid and defined structure, leaders have historically had greater control over their country in societies dominated by Confucian thought.

Governments promoted the Confucian structure of society because it instilled loyalty to the government among all of the subjects. The Han Dynasty was instrumental in diffusing Confucianism, and the values and lessons diffused hierarchically, primarily to the upper class (Figure 8.30). 1,500 years later, in 1315, Mongol rulers in China integrated Confucius's writings into the written exam they used for entry into China's civil service. In South Korea, the Choson Dynasty (1392–1910) replaced a bilateral family system where a married couple could live either with the woman or man's family and where "male and female offspring could both inherit their parent's property" with a patriarchal family system based on Confucianism (World Bank 2007, 2). Confucian teachings held that in order to build a strong authoritarian state, which the Choson Dynasty sought to do, establishing a patriarchal family system was necessary. The dynasty worked to undermine Buddhism on the Korean Peninsula and instead establish Confucianism because Buddhism focused on individualism and self-realization instead of on "loyalty to family and state" (World Bank 2007, 3).

With such a strong societal preference for sons, **female infanticide** was common in parts of East Asia over the centuries. Even so, sex ratios in East Asia were much closer to normal before the 1980s. In the 1980s, ultrasound technology diffused to East Asia. Pregnant couples began to use ultrasound to detect gender, and an increasing number chose to selectively abort females. The sex ratio by country over time (Figure 8.31) demonstrates the spike in female abortions that occurred in the 1980s and early 1990s in South Korea. In response to the marked rise in the number of female abortions, China banned sex-selective abortions in 1995 after neighboring India passed the same ban in 1994.

Meeting between Confucius (551–479 BC) and Lao-Tzu (c.604–531 BC) from 'Recherche sur les Superstitions en Chine' by Henri Dore, 1911-20 (coloured engraving), Chinese School, (20th century)/Private Collection/Archives Charmet/The Bridgeman Art Library

FIGURE 8.30   Confucius among the upper class.   Henri Dore produced a colored engraving of a meeting between Confucius (551–479 BCE) and Lao-Tzu (c. 604–531 BCE) in *Recherche sur les Superstitions en Chine.*

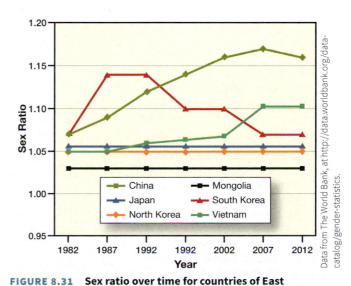

Data from The World Bank, at http://data.worldbank.org/data-catalog/gender-statistics.

FIGURE 8.31   Sex ratio over time for countries of East Asia.   Ultrasound to determine the sex of a fetus became available in 1980. In 1995, the government of China banned selective abortions based on gender discovery through ultrasound. Despite the prohibition, the gender imbalance rose in China after 1995.

CHINA: ONE-CHILD POLICY

China is famous for its **one-child policy**, which began in 1979 and encouraged or financially coerced couples into having only one child. The one-child policy has changed over time so that certain groups, including members of China's 55 minority groups (who comprise 6 percent of the country's population and live mainly in rural areas, including Xinjiang and Nei Mongol Autonomous regions) and couples living in rural areas who have a daughter first, are allowed to have more than one child. The one-child dictate, which applies mostly to couples in urban areas, is estimated to affect two-thirds of China's couples today. The Chinese government contends that because of the one-child policy, 400 million fewer children have been born in China since

1979. Some demographers argue that the smaller number of children born is not *only* a result of the one-child policy because the Total Fertility Rate (TFR; see Chapter 9) in China actually declined in the 1970s—from 5.8 in 1970 to 2.7 in 1979—before the one-child policy even began. After the one-child policy, TFR fell at an astonishingly fast pace to 1.7 in 2010, according to Chinese government estimates.

In view of Chinese culture's strong preference for sons, one may guess poorer families would be more likely to abort females and have their one child be a boy. However, within China, the less economically developed western provinces have sex ratios that are at or near normal. In fact, Tibet is the only province with a normal sex ratio. The most imbalanced sex ratios in China are in the south and east (Figure 8.32),

Data from "China's excess males, sex selective abortion, and one child policy: analysis of data from 2005 national intercensus survey." BMJ 2009;338:b1211, at http://dx.doi.org/10.1136/bmj.b1211.

Sex Ratio by Province

- Greater than 130
- 120–129
- 110–119
- Less than 109
- ✪ National capital
- ◉ Provincial capital
- ● City

FIGURE 8.32   China: Sex ratio by province.   The greatest imbalance between males and females is in Guangdong in the south and in the east, especially in the inland provinces. The western and northern parts of China have sex ratios closest to normal.
All the data in Figure 8.31 are from the World Bank, and the data about sex ratios in China in the text (unless otherwise noted) and Figure 8.32 are from a *British Medical Journal* study. Figure 8.21 reveals trends by country over time, but the data for current sex ratios from the *British Medical Journal* in Figure 8.32 are more accurate.

some of the wealthiest provinces in the country. The provinces with the highest gender imbalances have sex ratios above 130!

The impact of the one-child policy and of female abortions on sex ratios is unmistakable in these regions, but, ironically, the deleterious impact of the policy on sex ratios is even more extreme when we look at places where couples are allowed to have more than one child. Provinces in south and central China have exceptions to the one-child policy. If the first child is a girl, then couples are allowed to have a second child. After they have one child, many couples feel an extreme pressure that their second child must be a son. *The Economist* (4 March 2010) reported that in Guangdong Province, the most populous province in China, the sex ratio changes, becoming even more imbalanced, with birth order. For first children, the sex ratio is closer to normal at 108 boys to 100 girls. For the second child in a family, however, the ratio "leaps to 146 boys for every 100 girls. And for the relatively few births where parents are permitted a third child, the sex ratio is 167" (*The Economist* 4 March 2010).

As a result of son preference, female abortion, and female infanticide, the Chinese Academy of Social Sciences predicts that by 2020 between 30 and 40 million young men of marrying age will be unable to find a wife. "So, within the next ten years, China faces the prospect of having the equivalent of the whole young male population of America, or almost twice that of Europe's three largest countries, with little prospect of marriage, untethered to a home of their own and without the stake in society that marriage and children provide" (*The Economist* 4 March 2010, 2).

SOUTH KOREA: VALUING GIRLS

While South Korea never had an official one-child policy, the government did have a series of programs encouraging fewer children, dating back to its national family planning campaign in 1962. The Total Fertility Rate (TFR) in South Korea was 6.1 in 1960, which the government deemed too high. The family planning campaign encouraged couples to have "small and prosperous" families. In 1981, the government set the goal of two children for each couple and used economic incentives to encourage couples to comply. By 1984, the TFR in South Korea fell well below replacement level of 2.1 to 1.74.

In 2005, the TFR reached a record low of 1.08 in South Korea, but it has risen back to 1.2. Since 2005, South Korea has changed to a **pronatalist policy**, encouraging couples to have more children by providing "tax incentives, priority for the purchase of a new apartment, support for child care including a 30 percent increase in facilities, childcare facilities at work, support for education, and assistance to infertile couples" (Population Reference Bureau 2010). In 2005, the Supreme Court in South Korea also recognized the rights of women to remain a part of their parents' family in name and lineage and to claim their children as part of their lineage as well.

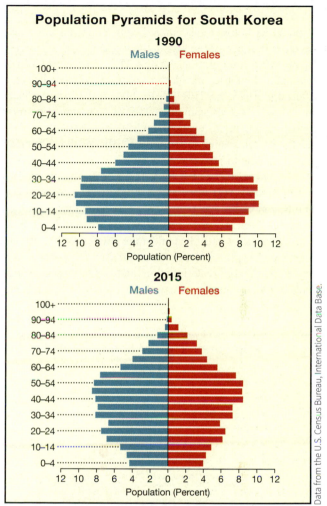

FIGURE 8.33 **Population pyramids for South Korea in 1990 and 2015.** The extreme gender imbalance and strong preference for boy children at birth was evident in South Korea's population pyramid in 1990. In 2015, a gender imbalance still exists, but the 2015 population pyramid reflects how much the preference has weakened in 25 years.

Data from the U.S. Census Bureau, International Data Base.

What is most remarkable about South Korea is not its shift to a pronatalist policy but rather the way society has shifted away from Confucian-based son preference to women feeling less urgently that they "need a son." Strong son preference in South Korea led to a gender imbalance that peaked at 117 boys for every 100 girls in 1990 (Figure 8.33). Today, the sex ratio in South Korea has fallen to 107 boys for every 100 girls. Much closer to the normal range, the marked change highlights a societal change in South Korea. A 2007 study by the World Bank analyzed South Korean fertility surveys in 1991 and 2003. Each survey asked women whether they preferred a son. The study found the group who felt the need to have a son had declined sharply across all age-groups and socio-economic groups in the population" (2007, 6). Overall, the study found that "son preference declines with increasing socio-economic status, lower parental control, younger birth cohort, and older age at marriage" (2007, 9).

JAPAN: WORKING MOTHERS

With TFRs below replacement levels, countries can make up the difference or the loss in the working-age population by allowing immigration. Japan's TFR after World War II was 4.54. Japan legalized abortion in 1948, and between 1948 and 1973, its TFR fell to 2.04, just below replacement level. Today, Japan has a TFR of 1.26, which is so far below replacement level that, when coupled with Japan's relatively strict immigration policies, the result has been an actual decline in population.

Between 1973 (2.04) and 1995 (1.42), the precipitous fall in Japan's TFR correlates strongly with a massive rise in inflation within Japan. In 1971, the United States released the dollar from the gold standard, which released Japan from a fixed exchange rate it had under the Bretton Woods Agreement (see Chapter 2) (*Japan Times* 2008). In response, Japan experienced large-scale inflation, and the economic crunch led to families choosing to have fewer children (Figure 8.34).

Like South Korea, Japan has several pronatalist policies. Legislation in Japan allows for women who work full time to take one-year maternity leave at partial pay. The government spends so much on child care that "the cost per month incurred by the government to fund day-care services in Tokyo for one infant currently exceeds the average monthly wage of a male worker in the capital" (*Japan Times* 2008, 1). The most recent Japanese program to encourage women to have multiple children is called the *Angel Plan*, which is designed to help women continue to work after having children and to encourage men to increase their workload at home to help raise children and maintain households.

CONCEPT CHECK

1. **What** is gender imbalance?

2. **How** does Confucianism help create gender imbalance in East Asia?

3. **Why** are population policies in China, South Korea, and Japan different?

HOW DO COMMODITY CHAINS LINK EAST ASIA AND THE WORLD?

LEARNING OBJECTIVES

1. **Describe** global sourcing.

2. **Understand** the reforms Deng Xiaoping made in China in 1979.

3. **Explain** how major ports and smaller ports are networked in East Asia.

Economic development theorists analyze how East Asian countries have successfully grown their economies in search of practices that other countries might apply. Some theorists focus on the openness of East Asian countries to foreign direct investment (FDI). Others examine the role of governments in encouraging investment. Some theorists study how regional growth helped spur national growth. Researchers propose that cultural practices that inform business management helped Asian countries develop their economies, while others attribute China's average annual growth rate of over 10 percent from 2000 to 2012 to "cheap" labor. Still other theorists imagine that it is simply Asia's "turn" in history (Rigg 2009, 30). Each theory attempts to reduce a multitude of factors to one "best practice," or set of guiding principles to instruct other countries how to develop. As geographer Jonathan Rigg explains:

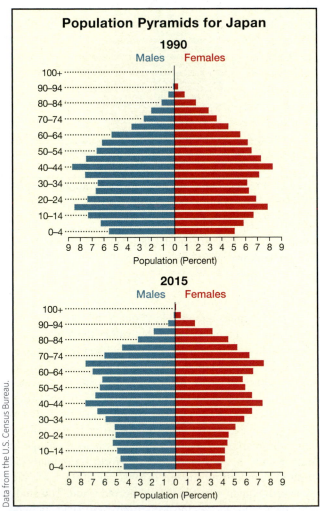

FIGURE 8.34   Population pyramids for Japan in 1990 and 2015. The total fertility rate (TFR) fell in Japan after 1948 and again after 1971. With restricted immigration and low TFR, each smaller generation of Japanese has fewer children than the generation before it, which has resulted in a sustained population decline.

Data from the U.S. Census Bureau.

The debate over what has made (parts of) Asia so economically successful is important, but ultimately we take away from the discussion barely more than we arrived with: geography matters; culture and society matter; policies and the quality of government matter; and the opportunities afforded by history matter (2009, 30).

Each theory, in its own way, oversimplifies what has happened in East Asia since 1960 and what has led to the phenomenal growth in the region's economy since 1980.

GROWING CHINA

No one factor has enabled China to grow at unprecedented rates. Its growth since 1990 stems from its geography, state investment and policies, and opportunities afforded by history. China's growth was not an accident or a miracle; its growth was purposeful. Since 1979 four players—the state, multinational corporations (MNCs), local firms, and universities—have worked in tandem to make China a global sourcing specialist and leading manufacturer in international commodity chains (Figure 8.35).

A **commodity chain** is a series of links connecting the steps in production of a good or service (Figure 8.36). These links include production of raw materials, processing materials, designing products, sourcing (choosing

commodity chain steps in the production of a good from its design and raw materials to its production, marketing, and distribution.

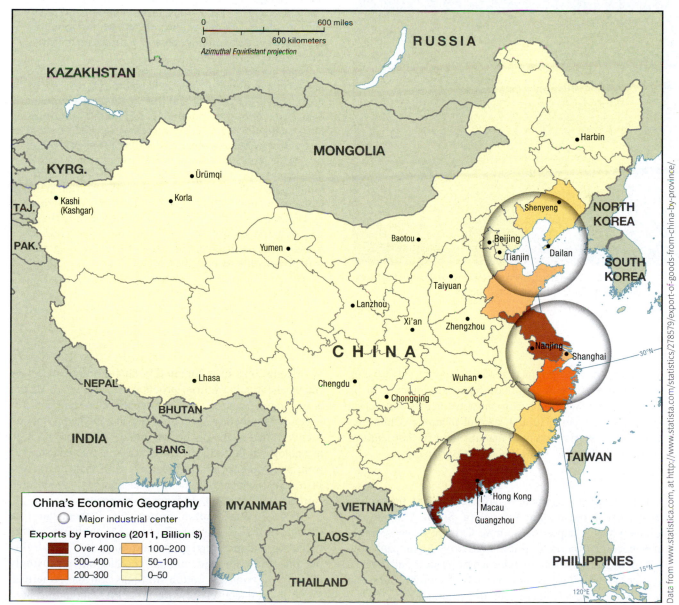

FIGURE 8.35 Manufacturing in China. Manufacturing in China is centered in three industrial clusters: Beijing, Shanghai, and Shenzhen. Exports per province are highest in the east where the industrial sectors are located and lower in rural, western China.

GUEST *Field Note* Chinese Link in the Commodity Chain

DR. JEAN-PAUL RODRIGUE
Hofstra University, New York

A value chain, also known as a supply or commodity chain, is a network of production, trade, and service activities used to gather resources, transform them into parts and products, and, finally, distribute goods to global markets. Each network is unique and dependent on the type of good, production factors (including resources and labor), market requirements, and the capacity of the transport system to effectively move raw materials, parts, and

FIGURE 8.36 Shenzhen, China. Pallets await shipping from this distribution center to the United States.

Courtesy of Jean-Paul Rorigue, Hofstra University.

finished goods. Value chains are inherently complex and involve a wide range of actors, including manufacturers, transport providers such as trucking and maritime shipping companies, transport terminals, distribution centers, and retailers.

This photo depicts apparel goods loaded on pallets in a distribution center in the major manufacturing hub of Shenzhen, China. When this photo was taken, I was visiting port and distribution facilities to get a better overview of the commercial relations brought by globalization. It was revealing to observe that many of the shipments were for large American retailers using the Pearl River Delta as a low production cost manufacturing facility. One shipment in particular was for a large, widely known fashion designer with the order address just a few miles away from where I lived. Thus, fashion design and retailing activities taking place in the New York metropolitan area were bound to production and distribution activities half a world away.

After leaving the distribution center, the pallets will be loaded into a container, which will be brought to a nearby port (most likely Yantian) and then loaded on a containership bound for the United States. From the port, most likely on the U.S. West Coast, the container will be carried by truck or by rail to a regional distribution center. There, the load is broken down into smaller shipments bound to individual retail stores. Such transactions underline the fundamental role of transportation and distribution in the global economy and the strategic and commercial interests behind them. The setting of global commodity chains has transformed the economic landscape of many regions of the world, particularly those that are highly integrated with global trade such as Shenzhen, China.

where and how to produce goods), shipping, advertising, storing, and selling. A commodity chain also includes many other intermediary and backroom steps such as accounting and legal services. Through careful planning, China has become the world's premier manufacturer of thousands of products and goods, from iPhones to Barbie dolls.

GLOBAL SOURCING

Much as Indian companies have established themselves as experts in outsourcing not only in South Asia but also globally (see Chapter 6), so have Chinese-owned companies become experts in **global sourcing**. A company anywhere in the world can imagine a new product and create a prototype, take it to one of thousands of sourcing fairs held worldwide, hire a Chinese sourcing agent, and hand over a large proportion of a particular good's commodity chain to a Chinese company.

Along links of a commodity chain, not all places or steps benefit equally from production. How much wealth is generated by production at a particular step, link, or node in a commodity chain depends on how the production occurs at that step. **Core processes**, which rely on high technology, highly educated workers, and financial support from governments, create greater wealth than **periphery processes**, which rely on low technology, uneducated workers, and little support from government.

For instance, cotton growers in Uzbekistan (see Chapter 10) use peripheral processes to grow cotton, including outdated irrigation systems that are drying out the landscape and making production increasingly difficult. In contrast, cotton in Texas and the American South is produced using core processes, including high technology fueled by massive research and development from agricultural implement companies making planters and harvesters specifically designed for cotton production and from seed and chemical manufacturers helping "guarantee" an output for cotton farmers. Designing technology and running it to produce cotton require high levels of education, which also helps churn up the regional economy through additional core processes. When an expensive piece of farming equipment is used in creating a commodity, many people "behind" the farming equipment receive high wages from engineering to producing to selling the equipment. Core processes, including higher technology and education, help create greater wealth than periphery processes, through designing, making, and using higher technology.

In East Asia, the growth of global supply companies has enabled China, South Korea, Japan, and Taiwan to gain a greater margin of the core processes in manufacturing. Global supply companies specialize in being the "brains" of production.

CENTRALLY PLANNED ECONOMY

To understand how radically China's economy changed between the 1950s and today, we need to understand how communism developed in China in the 1900s; how central planning worked; and how economic reforms in 1979 altered the path of Chinese economic development.

Development of Communism in China

Mao Zedong officially declared the establishment of the People's Republic of China (PRC), a communist country, on October 1, 1949, having defeated the Kuomintang (KMT), or Nationalist Party, after decades of civil war. At that point, KMT leaders fled to Taiwan, where they continued to govern a much-reduced Republic of China, which evolved into a multiparty democracy. Mao, whose portrait hangs outside the front gates of the six-century-old Forbidden City in Beijing (Figure 8.37), grew up in rural, central China. As a young man working in a university library in Beijing, Mao read Marx, but he, like Stalin in the Soviet Union, added his own interpretations to Marx's writings. In the 1950s, Mao recognized the need to industrialize China and also believed he could answer the decades-long call for a move away from feudalism and toward land reform in rural China by collectivizing agriculture.

Similarly to the Soviet Union, China instituted five-year economic plans, which continue today. Soviet planners helped the Chinese craft their first five-year plan, which dated from 1953 to 1957. The first five-year plan focused not only on collectivizing agriculture, but also on transferring privately owned industries to state ownership or joint state–private ownership, building new industrial parks, and centrally planning the Chinese economy. Mao planned to use monies generated from collectivizing agriculture to fund the creation of more state-owned industries, especially heavy industries.

How Central Planning Worked

During the first five-year plan, Mao hoped to intensify agricultural production in Chinese collective farms. He believed peasants would see collectivization as land reform, but peasant farmers generally refused to cooperate. In 1955, Mao delivered a speech urging peasant farmers to follow the plan of collectivization, explaining the issues facing Chinese agriculture by saying:

> The situation in China is like this: its population is enormous, there is a shortage of cultivated land (only three mou of land per head, taking the country as a whole; in many parts of southern provinces, the average is only one mou or less), natural catastrophes occur from time to time—every year large numbers of farms suffer more or less from flood, drought, gales, frost, hail, or insect pests—and methods of farming are backward. As a result,

Courtesy of Erin Fouberg.

FIGURE 8.37 Beijing, China. A woman takes a photo of a man and child in front of the 15-by-20-foot oil painting of Chairman Mao on the Tiananmen gate. Portraits of two other Chinese rulers adorned the gate before the first portrait of Mao was erected in 1949. A new, hand painted portrait of Mao is erected each year. The official portrait of Mao has changed in subtle ways since 1949, as a committee of artists led by an official portrait maker meets to make changes to the portrait. One of the most recent changes, according to the *New York Times* (2006) was to make both of Mao's ears visible to demonstrate that he can hear the masses.

peasants are still having difficulties or are not well off. The well-off ones are comparatively few, although since land reform the standard of living of the peasants as a whole has improved. For all these reasons there is an active desire among most peasants to take the socialist road.

In the same speech, Mao outlined three steps to collectivization. First, farmers were to organize themselves into teams or collectives. Second, the farmers in the collective were to pool their land and manage it as a semisocialist unit. Finally, the collectives were to unite into larger agricultural cooperatives that would be fully socialist.

Statistically, the first five-year plan was a success. The Chinese achieved the goal of transferring ownership of industries away from private individuals. By 1956, the state owned 67.5 percent of industries, and the remaining 32.5 percent were in joint state–private ownership. The growth in industrial parks and manufacturing plants through state construction helped industrial production increase quickly, rising 19 percent during the period of the first five-year plan. At the same time, Chinese income grew by 9 percent.

Despite the growth in Chinese industrial output, Mao was disappointed in the relative failure of collectivizing agriculture in the rural areas. Mao urged China to take a Great Leap Forward. Between 1958 and 1962, China instituted forced collectivization into communes. Mao's rationale was that collectivized farms would generate greater food production and more workers could be released from agriculture to work in heavy industries such as steel. His plan backfired. Farming productivity did not increase, and in fact by 1960 farming productivity suffered extraordinary losses. The lack of agricultural productivity in China spawned a massive famine. Between 1960 and 1961, 20 million people starved to death. China's Ministry of Foreign Affairs' website describes the period between 1958 and 1978 as a time when "the peasants' enthusiasm for [agricultural] production was greatly dampened."

In 1966, Mao sought to squelch his detractors by rooting out opposing voices. He wanted to remove all the "bourgeoisie" resistance to communism. Mao established the Red Guard, which was made up of small groups of radical students who were charged with destroying the four "old" habits that resisted communism: customs, habits, culture, and thinking (Figure 8.38). Mao called his movement the Cultural Revolution, and it lasted until his death in 1976. Under the Cultural Revolution, the Red Guard "reformed" not only farmers, but also teachers, artists, and writers who were not towing the communist line or were speaking or writing against Mao. One aspect of the Cultural Revolution was printing and distributing 35 million copies of what came to be known as "The Little Red Book" (originally titled "The Quotations of Mao Zedong").

Mao's reign established the Communist Party of China as the unequivocal controller of China and granted

ChinaFotoPress via Getty Images.

FIGURE 8.38   Beijing, China.   This image from 1966 shows Chairman Mao Zedong interviewing the Red Guards at Tiananmen Square. The Red Guards were a revolutionary mass organization composed primarily of university and high school students who were crucial supporters of the Cultural Revolution.

incredible power to the leaders of the Communist Party. From the Politburo to the armed forces to the provinces and townships, the Communist Party calls the shots in China. Today, the Communist Party has more than 66 million members, the vast majority of whom are men over the age of 30. Those who are members have privileges, including access to jobs and better housing. The Party controls aspects of everyday life for all 1.35 billion Chinese by censoring the media (Figure 8.39), controlling (as much as possible) the access to sites on the Internet, and providing housing and jobs. The Communist Party uses fear to intimidate dissidents and will use force or brutality to make examples of what happens when people speak up against the party.

Chinese Economic Reforms of 1979

After Mao's death in 1976, Deng Xiaoping, who was labeled as a "capitalist" during the Cultural Revolution, gathered other opponents of Mao and gained control of China. Deng Xiaoping added his own pragmatic spin to communism. He also established five Special Economic Zones (SEZs) much like the ports the Qing established for foreign investment in the 1800s. Under Deng's plan, the SEZs were open to foreign direct investment (FDI), enjoyed low taxes, had relaxed importing and exporting regulations, and had simplified land leases. The newly formed SEZs in five coastal locations became the foundation of the modern Chinese economy. While the Communist Party still controls the government, army, media, and much of daily life as it did under Mao, the economic systems in China have changed markedly since 1979. Overseas Chinese,

FIGURE 8.39 **Hong Kong.** In 2014, students in Hong Kong, including this 18 year old, occupied the streets near the city's financial district for pro-democracy rallies. While the months of protest in Hong Kong made the news around the world, the Chinese government censored Western media coverage in the rest of China, and the press in China gave limited and critical coverage of the Hong Kong protests.

especially in Hong Kong, Singapore, and Southeast Asia, were the first investors in SEZs. Overseas Chinese brought such rapid manufacturing development to China after the 1979 reforms that the post–1979 period is often referred to as China's Second Industrial Revolution.

Deng Xiaoping fundamentally reformed agriculture in China by abandoning the commune system in 1979 and replacing it with the Responsibility System. His first step was reestablishing the household or family as the "basic unit of farm production in rural China" (Veeck 2005, 4). Even with the reforms, farmers or households do not own the land they farm. Under the Responsibility System, each farmer contracts with government authorities at the village or township level to farm an amount of land for a certain period of time, typically 20 to 30 years. Households are allowed freedom to choose what to plant, how to produce, and when to sell. Farmers are also allowed to subcontract their land to other farmers through private, nongovernmental negotiations. In 1985, in a second reform, China ended its monopoly on purchasing agricultural products and allowed farmers to contract to sell their production. Agricultural reforms have led to higher output in Chinese agriculture (Figure 8.40).

INDUSTRIAL AGRICULTURE IN CHINA

In 2001, China joined the World Trade Organization (WTO). Since then, China has concentrated on advancing the use of science and technology in agriculture in order to generate increasing yields. In the tenth, eleventh, and twelfth five-year plans, China has specifically sought to intensify agricultural production through the use of technology. With approximately 20 percent of the world's population and less than 10 percent of the world's arable land, China recognizes a need to generate higher agricultural output. In 2005, the Organization for Economic Co-operation and Development (OECD) described China's agriculture as having abundant labor with little mechanization. The OECD estimated that over 200 million tiny farms averaging 1.6 acres produce China's agricultural output. With an abundance of labor, China has an economic advantage in producing labor-intensive agricultural goods, including fruits and vegetables.

China has historically increased production by moving onto marginal lands, thereby increasing the amount of land being farmed, and more recently, China has increased the use of fertilizers to generate higher production. The OECD claims that the use of fertilizers in China per acre is one of the highest rates in the world. China, like neighboring Japan and Korea, provides price supports for commodities that face import competition; these commodities include "sugar, milk, sheep meat, cotton

FIGURE 8.40 **Jiaozhou City, Shandong Province, China.** A Chinese farmer harvests corn while another watches. China is converting to using high-yielding seeds and embracing modern technology to increase grain production.

and soybeans, as well as some export commodities such as maize and rice" (Figure 8.41).

In 2007, the Chinese government issued the Number One Document of 2007, also known as the Modern Agriculture Document, which "stipulates that modern equipment, science and technology, industrial systems, management and development ideas shall be used to improve the quality, economic returns and competiveness of agriculture" (Waldron et al. 2010, 479). In 2012, the Chinese premier announced a plan to allocate $192 billion to create more water-efficient agriculture and to subsidize the use of agricultural technologies.

In the interest of promoting industrial agriculture, the Chinese state is increasingly seizing farmland from small producers. Chinese who continue to farm on small plots supplement their income by working in township enterprises or by receiving remittances from family members who have migrated to work in cities. With rising income levels, Chinese consumers are eating "10 percent more meat than they did five years ago" (*New York Times* 20 April 2012). To meet demand, the Chinese government is making it easier for industrial agriculture to consolidate

FIGURE 8.42 **Huaibei, Anhui Province, China.** A Chinese farmer feeds pigs. Between 1997 and 2007, the consumption of pork, China's most popular meat, doubled. China produces and consumes almost half of the world's pork.

land and import agricultural technologies. Chinese farmers are importing breeding stock from the United States, including breeder pigs and breeder broiler chickens. In addition, Chinese are importing bull semen from the United States to increase beef and dairy production. Between 1999 and 2008, China increased its beef production by an average of 21 percent a year (Waldron et al. 2010) (Figure 8.42).

Incorporating American agricultural technologies enables Chinese producers to increase meat production. For example, a Chinese broiler chicken takes 120 days to grow to market weight, but an American broiler chicken takes only 41 days to reach market weight (*New York Times* 20 April 2012). A livestock genetics expert illumined one shift from switching to American livestock breeds. The Chinese will become increasingly dependent on corn and soybeans. He explained "'Genetics and nutrition go hand-in-hand,'" and "'the more they use our genetics, the more they're going to need to import corn and soybeans from the U.S. and elsewhere'" (*New York Times* 20 April 2012).

CHINESE COMPANIES

Private enterprise is awash in China. The role of government is either in partnering with private companies or in Communist Party members owning or investing in private companies. **Minying firms**, which are nongovernmental firms "established outside either central or local government budgetary channels," have been major players in the growth of China's economy since 1979. Some minying firms are owned partially by the state, but all of these firms are market oriented and have "autonomy from the direct supervisory power of the Chinese government" (Zhou 2008, 3–4).

Beijing is home to the Zhongguancun (ZGC) high-technology district, which is the Silicon Valley of China (Figure 8.43). Beijing University, the Harvard or Stanford of

China's Producer Support Estimate by Commodity

Producer Support Estimates (PSE) and Single Commodity Transfers (SCT)

	2002	2006	2010
Producer Support Estimates			
¥ million	203,004	436,230	994,780
Market Price Support	121,357	205,310	609,209
% of PSE in gross farm receipts	8	12	17
Single Commodity Transfers (¥ million)			
Wheat	-17.844	53,569	66,149
% of SCT in gross farm receipts	-19	34	28
Rice	12,980	-11,167	-14,026
% of SCT in gross farm receipts	7	-4	-3
Maize	25,349	43,598	68,079
% of SCT in gross farm receipts	25	26	23
Soybean	5,885	5,927	12,767
% of SCT in gross farm receipts	17	16	24
Cotton	15,836	33,434	54,002
% of SCT in gross farm receipts	34	37	51
Rapeseed	-162	2,326	8,122
% of SCT in gross farm receipts	-1	9	16
Pig meat	-10,803	-14,656	87,712
% of SCT in gross farm receipts	-3	-2	12
Sugar	6,245	1,617	10,715
% of SCT in gross farm receipts	41	7	29

Data from: "Agriculture and Trade Policy Background Note: People's Republic of China," the Food and Agriculture Organization of the United Nations. Data from the OECD Producer and Consumer Database, 2012.

FIGURE 8.41 **China's producer support estimate by commodity.** The Single Commodity Transfer (SCT) is a measurement of the value of annual gross transfers from government policies in support of production of that specific commodity. The percent SCT is the proportion of government support for the commodity relative to the total gross farm receipts for the specific commodity (basically the government support relative to the total income from sales).

FIGURE 8.43   Beijing, China.   The Zhongguancun Science Park, in northwest Beijing, is China's Silicon Valley. The area houses 33 universities, research and development firms, domestic Chinese firms, and multinational companies.

major nodes in the networks of global trade, as containers of goods are gathered at port cities (some directly from land and some from smaller ports), organized, and placed on massive container ships for distribution around the world.

Geographers Lam and Yapp (2011) found that large ports compete with each other for market share. The researchers demonstrated that shipping carriers are equally willing to use Busan, South Korea, and Ningbo, China, to accumulate containers from smaller ports and then ship globally. The two ports compete with each other for business by reducing port charges to shipping companies, providing monetary incentives to shipping companies, and investing in smaller feeder ports in the region. Small ports are linked as spokes to large hub ports in East Asia. The port of Shanghai invests in smaller coastal ports in China to help secure a network of small ports that feed finished goods to the large port of Shanghai, China, for global distribution.

China, is located in the ZGC and its researchers have been instrumental in forming and attracting domestic companies, including Lenovo, Baidu, UFIDA, Founder, and Datang, to produce high-tech goods including software and hardware, and to establish high-tech services, including Internet firms.

ZGC is the headquarters for dozens of domestic companies in China. Domestic companies have knowledge of Chinese markets and government that have helped them find financial success in the ZGC. Since the 1980s, more than 1,600 foreign-funded companies have set up offices in the ZGC to partner with domestic Chinese companies (300 Chinese-funded companies are also located in the ZGC) and tap their knowledge to create goods and services in demand by the growing consumer market in China.

PORTS IN EAST ASIA

The growth of manufacturing in China has impacted other parts of East Asia. The ports of Busan, South Korea, Kaohsiung, Taiwan, Ningbo, China, and Shanghai, China (Figure 8.44) are major container ports that accounted for 43 percent of all throughput container traffic in East Asia in 2009. Ports are connectors between production on land and distribution overseas. Containers of goods or materials are trucked or moved by rail or river from where the goods are produced on land to where they are shipped. Smaller ports amalgamate containers, and then goods are moved by ship to larger ports. Large ports are

CONCEPT CHECK

1. **How** does being a leader in global sourcing enable China to capture more of the core processes in the commodity chains of manufactured goods?

2. **What** are SEZs, and **how** did they change China's economy?

3. **Why** are Busan, South Korea, and Shanghai, China, networked with smaller ports in East Asia?

FIGURE 8.44   Port of Busan, South Korea.   The Korea Express Container Terminal is just one part of the massive port of Busan, which is the fifth largest container port in the world.

WHAT ROLE DO GOVERNMENTS PLAY IN MIGRATION IN EAST ASIA?

LEARNING OBJECTIVES

1. **Describe** how the Chinese government uses the hukou system to try to control rural to urban migration.

2. **Explain** why an increasing number of women and girls from North Korea are being trafficked into China.

3. **Understand** why Japan encouraged immigration of Japanese Brazilians and why Japanese Brazilians are taking the path of return migration.

Whether in dictatorial North Korea, communist China, or democratic Japan, governments play a large role in influencing and controlling migration within East Asia (Figure 8.45). In China, the government works to slow rural to urban migration through a system of registering Chinese people. In North Korea, the government does not allow its people to leave, but North Koreans are being trafficked into China as an unintended consequence of China's one-child and hukou policies (hukou is explained below). In Japan, the government has encouraged Japanese Brazilians to migrate to Japan to supplement the country's declining labor pool, but the slowdown in the Japanese economy has resulted in Japanese Brazilians returning to Brazil.

CHINA

Growth in China, both industrial and agricultural, has fueled a stream of internal migration from rural to urban areas, generally west to east (Figure 8.46). Migration flows in China are distinct from those in other countries because the Chinese government registers each citizen by his or her home region. Some trace the system of registration to 2,000 years ago during the Han Dynasty, but its current version dates to a Maoist policy established in 1958. The registration policy is designed to control migration within China and to keep Chinese living throughout the country, especially in rural areas. Known as the **hukou** system, it limits where Chinese live and affects the flow of migration in the country. Hukou delineates both the location where a Chinese person is registered and whether that location is agricultural or nonagricultural. The Chinese government uses hukou to keep track of pertinent information,

FIGURE 8.45 **Political geography of East Asia.** East Asia includes the countries of China, Japan, South Korea, and North Korea.

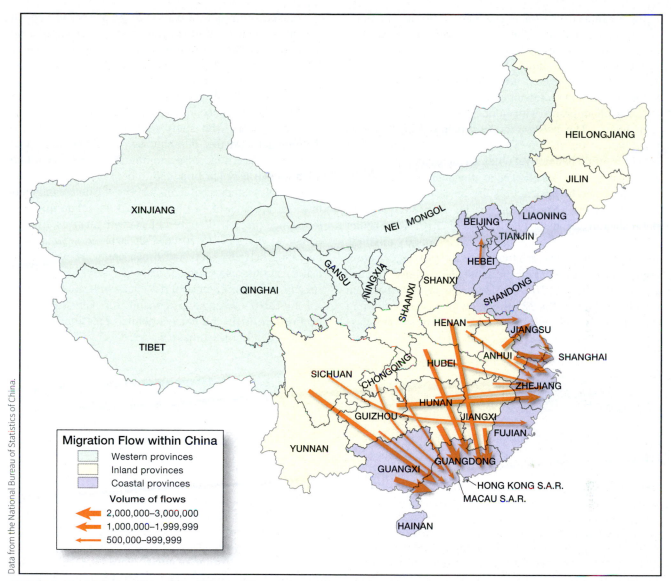

Data from the National Bureau of Statistics of China.

Migration Flow within China

- Western provinces
- Inland provinces
- Coastal provinces

Volume of flows
- 2,000,000–3,000,000
- 1,000,000–1,999,999
- 500,000–999,999

FIGURE 8.46 **Migration flow within China.** Chinese migrated from rural to urban areas between 2005 and 2010 despite the efforts of the Chinese government to stem internal migration through the hukou system.

including births, deaths, marriages, and divorces; to monitor its population, especially those who are politically subversive; and to "keep poor rural farmers from flooding into the cities" (Richburg 2010, A8).

Migration (see Chapter 11) in China follows two tracks. One track is *permanent* and occurs when migrants leave their homes and move to another region, typically urban, and then permanently changes their hukou to their new locale. The other track is *temporary* and occurs when migrants leaves their hukou for an urban area to work but never change their hukou to the new location.

The Chinese government provides housing support and other financial aids to Chinese who live in their hukou. When rural residents weigh the decision to migrate without government permission, they have to consider whether their potential earnings in urban areas outweigh financial benefits they receive by staying in the home province. The

growth in jobs and incomes in urban areas leads millions of Chinese to temporarily migrate, to move without moving their hukou, to urban areas for work. Temporary migrants are drawn to urban areas in part because the household per capita income for Han Chinese is 2.5 times higher in urban areas than in rural areas (World Bank 2011). Rural residents benefit financially from **remittances** sent to them by relatives working in the cities, and rural residents maintain the economic benefits afforded to them by their hukou, which includes access to leasing (and subcontracting) farmland.

Before 1980, population mobility in China was low, but by 2011 over 15 percent of Chinese lived in places other than their hukou (Sun and Fan 2011, 93). Millions of migrants in China move from rural to urban areas in order to earn higher wages. In 1990, 59.8 percent of migrants came from rural areas, and in 2000, 80 percent of internal migrants in China left rural areas (Sun and Fan 2011). Most of the migrants from rural to urban areas have not officially changed their hukou. When the government issues new hukou, it is "usually tied to home purchase, investment, age, education, and skills" (Sun and Fan 2011, 94). In the 1990s, Shanghai offered urban hukou to "investors, new homeowners, and professionals" (Sun and Fan 2011, 94).

Not all migrants with good jobs are able to change their hukou. The *Washington Post* reported in 2010 that the editor of the *Beijing News*, who had lived in Beijing for seven years had not yet obtained an urban hukou. As a result, his son was not allowed to attend public high school in Beijing, and he does not qualify for a subsidized apartment in the city. Nor are he and his family eligible for public health care in the city (Richburg 2010, A8). Having a hukou outside of the city means that temporary migrants must "travel to their home towns to get a marriage license, apply for a passport or take the national university entrance exam" (Richburg 2010, A8). Estimates are that between 7 and 8 million of the 22 million people living in Beijing do not have a Beijing hukou, making them second-class citizens in their city.

Hukou still acts as a barrier or gatekeeper to internal migration; Chinese who do migrate rely on **social networks** (Sun and Fan 2011). Reforms to the hukou system allow Chinese to change their hukou to that of their spouse. On Internet dating websites in China, some people who are looking for matches "will state upfront that they prefer a partner who has a Beijing hukou" (Richburg 2010, A8).

Geographers who study migration have established that migrants, whether international or internal, use social networks to determine where to migrate, to help them find a job in their new location, and to help them find housing (Figure 8.47). Studies based on surveys of Chinese migrants find that upwards of "70 percent of rural-urban migrants found their first job through contacts with relatives, friends, or fellow villagers" (Sun and Fan 2011, 96). Streams of migration between rural and urban areas are well established, and the use of social networks helps entrench the streams and connect internal provinces with coastal cities.

In 2010, when the demand for Chinese goods declined with the global economic recession, between 20 and 23 million Chinese workers were laid off. Most

 ## USING **GEOGRAPHIC** TOOLS

Surveys in Geography Research

Geographers use surveys to get a better overall picture of a situation and also to understand individual perceptions, interpretations, and motivations. Individual interviews give researchers anecdotes that can be used to paint a story of what is happening to an individual or family in a place, and surveys allow researchers to measure the magnitude of the issue and understand how many people are being affected by the same issue. In 2008, geographers C. Cindy Fan, Mingjie Sun, and Siqi Zheng surveyed people in 50 urban villages in Beijing to understand circular rural to urban migration in China. Rural to urban migration often involves families splitting temporarily with those who go to the city

earning money to send home to the rural area. The migration is circular because the migrants move from rural home to urban village and then back to rural home in response to opportunities and demands at home. Fan et al. found that family decisions regarding circular migration are made in the context of an extended family, "including those not living under the same roof." Through survey research, Fan et al. found that men were the first to migrate, but eventually couple and family migration became more common in China. Through survey research, they found that "rural Chinese are actively rearranging their household division of labor to take full advantage of migrant work opportunities."

Migration and Split Households in China

Forms of Households for Migrants Living in Beijing	Frequency	Percentage
Spouse and all children in Beijing (family migrants)	305	46.7
Spouse in Beijing, all children in the home village (couple migrants)	112	17.2
Spouse and all children in the home village (sole migrants)	104	15.9
Spouse and/or some children neither in Beijing nor in the home village	75	11.5
Some children in Beijing and some in the home village	51	7.8
Spouse in the home village, all children in Beijing	6	0.9
Total	**653**	**100.0**

C. Cindy Fan and Mingjie Sun, Migration and split households: a comparison of sole, couple, and family migrants in Beijing, China, *Environment and Planning A*, 201:43, 2164–2185. Used by permission of Pion Ltd., London, www.pion.co.uk and www.envplan.com.

FIGURE 8.47

Thinking Geographically

1. The data in the table are from 653 interviews. How might the data change if the researchers only had 10 interviews? Why is it important to have a large sample of people interviewed?

2. Figure 8.46 shows Guangdong as a major migration destination. How might the results differ if the researchers conducted the study in Shenzhen (a major city in Guangdong) instead of Beijing?

of the laid-off workers were temporary migrants. It remains to be seen what temporary migrants in urban areas including Beijing and Shanghai will do when they lose their jobs: Will they stay in the city to look for other employment, or will they return to their hukou region? When temporary migrants become unemployed, remittances decline to rural areas as well, which could accelerate the rural–urban wealth gap and undermine political stability in rural regions of China.

NORTH KOREA

At the end of World War II, the Allies considered how to handle Japan's former colonies in East Asia and the Pacific. The Allies divided the Korean Peninsula along the 38th parallel. The Soviet Union was to occupy North Korea, and the United States had South Korea. In 1950, North Korea attacked South Korea, which started a three-year war. Although exact numbers are not known, estimates are that more than 200,000 North Korean civilians, up to one million South Korean civilians, 150,000 troops from the South Korean side (including more than 36,000 Americans), and 800,000 troops from the North Korean side died during the Korean War.

When the Korean War came to a close in 1953, no official peace treaty was signed. The Demilitarized Zone (DMZ) divides North and South Korea on the 38th parallel (line of latitude) (Figure 8.48). Today, the DMZ is the focal point of the militaries of both countries. Between the two countries, a total of 1 million troops police the DMZ. In South Korea, two years of military service is mandatory for males. North Korea has a population of 23 million, and 1 million are in the military. Members of the military receive priority for food, which is vital in a country with frequent famines.

The leaders of North Korea have nearly absolute power and work to keep North Koreans in the dark about the realities of South Korea and the rest of the world. North Korea, more than any country in the world, tries to operate outside of the capitalist world economy. North Korea controls information in and out of the country, even more so than its neighbor China. North Korea teaches its citizens that the United States started the Korean War in 1950, although it is internationally agreed that North Korea started the war when it sent troops across the line in 1950. BBC reporter Sue Lloyd-Roberts reported that when she visited North Korea in 2010 her mobile phone was confiscated at the Pyongyang airport and that she did not have access to the Internet the entire time she was in the country. She described the country by saying "the entire country lives in a bubble of unreality, cut off from the outside world

Courtesy of Erin Fouberg.

FIGURE 8.48   Dora Observatory, South Korea.   View of Kijong-dong, North Korea, looking over the Demilitarized Zone (DMZ) from South Korea to North Korea. The DMZ divides North Korea and South Korea along a 160 mile (250 km) border. The zone is filled with landmines, fences, and open fields.

and watched by an army of informers" (Lloyd-Roberts 2010). The reporter described two mobile phone networks in the country. One is for nongovernmental organizations (NGOs) and diplomats who work in the country, and the other is for a small group of political elites. No commoners in the country have access to the Internet. Instead, the government created an internal intranet that it fully controls.

North Korea has a hereditary dictatorship. Kim Il-Song, called the Great Leader by North Koreans, is the founder of the country. He died in 1994 but is still officially the president of the country. His son, Kim Jong-Il, who was called the Dear Leader, ran the country until his death in 2011. After the death of Kim Jong-Il, his son, Kim Jong-Un, took control.

Relations between North Korea and South Korea remain tense. South Korea has discovered several tunnels under the DMZ that the North Koreans have dug in order to attack South Korea at some point (Figure 8.49). North Korean spies, some of whom the government abducted from Japan and trained as spies, are caught in South Korea with some level of regularity. Despite propaganda in North Korea about the evils of South Korea and the United States, North Koreans continue to attempt to defect to South Korea. With the DMZ, it is quite difficult to travel, even illegally, directly from North Korea to South Korea.

Between the end of Japanese occupation at the end of World War II in 1945 and the end of the Korean War in 1953, approximately 900,000 North Koreans migrated to the south. The Korean War split families, and the insularity of the North

Erin Fouberg.

FIGURE 8.49   DMZ, South Korea.   South Korea discovered four infiltration tunnels under the DMZ between 1974 and 1990. The third tunnel was discovered 27 miles (44 kilometers) from Seoul in 1978. The North Koreans designed the tunnel to allow their troops to attack South Korea from under ground.

Korean government since 1953 makes it nearly impossible for families split between North and South Korea to connect. The North Korean government strictly controls emigration and immigration. North Korea does not allow emigration, and those who emigrate without authorization are considered treasonous. If detained or deported back to North Korea, illegal North Korean emigrants are put in penal labor colonies where "prisoners serve anywhere between two to seven years" and "rates of torture and death are notoriously high" (Tanaka 2008).

During the Cold War, defectors to South Korea received a payment from the government of South Korea in exchange for information about what was happening in North Korea. The average payment by the South Korean government to a North Korean defector fell from $32,000 during the Cold War to $9,000 in 2005. The difficulty of finding one's way to South Korea from North Korea is reflected in the relatively few defectors who make it South Korea. The South Korean Ministry of Unification reports "a total of 8,661 North Korean defectors arrived between 1990 and 2006" (Tanaka 2008, 3).

Restrictions on mobility in North Korea, the heavily defended DMZ, and the North Korean penal code work together to encourage defectors to head to China first and hope to find their way to South Korea or elsewhere (Tanaka 2008). Estimates of the number of North Koreans who have illegally migrated to China range between 30,000 and 500,000. The migrants travel across the river border between North Korea and China, typically with the help of a smuggler or broker. "Brokers, who are usually

Korean, Chinese or former defectors and who work independently or in teams, charge anywhere between US $1,250 and US $19,950 per person to either accompany individuals across the border or to inform them of where and when it is safe to cross" (Tanaka 2008, 4). Defectors who do not use brokers typically bribe crossing guards, which is much less expensive, though more dangerous.

Since the North Korean famine of 1994–1998, the number of North Korean women and girls who have become refugees in China has increased markedly. In 2010, an aid worker estimated that "80 percent of North Korean refugees in China" are women and that more than "90 percent of North Korean refugee women" are trafficked in response to the gender imbalance in China (Kim 2010). As a consequence of the one-child policy in China, men under the age of 30 far outnumber women in China. Chinese men with hukou in urban areas are more attractive as marriage partners for Chinese women from rural hukous. Chinese men with rural hukous or who cannot find a Chinese bride increasingly look to North Korea and Southeast Asia for brides. NGOs that work against trafficking report an increasing number of North Korean women and girls trafficked into China as brides.

JAPAN

The Japanese government has historically been hesitant to allow immigrants to settle in the country. During European colonialism in East Asia, Japan refused colonizers except for the Netherlands and relegated the Dutch to the southern island of Dejima. During the Japanese occupation of Korea, approximately 2 million Koreans migrated to Japan. After World War II, approximately 500,000 Koreans remained in Japan. Migrants who came before 1945 are considered "old comers" by Japanese. In 1952, Japan passed the Immigration Control Act to restrict the number of immigrants. The government also uses an alien registration system to "monitor and control foreigners, whether newly arriving or already residing for years in the country" (Kashiwazaki and Akaha 2006, 2).

The TFR in Japan has been below replacement level since 1955, but the Japanese government generally opposes permanent migration that would fortify its labor supply. During the 1960s and 1970s, migration to Japan was relatively low, as Japanese industries focused on automating processes instead of encouraging the Japanese government to allow more immigrants into the country to work in factories (Kashiwazaki and Akaha 2006). During the 1980s, Japan's economy boomed, and the number of foreign workers grew to 300,000 by 1993. The immigrants who came after 1980, primarily from Southeast Asia, are considered "newcomers" by the Japanese. In 1993, Japan launched the trainee program, which allows foreign workers to work in agriculture, food production, and industries. Despite allowing foreign workers in certain sectors, Japan's economic recession began in 1997, and the lack of a working-age population has made it difficult for the economy to rebound fully.

Although Japan needs to increase its labor pool, the government controls the flow of immigrants and the duration of their stay. In 1989, Japan reformed the Immigration Control Law in order to "facilitate the immigration of professional and skilled personnel, while confirming its basic principle of not accepting 'unskilled' foreign labor" (Kashiwazaki and Akaha 2006, 2). Japan also opened its doors to Nikkeijin—descendants of Japanese who emigrated to other countries, including Brazil and the United States. An estimated 3 million people of Japanese descent live abroad, with half of the 3 million residing in Brazil. The number of Brazilians of Japanese descent living in Japan increased from 56,000 in 1990 to 286,000 in 2004 (Figure 8.50).

When the Japanese economy entered a recession in 2007, the lack of jobs in Japan and the growing economy in Brazil generated a **return migration** where Brazilians are leaving Japan and returning to South America. Between 2008 and 2012, Brazilians left Japan at an average rate of 8 percent a year.

The Japanese government encourages the immigration of highly skilled workers by separating immigration visas into two categories. One is for those who are married to or are children of Japanese nationals, and the other is for immigrants with skills in certain fields, especially science and engineering. Highly educated, well-paid workers are able to work in Japan for a certain amount of time. In order to prevent immigrants with skills needed in particular areas, such as nursing, from staying permanently, the Japanese government issues incredibly difficult exams in the subject areas, written in Japanese. Thus, a nurse from

© Aflo Co. Ltd./Alamy.

FIGURE 8.50 Tokyo, Japan. Brazilian Japanese hold placards and wave Brazil's national flag in Yoyogi Park to protest problems with the Brazilian government, including high taxes, inflation, corruption, rising prices, poor public services, inadequate hospitals, and weak schools. Demonstrators also denounced the Brazilian government for spending billions of dollars on the 2014 World Cup and 2016 Olympics.

Indonesia who has already graduated and passed exams in Indonesia can migrate to Japan on a work visa, but in order to stay beyond three years, she or he must pass another difficult nursing exam in Japanese. By recruiting workers from countries such as Indonesia, where people have little if any Japanese skills, the Japanese government is effectively preventing skilled immigrants from staying beyond three years.

CONCEPT CHECK

1. **How** does the hukou system in China work?

2. **Why** are women and girls from North Korea being trafficked into China?

3. **What** is return migration, and **why** are Japanese Brazilians following it?

SUMMARY

What Is East Asia?

1. Chinese civilization has existed for over 2,000 years, as reflected in the characters used in the language, in the records found in the loess caves along the Silk Road in western China, in the enormous building projects of the Great Wall and the

Terracotta Warriors, and in the attempts to control the Huang He over time.

2. The climate of East Asia is influenced by the monsoon in the southeast. In the west, the Tarim Basin is in the rainshadow of the Himalaya Mountains, making it incredibly dry. In the

northeast, the region's climate is dominated by the Siberian High in winter months.

3. Japan is an arc of volcanic mountains formed parallel to two plate boundaries, which are also subduction zones in the Pacific. Its proximity to subduction zones makes Japan vulnerable to earthquakes and tsunamis.

Who Are East Asians?

1. The Silk Road connected Xi'an China and points east, including Korea and Japan, with the Roman Empire in the west. Religions, languages, and diseases diffused along the Silk Road.

2. East Asia resisted, to a large degree, colonization by Europeans. China expanded and contracted its territorial control over dynasties, at different points controlling the Korean Peninsula and influence in the Korean language. After opening its economy in the late 1800s, Japan looked outward and colonized the Korean Peninsula, portions of coastal China, and parts of Southeast Asia and the Pacific in the decades leading up to World War II.

3. Daoism and Shintoism have their hearths in East Asia. Daoism in China and Shintoism in Japan found followers through the hierarchical diffusion of dynasties in the respective countries. Buddhism diffused to East Asia and is often mixed with or practiced with influences of Daoism and Shintoism.

How Do Confucianism and Political Policies Influence Population Growth in East Asia?

1. East Asia has a high rate of gender imbalance, where fewer girls are born within countries than should be if it were left to nature. Demographers report that in nature approximately 103 boys are born for every 100 girls. In 2010 in China, 123 boys were born for every 100 girls, and in South Korea in the same year, 111 boys were born for every 100 girls.

2. Confucianism is a political philosophy that strongly influences the cultural norms in East Asia. Confucianism stresses loyalty and diligence to family and government. Confucianism elevates and privileges sons over daughters.

3. China has had a one-child policy since 1979, which has resulted in much lower TFRs and gender imbalance. South Korea once encouraged small families, but the TFR dropped so low and the country experienced such a gender imbalance that the government now has pronatalist policies. Japan's low TFR coupled with its relatively strict immigration controls has led to a declining population. Japan has strongly pronatalist policies, which are having little impact to date.

How Do Commodity Chains Link East Asia and the World?

1. China is not only the manufacturing leader in the world, but the country also has a growing influence in global sourcing, which helps the Chinese economy capture more core processes in the commodity chains of manufactured goods.

2. As a communist country, China's economy was centrally planned and experienced collectivization in the 1950s and 1960s. After the death of communist leader Mao Zedong, Chinese leader Deng Xiaoping made reforms to the Chinese economy in 1979, including the creation of Special Economic Zones (SEZs), which drew manufacturing to the country.

3. East Asian cities serve as ports for the massive number of consumer goods manufactured in China, Korea, and Japan. Ports in Shanghai, China and Busan, South Korea, serve as hubs for smaller regional ports.

What Role Do Governments Play in Migration in East Asia?

1. China is experiencing massive rural to urban migration as a result of its growing economy. The Chinese government encourages people born in rural areas to remain or return to their home provinces through the hukou system.

2. The government of North Korea does not allow its people to emigrate. However, the gender imbalance in China has left between 20 and 30 million young men without a potential bride, which has resulted in increased trafficking of women and girls from North Korea and Southeast Asia into China.

3. Japan's declining population means the country has a shortage of working-age men and women. The Japanese government encouraged migration of Japanese Brazilians to Japan. As the Japanese economy has declined, however, many Japanese Brazilians have taken the path of return migration to Brazil.

YOUR TURN ▶ Geography in the Field

The golf course in this photograph is one of two reportedly located in North Korea. With a price tag of $74 million, the Ananti Golf and Spa Resort is a six-star golf course (on a scale that tops out at five stars) that boasts the longest hole in golf, a par 6 hole that is more than 1,000 yards long. The Ananti also has a par 3 funnel hole, where the green is designed as a funnel to help your ball reach the hole.

Kim Jong-Il, who was known as the "Dear Leader," died in 2011 holding what the government claims to be the world record golf score of 38 under par on North Korea's other course, the Pyongyang Golf Club. His record-setting round was reportedly the first he ever played and included 5 holes in one.

The Ananti, which was built as a joint venture between North Korea and South Korea to try to increase cooperation, hosted one professional golf tournament in 2007 and was opened for South Korean tourists to play golf. In 2008, a South Korean tourist was shot dead at the course and the border was closed. The golf course still stands, but it stands empty, as it is maintained but closed.

Thinking Geographically

- Search the Internet for aerial views of Ananti and of golf courses in your region. How does this course and the city around it look different from golf courses in your region?

© STAFF/X01095/Reuters/Corbis Images.

Mount Kumgang, North Korea. Built as a joint venture between North Korea and South Korea, this golf course no longer operates.

- Read the article by Michael Wray, "Golfing in North Korea: The Hermit Kingdom's Newest Pastime" published in *Time* on June 6, 2012: http://www.time.com/time/world/article/0,8599,2116565,00.html. The author describes the Potemkin Tourist Trail and his interactions with official tour guides in North Korea. Explain why the North Korean government has official tour guides and a prescribed tourist route.
- The new leader of North Korea, Kim Jong-Un, was photographed for the first time with his fiancé at a park in North Korea where he is planning to build a miniature golf course. Why is the North Korean government spending money on golf courses? What identity are they trying to project to North Koreans and to the world?

KEY TERMS

loess	Siberian high	gender imbalance	periphery processes
channelization	subduction	Confucianism	minying firms
humid subtropical climate	tsunami	female infanticide	hukou
humid continental climate	continental slope	one-child policy	migration
Siberian high	continental shelf	pronatalist policy	remittances
rainshadow effect	torii	commodity chain	social networks
monsoon	Buddha	global sourcing	return migration
orographic precipitation	bodhisattva	core processes	

CREATIVE AND CRITICAL THINKING QUESTIONS

1. The **site** of Gwnghwamun Gate in Seoul has changed slightly over time, and at the same time the **situation** has changed radically. Describe how the **situation** of Gwanghwamun Gate has changed over time.

2. Explain where the **hearth** of Buddhism is (use Chapter 6 to help) and then describe how and why Buddhism **diffused** to East Asia.

3. Describe how the Three Gorges Dam has changed the Yangtze River **network**. Determine how the Three Gorges Dam contributes to the concept of the **anthropocene**.

4. Use the U.S. Census website to create **population pyramids** for China and South Korea in 1990 and in the most recent year available. http://www.census.gov/population/international/data/idb/informationGateway.php. Explain how you can see evidence of the one-child policy and **gender** imbalance in the **population pyramids**.

5. In the **context** of the period between 1500 and 1950, determine what colonial power had the greatest impact on the **region** and describe what the impact is.

6. Describe how **gender** preference is tied to Confucian values. Explain how and why female infanticide and human trafficking are tied to the one-child policy in China.

7. Find something in your dorm, apartment, or home that was made in China. Imagine all of the steps needed to produce

this good and get it into your home. Write down each step as a link in the **commodity chain** of the good. At which steps/links in the **commodity chain** do you think **core** processes are most likely, and at which steps/links do you think **periphery** processes are most likely?

8. Explain what the hukou system is and how it is designed to slow rural to urban **migration** in China today.

9. Search for the Itsukushima Shinto shrine on the UNESCO site: http://whc.unesco.org/en/list/776. Describe the **cultural landscape** of the Shinto shrine, and explain the role of the torii in the sacred **site**.

10. Explain how **commodity chains** are **globalization** processes. The impacts of **globalization** are not evenly distributed. Use the example of a **commodity chain** to justify this statement.

SELF-TEST

1. What country colonized South Korea between the late 1800s and World War II?
   a. China
   b. Japan
   c. Great Britain
   d. France

2. Huang He means Yellow River. The river is yellow because the soil it flows through is:
   a. aridisol
   b. oxisol
   c. loess
   d. anthropoene

3. The Three Gorges Dam in China was built despite the concerns of environmental scientists. Building a dam is controversial because dams can:
   a. block migrating aquatic species
   b. trap sediment
   c. change water temperature
   d. all of the above

4. The islands of Japan are:
   a. arcs of volcanoes parallel to two subduction zones
   b. arcs of volcanoes perpendicular to two subduction zones
   c. volcanic islands formed by the Pacific plate moving west over a hotspot
   d. volcanic islands formed by the Pacific plate moving east over a hotspot

5. The tsunami in Japan in March 2011 hit Sendai, Japan, particularly hard because:
   a. the city of Sendai is built below sea level
   b. Japan does not have a tsunami warning system
   c. the coastline at Sendai is crescent shaped
   d. all of the above

6. In the fifteenth century, Koreans changed their language by:
   a. adopting the hangul lettering system for the Korean language
   b. adopting the Chinese character system for the Korean language
   c. creating a Creole language that combines the Korean and Dutch languages
   d. creating a Creole language that combines the Korean and French languages

7. Confucianism incorporates all of the following values except:

   a. matriarchal society where women are more important than men in the family lineage
   b. loyalty toward one's superiors in family and government
   c. diligence toward work
   d. benevolence (kindness) toward others in society

8. You can see evidence of European trade and influence in the cultural landscape of China today in the:
   a. Imperial Palace East Garden in Tokyo
   b. Forbidden City in Beijing
   c. Gyeongbokgung Palace in Seoul
   d. Bund in Shanghai

9. Daoism is a religion whose hearth is in _____, which was diffused hierarchically by _____.
   a. Japan/Tokuguwa Shogunate
   b. China/Han Dynasty
   c. South Korea/Choson Dynasty
   d. Hong Kong/British colonizers

10. In Japan, Shintoism is visible in the cultural landscape through ____ at entrances to Shinto shrines.
    a. mosques
    b. stupas
    c. torris
    d. statues

11. _____ Buddhism teaches that individuals can seek enlightenment through meditation and right acts, and it diffused in East Asia because it coincides well with the teachings of _____.
    a. Mahayana/Daoism
    b. Mahayana/Islam
    c. Theravada/Daoism
    d. Theravada/Islam

12. Gender imbalance in China is evident in all of the following cases except:
    a. for second children, the sex ratio is 146 boys for 100 girls
    b. for third children, the sex ratio is 167 boys for 100 girls
    c. the poorest provinces in the west have the most imbalanced sex ratios
    d. the wealthiest provinces in the south and east have the most imbalanced sex ratios

13. Mao Zedong is the father of communism in China. In this role, he:
    a. rejected Stalin's teachings and refused to collectivize farms
    b. designed Special Economic Zones in port cities to confine capitalism

c. distributed 35 million copies of *The Communist Manifesto* by Marx and Engels

d. established five-year plans and collectivized industries

14. China's global sourcing companies are capturing more of the core processes in the commodity chains of manufactured goods by:

a. designing prototypes and deciding where and how to manufacture and distribute goods

b. providing cheap labor to manufacture massive numbers of goods

c. extracting the minerals and raw material resources used in global manufacturing

d. lobbying the World Trade Organization to allow more goods to be produced in China

15. The hukou system in China is a(n):

a. universally taught Chinese lettering system adopted by the Koreans

b. registration system the government uses to limit where Chinese live

c. method of keeping track of births and deaths in the Confucian church

d. innovation created by Deng Xiaoping in his 1979 reforms

ANSWERS FOR SELF-TEST QUESTIONS

1. b, **2.** c, **3.** d, **4.** a, **5.** c, **6.** a, **7.** a, **8.** d, **9.** b, **10.** c, **11.** a, **12.** c, **13.** d, **14.** a. **15.** b

9

EUROPE

If you have had the opportunity to travel to Europe you may have noticed a scene like this flying into Paris's Charles de Gaulle Airport, Amsterdam's Schiphol Airport, or London's Heathrow Airport. Hanging from the ceiling is a sign listing who can go through this line for customs. If you are not from Europe, you may look for your line and realize it is quite a bit longer. If we could watch this scene on a security camera, you would notice the European Union line is also moving more quickly.

Study the picture a little more, and you'll notice that the sign for the line sports two blue squares with 12 gold stars in a circle, the symbol for the European Union (EU). Nearly everywhere you travel in Europe today, you will see symbols of the EU. The European Union flag can be found on signs like this one at the airport, hanging from hotels next to individual country flags, and even on umbrellas. Other symbols of the EU are evident in the money you exchange at the airport. In both France and the Netherlands, you will receive euro bills and coins, with the flag of the EU on one side of each coin.

The customs line in this photo is designated for people who hold passports from one of 28 European countries. The European Union does not issue passports, but 26 countries (including 23 EU countries plus Norway, Liechtenstein, and Switzerland) have agreed to the Schengen Accord, which allows free movement of people among those countries that sign the accord. Each citizen of Europe holds a passport from his or her own country, and holding a passport from a member country gives one the right to free movement within the border-free Schengen area, whether working, retired, touring, or studying.

This line symbolizes the great experiment in integration that the countries of Europe have been working toward since 1951. The European Union is a complex structure, and free movement of people is only one part. The European region has changed in many ways since World War II, as Europeans have developed the EU and migrants have arrived from former colonies around the world.

London, United Kingdom. Passengers queue in the customs line at an airport in the United Kingdom. Although the United Kingdom is not a signatory to the Schengen Accord, it is a member of the European Union. This customs line is for those passengers who are arriving and have passports from the UK or a Schengen country.

© david pearson/Alamy.

CHAPTER OUTLINE

THRESHOLD CONCEPT in this Chapter

Population Pyramid

WHAT IS EUROPE?

LEARNING OBJECTIVES

1. **Explain** how Europe came to be considered a continent.

2. **Apply** the geographic concept of region to Europe.

3. **Determine** whether traditional criteria used to define Europe still apply to the region.

Quickly—name the seven continents. Europe is somewhere on your list, right? Now, picture a map of the world. Physically, Europe is not divided from Asia. Most school children are taught that the world is divided into seven continents: North America, South America, Europe, Asia, Africa, Australia, and Antarctica. Yet, if we accept a common definition of **continent**, an extensive, contiguous, discrete landmass, Europe does not fill the bill. Europe has no clear eastern border separating it from Asia. The two are physically one continent, Eurasia. Europe is better defined as a peninsula of the Eurasian continent.[1] Why, then, are we taught that Europe is a continent?

EUROPE AS A CONTINENT

The idea of Europe as a continent stems from ancient Greek and Roman maps. The maps of the known world drawn by scholars from around the Mediterranean Sea about 2,000 years ago typically show the earth as divided into three continents: Europe, Africa, and Asia. Ptolemy, a Greek

---

[1] We must credit Terry G. Jordan-Bychkov, whose third edition (New York: Harper Collins, 1996) of *The European Cultural Area* convincingly establishes the description of Europe as a continent is human-created and not based on physical geography.

 USING GEOGRAPHIC TOOLS

Historic Maps

Over the last 2,000 years, European cartographers changed their world maps to reflect power fluctuations and influences in Europe. Ptolemy's map reflects the world as known to ancient Greece. A T in O map drawn by western European monks in the 1300s and 1400s demonstrates the world according to the Catholic Church in the Middle Ages. Al-Idrisi's map from the 1100s shows the eastern European and Arab view of the world during western Europe's Middle Ages.

Ptolemy's map, which dates to approximately 150 CE, balanced Europe and Asia with Africa and a hypothetical southern continent (Figure 9.1a). The T in O map, drawn by medieval monks and based on the Bible, breaks apart Asia, Europe, and Africa by rivers and the Mediterranean Sea (Figure 9.1b). The "T" of the rivers and sea symbolizes the cross and separates the three continents, which are surrounded by the "O" of the world ocean. Such T in O maps were popular

Bridgeman-Giraudon/Art Resource, NY.

FIGURE 9.1a **Ptolemy's map.** The map, recreated in the 1400s from Ptolemy's notes, shows the world known to ancient Greece around 150 CE along with places Ptolemy imagined existed, such as a great southern continent south of the Indian Ocean.

f.177v T-in-O map, with letterpress place names, from Isidore of Seville, 'Liber Etimologiarvm' (woodcut), German School (15th century). Newberry Library, Chicago, Illinois, USA/Bridgeman Art Library.

FIGURE 9.1b **T in O map.** Monks drew T in O maps, based on places and descriptions in the Bible, during the Middle Ages. Europe and Asia are completely separated on the map.

(Continued)

FIGURE 9.1c   Hereford mappa mundi.   The oldest surviving T–O map dates to 1300 CE and is found in Hereford Cathedral, United Kingdom. The size of the Don River is exaggerated to create a greater separation between Europe and Asia.

in medieval Europe. The T in O map from a 1472 manuscript of the *Etymologiae* reflects the power of the Catholic Church in the Middle Ages in Europe in that the monks who drew the maps rejected 2,000 years of geographical knowledge gathered by mapmakers in Greece, Rome, and Byzantium. Instead, the cartographers used the Bible and teachings of the Church as the primary source for the T in O maps. The Hereford Mappa Mundi is the oldest surviving T–O map, dating to 1300 CE and found in Hereford Cathedral, United Kingdom (Figure 9.1c).

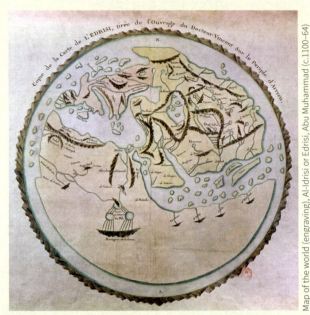

FIGURE 9.1d   Al-Idrisi Map.   Dating to the 1100s CE, Al-Idrisi's map reflects the knowledge from ancient Greece that the Arab world preserved during the Middle Ages. Compare this map to Ptolemy's, and note that Europe is still connected to Asia by a relatively small isthmus and the great southern continent is still hypothesized to exist.

Arab geographer Al-Idrisi's map, from the 1100s CE, draws from the Ptolemy map, continuing to show a great southern continent attached to Africa on the east (Figure 9.1d).

Thinking Geographically

1. Choose one of the maps in this box and imagine how your mental map of the world would have been influenced by this map if you lived at the time it was drawn.

2. Examine Europe in Figure 9.1d. What parts of Europe are accurate and what is missing or imagined on this map? What do these observations tell you about spatial interaction between Europe and the Middle East at this time?

geographer, astronomer, and mathematician, drew a map of the known world over 2,000, years ago. Ptolemy's map had relatively accurate details of the areas around the Mediterranean Sea, although he guessed the size and shape of the continents of Africa, Asia, and Europe (Figure 9.1a). In Ptolemy's map, the Indian Ocean is an enclosed sea, and a hypothetical great continent encircles the Southern Hemisphere.

After the fall of the Roman Empire, Christianity diffused into much of Europe. In western Europe, during the Middle Ages, from approximately 500 to 1500 CE, the Catholic Church rivaled and oftentimes exceeded the power of monarchs. The Church took control of science and math, and monks took on the role of cartographers. The maps of western Europe's Middle Ages are called T in O maps from the Latin *orbis terrae*, referring to the circle of ocean that enclosed the known earth (Figure 9.1b).

The T in O map used religion to perpetuate the division of Europe and Asia into two continents. Some T in O maps even sought to explain the varying pigmentations of human skin by labeling each continent for one of the sons of Noah, who the Bible attests built an ark and survived the great flood. Noah's sons and their wives, also on the ark, then traveled in three directions to repopulate the earth. Many T in O maps label the continents with the names of Noah's sons Shem (Asia), Japheth (Europe), and Ham (Africa). At around the same time, in eastern Europe, mathematicians and geographers advanced the field of cartography. Al-Idrisi, an Arab and eastern European cartographer and mathematician, mapped the known world, advancing from Ptolemy's map, as evident in the presence of the great southern continent (Figure 9.1d).

In the 1300s and 1400s, rejecting the T in O map and working from maps of the ancient Greeks and Romans that had been preserved and enhanced by Arab cartographers,

European explorers left the Iberian peninsula and rounded the tip of Africa to the east. Later, European explorers headed west across the Atlantic Ocean to the Americas. Eventually, European exploration and colonization revealed that Europe and Asia were not separate continents. However, the idea of a separate and distinct European continent persisted.

EUROPE'S PHYSICAL GEOGRAPHY

The peninsula of Europe is the western extension of the Eurasian continent. Europe is not its own lithospheric plate; it shares the same plate with Asia, creating the Eurasian plate, or Eurasian continent (Figure 9.2a). The European Peninsula is surrounded by more than 100,000 miles (160,934 kilometers) of coastline, with the Mediterranean Sea on the south, the Atlantic Ocean on the west, and the North Sea and Arctic Sea on the north.

Mountains in Europe run parallel to major plate boundaries. The east–west extending mountains of southern Europe, including the Betica Mountains in southern Spain, the Pyrenees between Spain and France, and the Alps stretching from France to Austria, run parallel to the boundary between the Eurasian plate and the African plate in the Mediterranean Sea (Figure 9.3). Even the north–south extending mountains of Europe run parallel to plate boundaries. The Kjolen Range in Norway runs parallel to the Mid-Atlantic ridge. The Apennines of Italy (Figure 9.2c) and the Dinaric

Reading the PHYSICAL Landscape

Plate Tectonics in Europe

The earth's crust is broken into approximately 15 major lithospheric plates, composed of either oceanic or continental crust, and several minor plates (Figure 9.2a). These rigid lithospheric plates move across the asthenosphere, which is the plastic layer of the upper mantle underneath the lithosphere. Where plates are converging, diverging, or transforming, the earth is marked by mountains, trenches, volcanoes, and earthquakes. Earthquakes and volcanoes in Europe occur mainly along plate boundaries in the Mediterranean Sea, southeast Europe, and Iceland (Figure 9.2b).

The L'Aquila, Italy, earthquake in 2009 did not occur along a plate boundary but instead, the earthquake occurred along a fault line or fracture within a lithospheric plate (Figure 9.2c). The L'Aquila earthquake was in the middle of the Richter scale, 6.3, but the destruction was dire, with over 200 deaths. Two fault lines run through Italy: one north–south and one east–west. Deaths and damage were high in part because of the high population density where the ground shook most. Many buildings in L'Aquila had been built since the 1960s and did not follow earthquake preparedness building codes. An Italian scientist predicted the earthquake, but his warnings were ignored. These two human factors contributed to the high damage and large death toll (Figure 9.2d).

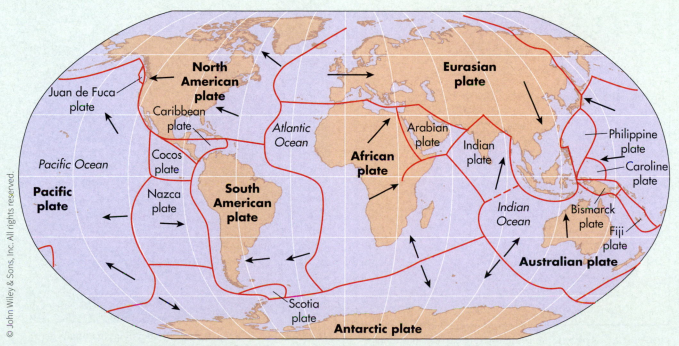

FIGURE 9.2a   **Tectonic plates.**   Earth's crust is broken into approximately 15 major lithospheric plates, composed of either oceanic or continental crust, and several minor plates.

(Continued)

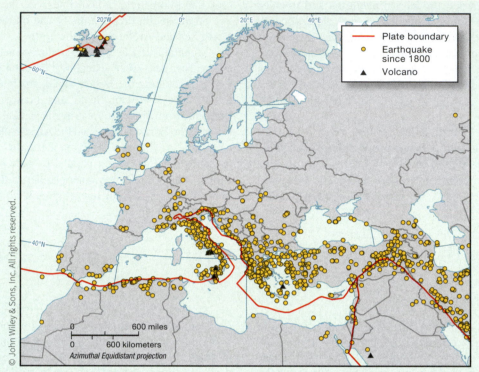

FIGURE 9.2b **Tectonic activity in the eastern Mediterranean.** Where tectonic plates converge, diverge, or transform, Earth is marked by mountains, trenches, volcanoes, and earthquakes. Earthquakes and volcanoes in Europe occur mainly along plate boundaries in the Mediterranean Sea, in southeast Europe, and in Iceland.

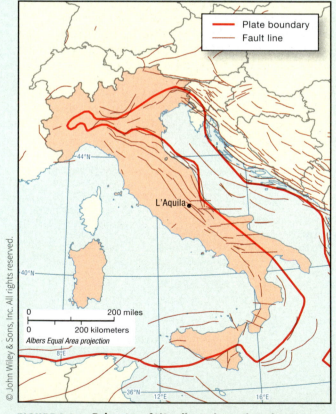

FIGURE 9.2c **Epicenter of L'Aquila, Italy earthquake, 2009.** The earthquake did not occur along a plate boundary. It occurred along a fault line or fracture within the lithospheric plate.

FIGURE 9.2d **L'Aquila, Italy.** The buildings in L'Aquila generally do not follow earthquake building codes, and more than 200 people died.

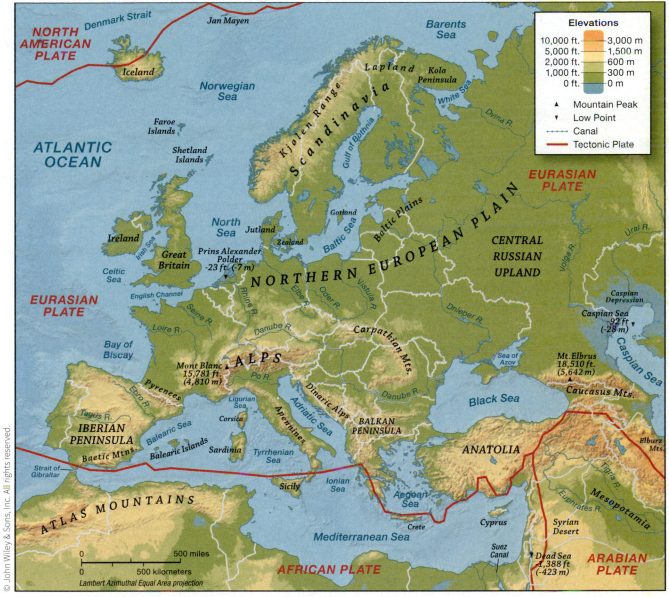

FIGURE 9.3 **Physical geography of Europe.** The mountains of Europe run parallel to plate boundaries in the Atlantic Ocean and the Mediterranean Sea. The lowlands in Europe have been either shaped by glaciers or carved by rivers over thousands of years.

Range of the Balkan Mountains are fold-thrust belts formed through tectonic activity—the Dinaric 50 to 100 million years ago, and the Apennines 20 million years ago.

The plains of Europe include lowlands shaped by continental **glaciers** that formed in Scandinavia and advanced as far south as Germany, Poland, and the Ukraine, ending about 15,000 years ago. In areas that glaciers did not reach, the flat plains have been carved over thousands of years by rivers, such as the Po River in Italy and the Danube River in the Hungarian Basin.

The climate of Europe is much warmer than we would expect for its latitude. For example, London is located at 51 degrees north latitude. In North America, Calgary, Canada, is located at 51 degrees north latitude. The climate of London, and Europe in general, is much warmer over the course of the year than the climate of Calgary and

other similar latitudes in Canada. Europe enjoys a warmer climate because of its location on the Atlantic Ocean and the North Atlantic Drift ocean current. A **high-pressure system** in the Atlantic pushes air and water along the equator, north along the coast of North America and then east to Europe, creating a warm influence on western Europe over the course of the year (Figure 9.4).

In Europe, the climates include Mediterranean in the south and marine west coast in the west. Coastal locations in Europe are warmer than inland locations not only from the North Atlantic Drift but also from the consistent warming effect of nearness to an ocean or sea. Water heats and cools much more slowly than land, so the coastline stays warmer in winter than a continent's interior. Eastern Europe's interior location, away from the Atlantic Ocean, creates a humid continental climate. Interspersed among

FIGURE 9.4 **Climates of Europe.** The climates of western Europe stretch from west to east, following latitudes. The climates of eastern Europe become increasingly dry in the east, away from the Mediterranean Sea and the Atlantic Ocean.

these climate zones are cold, high elevation climates, especially in the Alps. In a mountainous region, climate changes with elevation, and in the Alps, the highest elevations are in the tundra climate zone and includes glaciers.

EUROPE AS A REGION

Geographers use regions to categorize people and places in order to make sense of the world. When defining regions, geographers must determine what criteria and what set of characteristics to use. Regions are classified into three basic types: formal, functional, and perceptual (see Chapter 1).

The easiest way to define a formal region is to begin with cultural or physical characteristics and map it. Geographers who study Europe have traditionally used three cultural characteristics to define the formal region of Europe: language, religion, and race (Murphy, Jordan-Bychkov, and Jordan 2009).

Most Europeans speak a language that is part of the Indo-European language family. Historically, Europeans have largely been Christians, and traditional definitions of race distinguish Europeans as Caucasian (Figure 9.5). *Formal regions* are areas of peoples and places that are distinct from the rest of the world because of one or more shared cultural characteristics.

When these three characteristics are used, the formal region of Europe does not include Turkey because it is predominantly Muslim and Arab, and Hungary and Finland stand apart because their languages are not part of the Indo–European family. Choosing language, race, and religion as characteristics of Europe does not account for diversity within the region or for change in population or cultural traits over time. The racial and ethnic composition of Europe has changed since World War II as a result of migration and a rise in secularism.

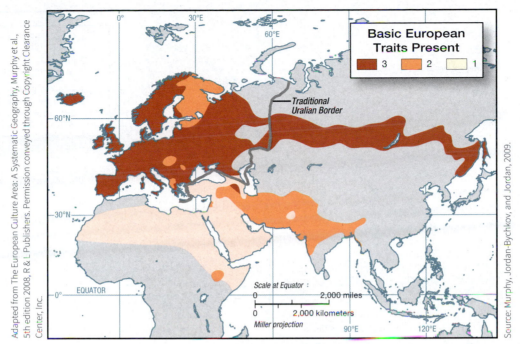

FIGURE 9.5 **The formal region of Europe.** Traditional cultural traits of Europe, including language, religion, and race, are mapped as three traits on this map. The darkest color on the map shows the places in Europe where all three traits have historically been present.

Europe can be classified, unlike most other world regions, as a functional region. The **European Union (EU)**, which began in 1951 as an economic community for the free trade of steel and coal, now functions economically, politically, socially, and culturally throughout its member countries in Europe (Figure 9.6). The EU is not a country, and it has not replaced the governments of the member states. However, the EU has grown in size and mission so much over the last six decades that it demonstrates an unprecedented level of state cooperation. The city of Paris is considered a functional region because it works politically to make laws that extend to the boundaries or limits of the city. Similarly, the European Union ties the region together to function across the 28 member states.

Europeans share a history of colonizing much of the world, fighting two world wars, taking sides during the Cold War, creating or joining the European Union, and leading efforts to advance human rights globally. In many of these instances, the region of Europe looked inward and saw an "us" and looked to the rest of the world and saw a "them." These experiences helped people within and outside of Europe perceive Europe as a distinct region. The fact that the European Union encompasses nearly the entire region helps entrench the perception that Europe is a cohesive region, one that will move forward, finding its way in the world together. As a result, Europe also meets the criterion of a *perceptual region* because people inside and outside Europe tend to see it as a region.

EUROPE DIVIDED AND UNITED

In the second half of the twentieth century, during the **Cold War**, Western Europe and Eastern Europe lived separately with an imaginary **Iron Curtain** dividing them (Figure 9.7). Today's Europe incorporates both west and east, although the Cold War is a not-too-distant memory. The Iron Curtain fell in 1989, but the division continues to have lasting impacts.

During the Cold War, the countries of Western Europe, Canada, and the United States formed the **North Atlantic Treaty Organization (NATO)** with the goal of defending themselves against the growing influence and power of the Soviet Union (Union of Soviet Socialist Republics, or USSR).

On the Soviet side of the Cold War, 15 different republics, including Russia, made up the USSR. Several of these republics were in Eastern Europe, including Latvia, Lithuania, Estonia, the Ukraine, Moldova, and Belarus. Each of these now independent states was a republic of the Soviet Union, which had a centrally planned economy. Leading members of the Communist Party in Moscow made decisions affecting all of the Soviet republics.

Other Eastern European states, including Poland, Hungary, Romania, East Germany, Bulgaria, and Albania, functioned as **satellite countries** of the Soviet Union, not physically a portion of the Soviet Union but strongly influenced by it. These satellite countries were all members of the **Warsaw Pact**, an agreement to unify the militaries of these countries under Soviet command.

Both Czechoslovakia (1918 to 1992) and Yugoslavia (1918 to 2003) were independent countries with their own

FIGURE 9.6 **The European Union.** Croatia was the most recent to join the European Union in 2013. Switzerland and Norway have chosen not to join the European Union, but both have trade relations with the EU. Not all European Union countries use the Euro, but the majority do.

communist governments—tyrannical in Yugoslavia's case. They were not part of the USSR, nor were they satellite countries of the USSR.

The countries of eastern Europe were not global colonizers, but they have histories tied to western Europe, and the people of eastern Europe have historically been considered "Europeans" racially and culturally. For example, many people in eastern Europe, including Austrians and Czechs, have cultural ties to Germany.

Since the late 1980s, the gaze of eastern Europeans has largely turned west. In his study of eastern European

maps in the post–Cold War era, geographer Donald Ziegler found that countries in eastern Europe re-centered their maps, changed names of areas or left names off of areas, and incorporated much more of western Europe in their maps of Europe. For example, the newly independent Czech Republic created a map that made the Czech Republic the center of Europe, encouraging the perception that the Czech Republic can bridge the divide between western and eastern Europe (Figure 9.8).

Since the Cold War ended in Europe in 1991, eastern European countries have worked to become member

FIGURE 9.7 The Cold War. During the Cold War, the states that were members of NATO were considered Western Europe, and the states that were members of the Warsaw Pact were considered Eastern Europe. The imaginary Iron Curtain divided them.

Donald J. Ziegler, "The Cartography of Independence: Maps of 'Eastern Europe.'" *Political Geography*, 21 (June 2002): 671–686.

FIGURE 9.8 The Czech Republic looks west. After the Cold War, countries in eastern Europe repositioned themselves politically. The new westward focus of the Czech Republic is reflected in this map produced and distributed by the Czech government, which positions the Czech Republic at the center of western Europe.

states of the European Union, and many have succeeded. Europe, divided during the Cold War, is united in the European Union.

CONCEPT CHECK

1. **Why** have western Europeans played a much larger role globally in defining Europe in the last 500 years than eastern Europeans?

2. **What** is your perception of Europe as a region?

3. **What** cultural characteristics can you use today to define Europe as a formal region?

WHO ARE EUROPEANS?

LEARNING OBJECTIVES

1. **Identify** the major religions and languages of Europe on a map.

2. **Understand** the growth of secularism and Islam in Europe.

3. **Question** how Europe is changing as a result of lowering fertility rates and increasing migration.

A woman walking down the street with a baguette peeking out of her shoulder bag . . . a man dressed in a suit and tie with a cell phone on his ear . . . women with kerchiefs covering their heads, selling onions and potatoes (Figure 9.9)—which of these people is a European? We have images, perceptions of who Europeans are, and many of our perceptions come from books, television, movies, or the Internet. Are all Europeans Christian? Do they all speak Romance or Germanic languages? Who are the newest Europeans, the recent immigrants? Are they, too, Europeans?

RELIGIONS OF EUROPE

Europe is largely thought of as a Christian region; however, Europe's history also includes nearly 2000 years of Judaism and over 1,200 years of Islam. Today, religion in Europe has increased in diversity to reflect the multitude of new migrants as well as the growth of secularism.

Although Christianity did not begin in Europe, the religion quickly diffused around the Mediterranean Sea, taking root in southern Europe. Christianity diffused from southwest Asia into eastern Europe and then into the west around the Mediterranean Sea. Paul of Tarsus converted to Christianity a few years after the death of Jesus, in approximately 36 CE. Paul traveled hundreds of miles around the Mediterranean Sea, diffusing Christianity to major port cities and writing letters that became books of the Bible.

In the early Christian church, the leadership was loosely organized among several patriarchs. One of the patriarchs was located in the west, in Rome, and others were located in the east. At the end of the Roman Empire (circa 350 CE), eastern and western Europe divided: The east became the Byzantine Empire, built upon the law, math, science, and architecture of the ancient Greco-Roman culture; the west centered its power on the patriarch in Rome.

From the 5th century CE to between the 13th and 15th centuries CE (the Middle Ages), the Catholic Church in Rome and monarchies in the west shared power over western Europe. The power of the Catholic Church during the Middle Ages in western Europe is reflected in the cultural landscape of cities built during that period (Figure 9.10).

Christianity evolved separately and distinctly in western Europe and the Byzantine Empire. In 1054 the Roman Catholic Church in western Europe split from the Eastern Orthodox Church in eastern Europe when the Catholic pope and the patriarch of Byzantine excommunicated (formally removed) each other from their respective churches.

In the sixteenth century, Christianity divided again. Reformers such as Jan Hus, Martin Luther, and John Calvin, who believed that Catholic leadership was corrupting the church, created distinct Christian faiths (called Protestant religions) based on their readings of scriptures, philosophy, and theology. Regions and countries in Europe divided between Catholic and Protestant. Southern Europe, the Mediterranean region, largely remained Catholic, and northern Europe predominantly adopted Protestant sects. Followers of each religious sect found themselves in countries and principalities where their religion was not the majority or where it was persecuted, which helped spur migration flows from Europe to the Americas.

Also starting around 1500, Europeans colonized first the Americas and later Africa and Asia. Religion became one

FIGURE 9.9 London, United Kingdom. Businessmen walk past the Bank of England in the financial district of London.

Chris Ratcliffe/Bloomberg via Getty Images.

© Perry Mastrovito/All Canada Photos/Corbis.

FIGURE 9.10 Krakow, Poland. With a central cathedral, a town hall, a marketplace, and a wall, Krakow, Poland, is a typical medieval city. Krakow's central square (Rynek Glowny) is home to the cathedral basilica of the Virgin Mary (built in the fourteenth century), the Town Hall Tower (built in the thirteenth century along with a town hall that was taken down in the nineteenth century), and a marketplace (Cloth Hall, built in the 1300s). The city of Krakow has remnants of the walls and gates that once enveloped it, though many of the walls were torn down in the nineteenth century.

way for Europeans to justify taking over territory and controlling people through colonization. Both Protestant and Catholic colonizers sought converts in their colonies, often through missionaries. Through colonialism, western European countries, especially Spain, Portugal, and France, helped diffuse Catholicism around the world. Today, more than 1.3 billion people adhere to the Catholic religion globally.

Each patriarch of the Eastern Orthodox Church developed a self-governing church, using a local language, including the Bulgarian Orthodox Church, the Georgian Orthodox Church, and the Serbian Orthodox Church. Whereas the Catholic Church universally used Latin at this point, the Eastern Orthodox churches used local languages for services and communication. Aside from Russia, eastern European countries did not engage in global colonialism between 1500 and 1900 (as western European powers did) because the Ottoman Empire, which began in 1301, expanded into eastern Europe during this time frame, taking control of part of the region. Nonetheless, some followers of Eastern Orthodox churches have migrated over time into North America, as other Europeans did (Figure 9.11). Today, the Eastern Orthodox Church claims about 250 million adherents worldwide.

Secularism in Europe

In today's Europe, churches are relatively empty during Sunday services. Whether Catholic or Protestant, church attendance is declining, and most who attend are over the age of 65. **Secularism** has risen markedly in western Europe since the end of World War II.

One sign of secularism in Europe is the number of people who report they "never" or "practically never" attend church (Table 9.1). In North America, an average 23.9 percent of people report never or practically never attending church. In comparison, 60.10 percent of French and 55.9 per cent of Dutch never or practically never attend church. In addition, 39.9 percent of the people surveyed in 11 western European countries and 21 percent of the people surveyed in eight countries in eastern Europe reported nonattendance.

In addition to low attendance at services, many European churches are struggling to find pastors or priests to minister to the church members and are also finding it difficult to maintain their properties with declining donations. In 2007, the Catholic Church only ordained 9 new priests in Ireland, and in the same year, 160 priests died. The *Irish Catholic* newspaper reported that by 2028 the number of priests in Ireland will drop to about 1,500 from 4,752 in 2008. The country saw a decline in church attendance from 85 percent in 1974 to 60 percent in 2004. Despite the growing shortage of priests, church attendance in Ireland rebounded between 2006 and 2011, partly as a result of the migration of Catholics from Poland to Ireland.

Academics question why secularism is on the rise in Europe. One theory is that today's affluent western Europeans do not struggle for survival as previous generations did or feel compelled to be "saved" by a church or cared for by an organized religion. Another theory holds that while European values have liberalized, churches have not kept up with people's new moral values. Still others cite sexual abuse by clergy as a more recent factor in the decline in church attendance.

Although some churches in western Europe have had to sell off real estate in order to keep going, in eastern Europe new churches are being built to keep up with demand. The Soviet Union had an official policy of atheism, which also applied in many satellite countries of Eastern Europe. Immediately after the fall of the Soviet Union in 1991, polls demonstrated a resurgence of attendance in the Eastern Orthodox churches. The resurgence in eastern Europe could be attributed to the freedom to publicly practice religion or to people turning to the church for solace during the uncertain transition of their governments and economies.

Recent reports show Christian missionaries (especially from the United States) actively seeking converts in eastern Europe, which is shaping a new religious mosaic in the region that incorporates a larger number of evangelical Protestants.

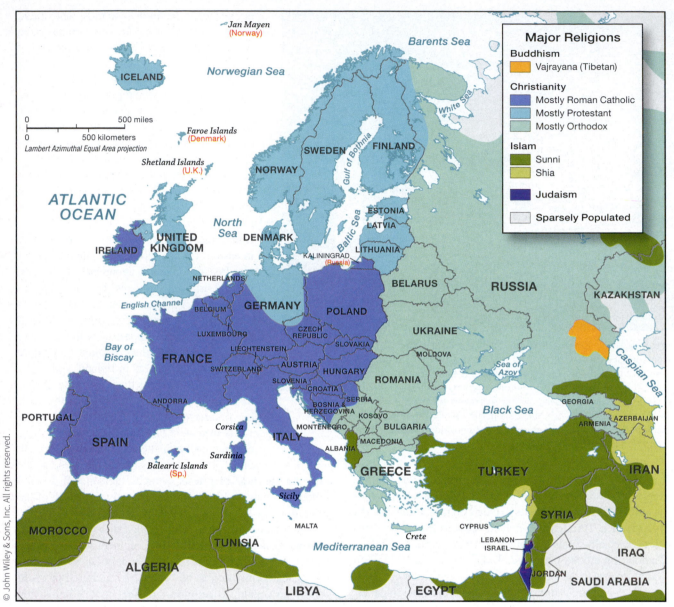

FIGURE 9.11   Religions of Europe.   Catholicism is found around the Mediterranean Sea, Protestantism is more prominent in northern Europe, and Orthodox religions are found in eastern Europe.

Islam in Europe

Among the many empty cathedrals and churches in the city streets of western Europe stand mosques built to serve the Muslim population. Islam first diffused into Europe when Muslim Moors took control of southern Spain in the 700s. This change in political control is reflected in the architecture and place names, or toponyms, of southern Spanish cities. The Moors built the Grand Mosque of Cordoba between the late 700s and the 900s. The Catholic Spanish crown pushed the Muslim Moors out of southern Spain in 1236. In the 1200s, the Catholic Spanish monarchy took control of the Grand Mosque and converted it into the Cathedral of Our Lady of Assumption in Cordoba, Spain (Figure 9.12).

A second wave of Muslims brought Islam to eastern Europe from the 1300s on with the rise of the Ottoman Empire in present-day Turkey. The Ottoman Empire expanded into Europe, extending to present-day Serbia and Bosnia and Herzegovina. Muslims continue to be in the majority in Albania, Kosovo, and Turkey, and Islam is one of three major religions in Bosnia and Herzegovina.

TABLE 9.1 Church Attendance in Europe (with North America for Comparison)

COUNTRY	PERCENT OF PEOPLE WHO "NEVER" OR "PRACTICALLY NEVER" ATTEND CHURCH, 2007
Andorra	61.90%
France	60.10%
Netherlands	55.90%
Sweden	50.60%
Spain	47.30%
Britain	46.60%
Germany	42.00%
Russia	34.30%
Turkey	32.90%
Switzerland	32.00%
Slovenia	30.50%
Finland	25.60%
Bulgaria	23.50%
Ukraine	22.30%
Italy	11.70%
Serbia	10.90%
Moldova	9.20%
Poland	5.20%
Romania	4.50%
United States	25.60%
Canada	34.70%
Mexico	11.40%

Note: The table shows countries in western Europe, eastern Europe, and North America (for comparison).

Source: World Values Survey, 2007.

FIGURE 9.12 **Cordoba, Spain.** Before the arrival of Muslim Moors, this site housed a small, early Christian church. Between 700 and 900, Muslim Moors controlled the region and built the Grand Mosque in Cordoba. In the 1200s, the Catholic Spanish monarchy took over the city, and eventually leaders converted the mosque into a cathedral, despite its hallmarks of Muslim architecture: the minaret tower, the arch-lined walls, and the intricate geometric designs. The site's multiple purposes over time are a good example of sequent occupance.

coming from former French colonies in West Africa. Between migration and conversion, Islam has gained a presence in southern Spain, eastern Europe, and major cities in western Europe.

Politics and Religion

The role of religion in government varies across Europe. In the north, where Protestantism diffused during the

Mosques from this period dot the landscape in southeastern Europe and in some places reflect a contentious history in the region. In 2003, Yugoslavia split into several countries, including Serbia and Bosnia and Herzegovina after a civil war that pitted Serbian Orthodox Christians against Bosnian Muslims and Catholic Croats. During their period of control, the Ottomans built over 200 mosques in Belgrade, Serbia. Today, only one remains, the Bajrakli Mosque (Figure 9.13). Only 2.3 percent of the people in Serbia today are Muslim, whereas 40 percent of the people in neighboring Bosnia and Herzegovina are Muslim.

The third and most recent wave of Islam began to diffuse into Europe in the 1950s and 1960s, as Muslims from former colonies and Turkey migrated to Europe, especially to cities. In France, Muslims now compose over 10 percent of the country's population, with 70 percent

FIGURE 9.13 **Belgrade, Serbia.** Bajrakli Mosque is the only remaining mosque in Belgrade, out of the 273 built during the reign of the Ottoman Empire.

© Jean-Pierre Lescourret/Corbis.

FIGURE 9.14 Borgund, Norway. Traditional Norwegian churches are called Stavkirke, named for the staves or wooden planks used to structurally support the church. The Borgund Stavkirke was built in the twelfth century, of trees felled from surrounding forests.

Reformation, most countries have a state church. The Danish National Church (Denmark), the Anglican Church (United Kingdom), the Church of Norway, and the Church of Sweden are all state churches whose official head is the monarch of the state (Figure 9.14). Traditionally, citizens of Sweden were automatically members of the state's Lutheran church and paid a 2–3 percent income tax to support it. Since 2000, Swedes have chosen whether to become a member of the state church and can opt not to support it.

In eastern Europe, most countries have a single prominent religion that is not necessarily supported by the state. Poland and Romania stand out as countries with large Christian majorities, with Catholicism dominating Poland and Eastern Orthodox the majority in Romania. Lithuania is predominantly Catholic, and Latvia and Estonia have larger Protestant populations. Eastern European countries in the Soviet sphere were officially atheist during that era (see Chapter 10), but many countries experienced religious resurgences after communism fell in the 1990s.

The south is traditionally Catholic, including France, Spain, Portugal, and Italy. Vatican City, the home of the Catholic Pope, is located within the country of Italy. The government of Italy is independent of the pope, but the pope weighs in on moral questions in the country. Despite the lobbying efforts of Vatican officials against RU-486, a pill that aborts a pregnancy between conception and seven weeks (or even later), the Italian government legalized it in 2009.

By contrast, religion and religious leaders are not part of the political process in France. The roots of secularism, the official policy of separating church and state, are in the French Revolution, which brought down the monarchy and the power of the Catholic Church in the late 1700s. In 1905, the French government made secularism the official law of the land. The French government holds that secularism means the government must be officially and strictly neutral on questions of religion.

In the 1990s, the French government instituted a ban on religious symbols as part of its secularism policy. In 2004, France banned the outward demonstration of religious symbols in public schools. The ban includes yarmulkes (skull caps worn by Jewish boys and men), crucifixes or crosses deemed to be too large (worn by Christians), and hijabs (headscarves worn by Muslim girls and women). Enforcement of the ban has made international headlines. One public school in France asked three Muslim girls to remove their hijabs and expelled two students for wearing Christian icons to school. France is not alone, as Belgium and several states within Germany have also banned wearing religious symbols in public schools. Two teachers in Belgium lost their jobs for wearing hijabs to school.

What a veil means to a Muslim woman and how it is perceived by others vary throughout the region of Europe and into Southwest Asia (Figure 9.15). In modern Turkey, women see the icon of the veil in very different ways. Some see it as a religious mandate, others see it as an expression of female virtue, and others regard it as oppressive. Still others view the veil as a symbol of class. The modern norm in Turkey has been for the urban elite not to veil and the urban poor to veil.

The Turkish Republic is a secular state founded by Mustafa Ataturk after World War I. The Turkish government discouraged Islamic political parties and prohibited women from wearing Muslim headscarves or veils in government-funded spaces, including campuses, courts, and parliament. Ataturk's dress code has been unevenly applied in Turkey. Geographer Anna Secor found that wearing a veil (or not) varies by the kind of space a Muslim woman is in (Secor 2002). Women are expected to wear a headscarf, either the traditional hijab or the modern interpretation, in certain areas of the city, such as the settlements where recent rural to urban migrants live. In other areas of the city, such as movie theaters, women are not expected to wear veils. In her spatial analysis, Secor found that older women in Istanbul were generally not supportive of

FIGURE 9.15   Istanbul, Turkey. Women wear hijabs outside of Topkapi Palace, a former home of Ottoman sultans in the historic area of Istanbul.

veiling. In the focus groups, older women expressed their belief in the value of Ataturk's secular policies.

Through Secor's fieldwork and focus-group interviews in Turkey, she found that the symbol of the veil impacted women who wore it as well as those who did not. Secor interviewed a female medical student who wore a veil and explained, "I am in a surgery class. I got ninety points and could have passed the exam, but the professor told the secretary to decrease my score twenty points because of my scarf" (Secor 2002, 14). The student described the general perception that "the ones who are not covered are probably cleverer and will be more successful." The medical student explained, "When I was uncovered I was just a student, now I am a symbol of something."

The perception of the veil in Turkey and the prohibition of outward signs of religion in schools in France and Belgium demonstrate how various individuals and governments perceive icons or symbols of a religion in different places. In Europe, guidelines on religious icons or symbols are typically instituted as part of a nationalistic identity based on secularism.

LANGUAGES OF EUROPE

Among the traits encompassed in a culture, religion and language are two of the most visible. We can walk through an unfamiliar place, study the cultural landscape, and immediately get a sense of the people's religions by their houses of worship, art, and architecture, and we can also easily see the language in the cultural landscape by the signs posted on buildings and placards.

Religion and language are spatially connected in Europe. In eastern Europe, where the Eastern Orthodox Church is

historically prominent, the languages are largely Slavic. In the southern/Mediterranean part of western Europe, where Catholicism has historically dominated, most people speak Romance languages, derived from Latin. In the northern part of western Europe, where the Protestant Reformation took hold and where most of Europe's Protestants live today, the people largely speak Germanic languages (Figure 9.16).

The distribution of languages in Europe gives us a glimpse of Europe's past cultures and their spatial interaction. Romance, Slavic, and Germanic languages are all part of the Indo-European language family. The languages that are not part of the Indo-European language family punctuate the areas of Europe where people, over centuries, lived apart from those around them and developed cohesive, internally directed cultures. A few languages in Europe, including Hungarian, Finnish, and Euskera (Basque), are not part of the Indo-European language family.

The Basques, who speak Euskera, live in the area between modern Spain and France. The Euskera language is not related to French or Spanish, demonstrating a high degree of isolation. Studies of the spread of the bubonic plague, which devastated Europe between the 1300s and 1700s, point to the Basque population as one that avoided the plague in large part because the Basque people had little interaction with those around them.

The languages that are identified with country names in Europe, including Romanian (Romania), French (France), Italian (Italy), and Greek (Greece), reveal a history of language used as a source of nationalism for Europe's states. Johannes Gutenberg's invention of the printing press in Germany in 1450 supported the connection of a language with a **nation** and that nation with a **state**. The printing press enabled a government to formalize writing in a national language and create an official language for the people of the state.

The languages on the western margin, denoted as mixed on Figure 9.16, are primarily Celtic languages whose native speakers were, at some point in history, politically marginalized. Celtic languages include Breton in Brittany, France; Irish (Gaelic) in the west of Ireland; and Welsh in Wales on the island of Great Britain. Linguists have recently found connections between Basque and Celtic languages. Although linguists do not universally agree that Basque is a Celtic language, its marginalization is similar to that of Celtic languages.

The number of native speakers of Celtic language is small; with 3 percent in Breton, 42 percent in Ireland, and 21.7 percent in Wales. Because language is a large part of each group's identity, the speakers and sometimes the government create incentives for people to learn and continue to speak Celtic languages.

The public school system in France is officially French speaking. While public school in Brittany is in French, private schools in Brittany offer bilingual instruction in Breton and French or exclusively Breton. France revised the constitution to recognize regional languages, including Breton, in 2008.

© Michael K Berman-Wald/Alamy.

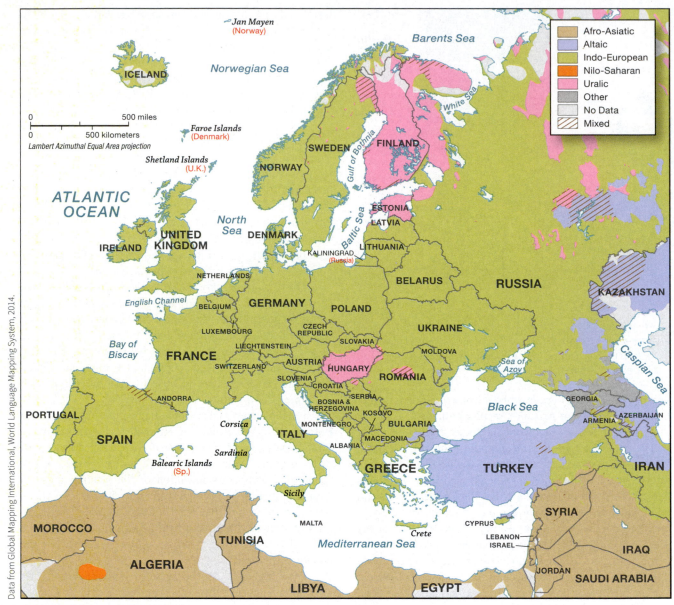

FIGURE 9.16   Languages of Europe.   The Indo-European language group includes Romance, Germanic, and Slavic languages. The languages in Europe are closely related, spatially, to the religions of Europe. Romance languages are in the Mediterranean region; Germanic languages are found in northern Europe; and most people in eastern Europe speak Slavic languages.

Wales promotes bilingualism in its public schools, teaching students both Welsh and English since 1990. Bilingual schools in Wales have helped Welsh children gain a sense of pride in being able to comprehend and speak Welsh as well as English and have reinvigorated the Welsh language.

Since its independence from Great Britain in 1922, Ireland's first official language has been Irish, and English has been its other official language. Also since 1922, the government of Ireland has had an Irish-language requirement for all government employees, in order to create an economic incentive for learning Irish. In 2005, Irish became one of the official languages of the EU.

The EU recognizes the right of every member to speak in an official language of their country. Translation services within the EU have grown with its membership. The EU now has 24 official languages and offers translation services in each of these languages. The official languages of the EU are expanded when a new country joins and requests a new official language or, if a state that is already a member, like Ireland in 2005, petitions the EU to recognize an official language. A democratic institution, the EU encourages all of its citizens to participate. Approximately 1 percent of the entire EU budget goes toward language and translation services to keep its citizens informed.

Today, the diffusion of the English language throughout Europe, both east and west, challenges the everyday use of national languages. In France, for example, the Académie française is charged with preserving the French language and

Data from Global Mapping International, World Language Mapping System, 2014.

translating everyday terms, such as "email," into preferred French terms, such as "communication electronique." By recognizing official languages and encouraging their use, the European Union helps stem the tide of English throughout Europe. European schools teach foreign languages much more than American schools. In Europe, 50 percent of people speak a second language fluently enough to hold a conversation, but in the United States, only 25 percent of people report enough fluency in a second language to hold a conversation.

AGING IN EUROPE

No single country in Europe, either west or east, has a fertility rate high enough to keep its population growing. The

Total Fertility Rate (TFR) is the average number of children born to women of child-bearing age (between 15 and 49). To stay at replacement levels, a country needs a TFR of 2.1 (Figure 9.17). In eastern Europe, total fertility rates dropped precipitously after the Soviet Union dissolved. Economic uncertainty and the rocky transition from centrally-planned economies to private ownership and enterprise convinced people to opt for fewer children. Low fertility rates coupled with the migration of eastern Europeans into western Europe in the 1990s caused the populations of eastern European countries to decline.

Poland's population fell by 500,000 from 2005 to 2010. In response, the Polish parliament passed legislation

Population Reference Bureau.

FIGURE 9.17 **Total fertility rates in Europe.** In 2014, not a single country in Europe had a TFR above replacement levels (2.1).

to provide cash payment for each child born. In western Europe, countries such as Sweden and France have offered generous maternity leave and subsidized child care in hopes of encouraging higher birth rates.

A **population pyramid** shows the age structure and the gender breakdown of a country's population at a glance (Figure 9.18). To read a population pyramid, start by looking at the horizontal axis and note that the left side of the pyramid shows the males in the total population and the right side shows the females. Then, look at the vertical axis and note that the age composition of the country is broken down into five-year groupings, starting from birth through age 4 and extending to a larger grouping of all people ages 80 and over at the top.

The shape of a population pyramid conveys the age structure of a country's population. Population pyramids for developing countries are typically wider at the bottom and narrower at the top (Figure 9.18). The wide base shows that a large proportion of the population is young, and the narrow top shows that a small proportion of the population is old. The pyramid shape reflects a number of possible factors that generate higher birth rates, including a lack of education for women, lack of access to birth control, a cultural desire to have large families, higher death rates that correspond with lower life expectancies, higher infant and child mortality rates, and a lack of access to medical care. Many of these factors are common in poorer countries.

Wealthier countries have access to birth control, medical care, and, as we see in Europe, a cultural desire for women to have fewer children. Smaller families and longer life expectancies are apparent in the population pyramids of wealthier countries (Figure 9.18).

Population pyramids also reveal differences in the gender structure of countries. If the majority culture in a country has a gender preference for boys, we can see that in a population pyramid by comparing the left and right sides of the pyramid. In India, for example, selective abortions of female fetuses has led to a sex imbalance, with men outnumbering women. Studies of this practice in India have found that selective abortions of females is most common in second and third pregnancies.

A gender imbalance in a population pyramid can be caused by factors other than gender preference. Mortality rates from war typically impact men, who are fighting the war in larger numbers than women. European population pyramids reflect reductions in the male population during the two world wars. Over 5 million military deaths in Germany alone (in addition to civilians and Holocaust victims) created a great gender imbalance during World War II (Figure 9.19). Europe, as a whole, suffered over 12 million military casualties, over 14 million non-Jewish

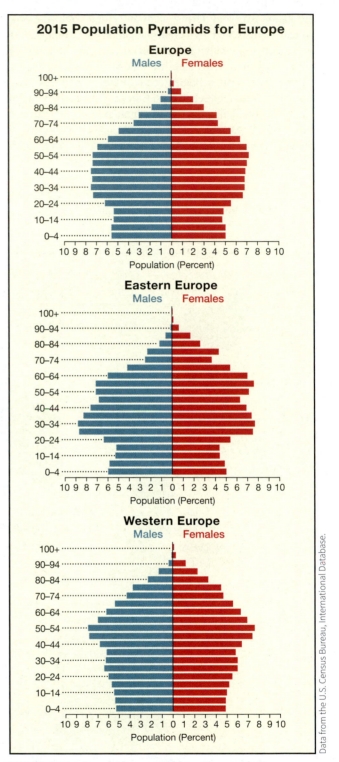

FIGURE 9.18 **Population pyramids.** The population pyramids for eastern Europe, western Europe, and all of Europe show the decline in Total Fertility Rates in eastern Europe after 1991 (look at the population under age 24) and longer life expectancies in western Europe.

civilian casualties (25 million including the USSR), and approximately 6 million Jewish civilian casualties.

Lower TFRs in Europe today cause concern for European governments because they create higher dependency

---

population pyramid a graphic representation of the age and sex composition of a population.

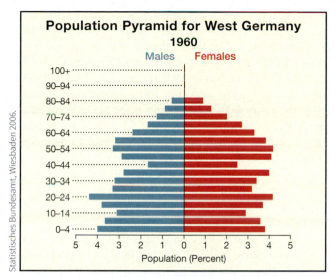

Population Pyramid for West Germany 1960

Males | Females

Statistisches Bundesamt, Wiesbaden 2006.

FIGURE 9.19 **Germany's population following World War II.** The population pyramid for Germany in 1960 shows the impact of World War II on the country. Men in the age range of 25 to 50 on this graph were most impacted by military deaths during World War II.

ratios, a measurement of the number of people under the age of 18 and over 65 that each working-age adult has to support with the taxes they pay. Without a large labor pool to work and pay taxes, governments in Europe have found it increasingly challenging to pay for all government services, including services to aid the young and old. In response to the decline in population since World War II, Europe has opened its doors to guest workers and to permanent immigrants in order to supplement the native-born labor supply.

MIGRATION IN EUROPE

In the last 60 years, Europe has become a region of immigration, but until 1950 Europe was largely a region of emigration. In the 1500s, European explorers and colonizers began to emigrate to the Americas and Africa. Europeans continued to leave the region for the next 450 years, migrating for opportunities abroad. Even immediately after World War II, emigration continued, especially of European Jews to what is today Israel. However, after 1950, Europe became a draw for immigrants from former colonies. Immigrants came largely from former European colonies in Africa and Asia. Each European country sets its own immigration laws, and each country handles immigrants, legal and illegal, in slightly different ways, although the European Union is working toward a common immigration policy.

Belgium and Germany had a severe shortage of working-age men after the horrors of World War II. Millions of young men who should have been the backbone of the postwar workforce perished. Both Germany and Belgium invited Turks, many of whom were young men eager to work, as **guest workers** into their countries (Figure 9.20). In both cases, most of the Turks stayed.

To chart the degree of integration of Turkish immigrants into the social fabric of Brussels, Belgium, geographers Kesteloot and Mistiaen (2005) studied restaurants in a Turkish neighborhood in Brussels. Through their fieldwork, they

---

GUEST *Field Note*

DR. CHRISTIAN KESTELOOT

Katholieke Universiteit
Leuven, Belgium

After World War II, both the Netherlands and Belgium recruited guest workers from Morocco and Turkey to help rebuild their economies. As the migrants arrived in Europe, they encountered different housing conditions and varying opportunities to become part of the community. Working with my colleague Cees Cortie, who lived in the Netherlands, we examined the integration of Moroccan and Turkish migrants into the cities of Amsterdam, the Netherlands, and Brussels, Belgium. We found that the housing situation in each country, as well as the place of origin of the migrant groups, factored into the level of assimilation of each group in the respective European societies. In Brussels, the government provided little public housing and guest workers found niches of subpar, low-cost housing. In Amsterdam, the city provided public housing, and guest workers tended to live together in public housing developments. The imprint of Turks and Moroccans in the cultural landscape of Amsterdam is more evident than in Brussels, as the concentrations of guest workers in certain areas of Amsterdam enabled migrants to establish ethnic neighborhoods.

Courtesy of Pascale Mistiaen and Y. Kazepov (ed.), 2005. "Visual paths through urban Europe," CD enclosed in the volume *Cities of Europe*, Oxford: Blackwell.

FIGURE 9.20 **Brussels, Belgium.** When a migrant group is large enough and concentrated in a single area, ethnic infrastructure appears, like this barber shop in Molenbeek, the heart of the Moroccan concentration zone in Brussels. Gradually, many ethnic enterprises, especially restaurants, open up for customers from outside their community. This barber clearly has similar ambitions.

found that Turkish restaurants in the Belgian capital began in the 1970s as small stands selling kabobs and Turkish food, often without a formal menu, to young, single Turkish men working in the city. Not long afterward, many young Turks married, often bringing their wives from Turkey to live in Brussels with them. Several restaurants began to cater to families, with tables and larger spaces.

At this stage, a few restaurants appealed to non-Turkish Belgians looking for new ethnic foods. With non-Turks frequenting the restaurants, many began to provide pictures of the dishes on the menus for their new clientele while continuing to cater to Turkish palates. In the final stage, in the 1990s, a few Turkish restaurants went upscale, catering primarily to non-Turks looking for something "different" but with an overt appeal to Belgian taste buds.

By tracking differences in restaurants in the study area, the researchers better understood the degree of integration of the Turkish immigrants in Belgian society and economy.

Immigrants in Europe typically have much higher fertility rates than native Europeans. The problems created by Europe's aging population may be answered by allowing higher levels of immigration. Migrants can lower the dependency ratio in the region by filling the gap in Europe's working-age population and paying their share of taxes.

Two migration flows affect Europe today: the first from former colonies, especially from Africa and South Asia into Europe after World War II, and the second from eastern European countries to western European countries after the Cold War. Many Africans and Middle Easterners living in Europe are second or third generation, meaning either their parents or grandparents migrated to Europe and they were, in fact, born in Europe. The flow of migrants from North Africa and the Middle East into Europe continues today, with 330,000 refugees fleeing strife in North Africa and Syria and finding asylum in Europe in 2012. Both refugees and illegal immigrants from North Africa follow similar paths into Europe, primarily through Spain and Italy (Figure 9.21).

In cities such as Paris, migrants typically live in low-income and public housing, which is in high demand and in relatively poor condition. The low-income and public housing in France is concentrated in the outskirts of the cities in neighborhoods called *les banlieues* (the outskirts). Second- and third-generation migrants, called *les jeunes* (the young) hope for a better life but find it elusive, as more than 20 percent of *les jeunes* are unemployed in France. A history of police brutality toward *les jeunes* coupled with the high unemployment rate sparked riots in *les banlieues* in Paris in

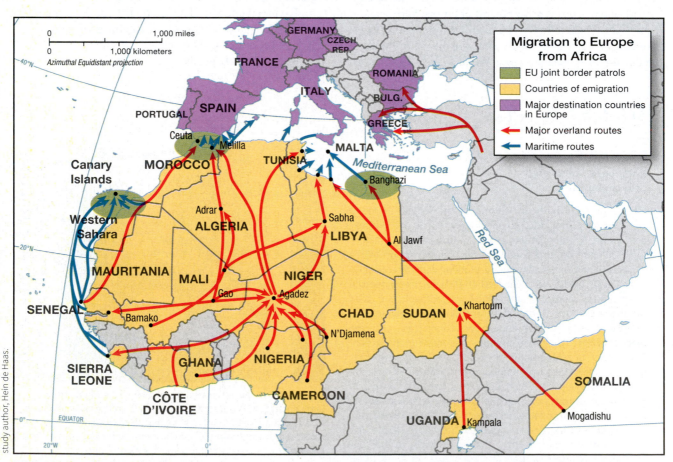

From The myth of invasion: Irregular migration from West Africa to the Maghreb and the European Union, published by the International Migration Institute, University of Oxford, October 2007. Used by permission of study author, Hein de Haas.

FIGURE 9.21 Migration to Europe from Africa. Africans who migrate to Europe follow a variety of routes from their homes in Africa to North Africa. From North Africa, two main paths are followed into Europe: in the west Africans migrate to Spain, and in the east Africans migrate to Malta. Once in Spain or Malta, they can travel to other regions in Europe with relative ease.

2005 and 2007. Unrest in the neighborhoods continues as poverty persists.

In 2005, the riots in *les banlieues* were in part a response to a French law that *les jeunes* believed would make it more difficult for them to get or maintain a job. In 2010, French workers throughout Paris, beyond *les banlieues*, demonstrated against a policy to change the retirement age in the country. The French government is working to find its way through a global recession while still providing the high level of government services the French political culture expects.

The flow of migrants from eastern European to western European states was precipitated by two factors: the end of the Cold War and the expansion of the European Union. The Iron Curtain fell with the end of the Cold War, which made movement possible. The Schengen Agreement, which governs much of the European Union, gives people within the Schengen region the right to free movement. Four of the countries in this region—Iceland, Norway, Liechtenstein, and Switzerland—are not members of the EU. Two EU member states, the United Kingdom and Ireland, have chosen not to join the Schengen Agreement. In 2014, Bulgaria and Romania, which became EU members in 2007, joined the Schengen region where labor can move freely. Cyprus, also a EU state, will not become a full member of Schengen until the EU Council determines they are ready.

CONCEPT CHECK

1. **Where** are the major religions and languages of Europe?

2. **Where** is secularism on the rise in Europe and **why**? **Where** is Islam on the rise and **why**?

3. **Why** is the population of Europe aging, and **what** are the implications of this aging?

HOW DO EUROPEANS SHAPE THE PHYSICAL ENVIRONMENT?

LEARNING OBJECTIVES

1. **Explain** how Europeans work to control the physical environment.

2. **Describe** how environmental issues in eastern Europe are affecting the region.

Europeans developed early towns and trading centers on coastal and river ports. An estimated one-third of the population of the EU live within 31 miles (50 kilometers) of coasts. Over the last 2,000 years, Europeans have used science and engineering to keep the ocean, seas, and rivers at bay.

In the interior, Europeans deforested much of the land to develop agriculture, industry, and cities over the same time period. During the 1800s, industrialization diffused throughout western Europe. In the 1900s, continued industrialization, urbanization, and two world wars increased demand for energy in Europe. As a result, Europeans harnessed energy from coal, wind, water, and nuclear power. Between the 1940s and 1990s, eastern Europeans undertook rapid industrialization. The rest of the world did not understand its environmental consequences, including industrial waste and air pollution, until communism fell.

KEEPING WATER AT BAY

The Netherlands and Venice have fought hardest against the encroachment of the seas. Sixty-five percent of the Netherlands is below sea level, kept dry by building dikes and pumping water. The city of Venice, Italy, floods more than 100 times a year from the weight of 1,500 years of living and building on the island, coupled with the depletion of groundwater used in manufacturing.

The Dutch have contended with a number of environmental issues for centuries. In addition to being below sea level, the global sea level rose .067 inches (1.7 millimeters) per year during the twentieth century. Second, the sand dunes along the coasts of the Netherlands, which function as natural barriers to flooding, are migrating eastward because of prevailing western winds, the same winds that create the North Atlantic Drift. Finally, the Netherlands is sinking from massive-scale tectonic change.

During the most recent ice age, ending about 14,000 years ago, a massive continental glacier covered northern Europe. As the global climate warmed, the glaciers receded (Figure 9.22). Glaciers weigh down continental crust, and when the glaciers melt, land previously weighed down by glaciers rebounds, or uplifts. Like a seesaw, when one side

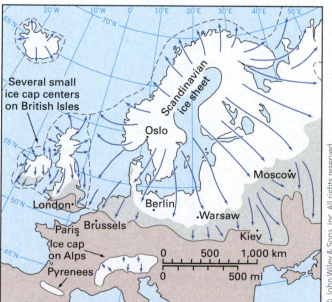

FIGURE 9.22 **Glaciers in Europe.** The most recent ice age, which ended about 14,000 years ago, brought continental glaciers into northern Europe and expanded alpine glaciers in the mountain regions of Europe, including the Alps and the Pyrenees.

of the fulcrum rebounds, the other sinks. Northern Europe is experiencing a rebound, but the Netherlands is sinking in this tectonic process, called **isostatic rebound**.

The Dutch began their fight with the sea as long ago as 300 CE. In response to fear of flooding in flat, agricultural lands, Dutch farmers constructed terps (small hills) and built their homes on top of them. By 900 CE, the Dutch were catching silt from rivers to build islands in the west to slow storms as they came ashore. Soon afterward, the Dutch built earthen dikes, and created the first **polders**, reclaiming land from the sea. By 1600, the Dutch were using windmills to pump water off polder land and begin draining interior lakes to create new polders (Figure 9.23).

Reading the **CULTURAL** Landscape

The Dutch Reclaim Land from the Sea

From the Baltic in the north to the Mediterranean in the south, seas surround the peninsula of Europe. Nowhere in Europe do people have a closer relationship to water than in the Netherlands. The Netherlands is geographically distinct because much of the country is part of the Rhine/Meuse River Delta. After the last ice age, massive melting glaciers deposited sediment frozen into the bases of the glaciers or pushed along by them. The outwash of glacial sediment formed the Rhine/Meuse Delta and the outwash of water carved the paths of the rivers themselves. The delta landscape lies only a few feet above sea level, which is why the name of the country literally means "low land."

Dutch efforts to protect land from the sea began as early as 300 CE, but efforts to reclaim land from the sea began centuries later. Since the ninth century, the Dutch have systematically protected their country from the sea, and, in many cases, reclaimed new land, by building a vast network of dikes and drainage ditches (Figures 9.23a and 9.23b). These landscapes include polders, large areas of drained or reclaimed land

FIGURE 9.23a   Zuider Zee, the Netherlands.   First, the Dutch built an 18-mile (30-kilometer) earthen dike across the Zuider Zee. Cut off from the North Sea in 1932, the waters of the Zuider Zee transitioned from salt water to fresh water over a short period of time.

© Planet Observer/UIG/age fotostock.

© Bettmann/CORBIS.

FIGURE 9.23b   Zierikzee, the Netherlands.   This photo from 1953 shows enormous pumps draining lowland tracts on an island south of Rotterdam.

FIGURE 9.23c   Beemster Polder, the Netherlands.   The land, literally taken from the depths of the Zuider Zee, is now productive agricultural land.

FIGURE 9.23d   Almere, the Netherlands.   Today, polder towns such as this one, located east of Amsterdam, are growing rapidly, in one of the most densely populated countries in the world.

bordered by protective dikes (Figures 9.23c and 9.23d). Windmills pump excess water from polders into nearby drainage ditches. The Dutch have reclaimed approximately 3,000 polders, more than in any other country in Europe.

Although draining water from polders is relatively effective, it causes formerly water-logged sediments to dry out, compress, and subside. As a result, much of the Netherlands is now below sea level. The Netherlands is not only a "low land" now; it is really a "below-sea-level" land.

The reclaimed lands are used for airports, housing, and farmland. Schiphol Airport is the third largest airport in Europe and is built on a polder. Some polders are used to hold urban growth around major cities such as Amsterdam. The Netherlands is one of the most densely populated countries in Europe, with 1,270 people per square mile. The land area of the Netherlands is about the size of Maryland, but Maryland's population density is 541 people per square mile.

The largest polders, the ones reclaimed from the Zuider Zee when the Dutch constructed a dike across the mouth of the sea in 1932, are used mainly for agricultural fields and planned agricultural towns. The fight against the encroaching sea is ongoing in the Netherlands. Dutch engineers are planning and building defensive systems along the coastline of the country, as predictions of rising sea levels from climate change grow.

THE SINKING CITY OF VENICE

Venice, founded in 421 CE, sits on a series of islands in the Venetian Lagoon. The lagoon waters meet the Adri- atic Sea at three major openings in the barrier islands that separate the lagoon from the Adriatic. Each day, with the tides, waters of the Adriatic wash into the Venetian lagoon and out again. The tides are the sewer system for Venice. Venice does not have a modern sewer system, and buildings without septic tanks dump sewage directly into the city's canals, which are cleared by the cleansing tides of the Adriatic.

More than 100 days a year, the tides bring high waters, or **acqua alta**, which flood the city. The increased flooding in the city has two sources: more frequent storms in the Adriatic Sea and the sinking of the city of Venice. Larger and more frequent storms in the Adriatic Sea bring a greater volume of water into the lagoon, raising the levels of Venice's canals (Figure 9.24). The weight of the city is compacting the layers of sediments beneath it, pushing the island downward—squeezing it like a sponge.

Since the foundation of the city, Venetians dealt with the easily compactable soils by driving wooden pilings

FIGURE 9.24 **Venice, Italy.** Tourists sit in one of the cafe terraces of the flooded St. Mark's Square, during acqua alta.

into the soft sediment under the city in order to create firm, level foundations for the city's buildings. But the pilings only compressed the sediments by increasing the weight of the city. The weight of the city is compacting the layers of sediments beneath it, pushing the island downward and squeezing it like a sponge.

In the 1500s, Venetians diverted major rivers away from the lagoon to keep it free of the silt they brought and clear for ships to enter the lagoon for trade. In the 1930s and the 1940s, industrial plants on another island in the Venetian lagoon drew down groundwater under Venice, further compressing the sediments on the island of Venice. Although Venetians stopped pumping groundwater, scientists now believe that the city sank 9 inches between 1900 and 2000.

Venetians have responded by either boarding up, building up, or moving out. The city that was once home to 170,000 people now has only 50,000 permanent residents. The number of tourists flocking to Venice is skyrocketing and, of the approximately 15 million who visit Venice each year, about 85 percent are day-trippers. More businesses are serving the needs of tourists, and fewer businesses are catering to the dwindling local population. To stem the tide of tourists, in 2002 Venice enacted a Venice card, which limits access to city landmarks.

In response to the historic city's precarious position, the Italian government decided in 2002, after several years of debate, to fund Project Moses, which is slated to be completed in 2014. Engineers designed a system of 78 enormous floodgates placed on the bottom of the sea at the three openings into the Venetian Lagoon, and the gates are to be lifted to block the flow of water into the lagoon whenever a storm is brewing on the Adriatic.

Environmentalists protest the building of the gates. If they are used frequently, the gates will alter the ecosystem of the Venetian lagoon, which is home to hundreds of species of birds and fish. Additionally, without regular flushing from the Adriatic, waste will accumulate in the lagoon.

ENVIRONMENT IN EASTERN EUROPE

On April 26, 1986, reactor number 4 of the Soviet Union's Chernobyl nuclear power station exploded. The accident occurred during a routine safety test. The radioactivity from the fallout killed fewer people initially than predicted; a surprisingly few 56 people died as a direct result of the accident. Thousands have since died from cancers, especially thyroid cancer, respiratory illnesses, birth defects, and other diseases caused by the radiation contamination. Estimates of the number of deaths range from 4,000 to 90,000, including people yet to suffer. The government of the Soviet Union built up its industrial sector and established its nuclear power program during the Cold War without regard for environmental costs.

After the accident, the Soviets built a sarcophagus around the reactor to encase it and keep the radioactivity inside (Figure 9.25). However, the sarcophagus was built hastily, and so the reactor has continued to leak radiation.

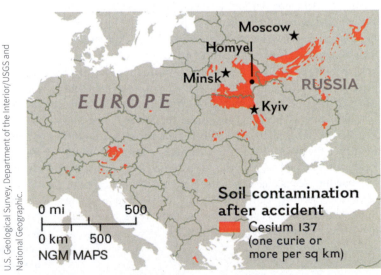

FIGURE 9.25 **Chernobyl disaster.** This map shows the area that was immediately affected by radiation from the Chernobyl disaster in 1986.

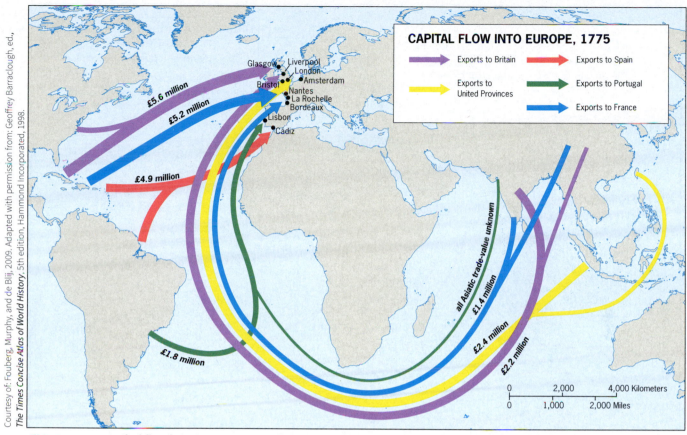

Courtesy of: Fouberg, Murphy, and de Blij, 2009. Adapted with permission from: Geoffrey Barraclough, ed., *The Times Concise Atlas of World History*, 5th edition, Hammond Incorporated, 1998.

CAPITAL FLOW INTO EUROPE, 1775

Exports to Britain
Exports to United Provinces
Exports to Spain
Exports to Portugal
Exports to France

£5.6 million £5.2 million £4.9 million £1.8 million £1.4 million £2.4 million £2.2 million all Asiatic trade-value unknown

Glasgow Liverpool London Amsterdam Bristol Nantes La Rochelle Bordeaux Lisbon Cádiz

0 2,000 4,000 Kilometers
0 1,000 2,000 Miles

FIGURE 9.26   Capital flow into Europe, 1775.   The arrows show the major flows of capital into Europe from its colonies in 1775. The capital helped fuel Europe's Industrial Revolution at the end of the 1700s and into the 1800s.

In 2011 a new steel structure, an arc-shaped coffin-like container, was being built to cover the entire reactor and contain the radiation by 2015.

The Chernobyl disaster was a wake-up call to Western Europe regarding the environmental problems hidden behind the Iron Curtain in Eastern Europe and the Soviet Union. With the expansion of the EU into eastern Europe since the end of the Cold War, the EU has helped with environmental cleanup and expanded the reach of its environmental policies into eastern Europe.

INDUSTRIALIZATION

Accidents such as Chernobyl are one of the consequences of living in an industrial society. The **Industrial Revolution** began in Europe, and the demand for energy, whether coal, oil, natural gas, or nuclear, is one of many outcomes of Europe's industrialization.

In the early 1700s, the British and the Dutch made major improvements in agriculture, including the invention of the seed drill, land consolidation, and livestock breeding. The diffusion of these technologies throughout the region increased the food supply in Europe. Crops from the Americas, including corn and potatoes, which were well suited to European soils, also improved western Europe's food supply. A more reliable food supply

enabled rural residents to migrate to urban areas to work in industry.

In addition to bringing crops to Europe to improve the food supply, the first wave of colonialism also brought a great deal of wealth (Figure 9.26). By the eighteenth century, the European countries that reaped the most financial benefit from colonization were primed for a massive change. The end of feudalism, worldwide trade, and the first centuries of European colonialism, which began around 1500, had all worked to ready Europe for the Industrial Revolution.

The Industrial Revolution diffused from United Kingdom into Europe in the 1800s (Figure 9.27). Most countries in eastern Europe, however, did not experience industrialization until the 1900s, under Soviet leadership or influence. During the Cold War, the Soviet Union encouraged rapid industrialization of Eastern Europe and in many cases, did so with little regard to the environmental impact (see chapter 10).

Similarly, in western Europe, industrialists in the 1700s, 1800s, and first half of the 1900s often paid little respect to the environmental impacts of mining and production. One twenty-first-century legacy is the parcels of land previously used for industry and difficult to develop because of industrial pollutants in the soil. Western Europe

FIGURE 9.27 Diffusion of the Industrial Revolution. The Industrial Revolution diffused eastward from the United Kingdom into Europe during the 1800s. Few pockets were labeled "major industrial areas" in eastern Europe by the 1880s. Industrialization in many Eastern European countries was led by the Soviet Union in the 1900s.

is dotted with these **brownfields** where the industrial economy has virtually collapsed and the economy is changing to a service basis (Figure 9.28).

The environmental movement, which encourages governments, companies, and individuals to pay heed to environmental impacts (and clean up brownfields and other remnants of industrialization), began in the 1960s in the United States and Western Europe. Pressure began to mount on countries, the newly forming European Union, and companies to cause less environmental damage. With the Cold War still in play, Western Europe

began cleaning up before Eastern Europe. Since 1992, the European Union has guided environmental policy and funded cleanups.

At the end of the Cold War, the desire of the newly independent states of eastern Europe to join the European Union became a major pull for environmental cleanup in the East. When eastern European countries joined the European Union, they agreed to the **acquis communitaire**, the body of all European Union regulations to date—including environmental regulations. However, the EU's principle of subsidiarity, which is the belief that decisions

WHAT ROLE DO EUROPEAN CITIES PLAY IN THE GLOBAL ECONOMY?

LEARNING OBJECTIVES

1. **Comprehend** how cities change to reflect societies, economies, and demographics.
2. **Understand** the transition of eastern European countries to the European Union.
3. **Explain** the role European cities play in the global economy.

European cities have undergone two major **urban revolutions**: first, thousands of years ago, when people began to live in cities, and second, just a few hundred years ago when the Industrial Revolution encouraged rural Europeans to move to urban areas for work. The Industrial Revolution is a good example of how social, economic, and demographic change is visible in the landscape of a city. With industrialization, the layout of the city changed to incorporate manufacturing zones, housing areas for factory workers, and stores and businesses to serve the workers.

EUROPEAN CITIES

Since 1900, cities in Europe have experienced several transitions, reflected in the urban landscape. The First and Second World Wars of the twentieth century destroyed entire cities in Germany and many other parts of Europe. Each war brought transitions in power, so that cities that were once part of empires were integrated into newly formed countries.

By 1950, the population of Europe became majority urban (Figure 9.29). Europeans continue to cluster in cities (Figure 9.30), and estimates are that by 2040 as much as 80 percent of Europeans will live in cities.

FIGURE 9.28 Bilston, England. This site in the West Midlands town of Bilston sits next to a canal that was used to move materials and products in and out of a metal processing works that once stood here. Abandoned in the 1980s, the site became a brownfield. The sign advertises a new housing development, Chestnut Walk, that has since been built on the site. A realtor advertises that the houses are next to a canal.

© Robert Brook Archive/Alamy.

and regulations should be made by the government body closest to the citizen, in this case local and state governments, makes it difficult for the organization to fully regulate environmental policy.

CONCEPT CHECK

1. **Compare and contrast** how and why the Netherlands and Venice battle the sea.
2. **How** has industrialization affected western and eastern Europe differently?

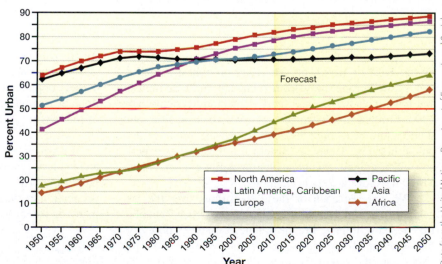

FIGURE 9.29 Percent urban by world region. North America is the most urbanized world region, followed by Latin America and Europe. Asia and Africa are less urbanized as of 2010. All world regions except the Pacific are slated to urbanize increasingly from now until 2050.

Data from the United Nations, Department of Economic and Social Affairs: Population Division, Population Estimates and Projections Section, at http://esa.un.org/unup/Analytical-Figures/Fig_1.htm.

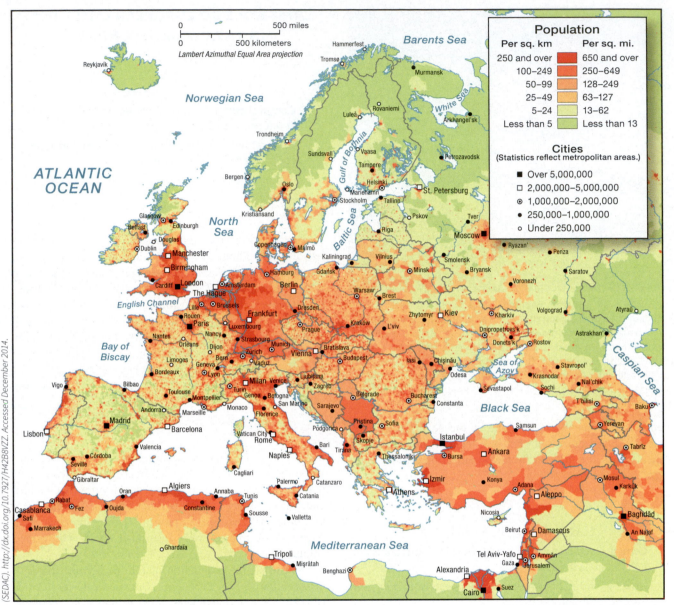

FIGURE 9.30 **Population density of Europe.** Europe is the third largest cluster of population globally, following East Asia and South Asia. Within Europe, population is clustered in cities, especially along coastlines and major rivers. Northern Scandinavia and the Alps, both of which are colder and more rugged than the rest of Europe, are less densely populated.

Data from CIESIN—Columbia University, United Nations Food and Agriculture Programme—FAO, and Centro Internacional de Agricultura Tropical—CIAT. 2005. Gridded Population of the World, Version 3 (GPWv3): Population Count Grid, Future Estimates. Palisades, NY: NASA Socioeconomic Data and Applications Center (SEDAC). http://dx.doi.org/10.7927/H42B8VZZ. Accessed December 2014.

With the onset of the Cold War, cities were influenced by the West or East. The year 1989 brought a transition in Eastern Europe away from socialism and communism and toward capitalism and more democratic systems. The 1990s brought rapid change, as former socialist cities and states addressed concerns such as private property and public housing. Today, as eastern European cities experience demographic change, including emigration and lower TFRs, the cities are again changing.

We can see each of these transitions over the last century in the landscape of Prague, in the Czech Republic. Like Krakow, Poland, Prague is a medieval city whose skyline is outlined with gothic spires and steeples. From the 1500s until the end of World War I, Prague was part of the

Austro-Hungarian Empire. The city's footprint expanded, and baroque architecture appeared but the city remained densely populated near the old town and Prague castle sections of the city. After World War I, Prague became part of the newly formed monarchy, the Kingdom of Czechoslovakia.[2]

Largely spared bombing during World War II, Prague's architecture remained relatively intact, but the demography of the city changed (Figure 9.31). At the time of Nazi occupation in 1939, more than 50,000 Jews lived in Prague, many of whom had arrived in the years prior fleeing Nazi

[2] Prague was part of Czechoslovakia from 1918 to 1993. In 1993, the Czech Republic and Slovakia separated.

FIGURE 9.31 Prague, Czech Republic. Built as the main marketplace in the Middle Ages, the Old Town Square in Prague survived World War II. The Tyn Church is the central structure in this photograph and is a classic example of Gothic architecture.

alle12/Getty Images.

control of Germany and Austria. The Holocaust emptied the old Jewish quarter in Prague and destroyed upward of 70 synagogues in the city and surrounding rural area. During the communist era in Czechoslovakia, another 105 synagogues were demolished. Since the fall of the Berlin Wall in 1989, some synagogues have been restored, and now 145 stand, with only a few holding regular services (Parik 1999).

The end of World War II brought a communist period to Czechoslovakia, and housing became a chief concern of the city's planners and the Czechoslovakian government. The goal was to eliminate homelessness by making housing near and in the central city accessible to all. The government built large, concrete-block tenements around the outskirts of the city of Prague. In the central city, historic dwellings were subdivided to fit multiple families. Under communism, housing in Prague was rent controlled and price controlled, which removed segregation among populations in the city because everyone had access to the same housing.

After the 1989 Velvet Revolution, when Czechoslovakia overthrew the communist government, Prague entered a new transition in urban form. The city quickly drew Westerners as tourists, overtaking Paris as the largest urban destination of tourists from western Europe by 1994. American, Canadian, and western European businesses that wanted access to markets in eastern Europe were drawn to Prague. The city changed rapidly. Land and buildings owned by the government under communism were opened to privatization.

At the same time, Prague and the rest of eastern Europe went through a transition in demography, in which women had fewer children and young adults migrated to western

Europe, creating a lower TFR in the East (in the Czech Republic, the TFR declined from 1.72 in 1992 to 1.16 in 1998). Private businesses bought up buildings, including housing, in the central City of Prague—locking out the renters from the Cold War era and turning buildings into high-end apartments and studios, especially for Westerners who came to work and live (even if temporarily) in Prague. In addition to creating high-end housing, the city of Prague was recast for tourists—pushing residences away from the central city. Between 1990 and 2000, residences in the outskirts of the city increased by 34 percent and residences in the city center declined by 21 percent. The loss of housing stock in the central city has led to higher rents and prices in the city center since 1990.

With declining total fertility rates since 1990, the family structure in Prague has changed, which is also reflected in the urban landscape. Prague, and the Czech Republic more broadly, now have many different household units, from single parents, to married couples choosing not to have children, to couples living together. New family structures create a higher demand for housing because fewer people in each dwelling unit means more dwelling units are needed. In Prague, the most rapid growth of housing continues in the outskirts. The demand for housing is so high that the old concrete block apartment buildings of the socialist era remain. They have simply been painted over in much happier colors than drab gray (Figure 9.32).

PRIVATIZATION OF CITIES IN EUROPE

In 2004, 10 eastern European countries (Cyprus, Czech Republic, Estonia, Hungary, Latvia, Lithuania, Malta, Poland, Slovakia, and Slovenia) entered the EU (Figure 9.6), Bulgaria and Romania entered in 2007, and Croatia in 2013. Aside from Malta, each of these countries was

© Profimedia.CZ a.s./Alamy.

FIGURE 9.32 Prague, Czech Republic. Since 1989, the drab, gray, unpainted concrete public housing in Prague has been revitalized with paint and architectural elements.

part of Eastern Europe during the Cold War and had to transition its economy in order to accede to the EU. In most of these countries during the Cold War, the government took land from private land owners in a process called *collectivization*.

After the Cold War, governments determined whether and how to *privatize* this same land, in some cases, returning it to individual land owners and corporations. East Germany was quick to do so and reunified with West Germany in 1990. The reunification treaty between the two countries required restitution of land to its pre-World War II (or pre-Cold War, in some cases) land owners, whether individuals or corporations.

According to a 2002 study of land restitution by geographers Mark Blacksell and Karl Born, governments in the Czech Republic, Slovakia, and Slovenia have also restored land ownership, but they have generally limited the calls for land restitution to the people who are current citizens of their states. In the Baltic States of Latvia, Lithuania, and Estonia, land restitution was mainly for people from these countries who were pushed into Russia or other parts of the Soviet Union during the Cold War and who wanted to return. Hungary opted to compensate people for their land loss instead of restoring ownership to them. Since Poland lost land to the Soviet Union in the east after World War II and gained land from Germany in the west, the Polish government has found land restitution nearly impossible for easterners who moved west. In Bulgaria and Romania, where Soviet-style collectivization of farmland was more popular, land restitution has lagged behind.

In addition to considering, and in many cases, implementing land restitution, the Eastern European governments also had to open their economies to investment from other EU countries and make their government actions, whether political or economic, transparent to all other members of the EU and to EU decision-making bodies.

EUROPEAN CITIES IN THE GLOBAL ECONOMY

In today's global economy, financial, capital, transportation, communication, and production networks crisscross the world, connecting some places to an unprecedented degree, while distancing other places that

lack access to these networks. European cities, along with the EU, are major players in the global economy. In fact, Europe shapes many of the financial and capital linkages and networks that pulsate through our world.

In the 1990s, a number of academic geographers and others formed the Globalization and World Cities Study Group to understand cities that make a major imprint on the global economy. The group defines a world city as "a key physical manifestation of contemporary globalization." A world city is a place that is integral to the world economy, a place where interactions with other world cities happen frequently and deeply. Through hundreds of studies of world cities and connectedness in the financial, advertising, accounting, and academic worlds, the Globalization and World Cities Research Network has classified cities throughout the world into different tiers. The two world cities most integrated into the world-economy are New York (United States) and London (United Kingdom). In addition to London, more than two dozen other European cities are world cities, meaning they are networked to other world cities in the flow of finance, advertising, accounting, and academia (Figure 9.33).

FIGURE 9.33 World cities in Europe. London is one of the two most important world cities, along with New York. Other European world cities are closely connected into the world cities network. Alpha cities are those most integrated into the world-economy through networks. Beta world cities link their country or region into the world-economy. Gamma world cities are important in regional networks but offer fewer producer services than alpha and beta world cities.

The world cities in Europe are major players in the world economy today; most of them became integral players in the world economy during the Industrial Revolution.

GLOBAL FINANCIAL SECTOR

When most Westerners think about banking in Europe, they think about Switzerland. Although banks in the United States have twice as many foreign deposits, Swiss banks have carved a niche for themselves as a major source of fuel for the world economy (Figure 9.34). Switzerland's banks are legendary for their bank secrecy laws, which were institutionalized by law in 1934. Under these laws, bank officials and the Swiss government could not divulge information about the bank's customers or their finances.

In 2014, Switzerland began complying with laws in the United States, disallowing bank secrecy for American clients so the US can track tax evaders who hide money in Switzerland. Switzerland now charges a tax on Swiss accounts owned by EU citizens. The EU pressured Switzerland to make the change because it believed EU citizens were using Swiss banks as a tax haven in the region.

Other countries in Europe and the Caribbean have copied Swiss bank secrecy laws. Luxembourg, one of the first six member countries in the EU, briefly followed the Swiss model, enacting secrecy laws but dropped them under pressure from the EU. Luxembourg is among the wealthiest countries in the world, and its financial institutions remain strong. Luxembourg is home to more than 7,600 investment funds worth approximately 900 billion euros, according to Luxembourg's VP Bank.

In addition to Switzerland and Luxembourg, banks in Germany are major financial players in the world economy. One of the largest global investment banks in the world is German-based Deutsche Bank. The company has a global presence that links together disparate places through global commerce.

SPACES OF GOVERNANCE

If you were driving around Washington, DC, and its suburbs, and you were studying the cultural landscape by paying attention to the signs on office buildings, you would notice that hundreds of organizations, representing an enormous range of concerns, have offices in the city. From the Association of School Bus Drivers to the Association of American Geographers, virtually every interest group in the United States has an office in Washington.

Likewise, if you travel in Europe, you can see in the cultural landscape the places where major government decisions take place (Figure 9.35). In Paris, the capital of France, French interest groups lobby the country's decision makers. The European Union has three capitals: Strasbourg (France), Luxembourg City (Luxembourg), and Brussels, (Belgium). Brussels is also the headquarters of the North Atlantic Treaty Organization (NATO).

FIGURE 9.34 **Zurich, Switzerland.** The Credit Suisse Bank stands on a corner of Bahnhofstrasse. Switzerland has more than 140 private banks and a history of bank secrecy laws that are now under scrutiny as countries increasingly accuse the banks of helping their citizens evade taxes.

FIGURE 9.35 **Political geography of Europe.** The borders of European countries are relatively recent. Germany and Italy did not exist as independent countries until the later part of the nineteenth century. After World War I, the borders of western European countries were redrawn, primarily by language. In eastern Europe, the countries that were formerly part of the Soviet Union and Yugoslavia achieved independence starting in 1991.

Dozens of intergovernmental organizations (IGOs) are also based in Europe. An IGO is set up by an organization of government entities that work together, such as the United Nations (the United Nations Environmental Program is based in Geneva, Switzerland) and the International Court of Justice (based in the Hague, the Netherlands).

Europe is also home to interest groups representing global concerns and with global outreach. Nongovernmental organizations (NGOs) are set up by individuals to advance a particular concern, many at the international or global scale. Some of the most powerful NGOs

in the world are based in Europe, including the International Committee of the Red Cross (Switzerland), Médecins sans Frontières (Doctors without Borders, France), and the World Economic Forum (Switzerland) (Figure 9.36).

The most powerful IGOs in the world give Europeans a strong say in the organization. For instance, the G-8 is an annual economic and political summit among eight of the most powerful countries in the world. Included in the lineup of the eight states are four European states (the United Kingdom, France, Germany, and Italy) in addition to a delegation from the EU. The

FIGURE 9.36 **Davos, Switzerland.** Christine Lagarde, Managing Director of the International Monetary Fund (IMF), listens to a speaker with Renault-Nissan CEO Carlos Ghosn and Facebook COO Sheryl Sandberg at the World Economic Forum in Davos in 2014. Some 40 world leaders gather in the Swiss ski resort Davos to discuss and debate a wide range of issues including the causes of conflicts plaguing the Middle East and how to reinvigorate the global economy.

ERIC PIERMONT/Getty Images.

remaining four states are the United States, Canada, Japan, and Russia.

The United Nations (UN) is headquartered in New York and has five permanent members on the Security Council, each with veto power. The Security Council is the most powerful body in the UN, and Europe holds two of the five permanent seats, or three of five if we include Russia as part of a larger Europe. The five permanent members are the United Kingdom, France, and Russia, in addition to the United States and China.

CONCEPT CHECK

1. **How** did Prague's central city change as it transitioned from a socialist city during the Cold War to the post–Cold War, modern city?

2. **What** did eastern European countries do with publicly owned land after the Cold War ended?

3. **Why** are some European cities world cities, and what roles do they play in the world economy?

SUMMARY

What Is Europe?

1. Europe is not a **continent**. Europe has no clear eastern border separating it from Asia. The two are physically one continent, Eurasia. In physical terms, Europe is better defined as a peninsula of the Eurasian continent.

2. The idea of Europe as a continent stems from ancient Greek and Roman maps. The maps of the known world drawn by scholars from around the Mediterranean Sea about 2,000 years ago typically show the earth as divided into three continents: Europe, Africa, and Asia.

3. In the second half of the twentieth century, during the **Cold War** Western Europe and Eastern Europe lived separately with a figurative **Iron Curtain** dividing them. Today's Europe incorporates both West and East, although the Cold War is a not-too-distant memory. The Iron Curtain fell, but the division continues to have lasting impacts.

Who Are Europeans?

1. Europe is largely thought of as a Christian region; however, Europe's history includes nearly 2,000 years of Judaism as well as over 1,200 years of Islam, especially in southern Spain. Today, religion in Europe has increased in diversity to reflect the multitude of its new migrants, as well as a growth of secularism among many Europeans.

2. Religion and language are spatially connected in Europe. In eastern Europe, where the Eastern Orthodox Church is historically prominent, the languages are largely Slavic. In the southern/Mediterranean part of western Europe, where Catholicism has historically dominated, most people speak Romance languages (derived from Latin). In the northern part of western Europe, where the Protestant Reformation took hold and where most of Europe's Protestants live today, the people largely speak Germanic languages.

3. No single country in Europe, either west or east, has a fertility rate high enough to keep the population growing. In eastern Europe, fertility rates dropped precipitously after the Soviet Union dissolved. Economic uncertainty and the rocky transition from state-planned economies to private ownership and enterprise convinced people to opt for fewer children. Low fertility rates, coupled with the migration of eastern Europeans into western Europe in the 1990s, caused the populations of eastern European countries to decline.

4. Historically a region of emigration, Europe has become a region of immigration. Two migration flows affect Europe today: the first from former colonies, especially in Africa and South Asia into Europe, and the second from eastern European countries to western European countries. Although the flow of migrants from Africa gains much more global attention, the flows of migrants from EU state to EU state and from other European states to EU states are even larger.

How Do Europeans Shape the Physical Environment?

1. Over the last 2,000 years, Europeans have used science and engineering to keep the ocean and seas surrounding the peninsula at bay and to extract and produce energy to fuel their economies. The most precarious environmental problems in Europe stem from human action—people have exacerbated flooding problems in the Netherlands and in Venice. Likewise, humans deforested much of Europe to fuel urbanization and industrialization and also built the Chernobyl nuclear power plant, which melted down in 1986, creating one of the worst environmental disasters in modern history.

2. The Industrial Revolution dates to the 1700s, beginning in Great Britain. The end of feudalism, the Age of Exploration, and the first centuries of European colonialism (which began around 1500) all worked to ready Europe for an Industrial Revolution. Prior to the Industrial Revolution, major improvements in agriculture, including the invention of the seed drill, land consolidation, and better practices in livestock breeding, increased the food supply in Europe. Western Europe's food supply was also aided by access to crops from the Americas (including corn and potatoes) introduced to Europe through exploration, which were well suited for Europe's soils. A more reliable food supply enabled rural residents to migrate to urban areas to work in industry.

What Role Do European Cities Play in the Global Economy?

1. Since 1900, cities in Europe have experienced several transitions, which are reflected in the urban landscape. The First and Second World Wars destroyed entire cities in Germany and other parts of Europe where fighting was heavy. Each war brought transitions in power as well, so that cities that were once part of empires were integrated into newly formed countries. With the onset of the Cold War, cities became either western or eastern. Eastern cities were either under the purview of the Soviet Union or they were under socialist governments of their own. The year 1989 brought a transition in Eastern Europe away from socialism and communism and toward capitalism. The 1990s brought rapid change, as former socialist cities and states had to address concerns such as private property ownership and public housing. Today, as

YOUR**TURN** ▶ Geography in the Field

The European Union flag flies at the same height as the Luxembourg flag throughout the small state of Luxembourg. In many areas of the city, the European Union flag flies more prominently than Luxembourg's. A 2006 poll by Eurobarometer (the European Union's statistical division) found that 76 percent of Luxembourg residents feel attached to Europe (63 percent for all residents of the EU), and 64 percent of Luxembourg residents feel attached to the European Union (50 percent for all residents of the EU).

Thinking Geographically

- Why do the residents of the tiny state of Luxembourg feel more attached to Europe and the European Union than do other people in Europe?

- What part does the European Union play in the identities of people in Luxembourg?

- If you were walking through the streets of Luxembourg City, what else could you see in the cultural landscape (in addition to EU flags) that would show you the importance of the EU in Luxembourg?

Read:
Van Gorb, Bouke, and Hans Renes. 2007. A European Cultural Identity? Heritage and Shared Histories in the European Union. *Tijdschrift voor Economische en Sociale Geografie*, 98, 3, 407–415.

DOMINIQUE FAGET/AFP/Getty Images/Newscom.

Luxembourg City, Luxembourg. European Union flags fly next to Luxembourg flags.

eastern European cities undergo demographic shifts, including migration out of eastern Europe as well as lower TFRs, the cities are again changing.

2. In today's global economy, financial, capital, transportation, communication, and production networks criss-cross the world,

connecting some places to an unprecedented degree, while distancing other places that lack access to these networks. European cities, along with the EU, are major players in the global economy. In fact, Europe shapes many of the financial and capital linkages and networks that pulsate through our world.

KEY TERMS

continent

glacier

high-pressure system

European Union

Cold War

Iron Curtain

North Atlantic Treaty Organization (NATO)

satellite countries

Warsaw Pact

secularism

nation

state

total fertility rate (TFR)

population pyramid

guest workers

isostatic rebound

polders

acqua alta

industrial revolution

brownfield

acquis communitaire

urban revolution

CREATIVE AND CRITICAL THINKING QUESTIONS

1. How does thinking of the **region** of Europe as a continent impact the **identity** of Europeans?

2. Go online to the Library of Congress Geography and Map Division website and look at its collection of maps from exploration and discovery: http://memory.loc.gov/ammem/gmdhtml/dsxphome.html. Choose a map to study and analyze its accuracy (are continents shaped correctly, are places where they really are?) as well as the toponyms (place names) on the map. Look at who the cartographer was and where he came from and determine what parts of the map are based on his **mental map**, and what parts are based on perceptions of places he has not been.

3. Go online to the NOVA website and examine the page for the video "The Sinking City of Venice": http://www.pbs.org/wgbh/nova/venice. Study the gate system under construction, and describe how these gates are part of the **anthropocene** and what their impact will be on the environment in the Venetian Lagoon.

4. The North Atlantic Treaty Organization (NATO) was created in response to the threat of the Soviet Union during the Cold War. The Cold War is over. The Soviet Union no longer exists. Thinking about the **context** of Europe, explain why you think NATO still exists and how you envision NATO's mission has changed in recent years?

5. Analyze the map of Europe that Czechoslovakia created after the Velvet Revolution (Figure 9.8). The Czech Republic (CZ)

and the Slovak Republic (SK) went through the Velvet Divorce in 1993, when each became independent states. Analyze the **scales** of the maps of the newly independent countries. How do you think each country's maps after 1993 reflected their newfound **identities**?

6. Analyze Table 9.1. Describe the hearth of secularism and determine how you think secularism has **diffused** through Europe.

7. Reread the definition of nation in the chapter. Using the concept of nation, explain why the European Union recognizes so many languages.

8. Visit the website http://www.census.gov/population/international/data/idb/informationGateway.php. Create a **population pyramid** on the website for one country in Europe. Analyze the population pyramid to predict what the impacts of Europe's aging population will be on European language and religion 50 years from now.

9. How could you use the field methods employed by Kesteloot and Mistiaen (who studied Turkish restaurants in Brussels) to study the economic and social integration of a group of **migrants** in the region where you live?

10. Where in Europe is the largest concentration of **world cities**, and why do you think it is in this area?

SELF TEST

1. Ptolemy's map of the world 2,000 years ago was relatively accurate for:
   a. China
   b. the Mediterranean Sea
   c. the Indian Ocean
   d. Subsaharan Africa

2. The T–O map shows the power of the _____ over knowledge during Europe's Middle Ages.
   a. monarchy
   b. feudal lords
   c. Catholic Church
   d. peasants

3. The European Union has replaced the governments of its member states.
   a. true
   b. false

4. Several states in Eastern Europe were not part of the Soviet Union during the Cold War, but they were heavily influenced by the Soviet Union and were called satellite states. Which of the following was a satellite state?
   a. Moldova
   b. Estonia
   c. Latvia
   d. Hungary

5. On the whole, the growth of Islam in Europe since the end of World War II has come largely by:
   a. the conversion of European Christians to Islam
   b. the migration of followers of Islam from North Africa to Europe
   c. the conversion of European secularists to Islam
   d. the migration of followers of Islam from South Asia to Europe

6. France banned the outward demonstration of religious symbols in public schools because:
   a. of the Paris riots in 2005
   b. the French government does not allow people who profess a religion to migrate to France
   c. the French government is officially secular
   d. of a European Union regulation

7. Religion and language are spatially connected in Europe. Which of the following combinations is found in Europe today?
   a. Slavic languages and Eastern Orthodox religion
   b. Romance languages and Protestant religion
   c. Germanic languages and Catholic religion

8. In their fieldwork in Brussels, Kesteloot and Mistiaen found that the level of integration of Turkish immigrants into the city of Brussels was reflected in:
   a. how the immigrants dressed
   b. Turkish restaurants
   c. mosque attendance rates
   d. public school attendance

9. Much of the Netherlands is below sea level because:
   a. the Dutch stopped building terps
   b. the sediment in the polders has compressed and dried out over time
   c. the sea gates in the Netherlands are always closed

10. In the 1989 Velvet Revolution, this country overthrew its communist government:
    a. Hungary
    b. Poland
    c. Romania
    d. Czechoslovakia

11. To solve its problem with acqua alta, Venice is:
    a. pumping groundwater into the island of Venice
    b. draining the Venetian lagoon
    c. building a series of giant sea gates
    d. moving people off the island and opening it only for tourists

12. Public housing in Prague is now mainly located:
    a. in the central city
    b. in the suburbs

13. Since 2004, ___ countries have joined the European Union.
    a. 10
    b. 13
    c. 14
    d. 16

14. Africans who migrate to Europe generally migrate through these two centers:
    a. Spain and Malta
    b. Portugal and Cyprus
    c. Italy and France
    d. United Kingdom and Germany

15. Guest workers are:
    a. immigrants who are welcomed into a country for a short time to fill a labor shortage
    b. immigrants who are welcomed into a country on a permanent basis to fill a labor shortage

ANSWERS FOR SELF-TEST QUESTIONS

1. B, 2. C, 3. B, 4. D, 5. B, 6. C, 7. A, 8. B, 9. B, 10. D, 11. C, 12. B, 13. B, 14. A, 15. A

NORTH AND CENTRAL EURASIA

The region of North and Central Eurasia is tied together through similar experiences in history, which helped connect cultures and economies in the region. During the twentieth century, every country in the region was an individual republic (similar to a state in the United States) within one country called the Union of Soviet Socialist Republics (USSR or Soviet Union). The period from 1922 to 1991 can be called the Soviet era because the Communist Party ran the government, which was based in Moscow (in present-day Russia), dictating where and how people lived.

Vladimir Lenin, the leader of the revolution that ushered in the Soviet era, died in 1924, after which Joseph Stalin took over the USSR. Lenin's body is enshrined in a mausoleum in Red Square in Moscow.

Moscow, Russia. The Red Square area of Moscow features the Cathedral of St. Basil the Blessed on the left and the Spasskaya Tower of the Kremlin on the right. At the front right is the Mausoleum of Vladimir Lenin encased in red granite. Once the Soviets realized the preservation of Lenin's body was long-lasting, architect Alexey Shchusev designed the stone, stepped, pyramid-like structure in 1930 to replace the wooden structure that encased Lenin's body previously.

In 1991, the Soviet Union dissolved, and eventually every country in the region of North and Central Eurasia today gained its independence. Each country has wrestled with how to handle Soviet toponyms and memorials. Geographer Benjamin Forest and political scientist Juliet Johnson determined that governments had three options in determining the fate of Soviet era memorials: coopt the memorial, disavow the site, or contest the symbol (Forest and Johnson 2002, 525).

In their research on how Soviet spaces are used today, Forest and Johnson explained that memorial spaces are important in building national identities and that elites have more power than the general public to determine how a space appears and interpret what a symbol means. In the 1990s Russian leader Boris Yeltsin pushed to remove Lenin's body from Red Square and bury his remains. Russians protested, and today Lenin remains enshrined in his mausoleum but a private charity pays for the preservation of his body.

Governments pick and choose how to remember their history, memorializing some leaders and casting others (and literally their statues) aside. By constructing memorial sites, governments construct the identity of their nation—how they hope the people of their country see themselves as part of a larger whole and the role of the larger whole, the nation, in the world.

© Hubertus Blume/age fotostock.

THRESHOLD CONCEPT in this Chapter

Identity

WHAT IS NORTH AND CENTRAL EURASIA?

LEARNING OBJECTIVES

1. **Explain** the distribution of mountains and plains in the region.

2. **Understand** what the steppe region is and why the soils and climate are suitable for growing wheat.

3. **Describe** the oil and natural gas resources in the region and why it is risky to extract oil and natural gas in the tundra region.

4. **Understand** why environmental degradation was so common during the Soviet region and **describe** some of the lasting effects of central planning on the environment in Central Asia.

North and Central Eurasia is dominated territorially by Russia, the world's largest country in land area. Russia is nearly twice the size of the United States. The distance from New York City to Los Angeles, California (2,462 miles or 3,962 kilometers) is slightly shorter than the distance between Moscow and Irkutsk on Russia's Lake Baikal (2,609 miles or 4,199 kilometers) (Figure 10.1). The countries in the region have an historical tie to Russia, as each was once physically part of a single country called the Soviet Union from 1917 to 1991 (Figure 10.2). Russia dominated the Soviet Union, which

some scholars argue was a massive colonization of the region by Russia.

PHYSICAL GEOGRAPHY

Approximately 20,000 years ago, at the peak stages of the last ice age, the Fenno-Scandian ice sheet covered the western extent of northern Eurasia, and tundra vegetation covered the east. The glaciers of the ice age carved the East European Plain, which extends through the Baltic States, Belarus, Ukraine, Moldova, and western Russia. Much like the Great Plains of North America, which were similarly glaciated, the East European Plain is productive farmland.

The Ural Mountains extend north to south in a range about 1,600 miles (2,500 km) in western Russia and divide the east European plain from the West Siberian plain. The Urals, much like the Appalachians in North America, are an ancient mountain range more than 250 million years old. The Urals were formed when the ancient Siberian and Baltic plates collided in the process of building Pangaea (Figure 10.3).

Lake Baikal, in eastern Russia, is the deepest point in the region and the deepest lake on earth. The Amurian plate, which extends from Lake Baikal south and east across the Korean Peninsula to the Pacific Ocean, is diverging from Eurasia (Figure 10.4). Lake Baikal has formed in the rift valley between the Amurian plate and the Eurasian plate. Geologists, tectonic physicists, and physical geographers are studying the rift valley because millions of years from now the Eurasian plate will be split along this area, creating

Former Soviet republics

Russia

Contiguous United States

FIGURE 10.1 **Size of Russia and the Soviet Union relative to the United States.** Taken together, the light and dark green areas of this map are the footprint of the Soviet Union. The dark green area is Russia as it stands today. The contiguous United States is overlaid on an equidistant projection to give a sense of how far Russia stretches from west to east.

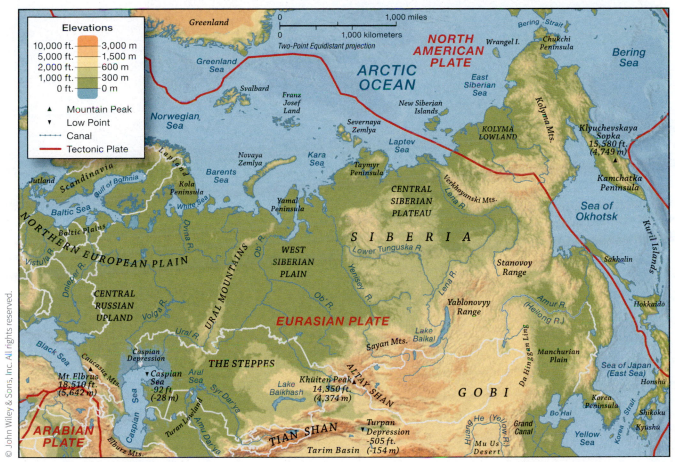

**FIGURE 10.2   Physical geography of North and Central Eurasia.   **The flat plains that dominate Northern Eurasia were carved by glaciers, and the hundreds of lakes and rivers in northern and eastern Russia were created by glacial melt water. Central Eurasia's mountainous terrain stands in contrast to the plains of the north.

new, smaller continents and building an ocean floor, much like the mid-Atlantic rift looks today.

CLIMATE

The climate of North and Central Eurasia is heavily influenced by latitude (Figure 10.5). Nearly the entire region is north of 40 degrees north latitude. The subsolar point migrates between 23.5 degrees south and 23.5 degrees north over the course of the year. At 40 degrees north and even closer to the North Pole, this region never receives direct rays (at a 90 degree angle) from the sun. Over the course of the year, the amount of daylight received in the region ranges from 14.8 hours on the summer solstice at 40 degrees north to 9.2 hours of sunlight on the winter solstice. Closer to the Arctic Circle, at 60 degrees north latitude, the amount of sunlight ranges from 18.6 hours on the summer solstice to 5.4 hours on the winter solstice. More than half of Russia is at or above 60 degrees north latitude.

**FIGURE 10.3   Ural Mountains.   **The Urals were built around the same time as the Appalachian Mountains in North America. However, the Urals have a higher average elevation than the Appalachians. The Urals are located at a higher latitude, which makes the climate colder than Appalachia and slows the erosion process.

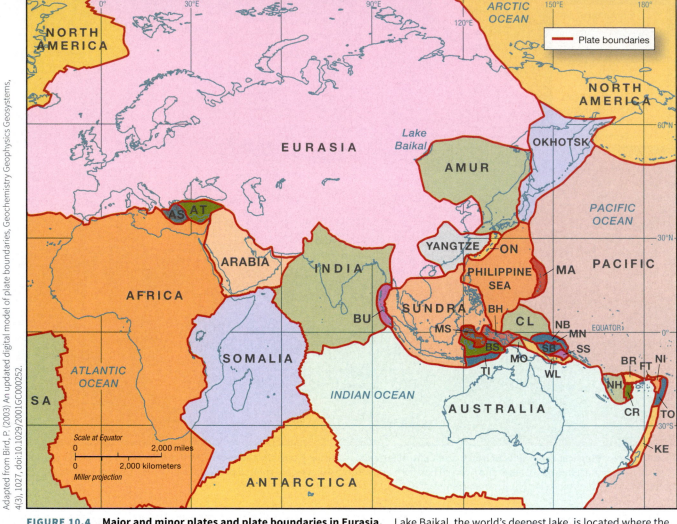

Adapted from Bird, P. (2003) An updated digital model of plate boundaries, Geochemistry Geophysics Geosystems, 4(3), 1027, doi:10.1029/2001GC000252.

FIGURE 10.4   Major and minor plates and plate boundaries in Eurasia.   Lake Baikal, the world's deepest lake, is located where the Amurian plate and Eurasian plate are diverging or separating, creating a rift valley.

How much **insolation**, incoming solar radiation, a place receives along with the angle of insolation helps determine the temperature. North and Central Eurasia have wide temperature ranges over the course of the year with high temperatures during the summer months of greater insolation and cold temperatures during the winter months of less insolation (Figure 10.5).

In addition to insolation, differences in high- and low-pressure systems over the course of the year also influence wind and weather systems in the region. In the winter, the lack of insolation creates intense cold in east Siberia, and a massive high-pressure system dominates the region (see Chapter 6). Air flows from aloft to the surface in a high-pressure system, and then air flows out from the high clockwise in the Northern Hemisphere (Figure 10.6). In the summer, the region is dominated by a low-pressure system in southwest Asia along with a high-pressure system in the northern Pacific Ocean and a weak high-pressure

system at the North Pole. The high-pressure system brings little moisture in the winter months, and precipitation increases in the summer months as air flows from oceans to land.

The climate zones in the region are dictated primarily by latitude and insolation (Figure 10.7). In the extreme north, the tundra climate dominates areas at and north of the Arctic Circle, about 10 percent of the area of Russia. The **tundra** climate is treeless and swampy. The rivers in Siberia flow north to the Arctic Ocean. During the spring and early summer thaw, the rivers melt in the south (closer to the tropics and warmer) first and in the north last. Meltwater flowing toward the Arctic hits ice that has yet to thaw, and rivers typically flood and fill in ponds and swamps in the region (Figure 10.8). The soils of the tundra region are wet and cold, which perennially freezes the soil in a condition called permafrost. Permafrost makes it impossible for trees to grow and makes it quite

FIGURE 10.5 **Climates of North and Central Eurasia.** Climate zones follow bands of latitude from west to east across the region.

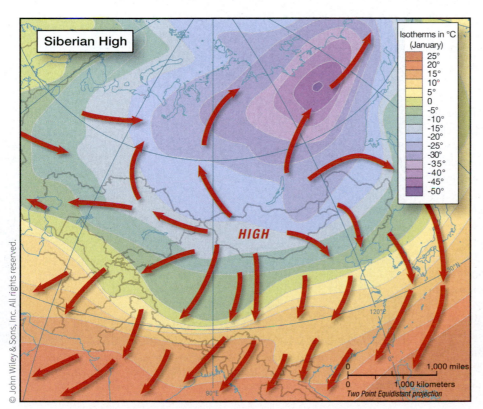

FIGURE 10.6 **High-pressure system in Siberia.** In the winter months when Siberia receives little to no insolation, the temperatures drop to massive lows and a high pressure system forms over Siberia. In a high pressure system winds flow from aloft to the surface and then in the northern hemisphere, surface winds flow clockwise out from the high.

FIGURE 10.7 Northern Russia. At noon during the winter months, the sun barely reaches over the horizon.

a challenge to build infrastructure, including roads and pipes.

Much like Canada (see Chapter 11), the **boreal forest** covers the massive area of the subarctic climate region known as the taiga in Russia. The subarctic climate region covers an area of North and Central Eurasia that is equivalent in size to the entire United States (Curtis 1996). Cold air masses form in the boreal forest during the winter months. Formed over the continent of Asia, the air masses are dry. Precipitation in the taiga occurs in small amounts each month, with an annual average of 40 inches (101 cm).

Eastern Europe and western Russia have a moist continental climate like the climate of the Midwest and east coast of the United States. The **humid continental** climate region is well suited for growing crops and covers much of Ukraine, Belarus, and western Russia (Figure 10.9). The humid continental climate receives moisture primarily from the Baltic and Black seas. The jet stream flows from west to east through the midlatitudes, attracting air masses to the region. In North and Central Eurasia, the jet stream draws moist air masses that form over the North and Baltic seas into the moist continental climate zone.

South of the boreal forest, a belt of the **steppe** or semiarid climate zone stretches from the northern coast of the Black Sea through the northern edge of the Central Asian states. Romantic and artistic renderings

of Russia often evoke the steppe climate that dominates the region around the southern Volga River where the Cossacks lived. (Figure 10.10).

A **midlatitude desert** extends from the east coast of the Caspian Sea through the southern side of Central Asia. This climate receives little, if any, precipitation because of its location. First, the climate zone is located on the interior of the continent, far from oceans. Second, the climate zone is located on the rainshadow of massive mountain ranges, including the Hindu Kush Mountains.

OIL AND NATURAL GAS

Over millions of years, heat and pressure created through the formation and erosion of the Urals have metamorphosized carbon into fossil fuels, including coal and natural gas. Heat and pressure also shaped precious gemstones, including emeralds and diamonds. Natural resources are crucial to the functioning of Russia's economy. "[O]il is the lifeblood of Russia's economy, providing two-thirds of its exports and half of federal revenue. It's not just oil: 85% of Russia's exports are raw materials or primary commodities, and their prices have also risen to unprecedented levels over the past 10 years" (Zakaria 2010, 21).

FIGURE 10.8 Tundra, Russia. When glaciers retreated from northern Eurasia approximately 20,000 to 10,000 years ago, they left behind massive amounts of water, which created a landscape of hundreds of rivers and lakes.

© Mykola/Moment Open/Getty Images.

FIGURE 10.9   Shidnytsia, Lviv region, Ukraine. The Stryi River Valley carves through the Shidnytsia, which features a spa resort area and village, located in the foothills of the Ukrainian Carpathians.

Although eastern Siberia has oil, little of it is explored or tapped currently. The oil boom in western Siberia began in the 1960s on these oil fields but has declined since the early 2000s.

When North and Central Eurasia were part of the Soviet Union, the communist government controlled energy prices, thereby "artificially" pricing "energy far below the level of world market prices," subsidizing the price of energy, and unintentionally encouraging "excessive consumption of energy" (Cooper, Energy, 1). Low energy prices helped the Soviet Union push growth in defense and heavy industries. By the demise of the Soviet Union in the early 1990s, the problems central planning had created in the energy sector were apparent, including "poor management of resources, underinvestment, and outdated technology and equipment" (Cooper, Energy, 1).

After 1992, Russia slowly deregulated energy pricing. Today, state-owned energy enterprises, including Gazprom and Trasneft, can control prices because of their monopolies on gas and oil, respectively. Natural gas energy company Gazprom licenses the rights to extract natural gas from the largest fields in western Siberia. Gazprom generates more than 90 percent of all natural gas output in Russia, and its subsidiary company, Gazexport, controls natural gas exports, most of which go to Europe.

Oil and natural gas extraction and refinement in Russia are controlled by domestically owned oil companies.

The Urals are not the only region with a high density of natural resources (Figure 10.11). Russia has the largest reserves of natural gas in the world. East of the Urals, western Siberia has three of the largest natural gas fields in Russia. The region accounts for more than 95 percent of Russia's natural gas extraction.

Western Siberia also has five major oil fields, and two-thirds of Russia's oil production comes from these fields.

Reading the **PHYSICAL** Landscape

The Russian Steppe

When people think about Russian landscapes probably the first they consider is Siberia because of its bleakness and intense cold. Another region in Central Asia that is often represented in paintings and photographs is the Russian steppe, which covers a broad corridor of land across the continent from the Ukraine to eastern Mongolia (Figure 10.10a). The word "steppe" means "flat and arid land" in the Russian language and is represented by a typical landscape shown in Figure 10.10b.

The Russian steppe is similar to the grasslands in the western Great Plains of North America. Both regions occur in semiarid climate zones where annual rainfall is between 10 and 30 inches (25 and 75 centimeters), summers are hot resulting in water stress, and winters are cool to cold.

Enough water is available to support thick stands of grass, with trees along streams.

The soils in the steppe are quite dark and are generally lumped into a category called Chernozems. This name literally means "black earth" in the Russian language. The rich, dark color occurs because the soils are extremely rich in decomposed organic matter, which largely comes from decaying grass. Like the genetically related Mollisols of the American Great Plains, soils on the Russian steppe are very fertile because of their high organic and mineral content. Without enough rainfall to grown water-intensive crops like corn and soybeans, the steppe is an excellent place to grow less water-intensive small grains like wheat (Figure 10.10c).

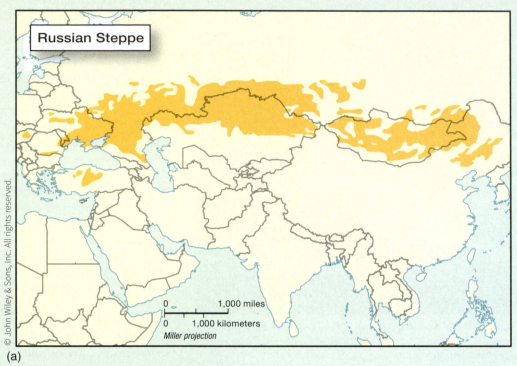

(a)

(b)

(c)

FIGURE 10.10 **The Russian steppe.** (a) The geographical extent of the Russian steppe. (b) **Tuva, Russia.** Tuvan horsemen ride past a burial mound, through the grasslands of the Russian steppe. (c) **Divnoye, Russia.** Wheat harvest on the Russian steppe. Wheat grows well in this environment because it is a domesticated grass perfectly suited to the climate and soils of the area.

International oil companies, including British Petroleum (BP), operate in Russia by partnering with Russian oil companies. In 2003, BP entered a contract with three Russian businessmen and formed TNK-BP. BP owns 50 percent of the company, and the Russian investment group owns the other 50 percent. TNK-BP is the third largest oil company in Russia, and in 2009, it opened the Uvat oil field. The largest oil producer in Russia is Rosneft, a domestically owned company that regularly partners with foreign oil companies. Rosneft is also Russia's largest oil refining company, operating Angarsk, the largest refinery in the country.

By controlling natural gas production and export, Russia has used state-supported Gazprom as a tool in its political relations with Ukraine, Belarus, Germany, the Baltic States, and the rest of Europe (Figure 10.12). In

2006, Russia cut off natural gas exports to much of Europe when it cut gas to the Ukraine over a dispute about energy pricing between the two countries. In 2006, Russia increased the price of Russian gas it charged neighboring Belarus. In response, Belarus levied a tax per ton of oil Russia shipped through Belarus by pipeline. Russia refused to pay the tax, and Belarus began siphoning gas off the Druzhba pipeline that flows west from Russia through Belarus. In early 2007, Russia's Transneft cut off the flow of gas along the Druzhba pipeline, which left Germany and Poland temporarily cut off as well. Controlling oil and natural gas exports to Europe enables Russia economic and political stature in the region (Figure 10.13).

Over 90 percent of the oil that Russia exports is transported by pipeline, and the state-owned company Transneft controls the vast majority of pipelines. Of the 7 million barrels per day of oil exported in

FIGURE 10.11 Berezniki, Ural Mountains, Russia. Natural resources in the Ural Mountains include gas, coal, precious gems, and minerals. The pile in this mine is potash, a potassium-based salt, primarily used in fertilizer. Potash is in high demand, and the company that owns this mine is worth over $14 billion.

FIGURE 10.12 Political geography of North and Central Eurasia. Despite the splintering of the Soviet Union into 15 independent countries after 1991, Russia remains the largest country in the world.

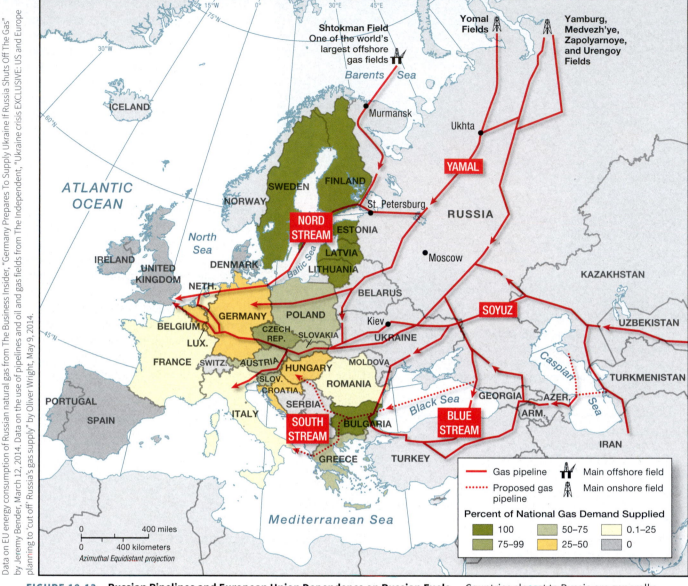

Data on EU energy consumption of Russian natural gas from The Business Insider, "Germany Prepares To Supply Ukraine If Russia Shuts Off The Gas" by Jeremy Bender, March 12, 2014. Data on the use of pipelines and oil and gas fields from The Independent, "Ukraine crisis EXCLUSIVE: US and Europe planning to 'cut off' Russia's gas supply" by Oliver Wright, May 9, 2014.

FIGURE 10.13 **Russian Pipelines and European Union Dependence on Russian Fuels.** Countries closest to Russia are generally more dependent on Russian natural gas, including Sweden, Finland, Estonia, Latvia, and Lithuania who receive 100 percent of the gas they consume from Russia. Ukraine is of particular strategic importance to Russia because Russian pipelines run through the country.

2009, more than 80 percent went to Europe, with Germany the single largest consumer of Russian oil. Another 5 percent of oil is shipped by rail, primarily west to the Baltic Sea or east to northern China or through Mongolia to central China.

Central Asia

Oil-rich countries in Central Asia include Kazakhstan, Turkmenistan, and Azerbaijan, all of which are located on the Caspian Sea and control 3.5 percent of world oil reserves and 6.8 percent of world natural gas reserves (Figure 10.14). During the Soviet era, Russia controlled oil and gas trade from Central Asia because all of these states were physically part of the Soviet Union. After 1991, the newly

independent Central Asian states still had to rely on Russia because Russia controlled oil and gas pipelines and also because Russia and its oil companies had the revenue to invest in energy extraction in Central Asia.

Since 2005, however, China has taken on an increasingly important role as consumer and producer of energy from and in Central Asia. The shift to China is partly motivated by a desire to move away from Russia. For example, in 2009, Russia was purchasing only one-third of what it had agreed to purchase from Turkmenistan. In fact, Russia turned off the taps on pipelines from Turkmenistan, which caused the pipeline to "explode" (BBC 2010, 2).

China's state-owned oil company, the China National Petroleum Company (CNPC), built a pipeline across

USING GEOGRAPHIC TOOLS

Mapping Oil and Natural Gas Resources

Fossil fuels form in sedimentary rocks that undergo heat and pressure over time. The carbon in fossil fuels comes from the live plant and animal material that slowly becomes oil and natural gas. The petroleum-dependency of the world-economy incentivizes finding new fossil fuel deposits and methods of extraction. Fracking (see Chapter 11) has gained widespread use to extract oil and natural gas in the United States. The possibility of tapping into larger, accessible fossil fuel reserves have encouraged geologists to use geographic information systems (GIS) to map large deposits of shale that contain accessible oil and natural gas. The U.S. Energy Information Administration (EIA) produced this global map in 2012, which caused countries to recalculate oil reserves and led to Saudi Arabia, Venezuela, and Canada topping the list of proved oil reserves with Russia in 8th place.

Shale is a sedimentary rock that forms where ancient water deposits, such as inland seas and massive lakes, leave behind algae, plants and animals that compact into rock under heat, pressure, and time. Geologists have known Western Siberia is home to a massive oil shale deposit called the Bazhenov for at least 20 years. The Bazhenov is equivalent in size to the area of Texas plus the area of the Gulf of Mexico (Helman 2012). New methods of fracking oil shale, many of which are used in the Bakken oil shales in North Dakota in the United States, now make it possible for fracking to begin in the Bazhenov. Russian owned oil company Rosneft is working with Exxon and Statoil to begin fracking the expanse that reportedly has enough oil to meet global demands and current rate for 64 years (Helman 2012).

Thinking Geographically

1. If shale forms where seas and lakes once stood, why are shale deposits found in the middle of continents?

2. What will happen to the global price of oil if Russia starts fracking in the Bazhenov?

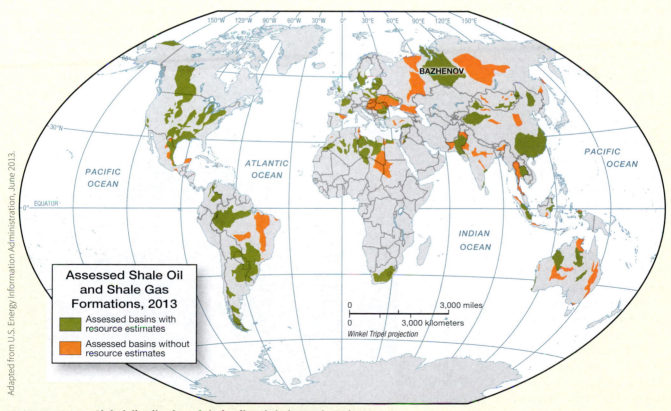

Adapted from U.S. Energy Information Administration, June 2013.

Assessed Shale Oil and Shale Gas Formations, 2013

- Assessed basins with resource estimates
- Assessed basins without resource estimates

FIGURE 10.14   Global distribution of shale oil and shale gas deposits.

Uzbekistan and Kazakhstan to Turkmenistan. Turkmenistan's South Yoloten gas field is one of the three largest in the world. CNPC is working with Turkmenistan to develop the South Yoloten field, which could be large enough to "supply gas not only to China but to Russia, Iran and even the European Union" (Roberts 2010, 2). Central Asian states with large debts, including Kazakhstan, are also drawn to China because China uses its deep pockets to fund new technologies and build pipelines to encourage higher production.

In 2010, Kazakhstan sent 200,000 barrels of oil a day into western China by pipeline. China and Kazakhstan share a lengthy border, and while more than one-third of the oil produced in Kazakhstan flowed east into China in 2010, other goods and also Chinese immigrants are flowing from China into Kazakhstan. CNPC bought into Kazakhstan at low prices beginning in 2009 because Kazakhstan had "over-borrowed," "suffered from falling oil prices," and was hit by the global economic recession (Demytrie 2010, 1). Since 2009, China has gained a "50–100% stake in 15 companies working in Kazakhstan's energy sector" (Demytrie 2010, 1). Relations between China and Kazakhstan have intensified quickly since the money began flowing in 2009. Kazakhs are leery of Chinese interests in renting farmland in Kazakhstan and fear a greater influx of Chinese energy workers and agricultural workers. With only 15.6 million people in the massive country of Kazakhstan, a relatively small flow of Chinese migrants could have a large impact on the ethnic makeup of the country.

ENVIRONMENTAL DEGRADATION

During the Soviet period (1922–1991), the communist government, which controlled the entire region of North and Central Asia, elevated industrialization over environment. Toward the end of the Soviet era, in 1986, Chernobyl, Ukraine, experienced a massive nuclear disaster (see Chapter 9). After Chernobyl, and especially after the fall of the Soviet Union in 1991 and the opening of the region, scientists, the media, environmental activists, and residents of the region came to recognize the magnitude of environmental degradation that had occurred during the Soviet years.

When the Chernobyl nuclear reactor accident occurred, the Soviet government did not reveal the accident until more than 48 hours after it occurred. The Soviet government did not report the extent of the accident until more than a week after the reactor broke down. The most egregious environmental issues were located not in Russia, but in the former Soviet Socialist Republics (SSRs), where the people had little say in their governance. In addition to Chernobyl, which took place in Ukraine and affected Belarus, many

other major environmental disasters were born from Soviet policies: agricultural land degradation in Uzbekistan, drying up of the Aral Sea in Central Asia, radiation exposure from a nuclear testing site in Kazakhstan, and extraction of energy resources on fragile lands in Siberia.

Uzbekistan

The desertification of agricultural lands in Uzbekistan stems from a Russian and then Soviet obsession with cotton production. In the 1800s, the tsars, and then in the 1900s the Soviets, encouraged and coerced Uzbek farmers to produce cotton. Uzbekistan has an arid climate; so, rivers, including the Amu Darya, irrigate the agricultural fields. The rivers originate from meltwater off of snow and glaciers in the Pamir Mountains. In order to intensify cotton production under Soviet leader Nikita Khrushchev's direction in the 1950s, Uzbekistani farmers drew increasing amounts of water from the river and expanded the irrigation canal system.

Overuse of agricultural lands has led to top soil depletion and soil erosion. Uzbekistani farmers are now struggling with salt deposits on 46 percent of irrigated lands, which makes growing crops nearly impossible, but "cotton is still king and the environmental destruction continues unabated, cutting into crop yields" (Tavernise 2008, 1) (Figure 10.15).

Once independent, the Uzbekistan government privatized collectivized agricultural lands but kept in place both price controls and quotas on cotton. Remedying the land degradation and salinization of Uzbekistan is a difficult

FIGURE 10.15 **Urgench, Uzbekistan.** This photo of women picking raw cotton in Uzbekistan reflects the low technology used in cotton cultivation during the Soviet era and still today. Technology has not advanced markedly, and the overuse of land has made production more difficult.

© JTB Photo/age fotostock.

USGS EROS Data Center.

FIGURE 10.16 The Aral Sea in Central Asia began disappearing in the 1960s because of the diversion of its two feeder rivers for agriculture. This series of images illustrates the unintended consequences of water management decisions. From left to right, the images were produced in 1977, 1998, and 2010. The darker color in 1977 shows the Aral Sea was still relatively deep. By 1998, enough water had been diverted that the sea had become two lobes, with the eastern more shallow than western. By 2010, the sea had shrunk to the point that the rivers no longer connected to the sea.

prospect because farmers still have little control over what to produce and are dependent on the state for subsidized fuel, fertilizer, and water.

Aral Sea

Through satellite imagery, the world became aware of the drying up of the Aral Sea (Figure 10.16) while the Soviet Union was still closed. The Aral Sea is an inland, saltwater sea that is a remnant of a vast inland sea (a body of water with no outlet) that covered the region millions of years ago. The saline Aral Sea is fed by fresh water flowing in from the Amu Darya River in Uzbekistan and the Syr Darya River in Kazakhstan.

In the 1950s and 1960s, the Soviet Union built dams across the rivers and developed massive irrigation canal systems. Water diverted for irrigation in both Uzbekistan and Kazakhstan resulted in less water reaching the Aral Sea, and so the sea level fell precipitously.

Over 30 years, the drying up of the Aral Sea became "one of the biggest man-made disasters in history" (Tavernise 2008, 1). The shorelines that dried up are now salt flats, and towns that were once located on the shore of the sea have lost their fishing and tourist industries (Figure 10.17). The water that remains in the rivers has high levels of pesticide runoff from cotton fields; the pesti-

cide residue and salt on dried lands regularly blow through the region; and the loss of sea, which moderated temperatures, has also changed the climate in the region.

Kazakhstan

Also in Central Asia, the Soviet Union used northeastern Kazakhstan as a nuclear testing site from 1949 to 1989. From 1949 until 1963, the Soviets conducted all nuclear testing above ground in the Semipalatinsk Oblast, which had a population of 1 million people. Between 1949 and 1989, the Soviet Union conducted 470 nuclear tests on the steppe of Kazakhstan. The above-ground tests created nuclear explosions that exceeded the total impact of "the Hiroshima nuclear bomb by 45,000" times, and the impact of the tests was grave, as

© PhotoAlto/Alamy.

FIGURE 10.17 Muynak, Uzbekistan. Once a town on the shore of the Aral Sea, the receding Aral Sea is now more than 62 miles (100 kilometers) from Muynak. The loss of water and the increasing salinity of the Aral Sea have destroyed 20 species of fish; in addition, with the loss of the sea and the fishing industry, ships have been abandoned.

"scientific research shows, in Kazakhstan approximately 2.6 million people fell victim to genetic mutation as a result of prolonged exposure to radiation" (UNESCO 2005). In 1977, 136 out of 1,000 children born had birth defects, and by 1993, the rate was 210 out of 1,000 live births in the four regions most affected by radiation (*Los Angeles Times*, June 16, 1996).

The movement to end nuclear testing in the steppe was born in Kazakhstan in the final years of the Soviet Union. In 1989, Olzhas Suleimenov, a Kazakh poet, called for citizens to demonstrate against the nuclear testing. Five thousand Kazakhs gathered at the National Writers' Union to demonstrate, and Kazakhs formed a nongovernmental organization (NGO) called Nevada-Semipalatinsk with the goal of ending nuclear testing and informing the public about the extent of radiation. Six months later, 50,000 Kazakhs gathered and threw rocks at the nuclear testing site, "following the Kazakh tradition of tossing stones at evil" (Nonviolent Action Database) (Figure 10.18). Between 1989 and 1991, Kazakh activists worked with antinuclear testing activists in Nevada, in the United States, to push for the closure of nuclear testing sites in both countries. In 1991, the president of Kazakhstan closed the nuclear testing site.

© Ustinenko Anatoly/ITAR-TASSPhoto/Corbis.

FIGURE 10.18   Almaty, Kazakhstan.   Kazakhs continue to use protest to bring political attention to problems. In 2012, this group of Kazakhs protested the violent treatment toward oil workers who had protested a month earlier as well as the results of recent elections.

Siberia

In the 1960s, the Soviet Union accelerated the extraction of oil and natural gas resources in the tundra climate and vegetative region of Siberia. The marshy soils of the tundra are particularly difficult to mine because soils shift easily when warmed. The wet conditions also make the region more susceptible to major impact from oil spills or natural gas flares. Marshes are connected, and spills move readily across linked water sources in the tundra. The cold climate and permafrost in Siberia make cleanup from either natural or human-made pollutants difficult. The "lower air, water, and ground temperatures slow natural self-cleansing processes that mitigate contamination in warmer regions, magnifying the impact of every spill and leak" in the permafrost zones (Curtis and McClave 1996).

CONCEPT **CHECK**

1. **Why** is Lake Baikal the deepest fresh water lake in the world?

2. **Where** is the steppe region, and **why** is it located there?

3. **Why** does North and Central Asia have so many oil and natural gas resources, and **why** is extracting the resources particularly difficult in the tundra?

4. **How** did Soviet central planning impact the environment in Central Asia?

WHO ARE THE PEOPLE OF NORTH AND CENTRAL EURASIA?

LEARNING OBJECTIVES

1. **Explain** how the Eastern Orthodox religion diffused into North and Central Eurasia and what happened to the religion during and after the Soviet era.

2. **Describe** what Russia is doing to try to increase its Total Fertility Rate (TFR) and **explain** why the Russian government is concerned about its low TFR.

3. **Understand** why the population density at northern latitudes is higher in North and Central Eurasia than at similar latitudes in other regions.

North and Central Eurasia covers a massive expanse of land, which includes hundreds of indigenous populations and a broad diversity of languages. The western part of the region is culturally tied to eastern Europe in many ways, including religion, where Eastern Orthodox churches are common. Central Asia is predominantly Islamic. The Soviet Union espoused an official policy of atheism, which resulted in pushing the Eastern Orthodox churches, especially the Russian Orthodox Church, underground. In Central Asia, however, the Soviet Union allowed Muslims to continue to follow and practice their religion.

The Russian tsars and the Soviets used the policy of Russification to move Russian speakers into the farthest reaches of their territories in the region. Through Russification, the Russian language diffused relocationally and supplanted other languages for purposes of government and business.

RELIGION

The predominant religions in the region of North and Central Eurasia are Eastern Orthodox in western Russia and eastern Europe; Islam in Central Asia and the Caucasus, Judaism in western Russia and eastern Europe, and traditional religions in Siberia and the far east of Russia (Figure 10.19).

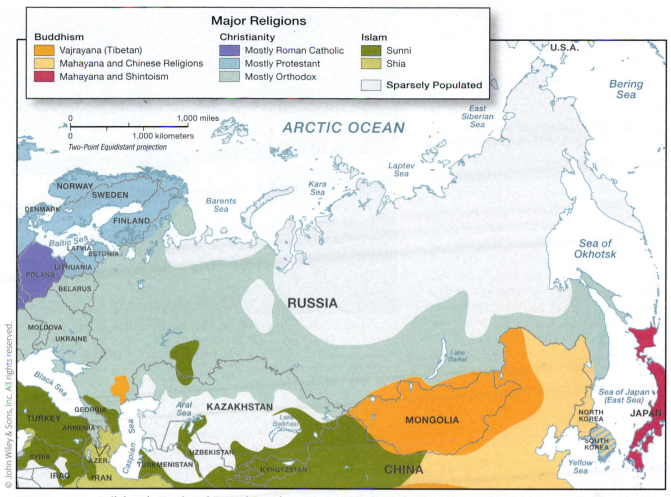

FIGURE 10.19 **Religions in North and Central Eurasia.** Eastern Orthodox and Sunni Islam are the predominant religions in the region.

Christianity

Christianity diffused hierarchically into Russia as members of the royalty converted. In the tenth century, Princess Olga of Kiev married Igor, who was the ruler of Kievan Rus, a Slavic kingdom in eastern Europe. Igor died when his son and heir was only 3 years old. Olga took over control of Kievan Rus until their son was old enough to rule. In 955, she converted to Christianity through baptism in Constantinople. In 988, Olga's grandson, Vladimir the Great, made Christianity the official religion of Kievan Rus and "ordered his people to be baptized in the Dnieper River" (Schemann 2009, 5).

Kievan Rus became Christian before the official split between western and eastern Christianity in 1054 (see Chapter 9). Some records in the western church contend that Olga was in contact with Otto the Great, the first emperor of the Holy Roman Empire, in Rome at the same time she was in contact with religious leaders in Constantinople. Russians hold that regardless of Olga's contacts, Prince Vladimir established the Christian church in the eastern tradition. Upon the split in Christianity in 1054, the western church, centered in Rome, became the Roman Catholic Church, and the Eastern Church, centered in

Constantinople, became the Eastern Orthodox Church. The Eastern Orthodox Church acknowledged Olga's pivotal role in diffusing Christianity to Russia by officially recognizing her as a saint in 1547.

Tsars established a Russian patriarch for the Eastern Orthodox Church, which came to be known as the Russian Orthodox Church. In 1589, Tsar Fedor officially established a separate Russian Patriarch for the Eastern Orthodox Church, which centered on Moscow. As the territory of tsarist Russia expanded, the Russian Orthodox Church diffused.

In the early twentieth century, the reign of the tsars ended, and the Soviet Union, in agreement with its official policy of atheism, converted Russian Orthodox monasteries and churches to other purposes and forced the church underground (Figure 10.20). Russian Orthodox adherents were one of the many groups to suffer at the hands of Stalin's purges—his campaign of rooting out opposition. Historians estimate that 30 million people in the Soviet Union died from Stalin's brutality and famines.

After the fall of the Soviet Union, the Russian Orthodox Church experienced a period of immediate resurgence, with the number of monasteries in Russia alone increasing from three in 1987 to 478 in 2009. In the same time frame, the number of Russian Orthodox churches rose from 2,000 to 13,000. The Russian Orthodox Church does not keep track of membership, but Russian historians of the church estimate that 60 percent of the people identify as Russian Orthodox, though less than 1 percent "actually enter a church at least once a month" (Schemann 2009, 4).

Islam

Islam diffused to Central Asia, the region east of the Caspian Sea, between 700 and 900 (see Figure 5.31). Central Asia is largely Sunni Islam, and to the south, Persia (modern Iran), is primarily Shi'a Islam. The Silk Road (see Chapter 8) traveled through Central Asia and helped diffuse Islam as far as China. As European exploration by sea expanded in the 1300s, 1400s, and 1500s, an increasing portion of trade between East Asia and Europe was conducted by sea routes instead of land routes.

FIGURE 10.20   Murom, Russia. Spassky Monastery dates to the eleventh century. The Soviets used the monastery as a barracks. In the late 1980s, the monastery was finally allowed to function again. Today, the Russian Orthodox Church has restored it to its former glory.

When Islam diffused to Kazakhstan, people blended the newly introduced teachings with traditional religions. Traditional religions in Kazakhstan uphold the belief that "separate spirits inhabited and animated the earth, sky, water, and fire, as well as domestic animals" (Country Studies, Kazakhstan, Culture). For centuries, Kazakhs practiced transhumance in the mountainous landscape, moving livestock from grazing lands at high elevations in the summer to low elevations in the winter (Figure 10.21). In the relative isolation of Kazakhstan in the period after the decline of the Silk Road, Kazakhs intermingled the teachings of Islam with the traditional belief in spirits. Professor Oraz Sapashev of the University of Wisconsin-Madison, who is Kazakh, explains that Kazakh and Kyrgyz Muslims:

> still recite prayers to the honor of our "aruaks," spirits of dead ancestors. But in classical Islam this is not envisioned. There is no understanding of reading the Qur'an for one's own dead as it's forbidden in Islam. According to Islam, a person recites prayers to ask for forgiveness, and to ask for protection and peace. But we read Qur'an to honor the dead (Sartbaeva 2012).

FIGURE 10.21   Itay, China. This picture, taken in 2012, shows Kazakh nomads herding their livestock by caravan across the plain in the far west of China's Xinjiang region. The plains are situated in the most northern part of Xinjiang, sharing a border on the east with Mongolia and on the west with Kazakhstan.

The combination of Islam with traditional religious practices is called **syncretic** diffusion because two different cultural attributes combine into something distinct.

Ethnic Russians were used as slaves in cotton production in present-day Uzbekistan in the 1700s and 1800s. In the 1800s, the British expanded their colonial empire to Pakistan and Afghanistan, to the south of Central Asia. Tsarist Russia continued to expand, and they set their sites on Central Asia in response to the enslavement of Russians, because of a desire to gain a source for cotton, and in response to the expansion of the British in South Asia. Tsarist Russia gained control over Central Asia in the second half of the 1800s. Russian migrants entered the region by railroad in the late 1800s. In response to the increasing presence of Russians, Central Asians worked to preserve their Islamic culture through **Jadidism**, a reform movement.

During the Soviet era, Muslims were allowed to practice their religion throughout the SSRs (Figure 10.22). Although the Soviet Union allowed Islam to operate, they also worked to control and coopt the religion through

official boards, which encouraged Muslims of Central Asia to acquiesce to Soviet control. The Soviet Union controlled Islam through a board called the Muslim Board of Central Asia. Operating out of Tashkent, Uzbekistan, the Muslim Board of Central Asia recognized 65 registered mosques in Uzbekistan. The board "carefully screened" Muslim clerics to ensure their allegiances were to the Soviet Union.

In 1997, Russia recognized five official religions: Russian Orthodox, Islam, Buddhism, Judaism, and several other non-Orthodox Christian denominations. Since independence, mosques and madrasas have been rebuilt, not only in Russia but in all of the former SSRs with large Muslim populations. Russia maintains a Muslim Spiritual Department, "which oversees the appointment of Islamic leaders" (Myers, November 22, 2005). The state limits the number of mosques that can be built and officially sanctions mosques. Russia fears political movements in Chechnya, Dagestan, and other Caucasus republics that would call for separatism, and has therefore acted multiple times to suppress any Islamic separatism from the Chechen War in the 1990s to operations to assassinate a militant leader in Dagestan in 2010.

In Central Asia, although all five independent countries continue to register mosques and madrasas, they have approached the practice of religion in different ways. Uzbekistan works to eradicate militants, including the radical group Hizb ut-Tahrir, which began in the Ferghana Valley of Uzbekistan and diffused within Tajikistan and Kyrgyzstan, parts of Kazakhstan, and part of Russia (Radio Free Europe 2005). In Turkmenistan, President Saparmurat Niyazov has worked to coopt Islam by "elevating himself to the level of a spiritual— as well as political—leader" (Radio Free Europe).

LANGUAGE

The most predominant language families in the region of North and Central Eurasia are Indo-European languages (especially Slavic languages) in Russia and eastern Europe; Altaic languages in Central Asia and central, far east and north of Russia; Chukotko-Kamchatkan in the far east of Russia (labeled 'other' on the map); Uralic languages in western Russia along the border with Finland; and North Caucasian languages in the Caucasus region (labeled 'other' on the map) (Figure 10.23).

Russification encouraged the movement of people across SSRs, diffusing and mixing languages. Intensified energy production during the 1960s also brought more non-Kazak migrants to Kazakhstan. By the 1970s, non-Kazakhs outnumbered Kazaks in the SSR. Kazakhstan did not have a written language until the mid-1800s, and the flow of political forces across Kazakhstan is well recorded back to the seventeenth century and is reflected in the language. In the 1860s, the first written form of Kazak was established using Arabic script. In 1929, Kazak was rewritten in Latin script, and in "1940 Stalin decided to unify the written materials of the Central

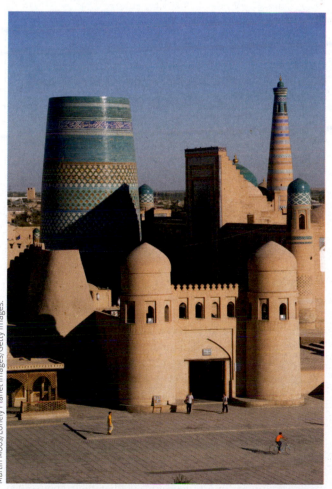

FIGURE 10.22 Khiva, Uzbekistan. The town of about 40,000 people today was once a busy trading city on the Silk Road. Ichon-Qala, meaning within the walls, is the old area of the city, featuring Muslim architecture built over more than 500 years, including these minarets.

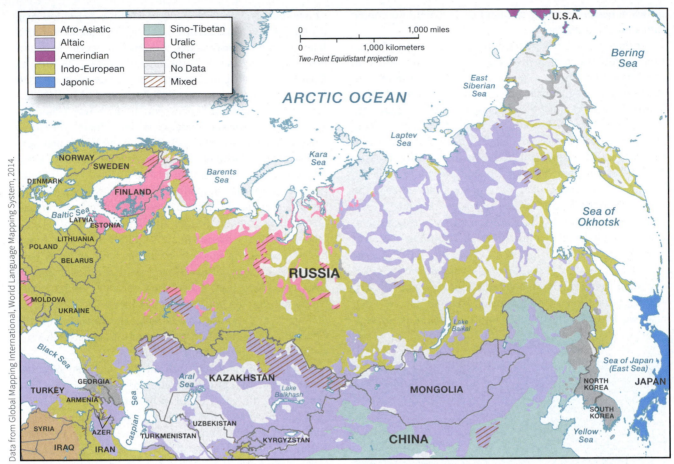

Data from Global Mapping International, World Language Mapping System, 2014.

FIGURE 10.23 **Languages of North and Central Eurasia.** Through the policy of Russification, the Russian language (an Indo-European language) diffused into many parts of the region.

Asian republics with those of the Slavic rulers by introducing a modified form of Cyrillic" (Curtis, Kazakhstan, 1996). After independence, Kazakhstan briefly considered returning the Latin script to the language, but determined that the idea was too costly. Instead, President Nursultan Nazarbayev, a Kazak, promoted the use of the Kazak language but recognized Russian as the country's **lingua franca**, the language of trade and business.

DEMOGRAPHIC ISSUES

Much as is happening in Europe and Japan, the population of Russia is aging. The dissolution of the Soviet Union had a major impact on Total Fertility Rates (TFRs) in Russia. In 1990, the TFR was 2.08, which was below replacement levels, and by 2004, TFR had plunged to 1.17. In 1989, Russia's population was 147 million; in 2002, it fell to 145.1 million, and in 2010, Russia's population fell to 142.9 million people (Figure 10.24). In western Europe and Japan, TFRs are declining primarily because women are delaying childbirth until after establishing their career. In Russia and the former SSRs of

Eastern Europe where TFRs are declining, women are giving birth to their first child at a relatively young age. However, birth rates for second and third children are quite low in Russia, which has "one of the world's highest abortion rates" (Weir 2006, 1).

The Population Council, a global population research nonprofit, conducted statistical research by surveying women over time to understand whether Russian women desire more than one child and whether they actually have more children after giving birth to their first child. The research went beyond macro-level issues such as economic uncertainty to study individual desires and decision making. The study revealed income through both formal and informal means was a major factor. The researchers found that in Russia women who were earning extra income through the informal economy desired to have more than one child. They found that whether women actually had more than one child hinged on whether their spouse was earning extra income in the informal economy.

Population change, which is based not only on births but also on deaths and migration, varies by region (Figure 10.25). The population has risen in the Caucasus region

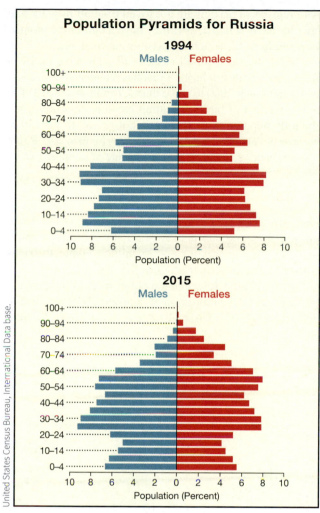

Population Pyramids for Russia

1994

Males   Females

Population (Percent)

2015

Males   Females

Population (Percent)

United States Census Bureau, International Data base.

FIGURE 10.24 **Population pyramids for Russia in 1994 and 2015.** Economic uncertainty in the post-Soviet period has led to a decline in fertility rates since 1991. Higher rates of suicide for men are apparent in the population pyramid, as men start dying at a higher rate than women after the age of 40 in the 2015 population pyramid.

of Russia and Central Asia, where Islam is the predominant religion. In western Russia, the population has experienced a decline of approximately 1 percent a year since the Soviet Union's demise. Unlike Europe and Japan, Russia's declining population is also impacted by relatively low life expectancy.

High Death Rates for Men

The working-age population is experiencing high mortality rates, especially among men whose "ranks have been decimated by alcoholism, war in Chechnya, AIDS, and accidents" (Weir 2006, 2). The number of working-age males who have committed suicide since independence is staggering relative to the number of women who have committed suicide (Figure 10.26). The high mortality rate for men is reflected in the much lower life expectancies for men, which was 60 years in 2008, as compared to female life expectancy of 73.

In June 2006, President Vladimir Putin remarked that the demographic situation created by low birth rates and high death rates is "the most acute problem facing our country today" (Putin 2006, 386). Even though Russia has experienced "large–scale net immigration (mostly due to the return of ethnic Russians from other republics of the former Soviet Union), the population in the last decade and a half has been shrinking: of late by some 700,000 persons per year" (Putin 2006, 385). Putin offered economic incentives, including monthly payments, extended maternity leave, and preschool child-care payments totaling about $9,300 to women for each child after their first. Prenatal care in Russia is free, and so are prenatal vitamins.

In addition to national financial incentives, provinces within Russia that are experiencing low birth rates are incentivizing higher birth rates. For example, the province of Ulyanovsk has declared September 12 the "Day of Conception" since 2005. Each September 12, women

© Igors Bolkovs/Demotix/Corbis Images.

FIGURE 10.25 **Riga, Latvia.** A woman uses a horn to demand improvements in pensions and housing allowances in a protest near the Ministry of Welfare. Low TFRs and an aging population pose challenges to governments who provide services to their older citizens. More welfare needs to be paid out to the relatively larger older population at the same time the relative size of the working age and tax-generating population shrinks.

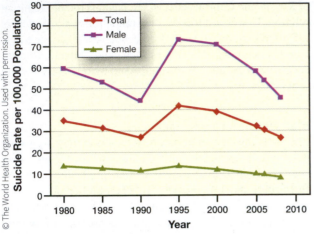

© The World Health Organization. Used with permission.

FIGURE 10.26   **Male and Female Suicide Rate in Russia, 1980 to 2006.**   The suicide rates for men are higher than for women. After the demise of the Soviet Union in 1991, the suicide rates for both men and women increased and peaked in 1995.

who register to participate are given time off of work and are eligible to win prizes if they have a child nine months later, on June 12, which is Russia's national day.

While deaths from all causes still outnumber births in Russia, the number of births a year has rebounded from a low of 1.2 million in 2000 to more than 1.7 million in 2010. The higher number of births is putting pressure on schools, but the total population of Russia is not rising, in large part because of the remaining high death rates for men, a TFR that remains under 2.1, and relatively low rate of immigration.

MIGRATION

One way to combat the declining population in Russia is to allow more immigration. President Putin has limited immigration to the approximately 25 million ethnic Russians who are living in former SSRs of the Soviet Union.

As part of the Soviet Union policy of Russification, the government forced migration of thousands of Russians into non-Russian SSRs (Figure 10.27). In addition to migration

Data from CIESIN—Columbia University, United Nations Food and Agriculture Programme—FAO, and Centro Internacional de Agricultura Tropical—CIAT. 2005. Gridded Population of the World, Version 3 (GPWv3): Population Count Grid, Future Estimates. Palisades, NY: NASA Socioeconomic Data and Applications Center (SEDAC). http://dx.doi.org/10.7927/H42B8VZZ. Accessed December 2014.

FIGURE 10.27   **Population density of North and Central Asia.**   People cluster in cities, especially in the western side of region. Eastern Russia and central Kazakhstan are sparsely populated.

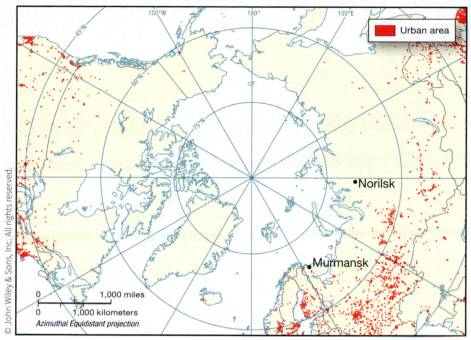

FIGURE 10.28   **Population clusters in the Arctic Circle.**   The Soviet policy of populating northern and eastern Russia is evident on this map where more urban areas are located in the Arctic in Russia than in other countries at the same latitudes.

to the non-Russian SSRs, the Soviet Union sought to populate northern and eastern Russia. The Soviets employed a policy called **development through resettlement** to assert their territoriality and physically control the far reaches of the country.

As a result of Soviet policies to populate the region north of the Arctic, the region's north is more densely populated than similar latitudes of landmasses around the world (Figure 10.28). North of the Arctic Circle, the cities of Murmansk and Noril'sk are located in the tundra climate zone, whose moist soils that repeatedly freeze and thaw over time make building difficult.

Both because it encompasses such a massive area and as a result of Soviet settlement policies, approximately 33 percent of Russians live in the boreal forest zone. Although the development through resettlement policy successfully increased the population of the far north, "the north was never evenly inhabited. Rather, settlements concentrated around the natural resource deposits often remained 'isolated islands of Soviet civilization'" (Spies 2009, 260).

The dissolution of the Soviet Union changed how Russians perceived the north and Siberia. The Soviets used subsidies and governmental programs to encourage people to migrate to the north, but post–Soviet Russia has not. Through the centrally planned economy, the Soviets guaranteed jobs and housing in the north. Without financial incentive to live in the north, Russians came to realize that "the population in this part of Russia is too large" (Spies 2009, 257). In post–Soviet Russia, residents in the north have migrated out of the region to cities in western Russia.

Russia, as a whole, experienced a population decline between 1991 and 2009 (Figure 10.29). Between 1990 and 2006, the Russian North lost 17 percent of its population. The Magadan Oblast and Chukotka Autonomous Okrug lost 57 and 68 percent of their population, respectively, in the time frame (Spies 2009, 260).

The oil and natural gas companies operating in the north continue to employ hundreds of Russians. However, many Russians working in the north today commute long distance between the oil and gas fields and their homes. One company in the north had 589 employees in 2004, and only 166 of them lived in the province where their job was located. The remainder had "their homes spread around Russia and commute from their place of residence to Usinsk for work duties" (Spies 2009, 258–259). As a result of migration and long-distance commuting, the average age of residents in the north is increasing.

CONCEPT CHECK

1. **Why** are the two major religions in North and Central Asia Russian Orthodox and Islam?

2. **How** have men been affected by the changes in the economy after the Soviet Union dissolved?

3. **Why** is the population density at northern latitudes higher in North and Central Eurasia than at similar latitudes in other regions?

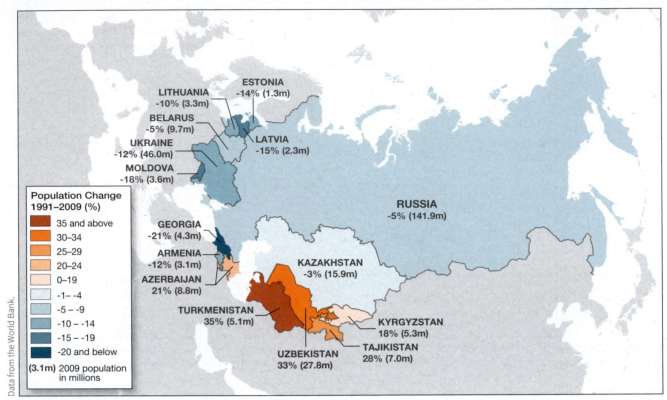

FIGURE 10.29   Population Change 1991 to 2009.   Population has fallen in all but five of the countries that were once part of the Soviet Union. Populations rose in the Central Asian countries of Kyrgyzstan, Azerbaijan, Tajikistan, Uzbekistan, and Turkmenistan. Georgia, Moldova, Latvia, and Estonia lost the most population in the same time frame with declines of 21, 18, 15, and 14 percent respectively.

HOW HAVE IDENTITIES FORMED AND BEEN CONTESTED IN THE REGION?

LEARNING OBJECTIVES

1. **Understand** what identities are and how they are constructed.

2. **Describe** the policy of Russification and **understand** how it impacted identities in the region.

3. **Explain** how independence has impacted identities in Belarus and Kazakhstan.

The Mongols (including Genghis Khan and his descendants) ruled a large expanse of land in what is now Russia starting in the 1200s for nearly two centuries (Figure 10.30). From their base in Moscow, a **Slavic** people (the Russians) organized under a leader in the 1300s and expanded first northwest to Novgorod and then east to the Volga River, outside the area of Mongol control.

By 1462, Slavic leader Ivan III had gained enough power to cast off the Mongols, and in 1480 he expanded his territorial control over the lands once ruled by the Mongols (Figure 10.31). Note the expansion of Russian

lands until 1505, when Ivan III's reign ended. During the 1500s and 1600s, Russian tsars, including Tsar Ivan IV and Tsar Peter the Great, expanded Russia's land area east to the Pacific Ocean, across Siberia. The land expanse was great, but the region was not densely populated. As late as 1720, Siberia's population numbered only 400,000. Russia did not expand westward into Europe between the 1500s and 1700s because Lithuania and Poland fought to retain their ground and actually gained some Russian lands themselves.

The 1700s marked the beginning of modern Russia, with Peter the Great naming St. Petersburg the capital in 1703. In wars with Sweden, Russia gained Estonia, Livonia, and part of Karelia, the region due east of Sweden. In the 1700s, Russians migrated to present-day Alaska and Canada. During Catherine II's rule, Russia expanded control to the south, to the shore of the Black Sea. Russians founded the city of Odessa in 1794, and the city became an outlet for Russian exports from the drainage basins of rivers that flowed into the Black Sea. In the 1800s, Russia continued to expand its territorial control to the south into Central Asia (see Figure 10.31).

The powerful tsars derived most of their wealth from a feudal system of agriculture, where serfs paid tribute to feudal lords, who in turn paid tribute to the tsars. In 1861, after losing the Crimean War, the tsars emancipated the

FIGURE 10.30 **Mongol lands.** The Mongol lands stretched over a massive expanse, including much of what is now Russia from the thirteenth to fifteenth centuries. The Mongol lands, which were divided into four khanates in 1294, are shown in orange on the map. The green territories are vassal states, which paid tribute to the Mongols. The green territory the farthest northwest were the Russian principalities.

serfs, releasing them from feudal control. Some serfs in western Russia migrated to towns in the same region, and others migrated as far east as Siberia. In the late 1800s, the tsars pushed a policy of rapid industrialization in western Russia, with especially intensive industrialization periods between 1893 and 1904 and again between 1909 and 1913. The new factories in urban areas were run to benefit the tsars, and the working conditions were horrendous.

IDENTITIES IN THE SOVIET UNION

As the region of North and Central Eurasia shifted from Kievan Rus and Mongol control to Russian tsarist control and then to the Union of Soviet Socialist Republics (USSR), people throughout the region negotiated their local cultures with the larger empire or country in which they were located. People do not take their identities on and off like masks. Rather, people negotiate their identities through experiences in places as well as through interactions with and perceptions of others.

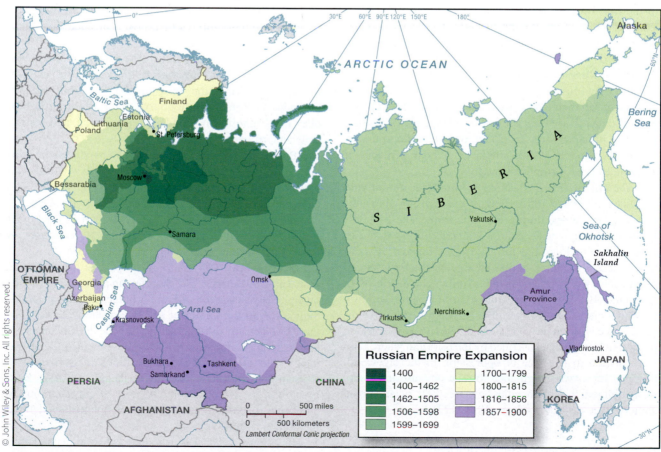

Russian Empire Expansion

1400	1700–1799
1400–1462	1800–1815
1462–1505	1816–1856
1506–1598	1857–1900
1599–1699	

500 miles

500 kilometers

Lambert Conformal Conic projection

FIGURE 10.31 **Expansion of Russia and Soviet Union.** Tsarist Russia and the Soviet Union expanded to encompass all of North and Central Eurasia over a period of more than seven hundred years.

Identity is how we define ourselves, how we explain who we are at a given scale or in a particular context. Identity includes common markers such as gender, race, ethnicity, and nationality. One person may have different identities at the scale of the world, the nation, locality, or family (see Figure 3.5). People create their identities by negotiating how they see themselves, how others define them, and how they define themselves relative to others. Identities are dynamic, negotiated, and contextual.

Countries seek to form a common identity for the people within their territory so that the people will feel an allegiance to the country, the state, which makes both governing and exerting territorial integrity easier. The rise of nationalism, or of states seeking to build nations, in the 1700s helped lead to the American and French Revolutions. In the 1900s, both World War I and World War II took place in the context of nationalist movements that undermined the territorial integrity of other states.

A **nation** is a group of people who see themselves as connected (often culturally), with a shared past and a common future, and who have a political goal. A nation is a constructed identity that governments often use to strengthen the allegiance of their citizens. People are born into nations, but no nation is "natural." Every nation is actively created either by a group of people or by a government.

RUSSIFICATION

In the late 1800s, the Russian tsars instituted a policy of **Russification** in an attempt to build a Russian national identity throughout tsarist territory. How Russification was manifested under the tsars varied from place to place within the greater Russian region.

In Poland in the later 1800s, Polish people were only allowed to sell their land to Russians, and schools in Poland were allowed to use only Russian as a language of instruction. In Lithuania, the tsars geared Russification to undermining the strength of the Catholic Church and replacing it with the Russian Orthodox Church.

The policy of Russification also impacted Jews living in Russia in the late 1800s. In the context of growing anti-Semitism in Europe during this time frame, the tsars targeted Jewish language, culture and religion—seeking to replace it with Russified forms. Tsar Alexander III controlled access to high schools and universities by Jews and moved Jews westward and then into urban ghettos. In 1890, Alexander III "issued a sweeping decree against the Jews: all who remained in the interior of Russia were now to emigrate to the western provinces." In the Jewish Pale, a region of eastern Europe that was ruled by the tsars and where Jews were invited to live free of oppression in the eighteenth century, Jews "were forbidden to own or lease lands and were obliged to live in cities" where their occupational choices were restricted (Hayes 1916, 472).

The Soviet Union continued the policy of Russification by moving people from one SSR to another and by working to populate the far expanses of the Soviet Union with Russians who had allegiance to the country.

IDENTITIES IN FORMER SOVIET SOCIALIST REPUBLICS

When you see a map of all the countries in the North and Central Eurasia, you may think each country (state) has its own nation. For example, Belarus has a Belarusian nation and Kazakhstan has a Kazak nation. However, the current boundaries of Belarus, Kazakhstan, and other SSRs are relatively recent creations. The indigenous populations of each state are diverse, and Russification changed the ethnic and linguistic composition of the countries. Since becoming independent in the 1990s, each country has had to consider questions of language, religion, and government, and also negotiate how to interpret the Soviet period under which they had little say about their place in the world.

Belarus

In Belarus, approximately 78 percent of the population is Belarusian, 13 percent are Russians, 4 percent Poles, and 3 percent Ukrainians. Before Belarus was part of the Soviet Union, the territory of Belarus was part of the "Kievan Rus (900–1200), the Grand Duchy of Lithuania (1300s–1569), the Polish-Lithuanian Commonwealth (1569–late 1700), and the Russian empire (1790s–1917)" (Buhr et al. 2001, 430).

The different empires that have controlled the region are reflected in the cultural composition, in terms of religion and language, in Belarus. The language of government changed from Belarusian to Polish and then to Russian. Likewise, the religion changed from Eastern Orthodox to Roman Catholic to a combination of Orthodoxy and Catholicism called a Uniate church, and then to Russian Orthodox.

Before World War II, Belarus had a large Jewish population, primarily in major cities where they comprised upwards of 50 percent of the urban population. The Germans targeted Belarus because the Jewish Pale ran through the country and because it was part of the Soviet Union. Stalin led Soviet troops to fight the Germans on Belarussian soil in order to stop Nazi occupation.

After World War II, Stalin's purges furthered the destruction of Belarus. The Belarussian population dropped precipitously as a result of Stalin's purges and deportations. Stalin ordered the killing of any Belarusian who was captured by Germany during the war; forbade the official use of the Belarusian language and made Russian the official language; and filled government positions in Belarus with Russian migrants from other SSRs, including Russia. Both because of the Holocaust and because of migration, in the 1989 census, only about 1 percent of Belarusians defined themselves as Jewish.

In August 1991, Belarus declared its independence. In the post-Soviet era, the first priority of the newly

---

 identity how we make sense of ourselves.

independent state was to define the Belarusian nation. The various peoples' shared experience of World War II, Soviet control, and Russification would be expected to bind them into a nation and to see their nation as cohesive enough for them to experience the future together. In the process of building a national identity, Belarus initially distanced its iconography from Russia by creating a distinct flag and declaring Belarusian the official language.

But today Belarus remains tied to Russia, both economically and militarily. Economically, in 1994 and 1995, Belarus and Russia created a trade union between them and eliminated customs check points. Belarus depends on oil and natural gas from Russia. Many of the officers in the Belarus military are Russians, and the Russian military trains in Belarus.

Since independence, several efforts and overtures toward formally uniting with Russia have come from conservatives and ethnic Russians in the government. In a 1995 referendum, Belarusians voted overwhelmingly (83.1 percent) to adopt a flag, language, and official holidays that tied it more closely to Russia. The new flag was a "modified version of the soviet flag," and they added Russian as an official language (Buhr et al., 2011, 425) (Figure 10.32).

In establishing a Belarusian nation after 1991, the country, under the leadership of President Alexander Lukashenka, has been influenced by three schools of thought. The first is a desire to develop a Belarusian nation that identifies with pre-Soviet Belarus by reinvigorating the cultural practices of the Grand Duchy and the Polish-Lithuanian Commonwealth. Followers of this school of thought promote the use of Belarusian in government and schools. President Lukashenka has rejected the pre-Soviet sense of nationalism by arguing that those who take an anti-Russian stance are ignoring the fact that the Belarusian economy developed during the Soviet era and that most Belarusians who speak Belarusian speak *trasianka*, which is a hybrid of Belarusian and Russian.

The second school of thought on Belarusian nationalism is to develop a Belarusian nation that identifies with Russia and Ukraine and recognizes the importance of the Soviet-era link among the three countries. Elites who espouse tying Belarusian nationalism to the Soviet Union uphold Russian as a language of "culture and civilization" and argue the Belarusian language should be abandoned (Buhr et al. 2011, 428). President Lukashenka contends that the Soviet school of nationalism is realistic in that it acknowledges the political, economic, linguistic, and social ties among Belarus and Ukraine.

The third school of thought on Belarusian nationalism is that the nation has a **Creole** identity or is a hybrid of Belarusian and Soviet experiences and identities. This school recognizes the economic development and the common experience of World War II during the Soviet era. It also "pays homage to Belarusian culture, part of which is language and part of which is a distinct, if difficult to define, sense of 'localness,' as well as a strong stance on Belarusian self-determination" (Buhr et al. 2011, 429). The Creole sense of Belarusian nationalism is gaining popularity because it straightforwardly defends Belarus's independence, recognizes both the Russian and Belarusian languages, and makes sense of the Soviet era and its meaning for Belarusians.

The national identity of Belarus, which Buhr et al. describe as Creole today, is not etched in stone. National identities are constructed, and the parties responsible for the story of the nation change over time. Belarusians elected President Lukashenko in 1994, and he can best be described today as an authoritarian dictator. In a 2004 referendum, Belarusians voted to allow Lukashenko to stay in power beyond the restrictions of the country's constitution. He has centralized control of the state and operates it as a centrally planned economy, much like the Soviet Union. Lukashenko retains close ties with Russia, which helps him stay in power. Polling data in Belarus demonstrate that since 1994, Belarusians have abandoned the idea of restoring the Soviet Union and want to remain independent (Figure 10.33). Lukashenko has used his power to create a government that depends upon and is subservient to him. This government defines the Belarusian nation as independent and yet linked to Russia.

© VASILY FEDOSENKO/Reuters/Corbis.

FIGURE 10.32 Minsk, Belarus. A man stands with the post–1995 Belarusian national flag in Independence Square in front of the government building. During the Soviet era, the Belarus flag looked very similar but included the Soviet hammer and sickle in the upper corner of the red stripe.

Response in Belarus to the Question:
"Would you Like the Soviet Union to be Restored?"

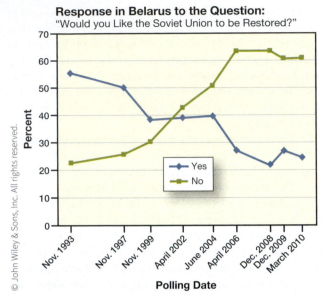

FIGURE 10.33   Polling data from 1993 to 2011 in response to the question "Would you like the Soviet Union to be restored?"

Russia's expansion into Ukraine and annexation of Crimea in 2014 has made it more difficult for Lushenko and Belarus to walk the tightrope between independence and Russia. The Belarusian government first condemned Russia's expansion into Ukraine and then stepped back to give Putin and Russia space to justify their territorial takeover. In fall 2014, a soccer match between Belarus and Ukraine ended with between 15 and 100 fans from both countries being arrested for "mild hooliganism" and other charges when Belarusian fans cheered for Ukraine's independence with a song used to protest Russia's expansion into Crimea. In response, Ukrainian fans cheered "Long Live Belarus" to support the continued independence of their opponent's country. A mixture of both fans then broke into a parody song popular in both countries titled "Putin is a prick." The display by soccer fans at a stadium in Belarus and the arrest of soccer fans by the Belarusian authorities demonstrates the pull Belarus feels between independence and allegiance to Russia.

Nations are constructed identities, and the myriad ethnicities, languages, and religions, the experience of being part of the Soviet Union, the experiences of World War II, the level of economic connections between newly independent states, and the sense of what characteristics make a nation, all factor into how each former SSR defines its nation in the post-Soviet era.

Kazakhstan

Kazakhstan is the only former SSR where the indigenous population, Kazaks, accounts for less than 50 percent of the total population. Stalin's collectivization policy in agriculture resulted in the death of 1.5 million Kazaks who suffered famine between 1930 and 1933 in large part because Kazak farmers rebelled and slaughtered their livestock in protest of collectivization. During and after World War II, Russification was particularly strong in Kazakhstan, with massive migration from Russia and Europe into the SSR as the Soviets relocated industries to Kazakhstan.

Between 1953 and 1965, Khrushchev waged a **Virgin Lands Campaign**, urging Soviets to turn grazing lands into croplands. In Kazakhstan, a traditionally nomadic country where livestock graze on huge expanses of land, the Virgin Lands Campaign affected thousands of people and livestock.

In 1994, Kazakhstan's population was 44 percent Kazak, primarily in the southern provinces; 36 percent Russian, primarily in the northern provinces; 5 percent Ukrainian and 4 percent German, primarily in the city of Almaty. In their 1995 constitution, Kazakhstan declared Kazak and Russian as its official languages (Figure 10.34). With multiple ethnic groups and nations each associating with a certain territory within the country, the newly independent government of Kazakhstan feared the country would split or **balkanize** along national lines.

Kazak President Nazarbayev chose to create a forward capital, moving it from Almaty to the existing town of Akmola (called Tselinograd, which means Virgin Lands City,

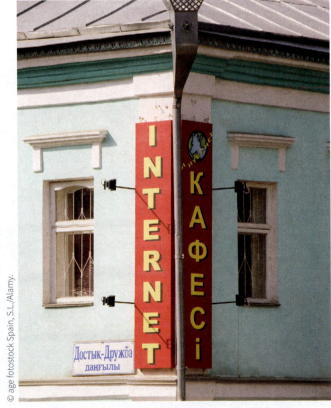

FIGURE 10.34   Uralsk, Kazakhstan.   Kazak and Russian languages are both used on the blue and white street sign attached to this internet cafe in Kazakhstan.

during the Soviet era). He renamed it Astana (meaning capital) in 1997, to help divert an **irredentist movement** in the Russian north.

CONCEPT CHECK

1. **Why** can nations be thought of as constructed identities?

2. **How** did the Soviet era impact identities in North and Central Eurasia?

3. **What** is an irredentist movement, and **how** did Russification (and development through resettlement) later lead to irredentist movements in the region?

WHAT ARE THE SOCIAL AND ECONOMIC LEGACIES OF THE SOVIET ERA?

LEARNING OBJECTIVES

1. **Describe** what collectivization was and how it impacted farming areas in the Soviet Union.

2. **Explain** what oligarchs are and how they gained power in the region after the Soviet Union dissolved.

3. **Compare and contrast** mikroyan and dachas.

Centuries of feudalism, decades of rapid industrialization, and an escalating policy of Russification made Russia ripe for a revolution. The story of the Bolshevik Revolution is difficult to condense, as the revolution actually started in several places and manifested differently among people. Historians refer to the revolution as the Bolshevik Revolution because the leader of the Bolsheviks (one of the revolutionary groups), Vladimir Lenin, became the first ruler of the Soviet Union in 1917, although the country was not officially called the Soviet Union until 1922. Lenin drew his revolutionary fervor and political ideology from the writings of Karl Marx and Frederick Engels.

Lenin gained support for the revolution both from the landless former serfs, many of whom became a rural peasant class in western Russia, and from serfs who migrated to become jobless peasants in urban areas. Even the working-class people who had jobs in Russia's industrial cities quickly became disheartened because of their horrific working conditions, and so many found Lenin's revolutionary ideas attractive.

GEOGRAPHY OF THE BOLSHEVIK REVOLUTION

The geography of the Bolshevik Revolution is complicated by time and space. The revolution really started with peasant unrest in western Russia in 1905, which was followed by years of economic uncertainty. In 1917, two more revolutions, one in February and one in October (this is the one typically called the Bolshevik Revolution), came both from a discontented urban working class and peasants in rural Russia. The industrial working class sought 8-hour workdays and safer working conditions, and peasants in rural areas claimed farmland.

At the same time these revolutions occurred, a class of politicians called the Bolsheviks led political movements against the tsars in cities from western Russia to east of the Ural Mountains. Communicating with the Bolsheviks from a hiding place in Finland, Lenin urged them to overthrow the provisional government in Russia, which led to the October Bolshevik Revolution.

As the dust settled, Lenin became leader of the newly formed Soviet Union, holding the position until his death in 1924. After Lenin's death, Joseph Stalin, who was born in Georgia, became the Soviet leader. Stalin established a powerful and centralized government in the USSR. During his reign, Stalin led purges—systematic efforts to destroy his opposition. Through Stalin's purges and as a result of famine that grew out of his agricultural policies, Stalin's rule was responsible for killing 30 million people. Stalin instituted two economic policies that fundamentally changed the Soviet Union, both of which had long lasting impacts on the region. Starting in 1928, Stalin **collectivized** farms and instituted a five-year plan.

The idea behind collectivization was Marxist: Marx asserted that the working class, the **proletariat**, should wrest control over the means of production from the rich, the **bourgeoisie**. In the process of collectivization in the Ukraine in 1928, Stalin took land from the wealthier, landholding peasants, a group he defined as the Kulaks, and encouraged the very poorest of the peasants to rise up with him in collectivizing lands. The most successful peasant farmers, in areas including the fertile soils of the Ukraine, resisted collectivization. In response to their resistance, Stalin intensified a campaign of terror by killing thousands of Kulaks and successful farmers who resisted and banishing others to Siberia.

While Lenin attempted to squash the Kulaks during his term, Stalin sought to annihilate the entire class of people he defined as Kulaks. The terror employed in the process of collectivization bred a deep-seeded hatred toward Stalin and Russia among peoples in the Ukraine and other areas where collectivization and terror went hand in hand (Figure 10.35). In fall 1932, the Soviet government took all grains produced by peasants and collectives, creating a human-made famine known as the Holodomor in 1932 and 1933. By 1933, Stalin had collectivized 75 percent of the land in the Ukraine, along with prime agricultural lands throughout the Soviet Union, and approximately 7 million had died from the human-made famine in the Ukraine alone.

Stalin used the monies gained by selling grains produced through collectivization to fuel a series of five-year plans beginning in 1928. He created the first five-year plan to reindustrialize the Soviet Union. Through these plans, the government determined what industrial goods would

FIGURE 10.35 **Vilshanka, Ukraine.** This photo from the 1930s shows the district newspaper *Collective Farmers* being produced on the flatbed of a truck in the fields of the "Lenin's Way" collective farm, while workers harvest hay in the background.

be produced and where they would be produced. Stalin found "slave" labor for the newly built Soviet industries by imprisoning Soviet citizens under false pretenses and placing them in **gulags**—prison labor camps. Stalin chose to **site** the gulags in remote regions such as Siberia and the Far North in order to avoid further resistance. Gulag prisoners constructed a major part of the Soviet infrastructure during Stalin's reign, including canals and railroads.

Subsequent five-year plans under Stalin continued both of these programs: industrialization fueled by gulag labor camps and collectivization. Soviet archives report that by 1940, 97 percent of peasant households were collectivized.

CENTRALLY PLANNED ECONOMY

Although the Soviet Union was officially communist and the economy was **centrally planned**, free market economies survived on the street as part of the informal economy. Central planning gave low priority to commercial goods and high priority to security sectors of the economy, including defense and space exploration. High-priority sectors received the natural resources they needed, including energy, and low-priority sectors had to swap, barter, and get by in order to produce goods. Because commercial goods, including cars, appliances, and clothing, received low priority, an underground market economy focused on trading commercial goods from the West grew during the Soviet era.

MARKET ECONOMY

With the transition to a **market economy** during the 1990s, the underground market in Western goods flourished. In **shuttle trading**, Russian traders brought Western goods from Turkey, the United Arab Emirates, and China into Russia and sold them. In 1995, the Russian minister of economics estimated that the shuttle trade brought $11 billion in goods into Russia that year (Cooper 1996). The shuttle trade in the 1990s was part of the informal economy, which means the government did not regulate or tax it. The Russian mafia gained a great deal of wealth by operating a large part of the shuttle trade.

The transition to a market economy in the 1990s also helped to create a number of oligarchs, well-connected businesspeople who own natural resources, media agencies, and industries.

GOVERNMENT CORRUPTION

During the 1990s, President Boris Yeltsin privatized the Russian economy but created several **oligarchs**, elites with political and economic interests in the oil and gas industries as well as the media industries.

In 2000, Russians elected Vladimir Putin president. He served in that post until 2008, when he traded positions with his prime minister, Dmitry Medvedev. Putin then served as prime minister and Medvedev as president until 2012 when Putin ran for and won a third term as president. In 2012, the Russian parliament changed the length of a presidential term to six years, so Putin's term will not be up until 2018.

In 2012, President Putin criticized his own government for corruption in a column published in *Kommersant*, a newspaper owned by a Gazprom executive: "The problem [Putin said] is much more profound [than that of individual corruption]—it comes from the lack of transparency and accountability of government agencies to society" (Zakaria 2012, 21). Putin explained that teenagers now dreamed of becoming state officials so that they could make "fast and easy cash" by receiving bribes (Zakaria 2012, 21).

The Russian Investigations Committee (RIC) is an agent of the Russian government that tracks corruption cases in the country. In the first 9 months of 2012, the RIC received 33,595 complaints about corruption by Russian officials, and 16,603 of the cases were sent to court. In 2011, the RIC sent 11,137 cases to court. Many of the officials charged are high ranking. For example, the former governor of the Tula region is charged with accepting a multimillion dollar bribe. In 2012, the Russian government fired the head of the Russian Space Systems company after Russian auditors realized $217 million in funds designated for Russia's space industry were missing. The government also fired the Russian Defense minister after finding more than $200 million had been embezzled by his subordinates and associates

UIG via Getty Images.

(*Los Angeles Times* 2012). The media also suffers from being agents of propaganda and in some cases of being corrupt. Two of the three largest television channels are owned directly by the Russian government, and the third is owned by Gazprom (BBC 2012).

Internet usage in Russia is relatively unregulated. Opposition political parties use blogs and social media sites to garner support. Blogs are particularly popular in the region and have been since the early 2000s when Russia increased government control over the formal news media. Free blogging sites that enable use of the Cyrillic alphabet, with which Russian is written, enable more than 40 million blogs to operate in Russian. Of the 40 million Russian blogs, less than 10 percent are active with posts more frequently than once a month (BBC 2011). The active blogs have the potential to impact Russian oligarchs because they have an audience of about 15 million people, and popular bloggers have become dependable alternative media sources for their followers.

Corruption levels in North and Central Eurasia rival those in Southeast Asia. Azerbaijan, which is a relatively wealthy oil country, is one of the 40 most corrupt countries in the world. Of the 183 countries ranked by Transparency International for perceived corruption, Azerbaijan is tied with Russia at the rank of 143; the first ranked country, New Zealand, is the least corrupt (see Figure 6.33).

In Russia, government and oil are directly linked, and government has cleared the way for Putin to continue to serve as president. Similarly in a 2009 referendum, Azerbaijan voted to allow President Ilham Aliyev to "run for office as many times as he wants" (BBC 2011 super-rich oil elite, 2). Aliyev's father was president from 1993 to 2003, and Aliyev has served since (Figure 10.36). In 2010, the *Washington Post* reported that President Aliyev's son, who was 11 years old at the time, purchased nine waterfront mansions on the Palm development in Dubai for a whopping $44 million in a two-week period of time (Figure 10.37). President Aliyev's annual salary of $228,000 is "far short of what is needed to buy even the smallest Palm property" (Higgins 2010). Between his son and two daughters, Aliyev's children own more than $75 million in real estate in Dubai.

FIGURE 10.37 Dubai, United Arab Emirates. The Palm Jumeirah, the smallest of the human-made Palm islands off the coast of Dubai, is the location of several properties purchased by the children of Azerbaijani President Aliyev. Mansions on the Palm islands typically sell for between $2 and $8 million. The Aliyev children have purchased mansions in Dubai valued at a total of $75 million.

Sasha Mordovets/Getty Images.

© ALI HAIDER/epa/Corbis.

FIGURE 10.36 Shanghai, China. Azerbaijani President Ilham Aliyev and his wife Leila Aliyeva arrive in style during a welcoming ceremony at an Asian summit in 2014.

RICH–POOR GAP

Since independence in 1991, the wealthiest people in Russia have doubled their wealth, "while almost two-thirds of the population is no better off and the poor are barely half as wealthy as they were" in 1991 (Parfitt 2011, 1). The rise in income inequality in Russia is "eight times more than in Hungary and five times more than in the Czech Republic" (Parfitt 2011, 1). In the post-Soviet period, the best connected members of the Communist Party used their wealth and access to purchase the most valuable industries and mineral and energy resources.

The rich–poor gap is evident in the cultural landscape in urban Russia. Government-owned housing was provided free to citizens during the Soviet era. In the 1950s, the government hastily built five-story-tall apartment buildings to house urban migrants. From the 1960s to 1980s, the norm was to build **mikroyan**, which are tall, prefabricated apartment buildings skirting the periphery of the city and were envisioned to be self-contained neighborhoods. The poor in Russia tend to live in the mikroyan apartments that date to the Soviet era, while the wealthiest live in privately owned houses and apartments in urban areas and spend their weekends in country homes (Figure 10.38a and b).

● Reading the **CULTURAL** Landscape

Housing in Soviet and Russian Styles

Soviet-style mikroyan (Figure 10.38a) were prefabricated apartment buildings designed to house thousands of Soviets. The Soviet government saw housing as a basic necessity that the government should provide, and the mikroyan were the simplest way to achieve this goal. Soviet-era mikroyan are visible in the cultural landscape of countries that were SSRs during the Soviet Union and also in the satellite countries of Eastern Europe, where the Soviet Union played an influential role in the economy and government (see discussion of Prague in Chapter 9).

Since the end of the Soviet era, developers have built apartment buildings primarily in Moscow and other major cities (Figure 10.38b). Luxury apartments in complexes like the Scarlet Sails sell for 750 times the average annual wage of Russians (*New York Times* 2002). Government officials and corporate executives reside in luxury apartment complexes around Moscow. The complexes compete for residents by providing yacht clubs, indoor water parks, and concierge services. Mikroyan are often razed in order to clear land to build luxury housing developments.

© Iain Masterton/Alamy.

FIGURE 10.38a **Kamchatka, Russia.** Soviet-style mikroyan buildings stand in far eastern Russia.

© Serguei Fomine/Global Look/Corbis.

FIGURE 10.38b **Moscow, Russia.** The Scarlet Sails apartment complex is privately owned, and residents have access to a yacht club.

RUSSIAN CITIES

In the United States, where localities compete for tax revenue from businesses and from property taxes, cities have experienced massive sprawl as outlying cities and counties encourage development of residential and business real estate. Cities in eastern Europe and Russia have the imprint of central planning from the Soviet era. Sewers and running water were not readily available in suburbs, so Soviet cities, unlike American cities, experienced little sprawl between World War II and 1991.

During the Soviet era, instead of sprawling suburbs, major cities were satellite cities with 20,000 to 100,000 residents developed that had their own "urban infrastructure, industries, and institutions" (Mason and Nigmatullina 2011, 318). Several satellite cities of Moscow were enveloped into the city in the twentieth century and became suburbs as large-scale urban to rural migration took place.

Before the 1950s, wealthy Soviets often had country homes where they retreated from the cities on weekends and in the summers. The first country homes were typically two stories tall and were originally built on lots between 0.3 and 1.25 acres (Mason and Nigmatullina 2011, 321). The country homes were known as **dachas**, meaning gift.

In the 1950s, the government established collective orchards and gave connected workers in the government and state-owned industries small plots of land on the orchards with the initial goal of encouraging gardening. The government's concept was that Soviet citizens living in cities could go to the country during the summer to grow their own food. Families consumed most of the food they grew, or they used the food in bartering for other goods or services (Round et al. 2010).

In the 1960s, the government loosened restrictions and allowed people to build small cabins on their orchard plots. The people referred to their small cabins without electricity as dachas, even though they did not resemble the original dachas. By the 1980s, 648,000 families in the Moscow region had orchard plots (Mason and Nigmatullina 2011, 321). In the summer months, Muscovites moved to their dachas, typically 5 to 25 miles from the central business district (CBD), and the city was relatively empty.

After independence, the government privatized public housing stock in the cities. Dachas went through a similar process of privatization, and Muscovites purchased private homes in the countryside and in orchard plots that became the suburbs of a sprawling Moscow.

GUEST *Field Note* Elite Landscapes in Moscow

DR. ROBERT J. MASON

Temple University with assistance from Liliya Nigmatullina

As socioeconomic divisions have deepened in recent years, elite landscapes have proliferated in Moscow. In stark contrast with the more egalitarian Soviet model, with its densely populated "satellite cities" of high-rise apartment towers situated in the metropolitan periphery, these new housing estates are exclusive and highly land consumptive. The house pictured here is located in the Rublevo Uspenskoe Highway development, to the west of central Moscow. This is Moscow's most expensive suburban locale, often referred to as the "golden ghetto." Here Moscow's nouveau riche live in homes guarded by fences, walls, and elaborate security systems. Public transportation is eschewed in favor of automobile travel, contributing to Moscow's legendary traffic jams. While the vast majority of Moscow's residents cannot afford to live in such exclusive developments, these landscapes of wealth—with their sprawling homes, clubs, fitness centers, and shopping areas—are a very visible manifestation of the dramatic changes taking place in post-Soviet Moscow.

Courtesy of Robert J. Mason and Liliya Nigmatullina, Temple University.

FIGURE 10.39 Moscow, Russia. Private home in a western suburb.

The suburbs have increasingly become the domain of the elite, with large estates and gated communities (Mason and Nigmatullina 2011, 323) (Figure 10.39). In response to the growth of Moscow suburbs and the growing number of elites living in the suburbs, businesses have moved to the suburbs, which has furthered sprawl in the region.

CONCEPT CHECK

1. **How** did central planning work under the Soviet Union?
2. **Who** are the oligarchs in the region, and **how** did they gain power?
3. **What** was the role of dachas in the Soviet era and what role do dachas play now?

SUMMARY

What Is North and Central Eurasia?

1. During the last ice age, a massive continental glacier covered the western part of North and Central Eurasia, creating the fertile East European Plain. The Ural Mountains are similar in age to the Appalachian Mountains in North America, but the higher latitude of the Urals slows the erosion of the mountains (less frequent freezing and thawing), which is why the Urals are a higher elevation than the Appalachians.

2. The tundra climate dominates the northern part of the region. The tundra and subarctic regions are swampy because rivers flow north to the Arctic Ocean, and in the spring the rivers flood when ice dams remain in the north while the southern ends of the rivers thaw. The soils in the tundra also have a layer that is perennially frozen, called permafrost.

3. The Soviet Union pursued a program of industrialization with little regard for environmental impact. In Central Asia, the Aral Sea shrank and agricultural fields in Uzbekistan became increasingly saline. In Kazakhstan, the Soviet Union conducted 470 nuclear tests between 1949 and 1989, exposing millions of people to nuclear radiation. Nuclear testing ended in 1989 in response to protests by Kazakh activists.

Who Are the People of North and Central Eurasia?

1. The two primary religions in the region are Eastern Orthodox and Islam. Eastern Orthodox Christianity diffused to present-day Ukraine in the tenth century. Centuries later, the tsars established a Russian Patriarch for the Eastern Orthodox Church, establishing the Russian Orthodox Church. The Soviet government espoused an official policy of atheism, but after independence in 1991, the Russian Orthodox Church reactivated and gained followers.

2. Islam diffused into Central Asia between 700 and 900 CE. The sect of Islam found in Central Asia is primarily Sunni. In some parts of Central Asia, people combined traditional religions with Islam, creating syncretic diffusion.

3. Total Fertility Rates (TFRs) in Russia fell after independence. The average life expectancy for men has fallen since 1991. High rates of suicide, alcoholism, and HIV/AIDS have resulted in a life expectancy for men of 60 years, while the life expectancy for women at the same time is 73 years.

4. The population distribution in North and Central Eurasia was shaped by the Soviet policy of development through resettlement. The government moved people to extreme reaches of the country. Still today, the population density in high latitudes is higher in this region than in other of high-latitude areas.

How Have Identities Formed and Been Contested in the Region?

1. An identity is how we define ourselves, how we explain to ourselves and others who we are at a given scale or in a particular context. A nation is an identity, typically constructed by governments, where a group of people see themselves as connected with a common past and common future.

2. The Soviet Union encompassed many countries and nations but sought to connect the people through a policy of Russification. The government moved Russians into other SSRs to spread Russian culture and language through relocation diffusion.

3. After independence, countries had to determine how to set up their governments and the place the Russian language has in their new countries. Belarus has a small Russian population of 13 percent. Belarusians have forged a Creole identity, which is a hybrid of Belarusian and Soviet experiences and identities. In Kazakhstan, Kazaks account for less than 50 percent of the population. Russians are concentrated in northern Kazakhstan, and the government is concerned the country may balkanize, or break up into smaller countries along ethnic lines.

What Are the Social and Economic Legacies of the Soviet Era?

1. Countries in North and Central Eurasia have transitioned from centrally planned economies, where the government owns the land and industries and makes five-year plans for production, to market economies, where privately owned businesses own land and industries and produce goods in response to the market.

2. In the process of moving from centrally planned economies to market economies, many of the government leaders, who were typically members of the Communist Party, invested in the industries and resources being privatized. Those who have gained wealth and power in the process are considered oligarchs.

3. Two visible elements of Soviet city planning in the landscape are mikroyan and dachas. Mikroyan are tall, gray, cinder-block apartment buildings designed to house people efficiently in small apartments. Dachas are suburban community farms designed during the Soviet era so that individuals could keep a garden. Dachas are now being privatized and land has been consolidated into large plots where mansions have been built.

Geography in the Field

Tyumen, a city of 600,000, is the capital city in the Tyumen Oblast (region) of Russia. Tyumen is located on the site of a small trading port on the Tura River tied into a Central Asian trade network. In 1586, Russia annexed the territory, sited the city, and named it Tyumen. Tyumen became an important node in the trade between Central Asia and the Moscow/Volga River region of tsarist Russia. Today, Tyumen is a relatively wealthy city because of the tax revenue generated from massive oil extraction in the region. The Tyumen region produces the majority of all oil in Russia and more than 80 percent of all natural gas in the country. Layers of history are evident in the cultural landscape of the city. New buildings and homes in the central region reflect both the oil wealth and the Soviet era with the mikroyan high-rise apartment buildings in the background.

Thinking Geographically

- The church in this photograph is in the middle of renovation. Based on what you learned in this chapter, why do you think the church is being renovated now?

- Most of the residential buildings in this photo were built since 1991. What do you imagine stood in this area of the city before then?

Read:

Martinez, K. 2012. Tyumen: The quiet powerhouse of Russia. *Moscow Times*. 12 February. http://www.themoscowtimes.com/beyond_moscow/tyumen.html.

Sergei Butorin/Shutterstock.

Tyumen, Russia.   Historically a trading site between Central Asia and Moscow and today the center of a large oil and natural gas production region.

KEY TERMS

insolation	development through resettlement	Balkanization	market economy
tundra		irredentist movement	shuttle trading
boreal forest	Slavic	collectivization	oligarchs
humid continental climate	identity	proletariat	mikroyan
steppe	nation	bourgeoisie	dachas
midlatitude desert	Russification	gulag	
syncretic	creole	site	
Jadidism	Virgin Lands Campaign	centrally planned economy	

CREATIVE AND CRITICAL THINKING QUESTIONS

1. The land in Siberia gets particularly swampy when snow thaws in the spring. Describe how the river **networks** in Siberia work and explain why the directional flow of the rivers causes seasonal flooding.

2. Determine whether you can assess and measure **development** using the same criteria in a centrally planned economy as you do in a market economy.

3. Use ArcGIS Online to look at the Aral Sea and the rivers that flow into it. Describe the **site** of the Aral Sea and explain how the sea is fed. Then, describe how the **situation** of the Aral Sea during the Soviet era caused the sea to diminish.

4. Describe the nuclear testing that occurred in Kazakhstan between 1949 and 1989. Then, explain how the environment in Kazakhstan today is a good example of **anthropocene**. Thinking about the **context** of the Soviet Union, describe how the **identities** of Kazakhs may have changed through the process of protesting nuclear testing in 1989.

5. People in the countries of Central Asia were allowed to practice Islam during the Soviet Union. Imagine how the religious **identities** of Central Asian Muslims may have been affected after independence in 1991.

6. Examine the **population pyramids** of Russia and two other countries in the region (you choose which ones) in 1980 and then in 2010. Describe any differences you see in the population pyramids by place, **gender**, and **context**.

7. The population distribution in North and Central Eurasia was shaped by the Soviet policy of development through resettlement. Using Belarus and Kazakhstan as examples, describe how the national **identities** in these places changed as a result of relocation **diffusion** from Russian **migrants**.

8. Determine whether **unequal exchange** happens only in a market economy or if you can find unequal exchange in a centrally planned economy.

9. Explain how shuttle trading in the 1990s was a process of **globalization** (both in process and in effects) in the region.

10. Compare and contrast mikroyan and dachas. Describe how you can see each in the **cultural landscape** in the region today.

SELF-TEST

1. The East European Plain is flat, fertile land in large part because:
   a. rivers drain north in this region
   b. it was covered by a massive continental glacier in the last ice age
   c. wind-swept loess soil covers the region
   d. the Soviet Union did not centrally plan agriculture in this region

2. The tundra region of North and Central Eurasia is swampy in the spring thaw because:
   a. rivers drain north in this region
   b. it was covered by a massive continental glacier in the last ice age
   c. wind-swept loess soil covers the region
   d. the Soviet Union did not centrally plan agriculture in this region

3. The Soviet Union elevated industrialization over environment. In Kazakhstan, the Soviet Union:
   a. tested more than 400 nuclear weapons, exposing people to radiation
   b. extracted almost all of the oil resources, leaving the country resource-poor
   c. engaged in mountaintop mining to expose and mine rare earth elements
   d. built the Three Gorges Dam across the most Aral River

4. The two primary world religions found in North and Central Asia are:
   a. Buddhism and Hinduism
   b. Hinduism and Islam
   c. Islam and Eastern Orthodox
   d. Eastern Orthodox and Buddhism

5. When the tsars expanded their landholdings and built up Russia, they expanded first to the _____ because the _____ blocked expansion in the east.
   a. north/Chinese
   b. west/Mongols
   c. north/Mongols
   d. west/Chinese

6. How Islam is practiced in Central Asia differs from practice in other areas of the world because:
   a. Islam mixed with indigenous religions when it diffused to Central Asia
   b. Central Asians were cut off from interaction with other non-Soviet countries during the Soviet Union era
   c. The neighboring Muslim country of Iran is Shi'a while the Central Asian countries are Sunni Islam
   d. All of the above

7. Russia, like Europe and Japan, has very low Total Fertility Rates (TFRs). One major difference in population characteristics between Russia and Europe and Japan is that:
   a. Russian TFRs started declining in the 1800s
   b. life expectancies for men in Russia have fallen
   c. HIV/AIDS has not reached Russia
   d. the government of Russia promotes low TFRs

8. The population density is higher at northern latitudes in North and Central Eurasia than at northern latitudes in North

America. The high population density in the north is the result of this policy:

a. one-child policy
b. de-Stalinization
c. development through resettlement
d. all of the above

9. This country has constructed a national identity that is Creole or a hybrid of its own nation and the Soviet identity.

a. Kazakhstan
b. Ukraine
c. Belarus
d. Azerbaijan

10. In the 1990s, shuttle trading was part of the _____ and brought goods from _____ into the region.

a. informal economy/China
b. informal economy/Europe
c. formal economy/China
d. formal economy/Europe

11. The group of people who have become wealthy and powerful during the transition from centrally planned economies to market economies are the:

a. tsars
b. senators
c. oligarchs
d. all of the above

12. The tall, gray cinder-block apartment buildings from the Soviet era are called _____ and since independence have been _____.

a. dachas/torn down
b. mikroyan/torn down
c. dachas/lived in primarily by urban poor
d. mikroyan/lived in primarily by urban poor

13. Mansions and large estates built on the outskirts of Moscow are primarily being built on sites that used to be:

a. mikroyan
b. dachas
c. central business districts
d. none of the above

14. The government of Kazakhstan is concerned that the country may balkanize, which means:

a. return to a centrally planned economy
b. experience a coup that overthrows the government
c. split into smaller countries along ethnic lines
d. have an informal economy that is larger than its formal economy

15. Extracting oil in the Siberian region is environmentally dangerous because Siberia has:

a. permafrost
b. large continental glaciers
c. rivers that run south all the way to the Indian Ocean
d. desert sands

ANSWERS FOR SELF-TEST QUESTIONS

1. b, 2. a, 3. a, 4. c, 5. b, 6. d, 7. b, 8. c, 9. c, 10. b, 11. c, 12. b, 13. b, 14. c, 15. a

NORTH AMERICA

When you imagine Chinatown, you may conjure up a scene from a Jackie Chan movie or a photograph from a family vacation to San Francisco or New York City. This photograph is from British Columbia, on the west coast of Canada, and the picture is not from a Chinatown located in the center of the city; rather, it is from a newly built Asian mall located in a suburb of Vancouver called Richmond.

© Ei Katsumata/Alamy Limited.

Richmond, British Columbia, Canada. The new Chinatown in Richmond, British Columbia, looks different from Chinatowns in nearby Portland, Oregon, and Seattle, Washington. What is distinct about Richmond's Chinatown is that it is in a suburb, not the central city. While Vancouver still has a Chinatown, Richmond has a much larger Chinese-Canadian population than the city. Ethnic neighborhoods in suburbs are a growing trend in North American urban areas and reflect the greater wealth with which some migrants are coming to North America today. This suburban Chinatown includes indoor and strip malls with Asian stores and decorations.

Immigrants in Canada and the United States share a similar history of migrating to North American cities, living in low-rent neighborhoods in the central city, working and earning money, and within three generations moving to neighborhoods in the suburbs.

Since the early 1990s, however, cities and immigration flows have shifted as a result of two trends that have changed the geography of immigrant settlements. First, many neighborhoods near the central city have experienced gentrification or the refurbishing of older neighborhoods, which increases housing values. Gentrification encourages less wealthy immigrants to locate in some older suburbs just outside of the central city where rents are lower. A second trend is that many immigrant groups, including Chinese immigrants to Canada, now come with greater wealth than previous migrant groups, thus enabling them to afford to live in more affluent suburbs. As a result of both of these trends, immigrants are moving to and living in suburbs rather than central cities at an increasing rate.

The North American region has a shared history of immigration, both voluntary and forced. Indigenous peoples throughout the region suffered disease at first contact with Europeans and then lost land when Canada and the United States expanded westward. The shared history of migration and change is reflected in the cultural landscapes of cities in North America.

THRESHOLD CONCEPT in this Chapter

Migration

WHAT IS NORTH AMERICA?

LEARNING OBJECTIVES

1. **Describe** the physical geography of North America.
2. **Explain** how climates are distributed in North America.
3. **Explain** why northern Canada is sparsely populated.

American children grow up singing "America, the Beautiful," proclaiming the country's "purple mountain majesties" and its "shining sea" (Figure 11.1). In Canada, children sing "Oh Canada," describing the northern land as "far and wide." Both of these countries invoke a vision of their land as diverse and grand in scale. Land is an integral part of the national identity, as evidenced by these patriotic songs.

The region of North America includes the second (Canada) and third (United States) largest countries in the world in land area (Figure 11.2). The topography ranges from Death Valley at 282 feet (85 meters) below sea level in California to Mount McKinley at 20,320 feet (6,293 meters) above sea level in Alaska. The entire continent of North America can be divided into three major topographic regions or physical geography landscapes: mountains, the Canadian Shield, and lowlands.

MOUNTAINS

The mountains of North America are found in two belts that are elongated north–south: the Appalachians in the east and the Rockies and Sierra Madre in the west. The Appalachians begin in Alabama and stretch northward across eastern Canada, encompassing Newfoundland. The Appalachians are a much older and more eroded mountain range than the Rockies. The last major tectonic uplift or major mountain-building event in the Appalachians was 350 million years ago, and the last major uplift in the Rockies was 40 million years ago.

Geologists believe that the Appalachians were formed before the North American plate even broke off of Pangaea. The Appalachians formed in two steps. One billion years ago, they were an arc of volcanic islands along a subduction zone in the middle of what became Pangaea.[1] The subduction zone that formed the Appalachians stretched into Africa, and thus the Appalachians in North America and the Atlas Mountains in Morocco were once part of the same mountain chain.

The second step in the formation of the Appalachians created today's distinctive ridge-and-valley topography. 350 million years ago the region folded and faulted after a major uplift. Streams uplifted, eroding limestone and other rocks, which created valleys through the region. The uplift and erosion left behind sandstone and other erosion-resistant rocks, which created the distinctive snake-like ridges and valleys of the Appalachians (Figure 11.3).

In western North America, the Rocky Mountains stretch from British Columbia in Canada to New Mexico in

---

[1] A theory published in 2006 in the journal *Geology* argues that the Appalachians were formed by the process of subduction, but more recently—about 380 million years ago.

FIGURE 11.1   Aurora, Colorado, United States.   Home to the Colorado Rockies major-league baseball team since 1995, Coors Field is sited so that at sunset fans can see the "purple mountain majesties" of the Rocky Mountains, visible just over the stadium beyond third base.

Doug Pensinger/Getty Images.

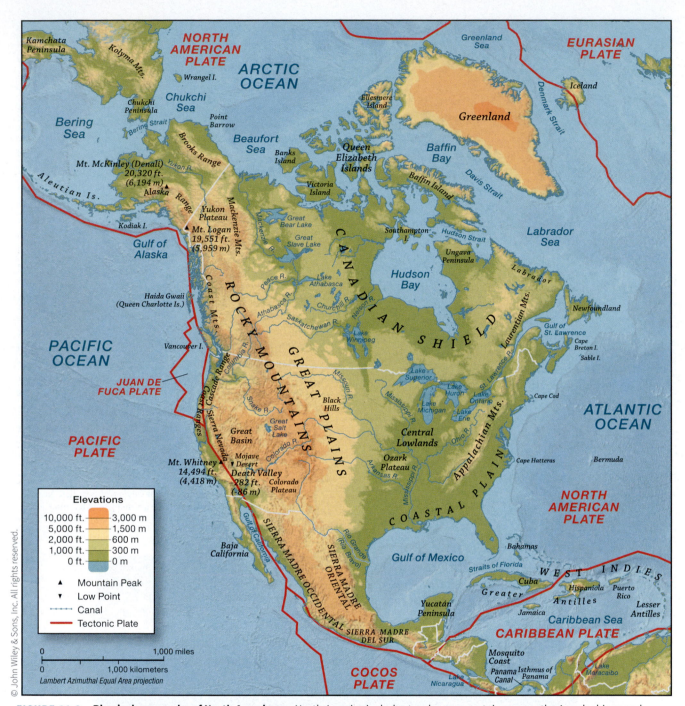

FIGURE 11.2   **Physical geography of North America.**   North America includes two large mountain ranges: the Appalachians and the Rockies (which extend into Mexico as the Sierra Madre). The region is flanked by the Atlantic on the east and the Pacific Ocean on the west, creating thousands of miles of coastal lowlands. Glaciers shaped the interior of the continent from Hudson Bay, through the Canadian Shield, through the Great Lakes and to the northern Great Plains.

the United States. The Rockies also emerged as a result of tectonic activity. Although the Rocky Mountains are hundreds of miles inward from the Pacific Ocean, they were actually created along a subduction zone. In this case, the angle of the subduction zone (the slope between the oceanic and continental plates) was more gradual than in subduction zones in other parts of the Pacific Rim (Figure 11.4). The gradual angle created a volcanic arc, the Rocky Mountains,

much farther inland from the plate boundary than is typical in other areas of the Pacific.

The coastal mountains in California, the Cascades in Washington and Oregon, and the arc of mountains in coastal Alaska were also created by subduction zones. Each of these landforms runs parallel to a plate boundary where an ocean plate is subducting under a continental plate.

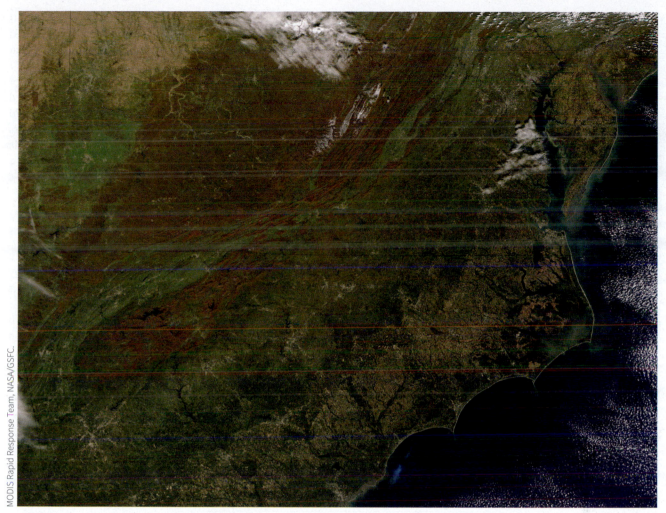

FIGURE 11.3   Appalachian Mountains, United States.   This satellite image of the Appalachian Mountains shows both their relatively low elevations and the snake-like ridges and valleys of the region's oldest mountain range.

The American Southwest and much of northern Mexico are dominated by basin and range topography, where the crust has stretched to about twice its original width. Along the normal faults in the crust, mountains (ranges) uplifted and valleys (basins) dropped downward, generally in a north–south elongation (Figure 11.5). After the faulting, snowmelt and rainfall eroded the mountains over time, creating **mesas** and **plateaus** throughout the region.

CANADIAN SHIELD

The Canadian Shield or the Laurentian Plain is the geologic center of the North American continent and covers

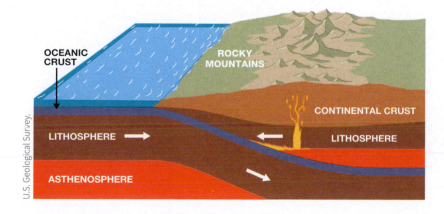

FIGURE 11.4   Subduction and the formation of the Rocky Mountains.   The angle of subduction for the Rocky Mountains is less sharp than the subduction zones in East Asia. As a result, the formation of the mountains occurred farther into the interior of the North American continent than along subduction zones in other parts of the Pacific Rim.

FIGURE 11.5 **Death Valley, California, United States.** Death Valley, the lowest elevation in North America, is a basin that dropped downward when the surrounding ranges uplifted. The straight line that cuts across the bottom of the alluvial fan in the background marks the fault line between the basin and range.

approximately 40 percent of Canada. The ancient rocks of the Canadian Shield metamorphosed into crystalline structures during mountain building approximately 2.5 billion years ago.

During the most recent glaciation, which began 2.6 million years ago and ended about 10,000 years ago (Figure 11.6), enormous continental glaciers scrapped off much of the topsoil from the Canadian Shield area, carried it south to the United States, and deposited it in the Great Plains and Midwest.

When the glaciers fully retreated, they left meltwater in lakes, formed new rivers, saturated aquifers, and filled the massive Hudson Bay. The soils of the Canadian Shield are thin, especially in places where continental glaciers pushed the soil southward. In many places the shield is bare and reveals the paths of glaciers in the **striations**, or elongated cuts, in the bedrock (Figure 11.7).

LOWLANDS

The Great Plains were formed millions of years ago about the same time as the Appalachians, with the melding together of continents to form Pangaea. Later erosion by glacial and fluvial processes gave the Great Plains their distinct, flat landscape.

Coastal plains surround the coastlines of North America. The Coastal Plain is a flat geological region stretching across the Yucatán Peninsula in Mexico, through Texas into Florida and north to New England (Figure 11.8).

The Coastal Plain formed when the North America plate split from the Eurasian plate, along the **mid-oceanic ridge** in the Atlantic Ocean. Deep under the Atlantic Ocean, a rift valley stretches from Iceland in the north to Antarctica in the south. A rift valley is a continental plate

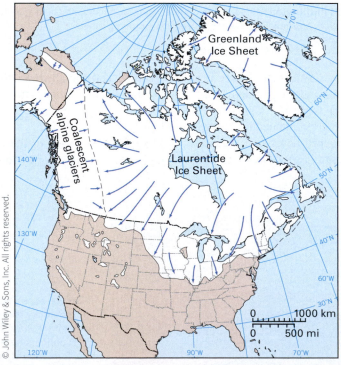

FIGURE 11.6 **Extent of continental glaciation.** During the most recent glaciation, called the Wisconsin Glaciation, which ended approximately 10,000 years ago, massive continental glaciers covered much of North America. The source region for the glaciers was Hudson Bay, Canada.

FIGURE 11.7 **Fox Bay, French River Provincial Park, Ontario, Canada.** Glaciers carve striations when rock frozen into the base of the glacier cut the path of the glacier in the bedrock under the weight and pressure of the glacier.

boundary where the two plates are diverging or splitting apart. As the North American plate actively moved westward, the edge of that plate, the Coastal Plain, trailed inactively behind. Sediment from the Appalachian Mountains covered the Coastal Plain and also settled on the seafloor of the newly forming Atlantic Ocean.

CLIMATE

The climate regions of North America stretch north–south in the western half of the continent and extend east–west in the eastern half of the continent (Figure 11.9). In western North America, the topography (Figure 11.10) greatly influences the climate, and in eastern North America, latitude has a much greater influence on climate.

Western North America

The combination of the Pacific Ocean, the prevailing winds from the west, and the mountain ranges along the western coast of North America creates a rainshadow effect along much of western North America, extending into the interior states of Arizona and New Mexico and into northern Mexico. Moist winds from the Pacific reach the west coast, and the air rises as it travels over the coastal mountains. When the air rises, the air pressure decreases and the air is cooled. When it reaches its dew point, water vapor in the air parcel condenses and

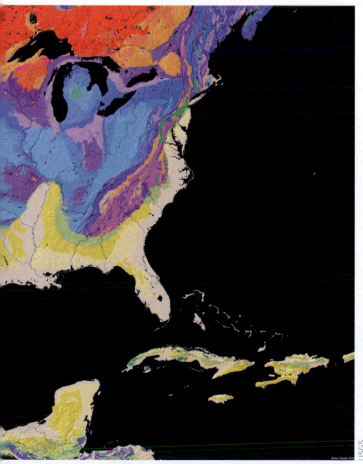

FIGURE 11.8a **Coastal Plain.** The North American coastal plain (the gray area on the image) stretches from the Yucatan Peninsula in Mexico, along the east coast of Mexico, around Texas and Florida, and along the east coast of the United States to Maine.

FIGURE 11.8b **Gulfport, Mississippi, United States.** The coastal plain is defined by flat, sandy beaches.

Reading the **PHYSICAL** Landscape

North American Precipitation Patterns

A major environmental variable that influences human activities in North America is the geographic distribution of precipitation. In some places, like the semiarid and arid western United States and Canada, people need to manage water carefully. In the more humid east, people generally worry less about water.

A physical geographer studies the map of precipitation shown in Figure 11.9a and sees the factors that influence the distribution of precipitation in North America. One factor is the origin of air masses that are drawn to North America as the jet stream moves from west to east across the middle of the continent (Figure 11.9b). The west coasts of Mexico, the United States, and Canada receive abundant precipitation

because high mountains along this margin, including the Sierra Nevada, squeeze water out of moisture-rich air originating from the Pacific Ocean through orographic precipitation. As a result, dense coniferous forests occur on the windward side of the mountains (Figure 11.9c).

In contrast, the northern part of Mexico and most of the American and Canadian west are dry because they lie to the east of the Cascades and Sierra Nevadas. Because air over the continent generally flows from west to east, these areas are in the rainshadow of the mountain ranges to their west. Deserts, including Death Valley and the Sonoran Desert, are found on the leeward side of the coastal mountains (Figure 11.9d).

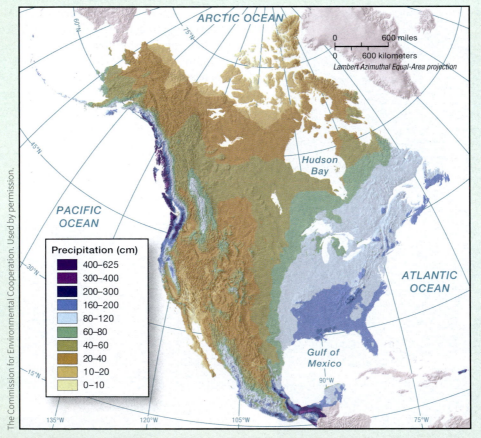

FIGURE 11.9a **Precipitation in North America.** The highest levels of precipitation annually are on the west coast and in the southern United States and Mexico near the Gulf of Mexico.

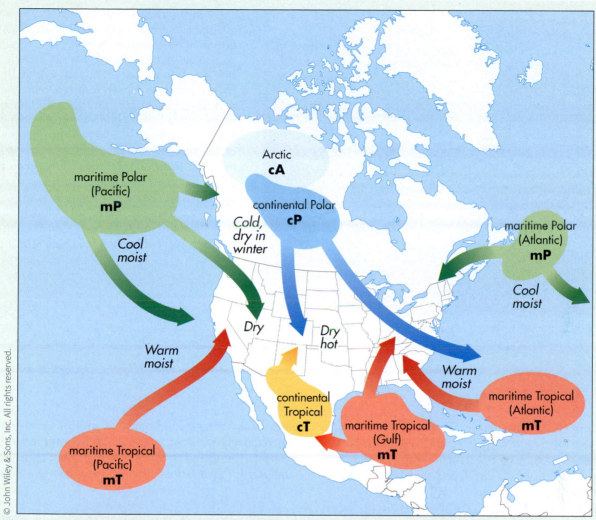

FIGURE 11.9b **North American air masses.** Air masses form where temperatures and moisture levels are relatively stable. Maritime (m) air masses have relatively high water vapor content, while continental (c) air masses are relatively dry. Tropical (T) air masses are warm, polar (P) air masses are cold, and arctic (A) air masses are even colder.

FIGURE 11.9c **Yosemite Valley, California, United States.** The elevation of the Yosemite Valley is well suited for coniferous trees to grow, and the location of the valley on the windward side of the Sierra Nevada Mountains provides plentiful precipitation.

FIGURE 11.9d **Sonoran Desert, Arizona, United States.** Located on the leeward side of the coastal range, the Sonoran Desert is home to the Saguaro cactus, an endemic plant.

(Continued)

The southeastern United States and eastern Canada have relatively high amounts of precipitation because these regions regularly receive air that is laden with water vapor from the Gulf of Mexico and the Atlantic Ocean. These coastal areas are densely forested, mostly with deciduous trees (Figure 11.9e). The central part of the United States, including the Great Plains, lies closer to the Gulf of Mexico and farther east of the rain shadow of the Rocky Mountains. Proximity to the Gulf of Mexico allows for greater precipitation, which enables grasses and crops to grow in the Great Plains in the Midwest.

The northern parts of Canada and Greenland are dry because the air over these areas is very cold and can thus hold little moisture.

Barrett & MacKay/© All Canada Photos/Alamy.

FIGURE 11.9e Capstick, Nova Scotia, Canada. An eastern deciduous forest covers Nova Scotia.

precipitation begins. On the eastern side of the mountains (called the leeward side or the rainshadow), air warms as pressure increases on its descent. Warm air can hold more moisture than cold air. This process is called orographic precipitation (Figure 11.10) (*oros* means mountain in Greek). As air descends on the leeward side, the air warms without any new moisture being added. The warm, dry air flows over the leeward side of the mountains, creating arid and semiarid climates in the west. A rainshadow is part of the process of orographic precipitation.

Along the west coast, a high-pressure system in the north Pacific brings cool but relatively moist air onto the coast. The climates in the west, marine west coast and Mediterranean, receive more precipitation in winter months than in other seasons (Figure 11.11). The intensity in moisture increases in the winter months because the polar jet stream strengthens owing to the extreme temperature difference between the cold northern polar region and the warmer region to the south. The strong jet stream pulls in varying air masses and the fronts that lead them, which creates conditions for greater precipitation in the winter.

Coastal climates are also called maritime because of the year-round influence of the ocean. Ocean water heats and cools much more slowly than land, and as a result oceans maintain relatively constant temperatures over the course of the year. The temperatures in the west coast climate are moderate for their latitude, lacking extreme high and low

© Kevin Ebi/Alamy.

FIGURE 11.10 Mount Baker, Washington, United States. Orographic precipitation occurs when a parcel of air rises over a mountain and cools. On the windward side, the cooling air brings rain, and on the leeward side the air warms and creates a more arid climate.

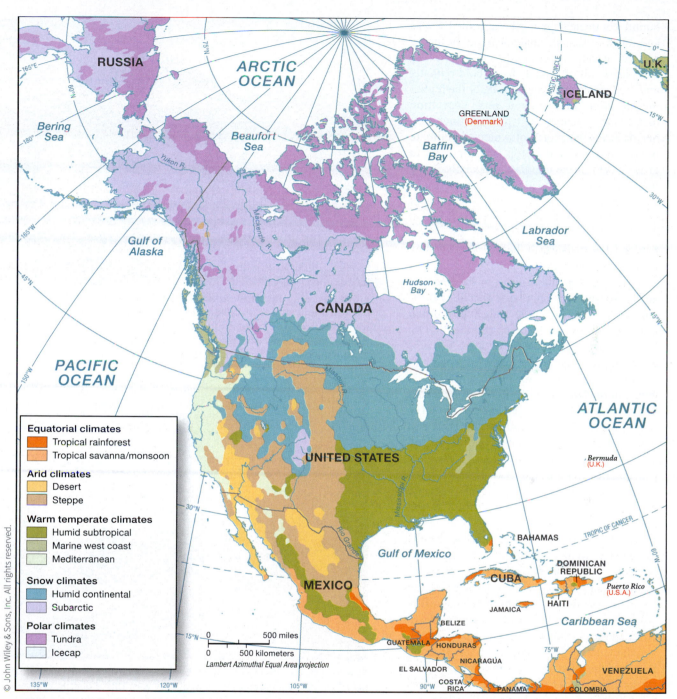

Equatorial climates
- Tropical rainforest
- Tropical savanna/monsoon

Arid climates
- Desert
- Steppe

Warm temperate climates
- Humid subtropical
- Marine west coast
- Mediterranean

Snow climates
- Humid continental
- Subarctic

Polar climates
- Tundra
- Icecap

0   500 miles
0   500 kilometers
Lambert Azimuthal Equal Area projection

FIGURE 11.11   Climates of North America.   Climate regions in eastern North America can be seen in west–east belts on the map, while mountains influence the climatic pattern in western North America.

temperatures, because of their proximity to the relatively warm temperature of the Pacific Ocean. The waters of the Pacific Ocean range, on average, from 46° to 56° F in Seattle, Washington, over the year and the average monthly high temperatures in Seattle range from 46° to 76° F. Farther inland, along the same latitude (47°N) as Seattle, Missoula, Montana, has an average monthly high temperature range of 32° to 84° F because Missoula does

not have the constant warmth of the Pacific to keep the temperature relatively consistent.

Eastern North America

The southern tip of Florida shares a tropical climate designation with neighboring Caribbean islands. The rest of eastern North America is humid subtropical in the south, humid continental in northern United States and much of Canada, and subarctic, or tundra and ice, in Canada's far north.

Nearly all of Canada has a cold climate, and the farther north the latitude or the higher the elevation, the colder it gets. Vegetation in Canada is demarcated by a distinct timber or **tree line**, north of which the climate is too cold and soil is too thin for trees to grow. North of the tree line, especially in the interior parts of Canada, the climate lacks summer warmth, which allows a permanent area of frozen subsoil called permafrost to develop. In areas of permafrost, some soil may exist at the surface, but it is too shallow for trees to take root.

Much of Canada is north of the tree line and is so sparsely populated that it is difficult to describe to people who live in urban areas or even in small towns (Figure 11.12). Gas stations are few and far between because towns are few and far between. For example, driving from Fort Simpson to Fort Providence in the Northwest Territories of Canada is 203 miles (328 km) and includes a ferry ride. Google Maps estimates that the trip takes 5 hours and 12 minutes, but there is no gas station between the two towns. Wrigley, Canada, lies 134 miles (215 km) north of Fort Simpson, and north of Wrigley, the roads are not paved and often not maintained by the government.

In the interior of North America, the climate varies by latitude with the coldest temperatures and greatest temperature ranges in the north and the warmer temperatures in the south. The drainage basin of the Hudson Bay covers one-third of Canada. When massive continental glaciers covered the region, they weighed down and compacted the Hudson Bay area the most, as it was the source region of the glaciers. The rivers and lakes in this region drain into the Hudson Bay because the elevation of Hudson Bay is the lowest in the region.

Courtesy of Dawn Bowen, University of Mary Washington, Fredericksburg, Virginia.

FIGURE 11.12a **Keno Hill, Yukon, Canada.** At the height of the silver boom in the mid-twentieth century, more than 30 mines, employing 800 workers, operated on Keno Hill. Today, the area, including the miner's cabin, is virtually abandoned. The nearby town of Keno thrived with a population of 2,000 in the 1960s but is now home to fewer than 30 people.

Courtesy of Dawn Bowen, University of Mary Washington, Fredericksburg, Virginia.

FIGURE 11.12b **Keno Hill, Yukon, Canada.** Only two silver mines remain open, and vast, empty, areas mark what once was. This photo shows several tailings (waste) piles and a sluice box that once carried water to the mine.

CONCEPT CHECK

1. **Where** are major mountain chains located in North America and why?

2. **Why** is Death Valley at such a low elevation, and why is the climate of Death Valley so dry?

3. **How** are climates distributed in the region?

WHO ARE NORTH AMERICANS?

LO

LEARNING OBJECTIVES

1. **Describe** where and how indigenous peoples have lived in North America.

2. **Explain** why migration is so important to North America.

3. **Discuss** how migration affects the countries of North America differently.

The prevailing theory in the physical and social sciences holds that the first humans lived in Central Africa. From Central Africa, people migrated around the world, first into southwest Asia and then east across Asia. From East Asia, some migrated south to Australia and Melanesia. Others migrated north to China and Siberia and then across the Bering Land Bridge to the Americas. A newer theory follows the initial steps of the first, but proposes that from the east coast of Asia, migrants sailed across the Pacific Ocean to North America, rather than crossing the Bering Land Bridge.

Studies of the indigenous languages of the Americas support the theory that the western part of the Americas was populated before the northwestern part of the continent. The western Americas had a much greater level of linguistic diversity in prehistoric times. This diversity reflects the thousands of years people lived in the region and developed their own languages. Some scholars estimate that the people of the Americas spoke over 2,200 languages before European arrival.

When Europeans arrived in the Americas in 1492, the continents were heavily populated with between 90 and 112 million people. In other words, "in 1491 more people lived in the Americas than in Europe" (Mann 1992). The Spanish priest Bartolome de las Casas described the Indies (the Caribbean) as heavily populated: "All that has been discovered up to the year forty-nine [1549] is full of people, like a hive of bees, so that it seems as though God has placed all, or the greater part of the entire human race in these countries."

European diseases, including smallpox, typhus, diphtheria, and measles, soon decimated the "beehive" of people in the Americas. In Europe, humans had thousands of years of history of interacting closely with animals from the Southwest Asian steppe, including cattle and horses. Asians and Europeans used animals to pull plows and used crops they grew to feed animals. All this interaction with animals exposed Asians and Europeans to animal diseases for thousands of years before contact with the Americas.

When Europeans came to the Americas in the sixteenth and seventeenth centuries, they brought diseases to which indigenous Americans had never previously been exposed. The diseases ravaged the population of the Americas until the early 1900s. Starting at points of first contact on islands and coasts, diseases diffused relocationally with migrants from Europe to the Americas, contagiously within populations, hierarchically across trade centers, and relocationally as migrants moved westward.

INDIGENOUS PEOPLES AND TITLE TO LAND IN THE U.S.

From the time of first contact, Europeans wanted lands from indigenous peoples in the Americas. Before the United States became a country, England and France both entered treaties with east coast American Indian tribes to gain land.

From independence in 1776 until 1871, the United States entered into hundreds of **treaties** to gain additional lands from tribes and expand the new country's territory westward. Reservations established by treaty are typically lands tribes retained after transferring vast tracts of farming, hunting, and gathering lands to the federal government.

After 1871 in the United States, Congress used official decrees, instead of treaties, to establish reservations by transferring land into trust. Despite the shift from treaties to decrees, the United States government has recognized American Indian tribes as sovereign since the country gained its independence. **Sovereignty** is an international legal term that means having the final say over a territory. In the world today, each country recognized as sovereign has authority over a defined territory. In the United States, the Supreme Court recognizes tribes as sovereign over their reservations but has finessed how they interpret tribal sovereignty, now calling tribes "domestic dependent sovereigns."

With more than 500 federally recognized sovereign tribes in the United States today, it is difficult to generalize about the American Indians because some lost all their lands, others suffered forced migration, some live in the poorest places in the United States, and others have newfound wealth through resources or gambling. Each tribe has its own story. Looking at a map of reservations in the United States today (Figure 11.13) reveals very few reservations on the east coast where European contact first happened and larger reservations on the Plains and in the West.

The Navajo Indian Reservation in the American Southwest has the largest number of resident Indians. The Navajo nation includes more than 250,000 members, and about three quarters of the members live on the reservation in small communities. Indian country in Oklahoma is more complicated than it looks on the map. The Choctaw, Chickasaw, Creek, Cherokee, and Seminole were removed from their homes in the east to Oklahoma in the 1800s. Tribal lands were allotted and reservations were dissolved when Oklahoma became a state. Western Oklahoma once had 13 reservations, but they were disestablished through federal allotment and opening of the reservations to non-Indian settlers. The only standing reservation on the map in Oklahoma belongs to the Osage, who preserved subsurface (mineral) rights and still produces oil and natural gas. Despite the loss of federally recognized reservations, tribes still denote the historical boundaries of their reservations throughout the state.

INDIGENOUS PEOPLES IN CANADA

The history of indigenous peoples in Canada is similar to that of American Indians in the United States. Both countries followed policies of land acquisition and cultural assimilation. The Canadian government groups indigenous peoples into First Nations (American Indians) and Inuit (historically called Eskimos). The Canadian government recognizes the sovereignty of the First Nations, but the First Nations have reserves owned by the Canadian government.

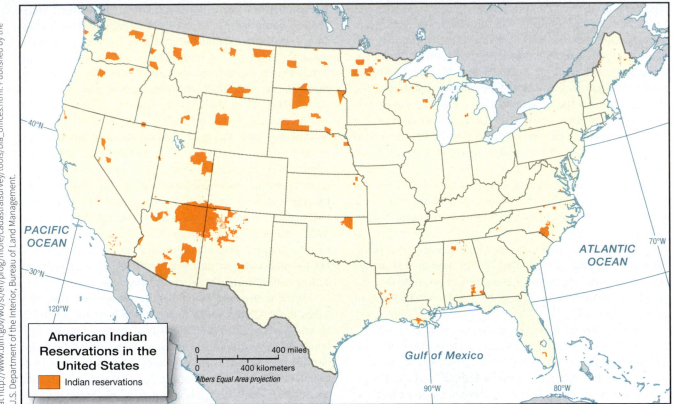

American Indian Reservations in the United States

Indian reservations

PACIFIC OCEAN

ATLANTIC OCEAN

Gulf of Mexico

0   400 miles
0   400 kilometers
Albers Equal Area projection

FIGURE 11.13 **American Indian reservations in the United States.** Reservations in the United States are concentrated in the Great Plains and the West.

The Canadian government defines a reserve as a "tract of land, the legal title to which is vested in Her Majesty, that has been set apart by Her Majesty for the use and benefit of a band." The lack of land ownership makes it difficult for First Nations to establish collateral in order to get loans because the value of homes they own is diminished by their not owning the land.

The map of First Nations reserves in Canada shows that reserves are primarily in the interior of the country, farther from the U.S.–Canadian border than nonindigenous Canadians (Figure 11.14). Globally, about 150,000 Inuit live in Alaska, Greenland, and Canada. In Canada's north, Inuit live primarily along coastlines and north of the Arctic Circle.

Inuit in Canada lived largely outside of Canadian influence until the 1950s and 1960s. The Canadian government believed that the Inuit should "retain their traditional way of life" and therefore generally did not encourage efforts to build missionary schools or fund federal education in northern Canada. Missionaries and American soldiers stationed in northern Canada were alarmed by the low literacy levels among the Inuit and condemned the government's policy of noninvolvement in Inuit education.

In the 1950s and 1960s, Canada encouraged Inuit to move to villages—primarily by providing health care, housing, and education. The Inuit became increasingly sedentary, but the Canadian government still saw the value in teaching Inuit language, survival skills, and stories. One barrier to educating children in their native language was that the Inuktitut language had no written form. In 1970, non-Inuit applied a version of the written form of the Cree (a First Nation located primarily in Quebec) language to create a written form of the Inuktitut language.

Despite the efforts to create a written form of the Inuktitut language and to teach it in schools, preserving the language became increasingly difficult with the introduction of English and French in the region. In 1972, the Canadian Broadcast Company erected a tower in Nunavut and began broadcasting television in English and French. Soon after, usage of the Inuktitut language began to decline.

Fearing a loss of Inuit culture, traditional ways, and language, 56 percent of the people of Canada's Northwest Territories voted in a 1982 plebiscite to divide the Northwest Territories and create a new territory with a majority Inuit population. The **toponym** for the new province is Nunavut, meaning "our land" in the Inuktitut language.

Over 85 percent of the people in Nunavut are Inuit (Figure 11.15), making it Canada's only territory or province with a majority indigenous population. The neighboring Northwest Territories has a majority nonindigenous population, with 52 percent categorizing themselves as nonindigenous on the most recent census. The neighboring Yukon Territory has an even larger nonindigenous population, with 80 percent reporting nonindigenous identities.

FIGURE 11.14 **First Nations reserves and Inuit lands in Canada.** Canada has a higher concentration of First Nation reserves in western provinces. The northern territories have relatively large Inuit populations, with the largest proportion in Nunavut and the Northwest Territories.

The economy of Nunavut is based on mining and oil, fishing and hunting, and ecotourism. Like many American Indian reservations and First Nations reserves, the relatively remote location makes it difficult to attract businesses, and unemployment is high—approximately 30 percent—in Nunavut. Teenagers in Nunavut have much higher rates of substance abuse and suicide than teenagers in the rest of Canada.

FORCED MIGRATION IN THE UNITED STATES AND CANADA

 The history of **migration** in North America includes both forced and voluntary movement. In **forced migration** people have no choice but to relocate. The removal of Indians and First Nations and the Atlantic slave trade are the most notorious examples of forced migration in North America.

---

migration purposeful movement from a home location to a new place with a degree of permanence or intent to stay in the new place.

Forced Removal of Indigenous Peoples

In 1838–1839, tens of thousands of Cherokee died on the forced march ordered by the United States government from the American South (primarily what is now Georgia) westward into Indian Territory (what is now Oklahoma). While the Trail of Tears is the most infamous example, dozens of other Indian tribes experienced forced removal. On the east coast, forced removal often happened after disease had obliterated most of a tribe's population. In the American Southwest, the government forced the Navajo to migrate eastward in 1864. During four years of internment, the Navajo demanded a homeland, and the federal government entered a treaty with the Navajo that recognized the first piece of their current reservation in 1868.

Governments justified the forced removal of tribes by claiming the Indians were "savage" and "didn't use the land." Both of these ideas were justified through a racist argument that European practices were "normal" and more "civilized" than tribal practices. The notion that indigenous populations were "savage" led to many

Percentage of Total Population Identified as Aboriginal and First Nations

Province/Territory	Percent Aboriginal	Percent First Nations
Alberta	6	3.3
British Columbia	5	3.5
Manitoba	15	9.0
New Brunswick	2	1.9
Newfoundland and Labrador	5	1.7
Northwest Territories	50	31.0
Nova Scotia	3	1.9
Nunavut	85	0.3
Ontario	2	1.4
Prince Edward Island	1	1.0
Québec	1	1.0
Saskatchewan	15	10.0
Yukon	25	21.0

Data from Statistics Canada, Census of Population, 2006.

FIGURE 11.15 **Inuit and First Nations proportions of population by province in Canada.**

unsavory and racist policies against indigenous peoples throughout North America. Both the United States and Canada forcibly removed individual members of tribes, primarily children, from their homes and homelands and also forbade indigenous children to speak their native languages in school.

Canada removed First Nations and Inuit children from their homes in a government policy that lasted from 1892 to 1969. The government primarily moved children from their homes on reserves to government-sponsored boarding schools, typically run by Christian churches. According to a study by the Australian government, "Many of the children in [Canada's] residential schools suffered physical and sexual abuse, as well as imposed alienation from families, communities and cultures, which have in turn led to a legacy of abuse and intergenerational trauma."

In the United States, the placement of American Indian children in federally owned boarding schools, beginning in the 1870s, was an integral part of the country's **assimilation** program. The federal government desperately desired that the American Indian learn to be like the "white man" and equated behaving like a "white man" with being "civilized." The United States awarded citizenship to American Indians based on their level of "civilization" as assessed by federal Indian agents on reservations, until 1924, when it extended U.S. citizenship to all American Indians.

The government-sponsored schools in both the United States and Canada not only separated students from their family and their heritage, but teachers and staff at the schools often used severe discipline and were too commonly guilty of physical and sexual abuse of the students. The schools operated under the dictum of Richard Pratt,

who founded the first federal boarding school in the United States; his guiding principle was "Kill the Indian in him, and save the man." Staff bathed Indian children in kerosene, cut their hair, gave them American names, and punished them for speaking in their native languages.

In 1928, the United States government issued the Meriam Report, based on a study of the consequences of assimilationist policies. The report was scathing, and it eventually led to Congress passing the "Indian New Deal" in 1934, which sought to promote the reestablishment of and self-government by tribal governments. In the section on federal boarding schools, the Meriam Report found the conditions deplorable, as the schools were overcrowded, toilets and facilities were not maintained, children were undernourished, and safety was ignored when fire escapes were locked in hopes of maintaining control over the children. The report also found that teachers and administrators inflicted "punishments of the most harmful sort" out of "sheer ignorance" of the "principles of human behavior" and that the amount and kind of work Indian children were forced to do would "constitute a violation of child labor laws in most states."

In 2000, the government of Canada issued a report on residential schools, titled "Restoring Dignity." In the report, members of Canada's First Nations sought "an acknowledgement of the harm done and accountability for that harm; an apology; access to therapy and education; financial compensation; some means of memorializing the experiences of children in institutions; and a commitment to raising public awareness of Institutional child abuse and preventing its recurrence."

Despite policies of land acquisition and assimilation, indigenous peoples maintained their culture and retained land. Indigenous peoples continue to live throughout the continent.

Atlantic Slave Trade

The Atlantic slave trade took place from 1502 until the 1860s, and during that time, somewhere between 12 and 28 million Africans were enslaved and taken from their homes, primarily in West Africa, to the Americas—especially the Caribbean and Brazil (Figure 11.16). The Spanish and Portuguese initiated the trade when they colonized Latin America in the sixteenth century, taking millions of slaves to Brazil and the Caribbean to labor on sugar plantations. The growth of the sugar trade, which was used both as a sweetener and to distill rum, in the seventeenth and eighteenth centuries propelled the slave trade. In the United States, cotton plantation owners in the South sought to acquire enslaved Africans for the labor-intensive industry.

Forced migration had vast impacts on both the regions of origin and the regions of destination. Enslaved Africans brought rice cultivation to the Yazoo Delta region of the American South, and geographers argue that the success of rice cultivation in that region was due to the skills of

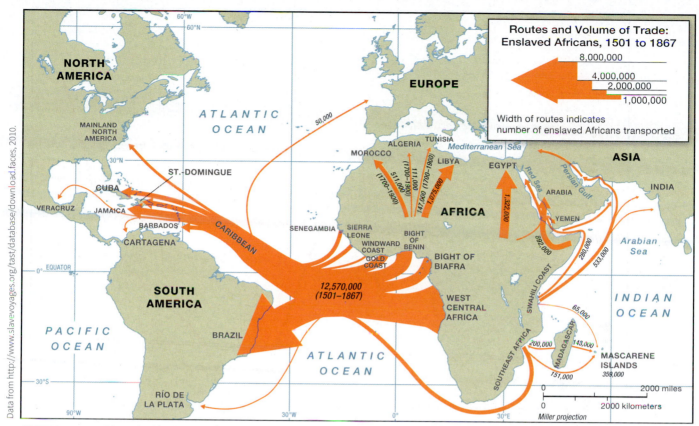

FIGURE 11.16   The Atlantic slave trade.   The majority of enslaved Africans who were forcibly migrated from Africa to the Americas in the 1700s went to the Caribbean where Europeans were intensively growing sugar on plantations.

enslaved Africans. The slave trade and the use of slave labor in plantation economies lay the foundation on which European colonizers grew their economies and established the Industrial Revolution. The sixteenth- and seventeenth-century American economy and the larger Caribbean and Latin American economies could not have grown at such an outstanding pace without the labor of millions of unpaid enslaved Africans.

Aside from these economic and cultural impacts, the Atlantic slave trade exacerbated a system of racism, which was entrenched through economic and political power relationships in the plantation system. People in positions of political and economic power used white supremacy to justify the forced migration and enslavement of millions until the United States outlawed the Atlantic slave trade in 1808 and ended slavery in 1865.

VOLUNTARY MIGRATION TO THE UNITED STATES AND CANADA

Voluntary migration occurs when a migrant weighs **push factors**, which encourage a person to leave his or her home place, and **pull factors**, which help the migrant decide where to go and make a decision to migrate. Voluntary migration is the common thread in the histories of the United States and Canada. From the sixteenth through the nineteenth centuries, immigrants to the United States and

Canada hailed from similar regions of Europe because the United States and Canada have similar colonial histories. Both were colonies of Great Britain, and both had regions that were colonized by France. French colonialism in the United States focused on the Mississippi River, from New Orleans northward. French colonialism in Canada focused on the St. Lawrence River, from its connection to the Atlantic inward to Quebec City.

Migration to the United States

Northern (Scandinavia) and western (Britain, Scotland, Ireland, and France) Europeans comprised the majority of immigrants to the United States from the early 1800s into the early 1900s (Figure 11.17). When the United States government conducted the first census in 1790, two-thirds of the Caucasian population in the new country claimed ancestry from Great Britain, Scotland, Wales, or Ireland. At that same point, about 20 percent of the people counted in the census were from Africa. Of the 757,208 "Blacks" counted in the 1790 census, 697,681 (92 percent) were slaves and 59,527 (8 percent) were free. The majority of immigrants in the peak wave between 1900 and the Great Depression came from southern and eastern Europe, especially the Mediterranean states of Portugal, Spain, Italy, and Greece.

The waves of immigration depicted in Figure 11.17 reflect several processes happening at different scales.

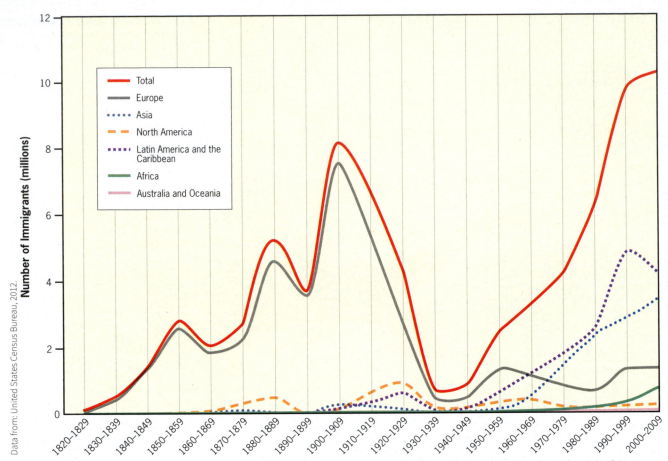

Data from: United States Census Bureau, 2012.

FIGURE 11.17 **Immigration to the United States by region, 1820 to 2010.** Regions of origin for migrants to the United States have changed from before the Great Depression, when most came from Europe, to post–World War II, with an influx of migrants from Asia and Latin America.

When migrants decide to leave their home countries, they decide based on global, regional, and local push factors, including economic growth and depression and civil wars. When they decide where to go, they consider pull factors, such as joining family, finding jobs, or feeling safe, but ultimately the receiving country has the final say as to whether they can legally migrate. All countries determine to whom they legally open their doors.

Between the onset of the Great Depression and the end of World War II, the United States received relatively few immigrants (the low point in the graph). The latest wave of immigration, beginning in the 1950s and stretching to the present, marks a change in the immigration flows in the United States. Europeans have become a relatively small proportion of the immigrant pool, and most legal immigrants now come from either Latin America or Asia.

Migration to Canada

Approximately 15,000 French migrated to Canada in the late 1600s, making an indelible mark on the landscape and laying the foundation for French Canada, a cultural region that is still visible in the Province of Quebec. The formal French influence in Canada lasted from the establishment of a French colony in Canada in 1627 until 1763, when the Treaty of Paris (the treaty ending the Seven Years' War, not the one ending the Revolutionary War in the United States) gave Britain control of all French colonies east of the Mississippi River, including French Quebec. Immigration to Canada peaked in the first decades of the 1900s with the opening and settlement of western Canada. In Canada's 1911 census, 49 percent of its immigrants were from England, Scotland, Wales, and Ireland. In the 1921 census, Canada's population of 8.7 million included 1.9 million migrants, 52 percent of whom were from these countries.

Immigration in Canada tapered off during World War I and dropped precipitously during the Great Depression and World War II. In the post–World War II era, Canadian immigration ebbed and flowed. The regions of origin of Canada's immigrants changed during this period, in large part because Canada changed its immigration laws in the 1960s to allow more non-Europeans to enter the country (Figure 11.18). As a result, Europeans now comprise a smaller portion of the immigration flow, and increasingly immigrants from Asia, Central and South America, and Southwest Asia and North Africa comprise the largest flows into Canada. Canada's relatively open immigration policies are designed to counteract Canada's falling birthrate. As in Europe (see Chapter 9), Canada's Total Fertility Rate (TFR) of 1.68 is below the replacement level of 2.1.

Step and Chain Migration

In both Canada and the United States, much of the voluntary migration follows **step migration**, **chain migration**, or both. Step migration is when a migrant follows a path of a series of stages or steps toward a final destination. Chain migration is when a migrant communicates to family and friends at home, encouraging further migration along the same path, along kinship links.

Geographer Robert Ostergren studied the patterns of European migration into the United States during the 1800s and up to World War I and found repeatedly that Scandinavians followed both step and chain migration into the upper Midwest. New migrants followed the paths of their friends and family in migrating from east coast ports to places in the Midwest and Great Plains. According to Ostergren, the Scandinavian migrants "made their way to the Dakota frontier quite directly or by way of acquaintances that shared their culture, and they settled down among people who had emigrated from the same agricultural regions of the homeland. The potential for cultural maintenance in these settlements was accordingly quite considerable" (1983, 58). In some cases, the Midwest became a step in the migration of Scandinavians to points further west, including Oregon and Washington. Pride in the Scandinavian heritage continues in Scandinavian festivals today (Figure 11.19).

Step and chain migration continue to be part of the present waves of immigration in Canada and the United States. In Canada, gateway cities including Toronto, Montreal, and Vancouver attract the majority of immigrants. Canadian geographers Harald Bauder and Bob Sharpe studied the segregation of visible minorities (most of whom are immigrants) in Canada's gateway cities and found that "space alone cannot explain residential configurations" (2002). Migrants may choose to live with similar migrants, but the options available to migrants on where to settle are determined by many factors over which they have little control. Thus, although voluntary migration is by choice, the range of options regarding where to settle in an urban area is not simply a question of choice.

Immigration to the United States and Canada from Mexico

In the United States, recent migrants from Mexico have followed both chain and step migration. The greatest flow

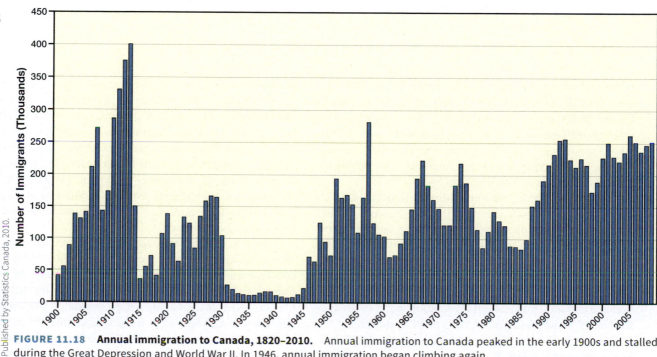

Data for 1871 to 2011 from "Cultural Diversity in Canada: The Social Construction of Racial Differences" by Peter S. Li, page 100. http://www.justice.gc.ca/eng/rp-pr/csj-sjc/jsp-sjp/rp02_8-dr02_8/rp02_8.pdf. Published by the Department of Justice Canada, 2000. Data for 1981–2013 from "Projections of the Diversity of the Canadian Population," by Éric Caron Malenfant, André Lebel and Laurent Martel, page 28, http://www.statcan.gc.ca/pub/91-551-x/91-551-x2010001-eng.pdf. Published by Statistics Canada, 2010.

FIGURE 11.18 **Annual immigration to Canada, 1820–2010.** Annual immigration to Canada peaked in the early 1900s and stalled during the Great Depression and World War II. In 1946, annual immigration began climbing again.

Reading the CULTURAL Landscape

Scandinavian Festival

In 1901, a Danish immigrant who was a real estate developer from Tyler, Minnesota, traveled west to Oregon and purchased a 1,600-acre ranch. He divided the land into small farms of 40 to 60 acres and began selling them to Danish migrants from the Midwest.

After World War II, the federal government concentrated on building an Interstate system to ease the transportation of goods and people. When the Interstate was platted through Oregon, it went around Junction City, and Highway 99, which went through the town, lost traffic to the Interstate. Dr. Gale Fletchall, a resident of Junction City, described the effect of the Interstate routing: "'When that highway [Interstate] opened it was just like you had a shut off valve'" (Scandinavian Festival). The town's economy quickly faltered. Fletchall recalls walking through downtown and seeing empty storefronts and a gutted theater. Fletchall, seeking to find a way to reinvigorate the community, proposed creating a festival to honor the city's heritage.

Every year since 1961, the community has held a Scandinavian Festival featuring foods, dances, plays, art booths, demonstrations, and presentations. Now in a four-day festival held every August, the community

© David L. Moore-OR/Alamy.

FIGURE 11.19   Junction City, Oregon, United States.   Adults and children dressed in traditional clothing enjoy the Scandinavian Festival.

celebrates a different Scandinavian heritage each day: Swedish, Danish, Finnish, and Norwegian. Approximately 100,000 visitors attend each day of the festival hosted by a town of only 4,500 people.

pattern of immigrants from Mexico to the United States is from central western Mexico to rural areas of the western United States to work in agriculture. This flow of immigrants is well established and dates back to the early 1900s when, according to sociologist Elizabeth Fussell, labor recruiters went to Mexico to seek migrants who could help build the railroads in the American West.

When the border between Mexico and the United States changed in 1848 following the Mexican War, it had a long-term impact on the peoples, cultures, and identities of the region. In the short term, people moved relatively freely across the border for decades after its establishment. The United States began a border patrol to control the flow of people in 1904. The border patrol received little attention and had little impact until Prohibition in the 1920s and the United States Immigration Acts of 1921 and 1924.

In 1942, in response to a labor shortage in the United States created by World War II, the American government established a formal policy, the Bracero Program, to encourage the flow of rural Mexicans to farm jobs in the western United States. Even though the program ended in 1964, Fussell argues that Mexican migration to agricultural areas in the United States has actually increased since 1964. In the same time period, Canada began welcoming thousands of legal Mexican immigrants each year, primarily to fill labor needs in agricultural production. The Canadian program encouraging migration of agricultural laborers from Mexico continues today.

In 1993, Canada, Mexico, and the United States signed the North American Free Trade Agreement (NAFTA) to allow for more free trade and flow of goods among the three countries. Since passage of NAFTA, interaction between Mexico and Canada has increased markedly. The ease of migrating to Canada is the main reason why the number of Mexican temporary workers in Canada has increased 68 percent since 1993. Mexicans can enter Canada simply by showing a passport, which enables thousands of Mexicans to enter Canada legally without the hassle of the visa the United States requires.

While NAFTA resulted in greater migration flows between Mexico and Canada, its passage had the opposite effect on flows between Mexico and the United States. Since the 1990s, the United States has ramped up its patrol of the Mexico border. The United States government enacted Operation Hold the Line in 1993 and Operation Gatekeeper in 1994. Both employed more border agents and technology to catch, detain, and deport immigrants crossing illegally. Since September 11, 2001, the border has become all the more tightly controlled, with over $10 billion going to U.S.–Mexico border security each year.

One of the most highly used border crossings between the United States and Mexico is that between San Diego, California, and Tijuana, Mexico. Despite the tightly defended border, the San Diego–Tijuana metropolitan area functions, in many ways, as a single urban system. People, goods, and services flow across the border. Both formal and informal institutions make it easy to move between Mexico and the United States (Fussell 2004). Mexican border residents can obtain border crossing cards that allow them to travel within 25 miles (40 kilometers) of the U.S.–Mexico border for short periods of time in order to visit friends or relatives or to shop. In the other direction, Californians continue to travel frequently to Baja California for "social visits, shopping, or tourism, thus contributing to the social and economic ties binding the two sides of the border together" (Fussell 2004).

LANGUAGE AND RELIGION IN NORTH AMERICA

The languages (Figure 11.20) and religions (Figure 11.21) in North America reflect its colonial histories and forced integration of indigenous and African populations. The French colonial regions of Canada (Quebec) and the United States (Louisiana) have maintained French language and culture, including a strong presence of the Roman Catholic Church. Similarly, Spanish Roman Catholics built mission churches (Figure 11.22) and influenced not only North America's religion but also its linguistic culture. Areas colonized by Great Britain have majority English speakers today, along with a diversity of Protestant religious sects (from Anglican to Southern Baptist).

All of the countries in North America are majority Christian today, with approximately 90 percent of Mexicans, 43 percent of Canadians, and 24 percent of Americans affiliating with the Catholic Church. Protestants compose 5 percent of Mexico, 29 percent of Canada, and 51 percent of America. The United States and Canada are experiencing similar changes in religious affiliations. The proportion of Protestants in both countries is declining, and a growing number of adults in both countries

are choosing a religious affiliation different from their birth religion. The Pew Forum on Religion and Public Life in the United States found in a 2008 study that the country is in the "midst of a period of unprecedented religious fluidity, in which 44 percent of American adults have left the denomination of their childhood for another denomination, another faith, or no faith at all" (Paulson 2008). Approximately 16 percent of Canadians and Americans claim no religion or no affiliation with a particular religion.

The diversity of religions in Canada and the United States has increased as immigration from non-European regions has risen. The number of Canadians espousing Islam, Hinduism, Sikhism, and Buddhism is on the rise. In the United States, Jews account for 1.7 percent of all Americans, and Buddhists, Muslims and Hindus still fall below 1 percent each in proportion of adherents. Most of Latin America is Christian, mainly Catholic, and since so many immigrants to the United States since World War II have hailed from Latin America, the proportion of Christians in the United States remains strong.

Migrants from western and northern Europe were largely Protestant because the Protestant Reformation diffused most strongly to these areas of Europe (except for France and Ireland) in the sixteenth and seventeenth centuries (see Chapter 9). While the United Kingdom became Anglican during the Protestant Reformation, the Anglican religion is not the prevailing religion either in the thirteen original colonies of the United States or the Hudson Bay region of Canada. The pull of free religious expression attracted migrants to the United States. For example, Baptists were not allowed to practice their religion in England in the 1600s, so they came to the colonies looking for religious freedom. They traveled first to New England and then to the Southern colonies in the late 1600s, bringing the Baptist faith so strongly associated with the South today to the region through relocation diffusion.

LASTING IMPACTS OF COLONIALISM IN CANADA

France controlled the colony of Quebec from 1627 to 1763. Despite its relatively short period of formal control, the province is steeped in French culture. The French Canadians, known as the Québécois, have worked to maintain their culture. Starting with the 1774 the Quebec Act, Britain recognized Quebec as being a French-language and Roman Catholic region. After Canadian independence, Québécois have sought continued government recognition of their French identity.

During the last few decades of the twentieth century, the Quebec separatist party campaigned for secession from Canada and nearly succeeded. Through a series of provincial laws, including Law 101 passed in 1977, the Québécois have firmly established the French Canadian language in business and education. The education laws

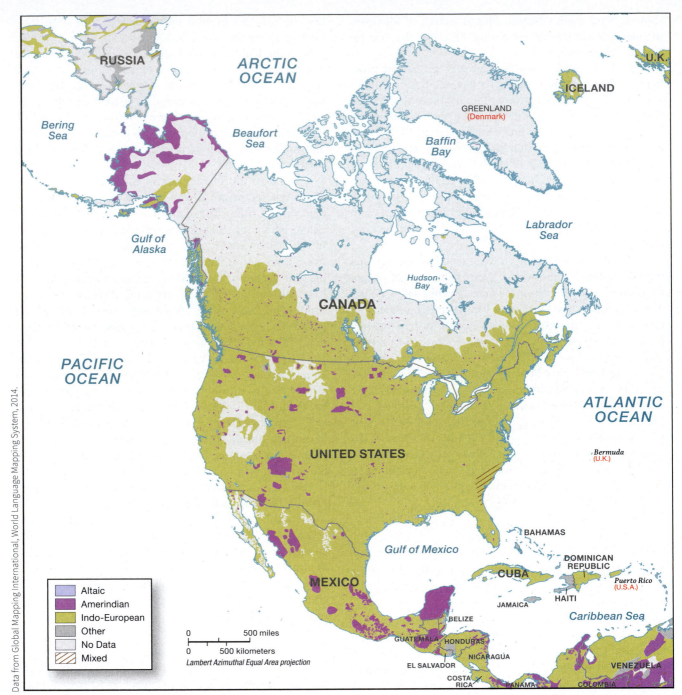

Data from Global Mapping International, World Language Mapping System, 2014.

Legend:
- Altaic
- Amerindian
- Indo-European
- Other
- No Data
- Mixed

0 — 500 miles
0 — 500 kilometers
Lambert Azimuthal Equal Area projection

FIGURE 11.20 **Languages of North America.** Indo-European languages are found throughout the region, with English being the primary language in both the United States and Canada.

in the province require that immigrants who come to Quebec be matriculated into French-speaking schools unless they can demonstrate that they came from English-speaking schools in their home country. In 1980 and again in 1995, people in Quebec voted on secession. The separatists lost both votes, but only by 1 percent in 1995. Although the province remains part of Canada, the Québécois continue to successfully protect French Canadian culture.

Demands for a separate Quebec have dissipated since the late 1990s in part because the government of Canada continues to recognize Quebec's right to be culturally distinct and yet remain part of Canada. Prime Minister Stephen Harper in 2006 proposed that Canada recognize Quebec as a nation within Canada. In the same year, the Parliament of Canada officially did so. Despite the success of the Parti Québécois, the separatist party, in 2012 provincial elections, the call for secession has lagged.

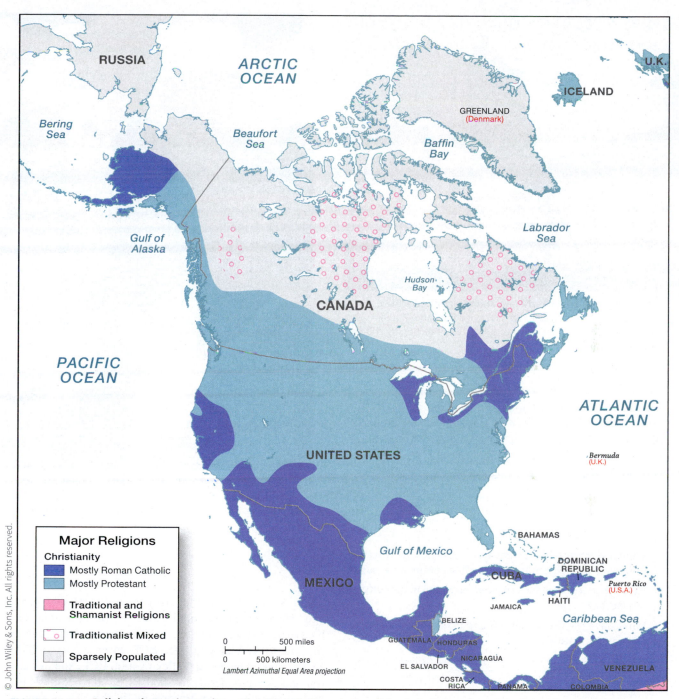

Major Religions

Christianity

- Mostly Roman Catholic
- Mostly Protestant

Traditional and Shamanist Religions

- Traditionalist Mixed

Sparsely Populated

0   500 miles
0   500 kilometers
Lambert Azimuthal Equal Area projection

FIGURE 11.21 **Religions in North America.** Christianity is the majority religion in both Canada and the United States. Catholics are concentrated in the St. Lawrence River Valley in Canada and in the Great Lakes and New England areas of the United States.

Lasting British Influence in Canada

British formal influence lasted two centuries longer in Canada than in the United States. In 1867, the British North America Act united Ontario, Quebec, Nova Scotia, and New Brunswick as a Canadian dominion within the United Kingdom. This status recognized Canada as an independent kingdom within the United Kingdom. Canada recognized the monarchy of Great Britain as the monarch of Canada. Canada grew in size over the next century, gaining increasing independence through acts passed in both Canada and Great Britain.

In 1982, Canada adopted a constitution, which acknowledges the close historical ties to the United Kingdom by recognizing the British crown as the sovereign of Canada. In these terms, according to the Canadian government, "the sovereign personifies the state and is the

FIGURE 11.22   Santa Barbara, California, United States. Franciscan Friar Fermin de Lasuen founded the Santa Barbara mission in 1786. The mission still operates as a functioning church and retreat center.

FIGURE 11.23   Calgary, Alberta, Canada. During their first months of marriage, the Duke and Duchess of Cambridge, William and Catherine, visited Canada as one of their first official duties.

personal symbol of allegiance, unity and authority for all Canadians." The Queen of England is not in charge of Canada, but the Canadian constitution recognizes the British monarchy as a symbol of the Canadian state (Figure 11.23).

CONCEPT CHECK

1. **What** evidence is there that North America was heavily populated by indigenous peoples when Europeans arrived?

2. **How** can you see evidence of step migration in the cultural landscape?

3. **What** have the Québécois done to maintain their French Canadian identity?

HOW ARE CITIES IN NORTH AMERICA ORGANIZED?

LEARNING OBJECTIVES

1. **Explain** why North American cities are located where they are.

2. **Describe** how North American cities have grown over time.

3. **Discuss** what urban sprawl is and how cities use planning to avoid it.

The first urban revolution began around 10,000 years ago when humans created cities in Mesopotamia and the Fertile Crescent region. Archaeologists date cities in the Nile River Valley to 5000 BCE, in East Asia to 2000 BCE, in the Indus River Valley to 4000 BCE, and in Central and South America to 2000 BCE. North Americans built larger cities than Europeans, in the Mayan region of Mexico (from 250 to 1500 CE) and in Cahokia, near present-day St. Louis (from 700 to 1400 CE). In several places in North America, including the Ohio River valley and the Rio Grande region, indigenous peoples lived in cities and farmed surrounding lands. However, none of these cities rivaled the size of the Maya and Cahokia cities (Figure 11.24).

The Cahokia cities declined before European contact. Archaeologists are not certain why. But the history of European contact with the Maya is relatively well documented (Figure 11.25). Succumbing to European diseases, Maya civilization broke down, and cities became historical markers of the empire.

CITY SITES

From 1500 on, new cities in North America were tied to the growth in trade on the Atlantic Ocean. North American cities were sited on rivers flowing into the Atlantic Ocean (Figure 11.26). Washington, DC, Baltimore, and Philadelphia are all on the **fall line**, which is the point along coastal rivers where rock softens and waterfalls form on the river. Below the falls, closer to the ocean, the rivers are navigable. Settlers often sited cities at or near these points to access coastal trade and ease inland trade. In the interior of the country, new cities were sited on waterways and connected by canals and later by railways to ease inland trade of agricultural products. Entrepreneurial founders sited other cities away from waterways for particular resources, whether the fur trade, coal, or gold.

Despite North America's vast tracts of farm and range land, the majority of North Americans live in urban areas (75 percent in Mexico, 79 percent in the United States, and 80 percent in Canada) (Figure 11.27). The population of North America clusters in cities for employment opportunities, cultural diversity, and a wide range of services and activities. The number of opportunities in cities is amplified because cities attract political processes, corporate headquarters, and cultural centers, thereby strengthening the pull to urban locations.

Today, 20 percent of the entire population of the United States and Canada lives in a relatively small area of land along a chain of urban areas from Washington, DC, north through Baltimore to Philadelphia to New York City and then to Boston. This urban agglomeration, tied together by Interstate 95, is called **Megalopolis**. Along the same lines, 18 million people (51 percent of Canada's population in the last census) live in a series of Canadian cities, including Windsor, Toronto, Ottawa, Montreal, and Quebec City.

The statistic that more than 75 percent of North Americans live in urban areas is somewhat misleading if we envision "urban" as "in the city." Most of the people living in urban areas of North America are not living in central cities (downtowns). Rather, the majority of people in the United States who live in metropolitan areas actually reside in suburbs. In fact, by the 2000 census, more than half of the United States population lived in suburbs (US Census, 2002).

In Canada, Montreal (Quebec), Toronto (Ontario), and Vancouver (British Columbia) are the three largest population centers. Each of these cities and their suburbs revealed the same trend in the 2006 census of suburban growth and downtown decline. The map of Montreal (Figure 11.28) shows the percent of the population under 14 years of age in the metropolitan area. A ring of young people in the suburbs surrounds an older central city. On the island of Montreal in the central city, only 1.2 percent of the population is under 14 years of age, whereas in parts of the suburbs, over 30 percent of the people are less than 14 years old (Census Canada 2006).

(a)

Source: Wood Ronsaville Harlin Inc./National Geographic Creative. Used by permission.

© Carver Mostardi/Alamy.

(b)

FIGURE 11.24a AND b   Cahokia Mounds, Illinois, United States. (a) Monks Mound is the largest human-made earthen mound in the United States. (b) This schematic shows the best estimate archaeologists have of what the Cahokia Indian City looked like at its height. Monks Mound is located near the center and toward the top of the diagram.

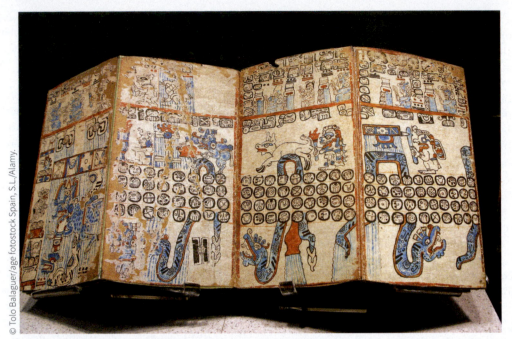

FIGURE 11.25 **National Museum of Anthropology, Mexico.** Grolier codex from the Maya civilization.

The government of Mexico has focused many of its economic development and growth policies on the Mexico City metropolitan area. Between 1959 and 1998, on average, 93 percent of all federal government investment in Mexico was concentrated on the greater Mexico City area (Aguilar and Alvarado 2008)[2], with much of the money dedicated to building Mexico City's road and

[2] The greater Mexico City area includes "the Mexico City Metropolitan Area, the Federal District, and seven metropolitan municipalities" (Aguilar and Alvarado 2008, 188).

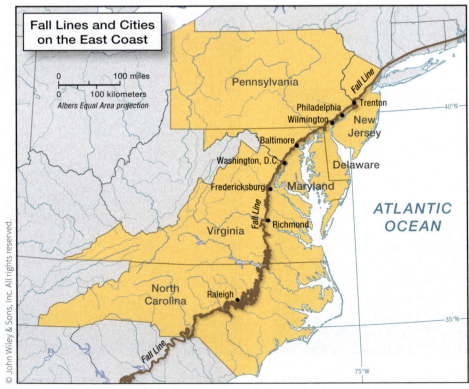

FIGURE 11.26 **Fall lines and cities on the east coast.** Major cities are located on rivers along the east coast, and trading ports that are now often suburbs of major cities were sited along the fall line of the rivers.

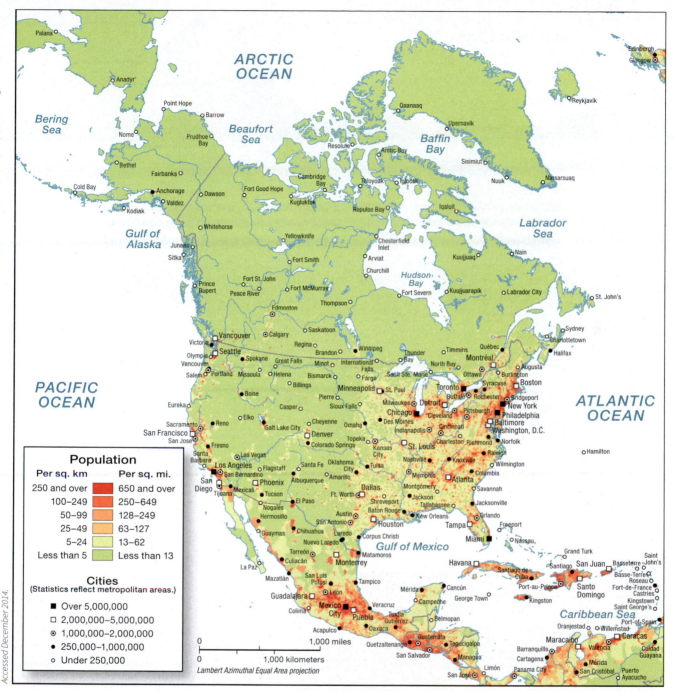

Data from CIESIN—Columbia University; United Nations Food and Agriculture Programme—FAO, and Centro Internacional de Agricultura Tropical—CIAT. 2005. Gridded Population of the World, Version 3 (GPWv3): Population Count Grid, Future Estimates. Palisades, NY: NASA Socioeconomic Data and Applications Center (SEDAC). http://dx.doi.org/10.7927/H42B8VZZ. Accessed December 2014.

Population

Per sq. km	Per sq. mi.
250 and over	650 and over
100–249	250–649
50–99	128–249
25–49	63–127
5–24	13–62
Less than 5	Less than 13

Cities
(Statistics reflect metropolitan areas.)

■ Over 5,000,000
□ 2,000,000–5,000,000
◉ 1,000,000–2,000,000
● 250,000–1,000,000
○ Under 250,000

0 — 1,000 miles
0 — 1,000 kilometers
Lambert Azimuthal Equal Area projection

FIGURE 11.27 **Population density of North America.** About 20 percent of the population of the United and Canada lives in the Megalopolis area of the United States and the stretch of Canada connecting Toronto, Ottawa, Montreal, and Quebec City.

transportation network. As a result, the city has spread outwards, and middle- and upper-class Mexicans have left central cities and moved to the sprawling suburbs. Part of the desire to leave the central city stems from the incredibly high air pollution rates in Mexico City (Figure 11.29). One newspaper reported, "The commuters are bringing broad changes to the face of the countryside for 100 miles on all sides of the capital. Land prices throughout the suburban region are rising fast, and the few patches of farm and forest land that remain near major highways

are quickly being bulldozed for strip malls and housing developments" (Dillon 1999).

MODELS OF NORTH AMERICAN CITIES

Suburban growth in Canada, the United States, and Mexico is just one signal of major changes in the geography of the North American city since World War II. Before World War II and before the automobile became commonplace on city streets, the typical North American city was laid out in

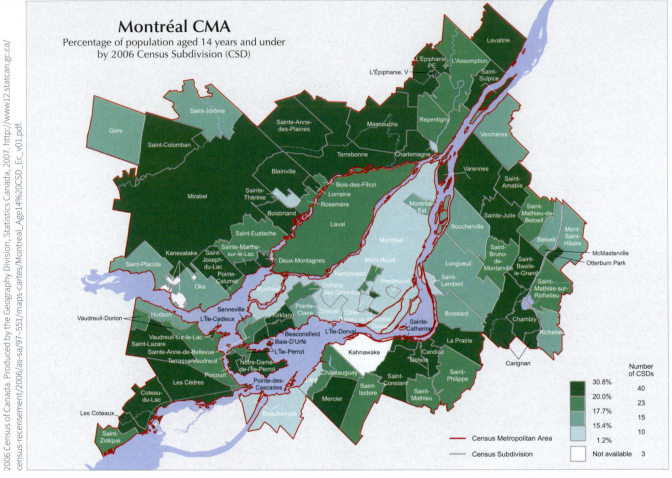

Montréal CMA
Percentage of population aged 14 years and under by 2006 Census Subdivision (CSD)

	Percentage	Number of CSDs
	30.8%	40
	20.0%	23
	17.7%	15
	15.4%	10
	1.2%	
⎯⎯ Census Metropolitan Area		
⎯ ⎯ Census Subdivision		
Not available		3

FIGURE 11.28   Average age of population in Montreal.   The central part of the city is populated primarily by older people, and the younger people in the city tend to be concentrated in a ring of suburbs around the city.

FIGURE 11.29   Mexico City, Mexico.   Pollution blankets Mexico City, with Ixtcihuatl in the background.

a *concentric zone model* or *sector model* (Figure 11.30). In both models, the **central business district (CBD)** is located at the center of the city, and the city is divided into zones of functions around the CBD. The wealthy lived in zones separate from those with lower incomes, and the lower income zones were close to the industrial zones of the city.

The **urban morphology** of the city was set up to serve the CBD. Because jobs were located in the CBD, transportation networks, including commuter rail lines and interstates, connected the outskirts of the city to the CBD. After World War II, with the growth of the automobile and the return of millions of veterans who qualified for Veterans Affairs (VA) loans, developers built tract homes in new suburbs around cities. Between 1944 and 1956, according to the Department of Veterans Affairs, the government guaranteed 5.9 million home loans at a value of $50.1 billion for

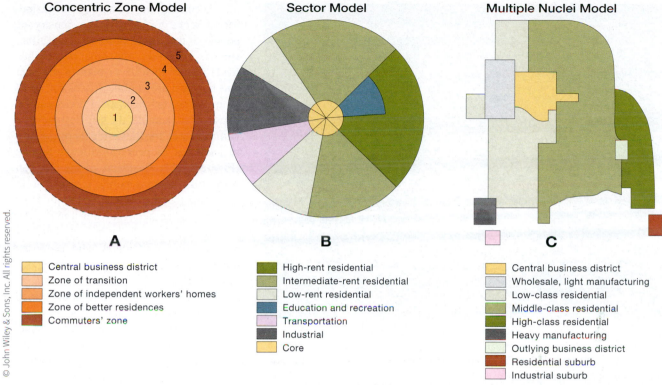

FIGURE 11.30 **Concentric zone, Sector, and Multiple Nuclei Models.** Concentric Zone, Sector, and Multiple Nuclei Models are three ways geographers have modeled the development and layout of North American cities.

returning soldiers. With the expansion of single-family homes afforded by support for veterans and the economic boom after World War II, the model of the North American city no longer looked like the concentric zone or sector model, and in the 1940s the *multiple nuclei model* was proposed.

Increasing use of the automobile and establishment of interstates through and around cities increased the central role of suburbs in the North American economy and encouraged the foundation and expansion of **edge cities**, which house office complexes, shopping malls, hotels, restaurants, entertainment venues, and sports complexes (see Figure 11.31).

With the growth of edge cities, commuters often travel from the suburb where they live to the suburb where they work, often an edge city. Another trend in North American cities is to rebuild or renovate older areas of the city that were low income, thereby increasing the housing values of the neighborhoods in a process called **gentrification**. A rise in occupied housing rates and in housing values in the center city does not necessarily attract jobs to the city. When city residents commute to edge cities for work, they create a new flow of traffic called a reverse commute.

One growing challenge of this new **functional zonation** for North American cities and suburbs is providing

transportation and services (such as police, fire, and ambulance) as new housing developments, jobs, and shipping districts continually change commuting flows.

URBAN SPRAWL

As the suburban population in the region increases, environmentalists and planners have become more concerned about **urban sprawl** and the amount of land that is becoming urbanized, suburbanized, or asphalted for infrastructure, housing, and business. Instead of buying an apartment in downtown Charlotte, North Carolina, a couple may choose to buy a house in a low-density suburb. The problem with this choice, according to anti-sprawl activists and planners, is the loss of farmland, the environmental impact of new housing, and the increase in pollution because of the greater commuting distances for the couple.

Portland, Oregon, is working to contain sprawl and is recognized as one of the world's most innovative and sustainable cities. Two resettled New Englanders founded the city of Portland in 1845 and named it after Portland, Maine. In 1979, Portland created an urban growth boundary, prohibiting expansion of the city beyond the boundary.

According to *Planning* magazine, Metro seeks to increase the density of people in the city in order to avoid urban sprawl. To do so, the organization has made it easy to walk or bike in the area and has designed public spaces for common use so that residential areas can be denser. For example, the city designated a common area for the residents of the city to gather at Pioneer Courthouse Square (Figure 11.32). Residents of the city can walk, bike, or take public transportation and come together to visit and enjoy their city and each other. Also downtown is the Tom McCall Waterfront Park, named after the governor of Oregon who proposed comprehensive planning for Oregon's cities in 1969.

One major way Portland has decreased sprawl is by encouraging public transportation. Portland has built a streetcar line, established bike lanes, and improved bus accessibility and routes (Figure 11.33). Portlanders are encouraged to hop onto a bus rather than into a car.

© Iain Davidson Photographic/Alamy.

FIGURE 11.31 **Scottsdale, Arizona, United States.** Scottsdale developed into an edge city after Joel Garreau wrote the book Edge Cities in 1991 (garreau.com). The Kierland area was built near the Scottsdale airport (to the left of the runway in this photo). Kierland Commons and neighboring Scottsdale Quarter are the central shopping districts in the new edge city, featuring Tommy Bahama, Nike, and an Apple Store.

Since the 1970s, Portland has embarked on a major community planning initiative to help shape and reshape the city's urban morphology. The city formed a regional planning organization called Metro whose goal is to encourage rebuilding and redevelopment of the area inside the urban growth boundary. The Metro region includes 24 cities that stretch across three counties. Metro is responsible for comprehensive, long-range planning and community planning in the region. Metro encourages public transportation to increase connectivity and to decrease automobile emissions.

CONCEPT CHECK

1. **Why** are the cities of Megalopolis sited where they are?

2. **How** do models of North American cities reflect the way cities in the region have grown over time?

3. **What** is urban sprawl, and **what** is Portland, Oregon, doing to try to avoid it?

Zack Frank/Shutterstock.

FIGURE 11.32 **Portland, Oregon, United States.** Designed as a common area for Portland residents, the Pioneer Courthouse Square is located in the middle of the redesigned downtown Portland.

USING GEOGRAPHIC TOOLS

GIS and Planning Portland

The Metro Regional Planning Organization and the city of Portland, Oregon, have used geographic information systems (GIS) as a tool to effectively plan development, track land use, and contain urban sprawl since 1991. The Metro GIS is a regional land information system (RLIS) that began in 1989. The first step in building the RLIS was to draw a basemap of tax lots built off a utility company's computer-aided design (CAD) maps. It took a subcontractor 16 months to digitally trace and geo-reference the tax lot map and bring it into an operating GIS. In 1991, Metro launched its GIS system, working with governments, emergency management systems, and private businesses to build layers of information and maintain the integrity of the data. Data are shared in both directions, from Metro to other agencies and businesses and back to Metro. Metro is better able to

"monitor land development and future growth capacity" by using aerial photographs to annually analyze the tax lots and determine whether they are "vacant, partly, vacant, or developed" (Luccio 2009).

Thinking Geographically

1. Why is the city of Portland concerned with tracking the development of tax lots?

2. If the Metro area did not have GIS, how would that inhibit the organization's ability to plan urban growth and control urban sprawl?

Sources: Luccio, Matteo. "Portland, Oregon, Trailblazes a Successful Regional GIS," *ArcNews*, Winter 2009/2010.

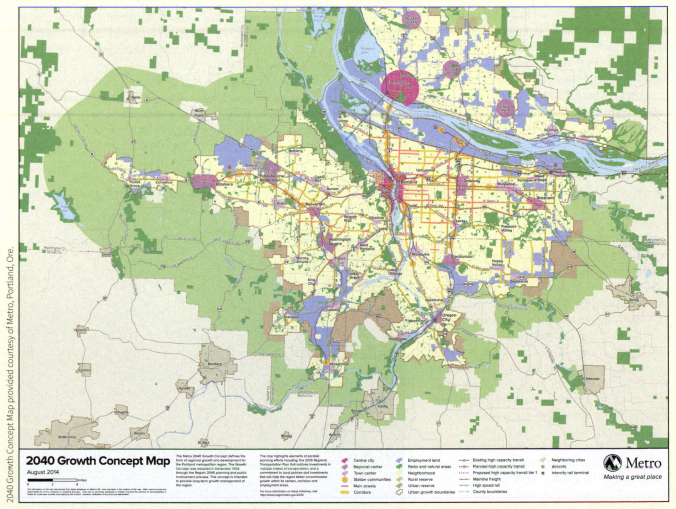

2040 Growth Concept Map provided courtesy of Metro, Portland, Ore.

FIGURE 11.33 Portland, Oregon, United States. Produced by the Metro Regional Planning Organization, this map charts a path for Portland's growth until 2040.

WHAT ROLE DO RESOURCES PLAY IN THE ECONOMIC GEOGRAPHY OF NORTH AMERICA?

AGRICULTURE

The greatest resource of North America is land, specifically land well suited for agriculture. The same glaciers that flattened the plains, formed the Hudson Bay, and scraped the soils from the Canadian Shield helped establish soils well suited for agriculture and filled the Ogallala Aquifer with water in the Great Plains and Midwest of North America. The agricultural lands in North America are some of the most productive in the world.

Agricultural production in Canada and the United States incorporates innovative technologies. In the American South, cotton is harvested using high-tech cotton-picking machines that separate the cotton seed from the burr, a process previously done by hand. Once cotton is stripped by machine and baled, bales of cotton are stored in modules on the fields until the farmer is ready to sell them.

In the Midwest, corn and soybeans are grown on large expanses of land (Figure 11.34). The United States Farm Program financially supports production of corn and soybeans, which encourages farmers to grow bumper crops of each, annually. In North America, corn is harvested using massive combines (costing more than $500,000) that cut stalks at the header, separate the ear from the stalk during threshing, shuck the corn kernels off the stock and then lift the corn to the trailer of a semi-truck. The demand for corn is so high right now, that farmers are removing tree belts and abandoned farmsteads to put more land into crop production.

The growing demand for corn and soybeans is driven by increased consumption by livestock in North America and China (Chapter 8), use of corn and soy products in processed foods globally, and corn and soy production marketed for ethanol and biodiesel fuels. Innovations in seeds, such as Roundup

Ready crops, enable farmers to grow corn and soybeans on marginal lands. Soils in drier climates and with lower nutrient values can now support crops as a result of seed and production innovations. North of the Corn Belt, wheat, barley, and canola are more commonly grown, as these crops require less moisture. Canola oil is used in processed foods and is considered a healthier alternative to other vegetable oils because the oil in new seed varieties remains stable in processed foods. If oil does not have to be partially hydrogenated, then the food produced will not contain trans fats. In Canada, provinces are opening previously forested areas for crop production (Figure 11.35).

FORESTS

Forests encompass much of the Canadian Shield and westward to Pacific Canada. This region where coniferous trees withstand the extreme cold of Canadian winters is called the **boreal forest**. This forest covers 35 percent of Canada and is harvested for paper products, railway ties, and other wood products. Similar climates across the oceans in Russia and Scandinavia create a global belt of boreal forest, of which Canada's boreal forest is one part. In addition to providing resources to Canada, the boreal forest produces oxygen and "absorbs and stores carbon dioxide and so plays a critical role in mitigating global warming" (*Atlas of Canada*).

Quebec, Canada, is home to Canada's first paper mill, established in 1805, and leads North American states and Canadian provinces in pulp and paper production, employing 91,000 people in the broader forest products industry, including lumber and forest harvesting. The pulp

© AgStock Images, Inc./Alamy.

FIGURE 11.34   Eastern Iowa, United States. Soybeans grow in rows in a field where corn grew the previous year.

GUEST *Field Note*   Expanding Farmland in Alberta, Canada

DR. DAWN BOWEN

University of Mary Washington
Fredericksburg, Virginia

Farming seems an unlikely pursuit in Canada's boreal forest, but in northern Alberta the amount of land being tilled is expanding dramatically. In the early 1980s, the province used a lottery system to open a 50,000-acre block of forested land to local residents, the overwhelming majority of whom are Mennonites. As a result, the landscape is dotted with dozens of new farms about 20 miles (32 km) southwest of the village of La Crete. When the lottery ended in 1984, the need for land had been met, albeit temporarily. The demand for land increased again when the next generation of Mennonites grew up and sought land for their own farms.

In 2011, the government of Alberta agreed to make more than 130,000 additional acres available, which would increase the amount of farmland in Mackenzie County (about the size of Maine in the far northwest of Alberta) by nearly one-third. Commenting on this process, one local resident declared that no other community in Canada would be willing to work so hard to bring raw land into production. But developing new farms does not happen quickly. It can take up to five years before a crop is planted. The

Courtesy of Dawn Bowen, University of Mary Washington, Fredericksburg, Virginia.

FIGURE 11.35 Mackenzie County, Alberta, Canada.   A farmer pulls roots from former forest land.

photo, taken in 2013, shows one stage in the transformation from forest to farm. The farmer has cut the trees, like those shown in the background, removed stumps, and is using a locally designed attachment on the tractor to pull roots from the ground. This is the final stage before a crop—wheat or canola—will be planted.

and paper industry in Canada generates 3 percent of the country's gross domestic product (GDP). Canada is a global leader in pulp and paper production, second only to the United States. Even though the United States produces more pulp and paper than Canada, Canada exports more pulp and paper products than the United States, primarily to the United States and western Europe. Quebec's pulp and paper industry primarily produces newsprint (40 percent of production), specialty paper, and Kraft paper (Figure 11.36).

The pulp and paper industry in Quebec is powered by Hydro-Quebec, which made a profit of $2.9 billion in 2013. Hydro-Quebec is owned by the Quebec provincial government and generates electricity using Quebec's many rivers that flow into the Hudson Bay, the Atlantic Ocean, and the St. Lawrence River. Of the $2.9 billion profit generated in 2013, Hydro-Quebec paid $2.2 billion to the Province of Quebec as a dividend.

The company works with Cree Indians in northwestern Quebec to generate electricity on

© Jacques Laurent/Design Pics /Corbis.

FIGURE 11.36 Mantane, Quebec, Canada.   The Tembec Pulp Mill, dedicated in 2012, is designed to use 90 percent less oil fuel than the previous generation of pulp mill. The mill includes an anaerobic treatment facility, which captures methane gas and uses it to dry pulp. A new boiler is fueled by electricity instead of heavy oil.

Cree land in the province. In addition to providing low-cost electricity to pulp and paper mills and Quebec residents, Hydro-Quebec sells electricity to residents of northeastern United States.

FOSSIL FUELS

Fossil fuels in North America are distributed in places where rich plant or marine life thrived in ancient times. Coal and oil come from the decomposition of organic material. Since the material, whether plant or animal, was once alive, it contained carbon. Under heat and pressure, the material became coal or oil, and these energy sources are burnable as a fuel because they contain carbon.

Coal often forms in areas that were once swampy and abundant in plant life. Over time, the plant life decays at the surface, forming peat. Sediment covers the peat and compacts it over time into lignite. Lignite can undergo enough heat and pressure to become sub-bituminous coal, then bituminous coal, then anthracite coal, and finally graphite.

North America boasts massive seams of bituminous and sub-bituminous coal, especially in the Appalachian Mountains in the United States and on the eastern side of the Rocky Mountains in Canada and the United States, with other, smaller seams in the Midwestern United States.

Coal

In the United States, coal mining is historically associated with West Virginia and the northern Appalachian region. Today, West Virginia is the second largest producer of coal in the United States, and Wyoming is the first. The most accessible coal in Appalachia has been tapped, and the remaining coal is costlier to mine. Wyoming coal is surface mined, which is less expensive than deep mining. In addition, Wyoming coal is abundant.

Over 90 percent of the coal used in the United States today is for generating electricity. The rising demand for electricity has encouraged mining companies to find cost-saving ways to extract the coal still in Appalachia. Since the 1970s, coal mined in Appalachia has increasingly been removed by a controversial practice called **mountaintop mining**.

In mountaintop mining, miners take off the top of a mountain (euphemistically called the overburden) in order to more easily access a coal seam lying beneath (Figure 11.37). Once mining is complete, the company regrades the now flat-topped mountain. Aside from the change to the landscape itself from a hill or mountain top to a flat meadow after regrading, mining companies blast and scrape the overburden into the surrounding valleys and streams.

According to the United States Environmental Protection Agency (EPA), the environmental impacts of mountaintop

© George Steinmetz/Corbis.

FIGURE 11.37   Kentucky, United States.   To access coal seams in Appalachia, mining companies blow off the tops of mountains and dump the sediment into river valleys, leaving behind sprawling, infertile mesas in what were once diverse temperate forests. The mountains in the background show the "before" image of the flattened mountains in the foreground. Seven units of overburden are moved for every unit of coal extracted.

mining include fully covering up streams, decreasing stream flow, and increasing the mineral content of streams, including sulfur. The United States Geological Survey (USGS) used remote sensing analysis to measure change in land use and **land cover** in the United States and found that mining had the greatest human impact on land-cover change. In their report, the USGS found that mountaintop mining has "ample documentation both supporting and criticizing it. Regardless of the arguments for and against mountain top mining, it is without dispute that a significantly altered landscape is the result of the practice."

Oil

In 2012, the United States was the second largest consumer of energy, behind China, and Canada was in ninth place. While North America is the largest consumer of oil, it is also a major producer of oil. The United States ranks first among the world's top oil producers, with Canada coming in fifth and Mexico ninth.

Fracking for Natural Gas and Oil

The methods for hydraulic fracturing, known as **fracking**, were established in 1947, but the practice was too costly and not widespread until the United States government loosened environmental regulations in 2005. Congress passed an energy bill, known as the Halliburton Loophole, that allowed oil and natural gas drillers to inject hazardous materials into or adjacent to known drinking water sources, thereby getting around the Safe Drinking Water Act and making room for fracking.

To frack oil and natural gas, companies drill wells vertically about 8,000 feet underground and then horizontally. Once the well is dug, the driller pushes down a mixture of water (400 to 600 truckloads per well), sand, and chemicals (a mixture of up to 600 chemicals). The water and chemicals open fractures along the horizontal well, and the sand keeps the fractures open. The natural gas or oil is released from the shale and is captured at the surface and trucked out. About half the water that is pumped in is later pumped back out, trucked out, and disposed of in holding ponds.

The environmental costs of fracking include air and water pollution and stress-induced earthquakes. Air is also polluted from the holding ponds, which contain water contaminated with fracking chemicals and methane. The 1,000 trucks that visit a single well while it is in production use massive amounts of petroleum and emit pollutants into the air. After evaporation, chemicals settle in the empty holding pond and are then blown by wind, causing people to breathe polluted air as well as drink polluted water.

Although oil and gas companies assert that fracking does not damage water supplies, widespread fracking in Colorado, Wyoming, Pennsylvania, North Dakota, and Texas has resulted in polluted wells, rivers, and groundwater supplies (Figure 11.38). Documentaries, news programs, and newspapers show landowners who can set their water on fire because of the natural gas and chemicals running into their drinking water supplies. Another environmental concern of homeowners is earthquakes and tremors, which

Adapted from the map, "North American shale plays (as of May 2011)", published by the U.S. Energy Information Administration, 2011. http://www.eia.gov/oil_gas/rpd/northamer_gas.pdf.

FIGURE 11.38a   Oil and gas shale deposits in North America.   The US Energy Information Administration mapped the location of shale plays (shale containing significant amounts of oil or gas) in North America. The expanse of shale from western Canada south to Texas was formed when this portion of the continent was the Western Interior Seaway during the mid-Cretaceous period (about 100 million years ago).

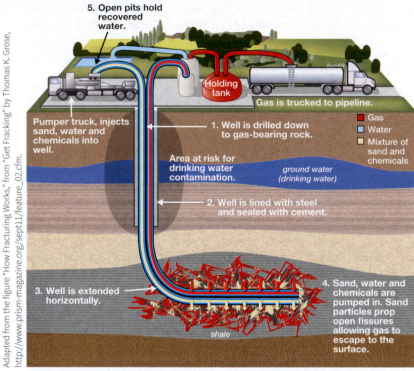

Adapted from the figure "How Fracturing Works," from "Get Fracking" by Thomas K. Grose, http://www.prism-magazine.org/sept11/feature_02.cfm.

5. Open pits hold recovered water.

Holding tank

Gas is trucked to pipeline.

Pumper truck, injects sand, water and chemicals into well.

1. Well is drilled down to gas-bearing rock.

Area at risk for drinking water contamination.

ground water (drinking water)

2. Well is lined with steel and sealed with cement.

Gas
Water
Mixture of sand and chemicals

3. Well is extended horizontally.

4. Sand, water and chemicals are pumped in. Sand particles prop open fissures allowing gas to escape to the surface.

shale

FIGURE 11.38b   How fracking works.   Fracking requires the drilling of deep wells and the use of chemicals and water to extract oil and natural gas.

can be generated when fracking induces pressure changes underground.

On the positive side, fracking can make the United States more energy independent, create jobs, and produce revenue. In 2001, shale provided 1 percent of the natural gas in the United States; in 2011, it provided 30 percent of the natural gas. Areas where large-scale fracking (for oil instead of natural gas) is occurring, such as the Bakken formation in North Dakota, are drawing in thousands of new residents. They drive trucks, drill, construct new housing, and provide goods and services. The state receives revenue from the new residents in taxes and from the companies drilling for oil or natural gas. The new revenue going into the state budget can be used to fund education, roads, or whatever the state needs.

Minerals

The *Wall Street Journal* lists Canada as the fourth and the United States as the second most resource-rich countries (Figure 11.39). The economies of the countries have benefited from vast agricultural lands, a wealth of minerals and fossil fuels, and the early diffusion of industrialization. In addition to fossil fuels, North America has a vast variety of minerals. The location of minerals is closely tied to the physical geography of the region.

In the west, copper is mined in open pits, and in the east, copper is mined underground. The most famous copper mine in North America (perhaps in the world) is the Bingham Canyon Mine in Utah. Kennecott Copper Corporation is owned by Rio Tinto, a mining corporation based in the United Kingdom and Australia. Rio Tinto mines copper, diamonds, and other minerals around the world

(Figure 11.40). Copper mining began at the Bingham Canyon Mine, 25 miles (40 kilometers) southwest of Salt Lake City, in 1903. The mine was once a small mountain, and today it is an enormous pit. One hundred trucks, each the size of a two-story house, travel the terraces of the mine to remove a half million tons of rock material from the mine each day. The rocks go through a lengthy process of being broken down physically and chemically in order to extract the copper, which is in high demand for wiring electronics from computers to cell phones.

INDUSTRY

When the Industrial Revolution began in Great Britain in the 1700s, the first industries were located close to water, either by rivers for power or by oceans for trade and transportation. During the 1800s, improvements in the steam engine and expansion of coke smelting to produce steel pushed coal ahead of water as the energy source of the Industrial Revolution. Manufacturers chose industrial sites close to sources of coal. With the advent of the railroad, industries grew in places without coal deposits, such as London, which imported coal along waterways from Wales and New Castle by barge.

When the Industrial Revolution diffused from Europe to North America, entrepreneurs sited industries on the east coast of the United States and in the St. Lawrence River valley of Canada (Figure 11.41). Proximity to coal resources, waterways (including canals), Atlantic Ocean ports, and cities with pools of immigrant laborers all drew industry to the eastern seaboard. Industrialization also required capital (monetary) input. In North America, Atlantic trade fueled industrialization, reinforcing the location of industry in the first manufacturing belts of the United States and Canada.

In the early 1900s, immigrants from southern and eastern Europe migrated to the Detroit, Michigan, area, along with immigrants from Southwest Asia and Canada and migrants from rural Michigan. Detroit is **sited** on the Great Lakes, with Lake Erie to the south and Lake Huron to the north. Detroit has an ideal **situation** for automobile production, which gives it access to iron ore deposits in northern Michigan and Minnesota; steel production in Pittsburgh; and coal deposits in northern Appalachia.

Early automobile manufacturing was concentrated in the Detroit region, with Ford Motor Company's largest plant in the western suburb of Dearborn. Ford, along with Chrysler and General Motors (also based in Detroit), make up the historic Big Three in North America's automobile industry.

Early immigrants to Detroit enjoyed relatively higher wages and more stable employment than workers in many other cities in North America. In 1914, Henry Ford, founder of the Ford Motor Company, promised to pay workers $5 a day, which was more than double the average manufacturing salary at the time. Ford's wages quickly drew one million

FIGURE 11.39 **Political geography of North America.** Many of the cities on the east coasts of Canada, the United States, and Mexico date to colonialism. Cities around the Great Lakes expanded with the diffusion of the industrial revolution. Many cities in the west are located close to mining or fossil fuel resources.

migrants to the city. The flow of migrants to the city included not only immigrants but also internal migrants, especially African Americans from the South. The number of African American migrants increased after the United States government tightened immigration laws in 1924. The flow of African Americans from Mississippi, Alabama, Tennessee, and Kentucky to Detroit was part of an era called the Great Migration (Figure 11.42). Immigrants and internal migrants alike moved to Detroit to live out the American Dream.

Detroit's Big Three built enormous manufacturing plants in the Detroit area, with laborers often living near the manufacturing plants. Laborers built single-family homes throughout Detroit and its suburbs, with neighborhoods concentrated near factories.

In 1917, Ford began production at the Rouge plant in Dearborn, a self-sustained factory complex that Henry Ford developed, which incorporated 93 buildings, over 100 miles (160 kilometers) of railroad tracks, and over 120 miles (193 kilometers) of conveyor belts, on over 2,000 acres of land (Figure 11.43). Raw materials, including iron ore and coal, came into the enormous factory complex, which included its own power plant.

FIGURE 11.40 **Salt Lake City, Utah, United States.** The Bingham Canyon Mine is the largest human-made excavation in the world and is managed by Kennecott Corporation. The mine is an enormous pit, 2.75 miles wide and 0.75 miles deep or 4.5 by 1.2 kilometers.

In its heyday in the 1930s and 1940s, over 100,000 Ford employees smelted iron, produced automobile parts, and assembled cars at the Rouge, which incorporated:

> ore docks, steel furnaces, coke ovens, rolling mills, glass furnaces and plate glass rollers. Buildings included a tire-making plant, stamping plant, engine casting plant, frame and assembly plant, transmission plant, radiator plant, tool and die plant, and at one time, even a paper mill. A massive power plant produced enough electricity to light a city the size of nearby Detroit, and a soybean conversion plant turned soybeans into plastic auto parts (The Henry Ford Museum).

Under this system, the 100,000 employees at the Rouge produced every part and assembled the car on site. This system of self-contained industrial production became known as the **Fordist** system and served as a model for industries worldwide.

The evolution of the automobile industry over the twentieth century reflects how manufacturing has changed in North America and globally. After World War II, automobile manufacturing diffused to Canada when the Big Three entered partnerships with Canadian manufacturers to produce

FIGURE 11.41 **Industrial belts in North America.** The manufacturing belt along the Great Lakes is often referred to as the Rust Belt today, as much of the industry in the region has declined. The newer manufacturing belt in the American South is referred to as the Sun Belt.

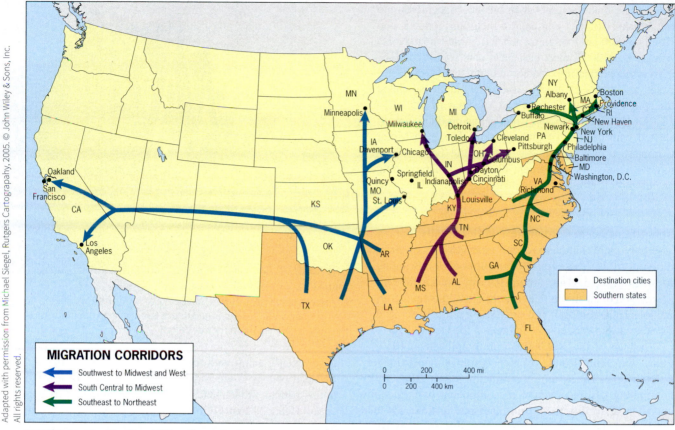

FIGURE 11.42   The Great Migration.   In the late 1880s and early 1900s African Americans followed three primary paths of migration as they were drawn to job opportunities in industrial centers in the north and west.

American cars in Canada. In 1965, Canada and the United States entered an Auto Pact under which new automobiles and automobile parts passed duty free between the two countries. The World Trade Organization overturned the pact in 2001, but the two countries' automobile industries remain linked. Although Canada does not house the headquarters of any automobile manufacturing company, automobile manufacturing rivals Canada's timber industry for the most important industry in Canada. The Ford Motor Company has a major presence in Windsor, Ontario, which is just across the river from Detroit. The region of Kitchener, near Toronto, Ontario, serves as a center of parts and tire manufacturing in Canada.

The decline of the Big Three automobile companies began in the 1970s with the rise of Japanese and then Korean automobile sales in North America (Cooney and Yacobucci 2005). In the 1980s, the United States placed quotas on the number of Japanese cars allowed to be imported, thus encouraging foreign producers to open plants in the United States. States outside of Michigan have lured foreign

FIGURE 11.43   Aerial photograph of the Rouge factory complex, circa 1950.   The Rouge factory complex included a power plant that fueled the manufacturing of all of the component parts and final assembly of automobiles on one site.

automobile manufacturers with business incentives and tax breaks, so that companies including Toyota and Honda, which are foreign owned, assemble cars in North America. By 2003, most passenger cars sold in the U.S. market were either imported or manufactured by foreign-based producers at North American plants (so-called transplant facilities) (Cooney and Yacobucci 2005, 1).

The automobile industry model and production itself shifted with the rise of globalization. The Ford Motor Company uses suppliers for automobile parts, which today accounts for 75 percent of the cost of producing an automobile (Rutherford and Holmes 2008). The component parts are made throughout North America and around the world. They travel **just in time** to the Ford Rouge Center, a modern, more environmentally friendly assembly plant that can produce nine vehicles on three different chassis (base frame of an automobile). Bringing in component parts from around the world just in time for production of an automobile means a loss in manufacturing jobs in Detroit itself and a massive decline in demand for storage and warehouse facilities in the Detroit area.

Instead of the 100,000 people employed in 1917, today the Ford Rouge Center employs approximately 6,000 people. The system of producing industrial goods by contracting the manufacture of component parts globally and then delivering them to a plant just in time, for final production and shipment, reflects the millions of interlinkages in the global economy and is called **post-Fordist** production.

The city of Detroit reflects the massive changes in automobile production over time. As manufacturing has changed and fewer people are employed in Detroit, workers have abandoned the residential areas of Detroit and its suburbs. The city's population fell from 1.8 million in 1950 to 688,701 in 2013, according to the United States Census (Figure 11.44). Without a job and unable to pay a mortgage or even sell their homes, thousands have left Detroit and abandoned their homes. The city of the American dream where everyone could afford a single-family home now has approximately 90,000 abandoned or vacant homes and residential lots.

The new production methods, including suppliers producing parts, just-in-time delivery, and transplant assembly plants (e.g., Japanese cars manufactured in Tennessee), have also impacted the labor unions and the role of governments in the automobile industry. Employees who work at parts manufacturers and transplants (10 foreign car companies own 16 factories in the United States) generally do not belong to labor unions (Healey 2012). Parts manufacturers are attracted to doing business in Canada, where employees have government-supplied health care paid for by their taxes. This lifts the burden of health care coverage from the manufacturer. The cost of health care for the companies that produce parts and the finished product per vehicle produced in Canada is about $30 versus between $1,000 and $1,400 in the United States (Rutherford and Holmes 2008, 530).

Dreading the further loss of jobs from economic recession in 2008 and 2009, the United States and Canadian

FIGURE 11.44 Detroit, Michigan, United States. Having lost 25 percent of the city's population between 2000 and 2010 alone, approximately 90,000 homes and lots are abandoned or empty in the residential neighborhoods of Detroit.

governments, along with the province of Ontario, offered a financial bailout to the Big Three. In exchange for a $50.5 billion bailout, the United States government owned 61 percent of General Motors (GM) until 2010, when it sold enough shares to reduce its ownership to 26 percent. In 2013, the government sold its final shares of GM stock, at a total loss of $11 billion, marking the official end of the GM bailout.

Changes in the automobile industry over the last century are a result of processes occurring simultaneously at multiple scales. The industry began as a local scale process—focused in places with the necessary resources (coal, iron ore) and transportation routes (waterways, canals, ports). Labor was either readily available or drawn into the manufacturing centers through migration. Manufacturing was centralized in one place—workers in one factory built parts, and nearby, workers in another factory assembled the parts into cars. Governments (national and state or provincial) had an impact on manufacturing with quotas on imports, business incentives to lure manufacturers, tax breaks, social welfare policies for employees, and bailout programs. Processes at the global scale, including World Trade Organization decisions, foreign investment, the growth of transplant companies, and regional trade agreements have also altered the location and methods of automobile manufacturing.

CONCEPT CHECK

1. **What** economic industries are made possible in Quebec as a result of the boreal forest?

2. **What** is the impact of mountaintop mining on the physical landscape of the Appalachian Mountains?

3. **How** has the automobile industry changed since World War II, and in what ways does Detroit reflect the changes?

(caption credit, vertical text) Trevor Collens/Photoshot/ZUMApress/Newscom.

What Is North America?

1. North America is a continent that includes mountains, the Canadian Shield, and lowlands. The Rocky Mountains extend north–south from British Columbia in Canada to New Mexico in the United States. The Appalachians are located in the east and are a much older mountain range, 350 million years old, compared to the Rockies, which are 40 million years old. The lowlands include the Great Plains, Midwest, and coastlines.

2. The climate regions in eastern North America extend east–west. In the west, climate regions are shaped in large part by the orographic effect created by the Rocky Mountains. Precipitation falls on the windward side of the Rockies, and the semiarid region is found on the leeward side of the mountains. In the east, climates extend east–west, impacted mainly by the amount of incoming solar radiation, which varies by latitude.

3. North of the boreal forest, above the tree line, Canada is sparsely populated. The soil is permafrost, which means an area of the subsoil is permanently frozen. Some soil may form at the surface, but it is too shallow to support trees.

Who Are North Americans?

1. North America is a continent settled thousands of years ago by peoples migrating from Asia across the Pacific. The cities and villages of the indigenous Americans created a hive of activity on the continent. In the late 1400s, Spanish explorers described the region as densely populated. European diseases killed large numbers of indigenous people.

2. Both Canada and the United States had assimilation policies, through which they sought to change the cultures of First Nations in Canada and American Indians in the United States. Both countries also had policies of removing indigenous children from their homes to federally sponsored boarding schools. The boarding schools in both countries were notorious for abuse, child labor, and low educational achievement.

3. Migration can be voluntary or forced. Both Canada and the United States were shaped by voluntary migration from Europe. Step and chain migration influenced the patterns of migration, helping to create a French Canadian region in Quebec and northern European regions in the Midwest and Great Plains.

How Are Cities in North America Organized?

1. More than 75 percent of people in North America live in urban areas. The cities of Megalopolis are sited on the fall line of eastern North America, the farthest inland point along a river that boats can travel. The biggest cities in Canada are sited along the St. Lawrence River and inland toward the Great Lakes.

Geography in the Field   **YOUR**TURN

Motor homes stand in a field that is devoid of trees. A muddy landscape is testament to the fact that the people working in this field want what is under the ground, not what is on top of it. The drill tells us exactly what they are hoping to find, and the trucks and small buildings surrounding the drill give us the impression that they have found it. This drill in Dimock, Pennsylvania (a town of 1,400 people), stands on the Marcellus shale formation, which is located in the northern part of Appalachia, and holds massive amounts of natural gas. The trucks are delivering water, sand, and chemicals to frack the natural gas out of the shale.

Thinking Geographically

- The average productive well is visited by more than 1,000 trucks a year. Knowing that this is one of hundreds of wells in the region, what would be the impact of fracking on the cultural landscape?

- In October 2009, the Pennsylvania Department of Environmental Protection acknowledged that the groundwater aquifer under Dimock had been contaminated with methane. How will this contamination affect house values for families living in the region?

- What will this cultural and physical landscape look like in 20 years when the well has run dry?

Read:
Bateman, Christopher. 2010. A colossal fracking mess. *Vanity Fair*. 21 June.

© Philip Scalia/Alamy.

2. The urban morphology, or the size, shape, and layout of urban areas, of North American cities was designed to serve the central business district (CBD). Transportation networks were designed to move people from suburbs into the CBD. The growth of edge cities, or suburban areas that house office complexes, shopping districts, sports complexes, hotels, and housing, have helped create urban sprawl, whereby the city is growing outward, expanding with low-density housing regions and massive new road systems.

3. Planners have sought to stem urban sprawl by encouraging high-density housing near the CBD. Portland, Oregon, recognized globally as an innovative and sustainable city, designated an urban growth boundary to encourage development closer to the CBD and contain urban sprawl.

What Role Do Resources Play in the Economic Geography of North America?

1. The boreal forest encompasses more than 30 percent of Canada's land cover. The land use of the boreal forest includes a pulp and paper industry that generates 3 percent of Canada's gross domestic product and massive hydrological power companies.

2. The demand for coal has risen as a low-cost option for generating electricity in the United States. As a result, surface coal mining in Wyoming and mountaintop mining in Appalachia are on the rise. Mountaintop mining is a method to cheaply extract coal by dynamiting the mountain top and removing the "overburden" to hillsides and streams in order to reach coal. A flat-top mountain is left behind.

3. Detroit reflects the multitude of changes that have happened in automobile manufacturing over the last century. Before World War II, Fordist manufacturing was centered in one place, with component parts and finished products created in one massive factory complex. Post-Fordist manufacturing is built on just-in-time production with component parts created around the world.

KEY TERMS

mesa	forced migration	central business district (CBD)	land cover
plateau	assimilation	urban morphology	fracking
striations	voluntary migration	edge city	site
mid-oceanic ridge	push factor	gentrification	situation
tree line	pull factor	functional zonation	Fordist
treaties	step migration	urban morphology	just in time
sovereignty	chain migration	urban sprawl	post-Fordist
toponym	fall line	boreal forest	toponym
migration	Megalopolis	mountaintop mining	

CREATIVE AND CRITICAL THINKING QUESTIONS

1. If you studied the Appalachians in North America and the Atlas Mountains in Morocco, what evidence could you look for to conclude that the two mountain chains, now separated by the Atlantic Ocean, were once part of the same physical region?

2. In North America, the prevailing winds are from the west. If there were a mountain range in the tropics, where prevailing winds come from the east, how would this different context affect the diagram of orographic (Figure 11.9) precipitation?

3. In 2006, the Canadian government issued an apology to First Nations and Inuits in Canada for the policy of assimilation. Would the United States government ever do this, and if it did, what might be the impact on American Indian identities?

4. Nunavut is a vast, sparsely populated province. What challenges does the situation of Nunavut pose for the Inuit government in terms of providing services to the Inuit people? How could the Nunavut government meet these challenges?

5. In the 1700s and 1800s, cities were sited on fall lines to link land and river trade. With current innovations in transportation, what is the new "fall line" for the most successful cities today?

6. In the 2010 census, more than half the population of North America lived in suburbs. What are the ramifications for American culture and the cultural landscape of urban areas with so many people living in suburbs?

7. How are mountaintop mining in Appalachia and fracking in North America evidence of the Anthropocene, and what are the long-term implications for development in areas where fossil fuel extraction is changing the landscape and environment?

8. North America can be described as a region of migration. How are voluntary and forced migration evident in the cultural landscape of the region?

9. Portland, Oregon, built new transportation systems and redesigned downtown to discourage urban sprawl. How can carefully planning the transportation network, urban morphology, and functional zonation of a city impact urban sprawl?

10. The movement to post-Fordist production in the automobile industry has meant a massive loss of jobs in Detroit. As a result, Detroit has approximately 90,000 vacant or abandoned houses and residential lots. What can Detroit do to encourage development and make the city vibrant again?

1. The Rocky Mountains are located farther inland than most mountain chains created along subduction zones. Why?
   a. the angle of subduction
   b. the angle of repose
   c. The continental plate is denser than the oceanic plate
   d. The continental plate and oceanic plates have the same density

2. The Appalachian Mountains are older than the Rocky Mountains. The easiest way for an amateur physical geographer to tell this is to notice:
   a. The Appalachians are at a higher elevation than the Rockies
   b. The Rockies are at a higher elevation than the Appalachians

3. The basin and range topography of the western United States is created when the crust is _____ and ranges ___ while basins _____.
   a. stretched/uplifted/dropped
   b. compressed/uplifted/dropped
   c. stretched/dropped/uplifted
   d. compressed/dropped/uplifted

4. In the western United States, orographic precipitation creates a rainshadow on the ___ side of the Sierra-Nevada Mountains, and the climate there is _____.
   a. eastern/moist
   b. eastern/dry
   c. western/moist
   d. western/dry

5. North of the tree line in Canada, trees do not grow for all of the following reasons except:
   a. the soil has permafrost
   b. the soil is too thin
   c. the climate is too cold
   d. the climate receives too much rain

6. European diseases killed millions of indigenous peoples in North America because:
   a. European diseases were animal-borne diseases to which North American indigenous peoples had never been exposed
   b. North American indigenous did not live in cities before Europeans arrived

7. The United States (and France and England before them) entered treaties with American Indian tribes that recognized tribes as:
   a. wards of the United States
   b. sovereign
   c. citizens of the United States
   d. protected minorities

8. True or False:
   Both the Canadian and United States governments forcibly removed members of indigenous tribes, primarily children, from their homes to boarding schools, where they were forbidden to speak their native language.

9. The Atlantic slave trade forced the removal of Africans to the Americas. The greatest proportion of slaves, according to Figure 11.17, went from Africa to:
   a. Central America and the Caribbean
   b. southern United States
   c. eastern Canada
   d. Brazil and the Caribbean

10. The Scandinavian Festival in Junction City, Oregon, is a good example of step migration because the town was primarily populated by Scandinavians from:
    a. Denmark
    b. Norway
    c. the Midwest
    d. California

11. True or False:
    Before 1500, cities in North America were larger than cities in Europe.

12. The 2006 census in Canada found a major difference in the populations of the suburbs and the central city in Montreal and other Canadian cities. The major difference they found was:
    a. The population of the central city was older than the population of the suburbs
    b. The population of the suburbs was older than the population of the cities

13. Megalopolis includes all of the following cities except:
    a. Washington, DC
    b. Charlotte, NC
    c. Philadelphia, PA
    d. New York, NY

14. True or False:
    More than 50 percent of Canada's population lives in an urban area stretching from Windsor, Ontario, to Quebec City, Quebec.

15. Urban sprawl means a city is growing:
    a. quickly
    b. slowly
    c. with low-density areas on the outskirts
    d. with high-density areas on the outskirts

ANSWERS FOR SELF-TEST QUESTIONS

1. a, **2.** b, **3.** b, **4.** b, **5.** d, **6.** a, **7.** b, **8.** True, **9.** d, **10.** c, **11.** True, **12.** a, **13.** b, **14.** True, **15.** c

12

LATIN AMERICA AND THE CARIBBEAN

This *campesino* (or peasant) from a small village in Guatemala is rolling cane to produce a bowl of fresh-squeezed cane juice, a delicious, sweet treat. In many ways he typifies a particular view of Latin America as a place dominated by rural landscapes and small-scale farmers. But this view does not reflect contemporary Latin American realities. Latin America and the Caribbean is the second most urbanized region in the world after North America, with 78.3 percent of the people living in urban areas.

Even Guatemala, one of the more rural countries in the region, had a majority of its population living in urban areas by 2013. By featuring a male farmer, the photo also fails to communicate another major shift in the region: the feminization of agriculture. As men migrate to cities in search of employment, women are increasingly taking on the responsibility of running household farms.

Finally, most of the remaining *campesinos* in Latin America and the Caribbean are not small-scale, subsistence farmers as the word "peasant" implies. Rather, they are more likely to be laborers on large farms who grow cash crops for international markets. They may also produce crops for their own consumption through subsistence farming on the side.

In many communities in Latin America, nonprofit groups have been working to develop agrotourism in order to diversify the economy beyond agriculture. Absent alternative sources of income, farmers may not be able to remain on the land.

Antigua, Guatemala. A man uses a machine to extract juice from sugar cane. The machine uses rollers to crush the cane, and the cane must be crushed within a day of harvesting. The juice is then purified and boiled. After boiling, the remaining liquid crystalizes into sugar once it cools.

© Bill Bachmann/Alamy.

CHAPTER OUTLINE

 THRESHOLD CONCEPTS in this Chapter

Race
Gender

WHAT IS THE PHYSICAL GEOGRAPHY OF LATIN AMERICA AND THE CARIBBEAN?

LEARNING OBJECTIVES

1. **Describe** the physical geography of Latin America and the Caribbean.
2. **Explain** Latin American agricultural systems.
3. Critically **evaluate** models of biodiversity conservation in the Latin American context.

The physical environment in Latin America is amazingly diverse, ranging from the world's largest rainforest area of the Amazon to the arid vegetation of the Sonoran, Atacama, and Patagonian deserts, to the high-altitude climates of the Andes and the Sierra Madres. Nearly all of the world's **biomes**, or major ecological zones, may be found in this region. These ecological variables, together with a mix of indigenous and European influences, along with striking differences in power, wealth, and access to resources, have shaped how Latin Americans use the land.

PHYSICAL GEOGRAPHY

The region of Latin America and the Caribbean is distinguished by its exceptionally great length, from north to south, stretching from the temperate zone in the Northern Hemisphere to the temperate zone in the Southern Hemisphere, a total of 5,500 miles (8,851 km) from the Rio Grande to Cape Horn (Figure 12.1). If we start at the

FIGURE 12.1 **Physical geography of Latin America.** The Andes Mountains stretch along the western side of South America, and the Amazon River basin creates a flat plain in much of the northern part of the continent. Mexico and Central America have mountains mainly in the central areas, generally formed by volcanoes. The islands of the Caribbean are primarily located on plate boundaries, especially surrounding the Caribbean plate.

equator and move poleward, the major climates vary by latitude from tropical rainforest to savanna to humid subtropical. The Andes Mountains in western South America shape climates by elevation instead of latitude. The influence of the cold water Peru current in the Pacific Ocean off the west coast of South America keeps climates cooler than they would be otherwise (Figure 12.2).

With the exception of northern Mexico and the Southern Cone region (Chile, Argentina and Uruguay), the majority of Latin America falls within the Tropics, between 23.5°N (the Tropic of Cancer) and 23.5°S (the Tropic of Capricorn). The convergence of the tropical trade winds at the **Intertropical Convergence Zone (ITCZ)** heavily influences the climates throughout the Tropics. The ITCZ brings rain and the wet season when it is overhead, because this zone of low pressure (where warm air rises, cools, and precipitates) draws moist air from the ocean over land (Figure 12.3). As the ITCZ moves away, it leaves clear skies

FIGURE 12.2 Climates of Latin American and Caribbean. Climates in Latin America vary by latitude and stretch across tropical, equatorial, and midlatitude climate zones. In southern Mexico and Central America, volcanic mountains create conditions for orographic precipitation. In South America, the Andes Mountains stretch from north to south along the western side of the continent. Different climates are found at each elevation along the windward and leeward sides of the Andes.

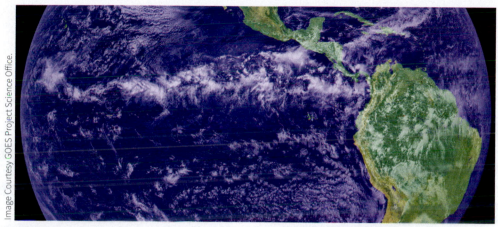

FIGURE 12.3 **ITCZ over the Pacific and South America.** A band of clouds forms along the ITCZ in the Pacific and over land in South America, just to the north of the equator. Between March and June, the ITCZ moves north from the equator toward the Tropic of Cancer.

and a dry season. The ITCZ moves seasonally, shifting south from June to December and north from December to June. The ITCZ passes over the equator at the spring equinox and at the fall equinox. The location of the ITCZ corresponds to the latitude of the subsolar point, where the sun's rays hit earth at a 90 degree angle.

The consistent presence of the ITCZ near the equator explains the year-round rainfall in the tropical rainforest of **Amazonia** and Central America. The Latin American rainforest is characterized by poor soils, as nutrients are quickly leached by rainfall from **ultisols** and **oxisols**, the most common soils in the rainforest; buttressed trees, which spread their roots broadly and horizontally in search of nutrients; and, lots of vine species, which depend on woody trees to support their vertical climb to sunlight in the upper canopy. The shifting ITCZ largely explains the

seasonality of rainfall in the tropical forest, tropical savannas, and, at higher latitudes, the midlatitude grasslands (Figure 12.4a, b, c, d). The near-constant subtropical high pressure cell close to the Tropic of Cancer largely explains northern Mexico's Sonoran Desert.

Altitude and ocean currents help shape the region's other climate regions. The Andes Mountains, running down the spine of western South America (stretching 4,000 mi or 6,437 km) form the longest mountain chain in the world. The Sierra Madre is a smaller mountain chain in Mexico and Central America.

One branch of the Sierra Madre, the Sierra Madre Occidental Mountains, dominates the landscape of western Mexico. The Sierra Madre is a combination of igneous (volcanic) and sedimentary rocks. Mayan and Aztec silver and gold resources came from volcanic rocks in the Sierra

FIGURE 12.4a **Amazon Rainforest.** Tropical rainforests are found along the equator across the world where the greatest amount of incoming solar radiation is received over the course of the year, and the largest expanse of rainforest is the Amazon.

FIGURE 12.4b **Rincon De La Vieja, Costa Rica.** A seasonal tropical forest has a brief dry season when the ITCZ is in the opposite hemisphere. Deciduous trees may lose their leaves during the dry season. Seasonal tropical forests are found in Mexico, the west coast of Central America, and around the Amazon in South America.

FIGURE 12.4c **Gran Sabana, Venezuela.** The Gran Sabana lies on a plateau at an altitude of 1,000 meters above sea level and is dotted with huge tabletop mountains called Tepuis, which rise from the surrounding plains.

FIGURE 12.4d **Chile.** Midlatitude grasslands are well suited for raising livestock, including sheep.

Madre. A subduction zone off the west coast of Mexico, near Acapulco, built a more recent belt of volcanic mountains in southern Mexico, near Mexico City.

The Andes and Sierra Madre Mountains influence climate and vegetation in at least two ways. First, with an increase in elevation, the average temperature declines, creating altitudinal climate zones. Near the equator, such as on the eastern slopes of the Andes in Ecuador, you can climb from the tropical rainforest to the temperate forest, and finally to tundra (Figure 12.5). In addition to cold climates being located at high elevations, temperatures also decrease at higher latitudes; so, higher elevations that

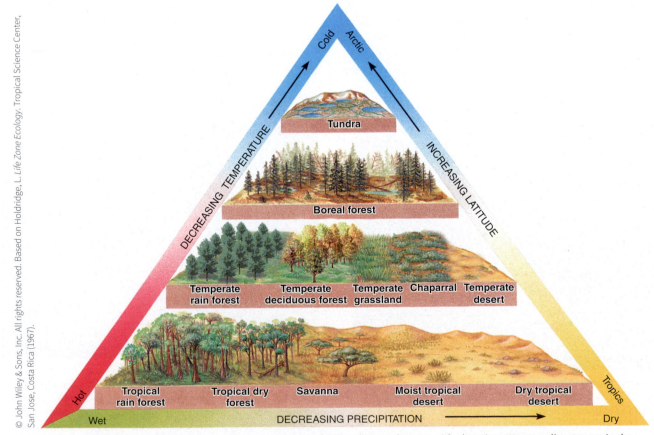

FIGURE 12.5 **Altitudinal climate zones.** In a mountainous area, the climate changes with elevation. Warmer climates are in the valleys, and cooler climates are at the highest elevations and latitudes.

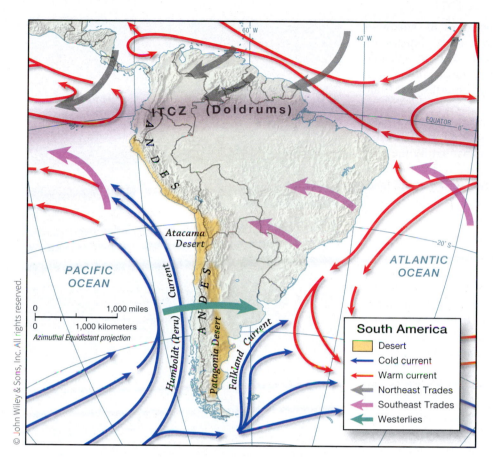

FIGURE 12.6 **Latitudes, prevailing winds, ocean currents, and deserts in South America.** The prevailing winds include the trades in the tropics and the westerlies in the south. The leeward side of the Andes is on the east where the westerlies dominate, creating the Patagonia Desert. Where the trade winds dominate, the leeward side of the Andes is on the west, creating the Atacama Desert.

are located at higher latitudes will experience even colder climates, including tundra and ice cap.

Mountains also lift and block winds, creating moist climates on the windward side and rainshadow, dry climates on the leeward side. Because the South American continent and the Andes mountain chain stretch far north–south across the tropical and midlatitude zones, the prevailing winds shift three times over the continent. North of the equator, the prevailing winds are from the northeast, intensifying the amount of moisture in the Amazon rainforest because moist winds from the Atlantic Ocean consistently blow onto the eastern side of the continent.

From the equator south to the midlatitudes, the prevailing winds are from the southeast, bringing moisture to the windward side, which is the eastern side of the Andes and includes northern Argentina, Paraguay, and southern Brazil. The leeward side, or the rainshadow of the Andes, is the west of the Andes in northern Chile where the Atacama Desert is located (Figure 12.6). The cold Humboldt or Peru ocean current, coming from the south near Antarctica, pushes north along the west coast of South America and brings cold, dry air, which

allows little to no chance of precipitation from the Pacific in the Atacama. The Atacama Desert is the driest place on earth, receiving virtually no rainfall (Figure 12.7).

From the midlatitudes (about 35 degrees south) south to the tip of the continent, the prevailing winds are from the

FIGURE 12.7 **Atacama Desert, northern Chile.** Rock formations in Death Valley in the Atacama Desert extend above an expanse of sand that has been likened to the surface of Mars.

west, making the western side of the Andes the windward side. Mediterranean and marine west coast climates are found in Chile, south of the Atacama, with precipitation falling mainly in the winter months. Grapes grow well in Mediterranean and marine west coast climates, and Chile is well known for its wines.

On the leeward side of the southern Andes, the Patagonia Desert sits in the rainshadow in southern Argentina. The dryness of the Patagonia is intensified by the cold Falkland current off the Atlantic Coast of South America. As cold air from the ocean moves over warmer land, the chance for precipitation declines further, creating a **cold water desert**.

AGRICULTURAL SYSTEMS

The world owes a debt to Latin American farmers for developing several plants it depends upon today for food and spices. Over centuries, Latin American farmers, by selecting seeds for certain characteristics, have **domesticated** corn (maize), potatoes, sweet potatoes, manioc, beans, tomatoes, and peanuts (Figure 12.8). The region has also contributed spices, including chili peppers, chocolate, and vanilla, and drugs such as quinine, tobacco, and coca.

The exchange of plants and animals across the Atlantic after 1500 CE profoundly influenced local and global food systems. Both Europeans and Africans were intermediaries in this process, known as the **Columbian Exchange** or the Triangular Trade Network (see Figure 2.2). The Spanish and Portuguese brought a number of crops with them, some native to their own countries and some from other parts of the world. The crops included wheat (which flourished in upland areas); grapes and olives (now produced commercially in temperate zones of South America); sugar cane (which became the primary cash crop of the Caribbean and the Brazilian tropical lowlands); and coffee (which emerged as an export crop that was cultivated in the upland areas of Central America, Colombia, Venezuela, and Brazil). Africans brought rice, millet, sorghum, coffee, okra, watermelon, and the "Asian" long bean to various parts of the Americas. Europeans sent a number of Latin American crops to Europe, Africa, and other parts of the world, including corn (maize), potatoes, manioc, hot peppers, tomatoes, pineapple, cacao (from which chocolate is made), and avocados.

The Columbian Exchange was not only about the transfer of plants and seeds, but also about knowledge of how to cultivate them (Figure 12.9). In slave ships, Africans brought rice, coffee, okra, watermelon, and other produce to the Americas. In addition to the crops that enslaved Africans were forced to produce, most grew foods for their own nourishment in

COURTESY OF ERIC CARTER.

FIGURE 12.8   **Tulum Quintana Roo State, Mexico.**   A Mexican farmer works a cornfield in the Yucatan Peninsula.

© National Maritime Museum, London/Image Works, The.

SLAVES IN BARBADOES.

FIGURE 12.9   **Barbados.**   The Columbian Exchange brought crops from Europe and Asia to the Americas, including sugar, which is native to Polynesia but had diffused to Persia (present day Iran) by the 7th century and to Mediterranean Europe as early as the 13th century. This painting demonstrates why the Columbian Exchange is marked as the beginning of globalization, as it depicts enslaved Africans enjoy time away from producing a Polynesian crop with a Dutch windmill in the background on an island in the Americas.

subsistence plots. The knowledge Africans had of how to farm now popular "American" crops was critical for their successful production in the New World (Carney and Rosomoff 2010).

Latin America supports a diversity of farming systems, built upon a rich tradition of indigenous agricultural practices, later influenced by colonialism and globalization. The Mayan, Aztec, and Incan civilizations of **Mesoamerica** and South America grew and thrived based on their highly productive farming systems. Geographers Tom Whitmore and Billie Turner (2002) describe the Spanish explorer Hernán Cortés's introduction to Aztec farming systems in Mexico. In *Cultivated Landscapes of Middle America on the Eve of Conquest,* Whitmore and Turner made a landmark contribution to **historical cultural ecology**, a subfield of geography that seeks to understand past farming systems and human–environment dynamics (Figure 12.10):

The penetration of the mainland yielded numerous marvels: coastal wetlands cultivated through the use of canals and raised-field networks and highland basins containing an array of complex land uses. The apogee of environmental transformation, however, awaited Cortés' descent into the Valle de México, where among the temples and material riches, the Spaniards encountered what could only be described as a sculptured landscape. Well-manicured, terraced hillsides, some with irrigated gardens, cascaded towards the valley

below. Neatly partitioned farms and orchards in the bottom lands were punctuated by villages and towns. A lacustrine system, controlled by dikes and sluices, separated saline and fresh waters, and sustained a vast network of *chinampas* (wetland fields and canals) rivaling in area the polders of the Netherlands. Impressive cities ringed the lakes, but aqueducts, causeways, and canoes led all eyes to the island capital of Tenochtitlán, a Venice in the New World (Whitmore and Turner 2002, 2).

This description of the complex agricultural systems in the Americas before 1500 debunks the myth that the region was a pristine wilderness, barely touched by humans before Europeans arrived.

The dense population of the Aztec Empire provided the labor required to build and maintain highly productive canal and terrace systems. Productive farming systems, in turn, allowed a large population to survive. Evidence collected by geographers and archaeologists suggests, for example, that the precolonial population of the Yucatán Peninsula in southern Mexico may have been many times what it is today.

Latin America's pre-Columbian demography sheds important light on a long-running debate in the social and natural sciences about the relationship between population dynamics and agricultural change. Based on the writings of the British parson Thomas Malthus, scholars had long

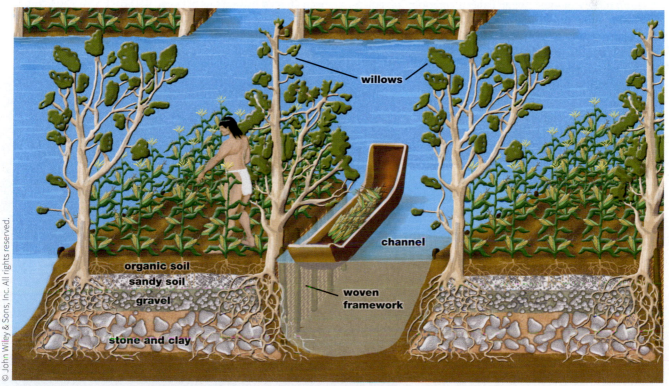

FIGURE 12.10 **Chinampa agricultural lands.** Mayans and later Aztecs created additional farmland by building raised beds called chinampas in lakes, ponds, and wetlands.

assumed that food production limited population growth. In his influential 1798 treatise, "An Essay on the Principle of Population," Malthus argued that agricultural production grew arithmetically, whereas population grew geometrically. Following this logic, population would inevitably exceed food production, leading to famine. For Malthus, famine was a "natural" check that rebalanced population to a sustainable level.

In 1965, the political economist Ester Boserup asserted that Malthus had it all backwards. It was not food production that set population levels, she said, but population that determined how much food could be produced. By looking at historical examples, such as the pre-Columbian situation in Mexico, Boserup argued that when population reached a certain density, food became scarce, and that this dearth drove people to innovate and intensify food production by building irrigation systems, applying organic fertilizers, and terracing in some cases (Boserup 1965).

Contemporary Latin American agricultural landscapes are typically divided into large-scale commercial farms, or **latifundias**, and small-scale subsistence farms, or **minifundias**. Many of the latifundias date from the colonial period when the Spanish crown granted **conquistadors** large landholdings called **haciendas**. Governments have made attempts at land redistribution in Latin America, and Chile is often lauded as having the most successful land redistribution program. Several countries in addition to Chile have attempted land distribution, but they continue to be hampered by highly inequitable access to land.

Latifundias include sugar cane plantations of northeastern Brazil and the Caribbean, wheat, potato, and dairy haciendas of the high Andes, fruit farms and vineyards of central Chile, and cattle ranches of northern Mexico and the Pampas of Argentina, Uruguay, and southeastern Brazil. *Minifundias* often punctuate *latifundias*. While minifundias are more numerous, they account for a smaller total area. Small farmers on minifundias work independently or spend part of the time as agricultural laborers on *latifundias*. Working for a larger farm while maintaining your own small farm, **agricultural dualism**, is a prevalent way of coping with inequitable access to land; this practice prevails throughout Latin America.

Farmers in less densely populated areas, most notably Amazonia, practice **slash and burn agriculture**, also known as shifting cultivation or swidden agriculture (Figure 12.11). With the slash and burn method, farmers clear an area of tropical rainforest and then burn the vegetative remains to release the plant nutrients back into the leached soils. A farmer will then farm the area for two to four years before abandoning it and moving to a new plot.

When such plots are small, the abandoned area regenerates with native vegetation. The farmer may return to farm the plot again in 20 to 30 years. Although some environmentalists dismiss slash and burn agriculture as wasteful, other scholars argue that it is a rational approach to farming when land is plentiful and labor is scarce. Slash and burn gives the land time to regenerate when a plot is abandoned, and in this way, farmers allow the soil to regenerate itself over long periods of time instead of relying on chemicals. If the population density is too high, farmers may allow less time for plots to regenerate. As population densities increase, slash and burn agriculture tends to transition into more intensive subsistence agriculture, employing more labor and fertilizer per unit area.

Relative to the rest of Latin America, the Amazon as a whole has a low population density (Figure 12.12). However, within the Amazon, areas of higher population density are found where pockets of people are practicing intensive gardening. Remarkably, the areas of intensive gardening date back to the pre-Columbian period and are located in patches of black soils, also known as terra preta or **black earth**. Scientists once thought that these patches of highly fertile soil in the midst of nutrient-poor swaths of land were naturally occurring, with the most common theory that the soils were deposited by volcanoes. Researchers now believe that humans created the black earth. Pre-Columbian farmers added a mixture of charcoal, bone, and manure to the infertile Amazonian soil over many years to make them nutrient-rich. Black soils owe their name to their very high charcoal content. Patches of black earth are thought to have been created by humans between 450 BCE and 950 CE, but there is an ongoing study seeking to determine whether the practice continued beyond 950 (WinklerPrins and Barrera-Bassols 2004).

© Edward Parker/Alamy.

FIGURE 12.11   Santarem, Brazil. Farmers used slash and burn to clear this field of Amazon rainforest near Santarem, Brazil to make way for planting soy beans.

Data from CIESIN—Columbia University, United Nations Food and Agriculture Programme—FAO, and Centro Internacional de Agricultura Tropical—CIAT. 2005. Gridded Population of the World, Version 3 (GPW3): Population Count Grid, Future Estimates. Palisades, NY: NASA Socioeconomic Data and Applications Center (SEDAC). http://dx.doi.org/10.7927/H42B8VZZ. Accessed December 2014.

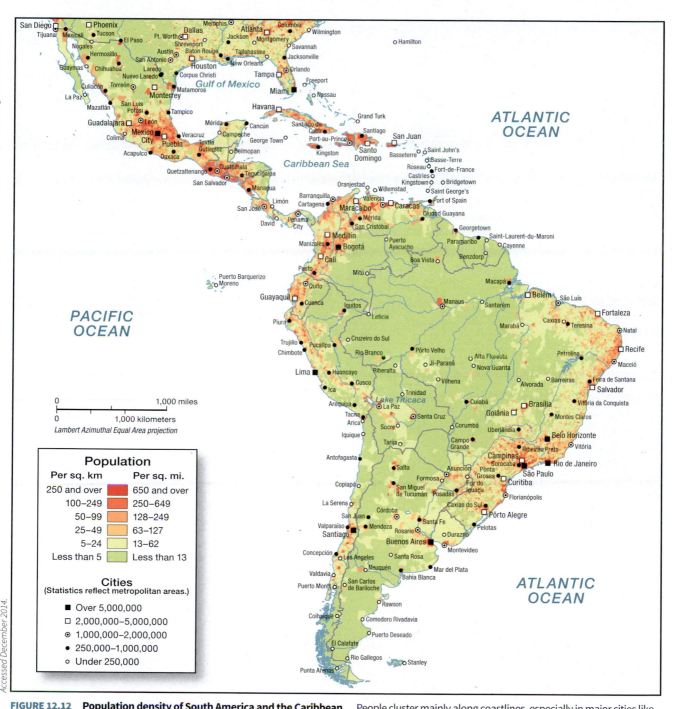

FIGURE 12.12 **Population density of South America and the Caribbean.** People cluster mainly along coastlines, especially in major cities like São Paulo and Rio de Janeiro, Brazil, which have more than 20 million and more than 11 million people, in the metropolitan areas respectively. Cities in Mexico are largely in the interior of the country, with the population of the metropolitan area of Mexico City topping out at over 21 million.

Intensive subsistence farmers occupy the mountainous areas of Latin America. In the Andes at lower elevations, between 4,500 ft (1,400 m) and 9,100 ft (2,800 m), farmers cultivate coffee, corn, and cotton. Between 9,100 ft (2,800 m) and 13,100 ft (4,000 m), farmers produce potatoes, wheat, and barley. Above the tree line, above 13,000 ft (4,000 m), land is used primarily for pasture land and for growing potatoes (Figure 12.13). Unlike swidden agriculturalists, these farmers will till a plot every year, occasionally letting it lie fallow. Where possible, some households have plots at varying elevations to allow for different crop mixtures.

LATIN AMERICAN BIODIVERSITY CONSERVATION

The region's high levels of biodiversity are what led Charles Darwin to conduct his research in Latin America and the Caribbean. Scientists conceptualize biodiversity at a variety of scales, and one of the most common scales is the total number of species found in a particular ecological region. South America has extensive humid lowlands because the continent is so wide at low latitudes near the equator. Within the humid lowlands, the region contains significant portions of tropical rainforest, which is the most biologically diverse climate region in the world (Figure 12.14).

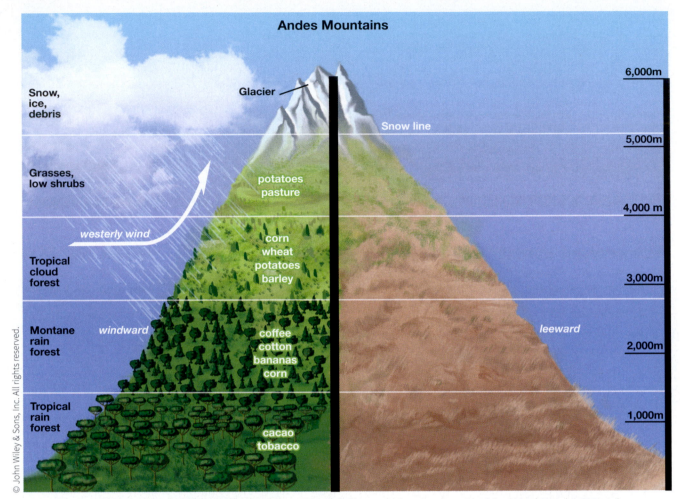

FIGURE 12.13 **Agriculture and altitudinal zonation.** The type of crop produced varies with altitude in the Andes Mountains. Coffee, bananas, and corn are grown at low elevations. Fruit, wheat, barley, and potatoes are grown at higher elevations, and livestock, including sheep, graze on pastures at higher elevations.

Reading the **PHYSICAL** Landscape

Jari Plantation

Misunderstanding of tropical ecosystems has often resulted in misguided management decisions. One particularly blatant example of this problem is the failed Jari Plantation in Brazil. This project began in 1967 near the intersection of the Jari and Amazon rivers. Anticipating a global shortage of pulpwood, American investor Daniel Ludwig established a plantation in the Brazilian Amazon based on the belief that this would be an enormously productive site for commercial soft wood production.

Ludwig cleared the indigenous vegetation and planted a fast-growing exotic tree species called Melina, even though the soils were never tested to see if they were

suitable for such a species. The company built a port on the Jari River, constructed a new town to house workers, and assembled a pulp processing plant out of parts shipped from Japan. By 1981, $1 billion had been invested in the project (roughly equivalent to $2.4 billion in today's terms, accounting for inflation).

It did not take long before problems began to surface. For starters, this massive **monoculture** (or planting of a single species) was particularly susceptible to pest infestations. Having destroyed the biodiversity of the original forest, which naturally controlled pest problems, the company used massive amounts of expensive and harmful pesticides to control pests. Serious erosion occurred between the time the forest was cleared and the newly planted trees took to grow to a sufficient height to protect the ground. This erosion, combined with the generally nutrient-poor soils (oxisols and ultisols) found in the Tropics, meant that the plantation was producing 40 percent below what had been projected.

Suffering huge financial losses, Ludwig sold the plantation in 1982 for $280 million ($720 million less than his investment). The lack of understanding of the physical geography of the rainforest and its soils resulted in a loss of financial resources and ecological destruction. Both could have been avoided if planners and investors had taken the time to understand the ecology of the Amazon rainforest (Jordan 1998).

Ken Gillham/Robert Harding Picture Library Ltd/Alamy.

FIGURE 12.14 **Jari River, Brazil.** The Jari project included a massive timber mill located at the intersection of the Jari and Amazon rivers.

Biodiversity is declining globally at unprecedented rates (Wilson 1999), which weakens ecological stability. Ecologists David Tilman and John Downing (1994) demonstrated that grasslands with more species diversity tolerated drought better than less diverse grasslands. Species in tropical rainforests have coevolved and depend on one another for their survival, a relationship known as **mutualism**. The loss of one species often leads to the decline or loss of another, which compounds the loss of biodiversity. A classic example of mutualism is leaf cutter ants and fungus (Figure 12.15). At least 47 species of leaf cutter ants are found in Central and South America. Leafcutter ants cut and process fresh leaves and grasses to serve as food for fungus they raise. The fungi need the ants to survive, and the ants feed the fungus they raise to their young.

Humans depend on and benefit from biodiversity. Approximately 45 percent of prescription medicines contain at least one product of natural origin. For example, the main ingredient in aspirin comes from the willow tree. By maintaining biodiversity, we safeguard a store of future, undiscovered medicines. Scientists monitor biodiversity in the rainforest through fieldwork on the ground and remote sensing from satellite images (Figure 12.16).

Biodiversity is also important economically, as **ecotourism** or nature-based tourism is a growth industry in Latin America. The use of parks or preserves to maintain biodiversity, or **preservation**, calls for humans to use biological resources not in a consumptive manner, but rather only for activities such as hiking or photography.

Christian Ziegler/© Danita Delimont/Alamy.

FIGURE 12.15 **Barro Colorado Island, Panama.** Leafcutter ants and the fungus where they live demonstrate the concept of mutualism because the ants process leaves that the fungi use as food. In turn, the ants feed the fungus to their young.

USING **GEOGRAPHIC** TOOLS

Remote Sensing of Deforestation Processes in the Amazon

Geographers use satellite images (often taken every 10 days at high resolution) in order to conduct studies of the patterns of land-cover change. This analysis, when combined with on-the-ground fieldwork, has deepened our understanding of the causes and consequences of land-cover change. Figure 12.16 is a satellite image of deforestation in the Brazilian Amazon. While deforestation is often blamed on poor peasants who practice slash and burn (or swidden) agriculture, the pattern in this image reveals a different process. What you see in the image are a series of logging roads. After dirt roads like

this are built timber companies log the forest around the dirt roads. Then, landless small farmers follow the paths of the dirt roads and move into the area to cultivate.

Thinking Geographically

1. If the Amazon did not have logging roads, what pattern would farmers follow in choosing new plots of land for slash and burn agriculture?

2. Notice the large dark brown area in the middle of the image, where vegetation has not grown back. Why do you think vegetation has not regenerated in this area?

NASA Earth Observatory.

FIGURE 12.16   Amazon, Brazil.   Deforestation spreads out from logging roads. The areas of the image where the brown and green create a checkerboard pattern are places where straight logging roads have been cut. Farmers move out from the roads to clear new swaths of forest to practice swidden agriculture.

Costa Rica has led the region and the world in its biodiversity preservation efforts. A centerpiece of these efforts is Costa Rica's well-developed national park system (Figure 12.17), which has made tourism Costa Rica's number one source of foreign exchange. The park system covers all of the major ecological regions in the country, often providing habitat for species that are only found in a particular area, highly **endemic species**. Yet, there are limits to the preservation approach in terms of maintaining biodiversity. Setting aside large tracts of land for parks or preserves often means that land

outside of these reserves must be used more intensively (Moseley 2009). The national parks in Costa Rica have sometimes been characterized as diamonds in a sea of devastation (Sanchez-Azofeifa et al. 2002) because of the concentrated land use around the parks.

To give a more specific example of this problem in a Costa Rican park, consider the Monte Verde cloud forest, which is a rainforest formed on the windward side of an upland area where moisture-laden air rises, cools, and precipitates. Monte Verde had been home to the Golden Toad (Figure 12.18), a highly endemic species. Each April,

FIGURE 12.17 **Tortuguera, Costa Rica.** The Tortuguero (turtle catcher) National Park on the east coast of Costa Rica is known for sandy beaches where turtles nest, along with sweeping views of rainforest.

FIGURE 12.18 **Monte Verde Preserve, Costa Rica.** The Golden Toad once inhabited a 4 square kilometer area of the Monte Verde Preserve. Tourists flocked to see the toads mate during the rainy season. Since 1989, the species has been extinct. Scientists are uncertain what led to the quick extinction of the species, but theories include pesticides, a fungus, and climate change.

tourists came to a certain area of the park where thousands of toads congregated to mate. Then the toads mysteriously disappeared, without a single individual being seen since 1989. While theories for their disappearance abound, one likely explanation is that, as amphibians have highly porous skin membranes, they succumbed to pesticides. Just below this cloud forest are large banana plantations where pesticides are used extensively. Some of these pesticides likely ascended the mountains in moisture-laden air and then precipitated out on to the Monte Verde forest. The broader concern really has to do with the limits of preservation, an approach that often results in a patchwork of land uses segmented by porous boundaries.

Some scholars argue that a broad application of **conservation**, which is different from preservation and calls for using land and species within their biological limits, or a sort of coexistence with nature, is the more effective approach to maintaining biodiversity. The biological limits of a renewable resource are defined in terms of a harvest that does not exceed the annual growth increment of the population. So, for example, if a fish population grows at 5 percent per year, then the annual harvest should not exceed this 5 percent growth.

Based on evidence from Latin America, ecologists Ivette Perfecto and John Vandermeer (2010) argue that agriculture need not be the enemy of biodiversity. For these scientists, an agroecological matrix composed of small-scale agricultural systems that use natural farming techniques holds the promise of maintaining biodiversity and improving rural livelihoods. This is quite different

from the large-scale, industrial agriculture currently seen outside of many parks in Costa Rica.

CONCEPT CHECK

1. **Why** does Latin America have a diversity of climate types?

2. **How** do the land ownership systems and agricultural practices set up during European colonialism still impact agriculture in Latin America?

3. **What** is the difference between conservation and preservation of biodiversity in Latin America?

WHO ARE LATIN AMERICANS?

LEARNING OBJECTIVES

1. **Describe** what racial and ethnic groups make up Latin America.

2. **Explain** why geographers call this region Latin America.

3. **Critically discuss** the roles of women and men in various areas of Latin America.

Latin America is composed of a mix of indigenous peoples, as well as people of Asian, African, and European descent. The ethnic and cultural mix of Latin America is notable on a number of fronts. Historians call this Latin America's **triple heritage**, referring to its main sources of people and culture: indigenous, African, and **Iberian**. Many Latin American countries still have sizeable indigenous populations. The region also has the greatest number of Africans outside Africa and Japanese outside Japan. In addition, it has the world's largest concentration of people speaking Romance languages, as well as the largest concentration of Catholics.

THE RACIAL AND ETHNIC COMPOSITION OF LATIN AMERICA

Latin America was the second to last region to be peopled by *Homo sapiens* (the South Pacific was last). Crossing the Bering land bridge from Asia about 14,000 years ago (or earlier), people pushed south into North America. They would soon make their way down through Central America, reaching the tip of South America about 12,000 years ago.

Scientists debate about the dates when humans reached different areas; a newer theory contends that migrants sailed from East Asia across the Pacific to the Americas, instead of crossing the Bering land bridge. Historians do know that many areas of Latin America were thickly populated before the Europeans arrived in the late fifteenth century. 90 to 112 million indigenous people lived in North America, Central America, South America, and the Caribbean before 1492.

The Maya and Aztecs of Mexico and Central America, as well as the Incas of South America, built complex and sophisticated civilizations, including highly productive farms, extensive road networks (especially the Incas), and cities (Figure 12.19). Within 150 years of contact with Europeans, indigenous populations in Central and South America declined by more than 90 percent to under six million. European diseases were the main culprit in this decline. Diseases including smallpox and measles, to which indigenous people had little immunity, decimated

Werner Forman/Universal Images Group/Werner Forman/Universal Images Group//Getty Images.

FIGURE 12.19 **Machu Picchu, Peru.** The Inca town of Machu Picchu at 7,970 feet (2,430 meters) above sea level in the Andes Mountains in Peru was never found by the Spanish. Rediscovered in 1911 by Hiram Bingham, a modern access road zig zags up the mountain so tourists can access the site.

the population. More than military power, disease and subsequent population collapse allowed the Iberians to control the entire region in about 40 years.

Today, indigenous people in Latin America comprise more than 12 percent of the population of Latin America, especially in Bolivia, Ecuador, Mexico, Peru, and Guatemala. The International Work Group for Indigenous Affairs (2014) estimates that 40 million people in Latin America belong to 600 different indigenous groups (Figure 12.20). The group found three commonalities among indigenous peoples. First, indigenous people live primarily in rural areas but are migrating increasingly to cities. Second, indigenous lands are threatened by resource extraction. Third, while most governments recognize indigenous rights, few follow through and protect indigenous rights.

With extensive loss of life among the indigenous people from diseases introduced by the explorers to the Americas, Europeans turned to slavery to provide labor on plantations

Reading the **CULTURAL** Landscape

Indigenous Peoples Along the Mexico–U.S. Border and in Mexico

The mix of indigenous peoples along the Mexico–United States border and in Mexico raises interesting questions about the nature of borders and the status of indigenous peoples. In the 1848 Treaty of Guadalupe Hidalgo, Mexico ceded half of its territory to the United States. The treaty set the border at the Rio Grande River, and former Mexican lands became

California, Texas, Arizona, western Colorado, Nevada, New Mexico, and Utah. The border between the United States and Mexico divided communities and created an artificial barrier between flows of people and goods (Figure 12.20a).

The cultural landscape of the border region reflects the complexities of the place. For example, the Yaqui people on the Yaqui Indian Reservation near Tucson, Arizona, and in ethnic neighborhoods in Tucson and Tempe, are indigenous people from the Mexican American border region. The border region switched hands among Mexico, Spain, and the United States, and the Yaqui ended up on the American side of the border. Thus, the Yaqui are considered indigenous Americans, but had the border been located farther north, they would be considered indigenous Mexicans.

The Yaqui settled in the region around 500 BCE. In 1500, the Yaqui began interacting with Jesuit missionaries from Spain. Today, most Yaqui speak Spanish, and many have blended traditional Yaqui and Catholic religions in a syncretic way (Figure 12.20b).

The Mexican government sees the status of its indigenous peoples differently from the United States or Canada. In Mexico, indigenous peoples are defined as a cultural group who have the right to preserve their culture and language under the Mexican constitution. The map of indigenous peoples in Mexico (Figure 12.20c) is based on speakers of indigenous languages. Indigenous peoples in Mexico do not have reserves or reservations, but the constitution protects their cultures.

Data used with permission from: Global Mapping International, World GeoDataSets.

FIGURE 12.20a   Mexican lands in 1824.   The region around the current border between the United States and Mexico changed hands several times between the sixteenth and nineteenth centuries. This map shows the states of Mexico in 1824. Note that the Rio Grande border, drawn in 1848, does not coincide with how Mexico was organized before that time.

(Continued)

FIGURE 12.20b   Hermosillo, Mexico. Members of the Yaqui tribe celebrate Holy Week, leading up to Easter. Jesuit missionaries brought Catholicism to the Yaqui in the 17th century. Yaqui males dress as Pharisees and wear masks as part of an extravagant pageant that culminates on Easter.

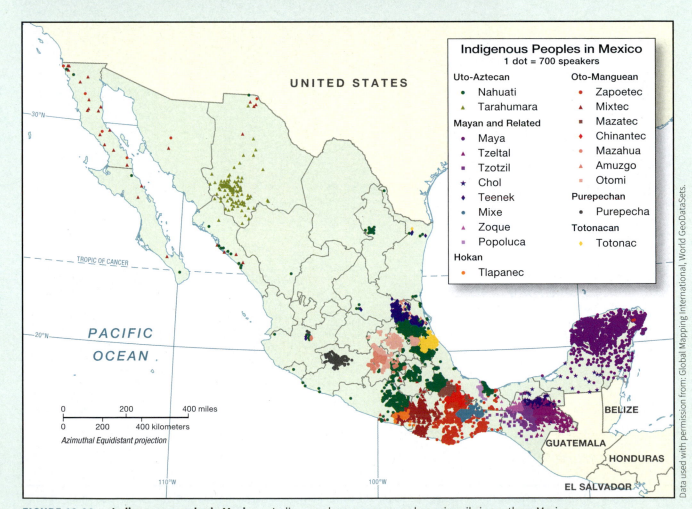

FIGURE 12.20c   Indigenous peoples in Mexico.   Indigenous languages are spoken primarily in southern Mexico.

(see Chapter 11). The Atlantic slave trade brought enslaved Africans to the New World from 1502 until the nineteenth century. European traders enslaved and transported between 12 and 28 million Africans and sold them primarily in the Caribbean, Brazil and Mexico. Cartagena, Colombia functioned as a major port in the slave trade. In South America, enslaved people were held in warehouses called factorias and then transported into the central part of the continent to labor in gold and silver mines (Brodzinsky 2013).

Great Britain banned the slave trade in 1833, and Portugal followed suit in 1836. However, the slave trade continued in the Caribbean and Brazil some 30 years after these bans, finally dying out in the 1860s. In the 1800s, Afro-Latin Americans helped fight in Latin America's revolutionary wars for independence from Spain and Portugal. Intermarriage and mixing among blacks, indigenous, and Iberians have been relatively common since the 1800s in Latin America. In 1810, Buenos Aires, Argentina, reported a population that was 20 percent black, but in the 2010 census only 0.4 percent of the country self-identified as black (Broadinsky 2014). Geneticists in Argentina estimate that 10 percent of the country is of African descent, but the population is not very visible because of intermarriage. In other countries such as Brazil, the African population is sizeable and visible. The 2010 census found that the majority of the population in Brazil was African or of mixed race for the first time since the census was conducted in 1872: 50.7 percent of the population identified themselves as black (7.6 percent) or mixed race (43.1 percent) (Phillips 2011). Despite intermarriage and the large population, people who are black or mixed race report significantly lower earnings than whites. A study by Data Popular Institute released in 2011 found that 82.3 percent of the wealthiest Brazilians are white and only 17.7 percent are African-Brazilians. Among the poorest in Brazil, the populations flip: African-Brazilians comprise 73.6 percent and whites 23.7 percent (Phillips 2011).

The number of Iberians who migrated to Latin America in the sixteenth and seventeenth centuries was relatively small compared to the size of the indigenous population, or to the enslaved African and their descendants. Nevertheless, Iberians held positions of political and military power and wielded considerable cultural influence. Roughly 240,000 Spaniards (mostly conquistadors) came to the Americas in the sixteenth century, followed by another 450,000 in the seventeenth century. About 100,000 Portuguese arrived in Brazil during a similar period.

British, Irish, French, Germans, Dutch, Chinese, Japanese, Koreans, and South Asians also migrated to the region in later periods. The more recent waves of immigration include about eight million Europeans from 1870 to 1930. These immigrants from Italy,

Portugal, Spain, and Germany mostly settled in the southern parts of Latin America. Many Asians (mostly Chinese, Japanese, and Korean) arrived in the nineteenth and twentieth centuries as contract workers or economic migrants. Many of the initial Asian immigrants crossed the Pacific from China and Japan to work on coffee plantations or formed agricultural colonies. More recent Korean immigrants settled in urban areas in Paraguay, Brazil, and Argentina.

 The idea of **race** is contested in the social sciences. From a genetic standpoint, genes do not tell us the color of our skin but do tell us where our ancestors came from. Race, then, is not biologically defensible, but racial identities enforced by society do often lead to differentiated social experiences. In other words, race is more of a social construction than it is a category backed up by real biological differences (Figure 12.21).

---

 race　social constructions of differences among humans based on skin color that have had profound consequences on rights and opportunities.

Four Different Racial Groups (oil on panel), Islas, Andres de (fl.1772)/Museo de America, Madrid, Spain/Bridgeman Images.

FIGURE 12.21　**Casta art.**　In eighteenth century Mexico, hundreds of oil paintings showing couples from different races (mainly Spanish, indigenous, and African) and their children helped define a new mixture of races in Latin America. Each painting shows a mother of a certain race, a father of another race, and then names the race of their child.

FIGURE 12.22　**São Paulo, Brazil.**　Japan Town is one of several ethnic neighborhoods in the cultural landscape of São Paulo.

Latin America is notable for mixing many peoples and cultures through generations of intermarriage among ethnic groups. People in the region are often referenced in terms of four broad categories: Blanco (European ancestry), **Mestizo** (mixed ancestry), Indio (indigenous or Indian ancestry), and Negro (African ancestry). Privileged during the colonial era, Blancos are often overrepresented in the elite classes of society and have dominated politics. Racism—the privileging of one racial group over another—persists in many Latin American countries.

Brazil is sometimes held up as an example of racial mixing and harmony, a place where race no longer plays a central role in determining social hierarchies (Figure 12.22). In contrast, anthropologist J. H. Costa Vargas (2004) has argued that Brazilian society has a hyperconsciousness of race that contradicts the "vehement negation of race" in public discourse. After decades defending the myth of "racial democracy," the Brazilian state admitted to racism in 2001. Shortly thereafter, the state endorsed a quota-based approach to affirmative action for Afro-Brazilians in government service and higher education (Htun 2004).

THE RATIONALE FOR LATIN AMERICA AS A REGION

Latin America, which commonly includes Central America, South America, and the Caribbean, refers to a contiguous region largely colonized by Spain and Portugal starting in the fifteenth and sixteenth centuries. The region stretches from Mexico and Central America in the northern hemisphere, through the islands of the Caribbean, and encompasses all of South America, largely found in the southern hemisphere. As a result of the region's colonial heritage, its dominant languages are Spanish and Portuguese (in Brazil) and the dominant religion is Catholicism. The British colonized Belize, Guyana, Jamaica, Trinidad and Tobago, and smaller island nations; the French colonized Haiti, French Guiana, and smaller isles; and the Dutch colonized Suriname and Aruba.

The Spanish were the first Europeans to colonize the region, beginning with Christopher Columbus laying claim to the island of Hispaniola (now split between the Dominican Republic and Haiti) for Spain in 1492. Spain then colonized the Greater Antilles (Puerto Rico, Jamaica, and Cuba), with its interests spreading to Central and South America. Spain established two capitals in the New World. The first was at the contemporary site of Mexico City, on top of the Aztec capital city of Tenochtitlán. Initially, the Spanish could not overcome Aztec defenses and captured the city only after disease had crippled its population. The Spanish referred to this seat of power as the Viceroyalty of New Spain (administering land between what is now Panama and California). The Spanish established a second seat of power at Cusco, Peru (the old Inca capital) and named it the Viceroyalty of Peru. At first, Spain was primarily interested in gold and silver, but conquistadors soon recognized the productivity of agricultural land in the Caribbean and Central America and became involved in crop production.

Unlike the highlands of South America or Mexico, which were densely populated by indigenous people prior to colonization, Amazonian Brazil had an indigenous population of only 2 to 5 million people prior to contact. Not finding gold or silver, the Portuguese initially harvested brasa wood (the eventual namesake of the country) for export. The Portuguese increasingly developed Brazil as a source of provisions and then (by the late sixteenth century) as a major center of sugar production.

The Iberians essentially invented a new form of **colonialism**, different from the colonial empires of antiquity (such as the Egyptians, Greeks and Romans). First, Iberian colonies were much farther from the power center of the colonizer than previous colonial ventures. Second, Iberian colonialism included supplanting the existing language and religion with those of the colonizer (Figure 12.23). Finally, the highly extractive nature of Iberian colonialism marked a distinct departure from previous colonial endeavors. The main purpose of the Spanish and Portuguese colonies in Latin America was to extract resources for the homeland, in keeping with **mercantilism**, a form of early capitalism in Europe that prioritized a positive balance of trade and the accumulation of gold and silver reserves. Indigenous economies were destroyed (or underdeveloped), and indigenous cultural practices and belief systems were aggressively dismantled.

In addition to granting haciendas, the Spanish ran their colonies by the **encomienda** system, in which the crown granted a person, often a conquistador (or former soldier or explorer), a specified number of indigenous people for whom he would take responsibility. He was to protect his charges from warring tribes, teach them the Spanish language, and instruct them in the Catholic faith. In return, the holder of the encomienda could extract payments from the indigenous people in labor, gold, or agricultural products.

By the second decade of the nineteenth century (1810–1826), Iberian colonists led a number of revolutions across Latin America leading to independence. In almost every instance, descendants of Spanish and Portuguese colonizers became the rulers of the new countries. The new governments did not allow indigenous people in the region to vote for at least another 100 years.

The Iberian influence that makes Latin America Latin is also seen in the dominant languages and religion of the region. The Portuguese began colonizing the region after the Treaty of Tordesillas (1494), in which the Pope divided the region at approximately 46°W longitude between the Spanish and the Portuguese. Although the British, Dutch, and French never recognized the Treaty of Tordesillas, the line is apparent in the language map of the region, with Spanish the dominant language west of the line and Portuguese the dominant language east of the line (Figure 12.24). Approximately two-thirds of the Latin American population speaks

FIGURE 12.23 **Cusco, Peru.** A Society of Jesus (Jesuit) Catholic Church is sited prominently on the Plaza de Armas in the center of Cusco, which was once the capital of the Incan empire.

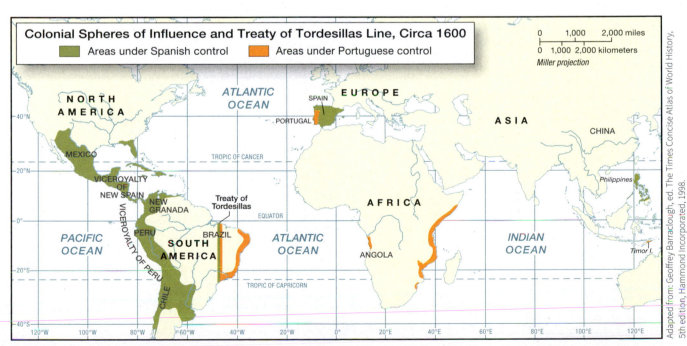

FIGURE 12.24 **Colonial spheres of influence and Treaty of Tordesillas line.** In 1494, just two years after Columbus landed on Hispaniola in the Caribbean, the Treaty of Tordesillas established a line of longitude a certain distance west of the Cape Verde Islands off the coast of Africa. The treaty designated Spain would colonize what was west of the line and Portugal would colonize lands east of the line.

Spanish, and one-third speaks Portuguese. Numerous Amerindian languages are still spoken in more remote places (Figure 12.25).

The majority of the Latin American population practices Roman Catholicism (Figure 12.26), although **syncretic religions** (a mixture of religious practices) have been formed to include traditional beliefs into the Catholic faith. Protestant denominations are also becoming more popular.

GENDER ROLES IN FLUX

Whereas sex is biologically determined, **gender** refers to the roles, behaviors and activities that a society considers appropriate for men and women, boys and girls. Social constructions of gender affect a person's identity, development, livelihood, and relations with the natural world.

> **gender**   socially constructed notions of roles, behaviors, and activities that are appropriate for men and women.

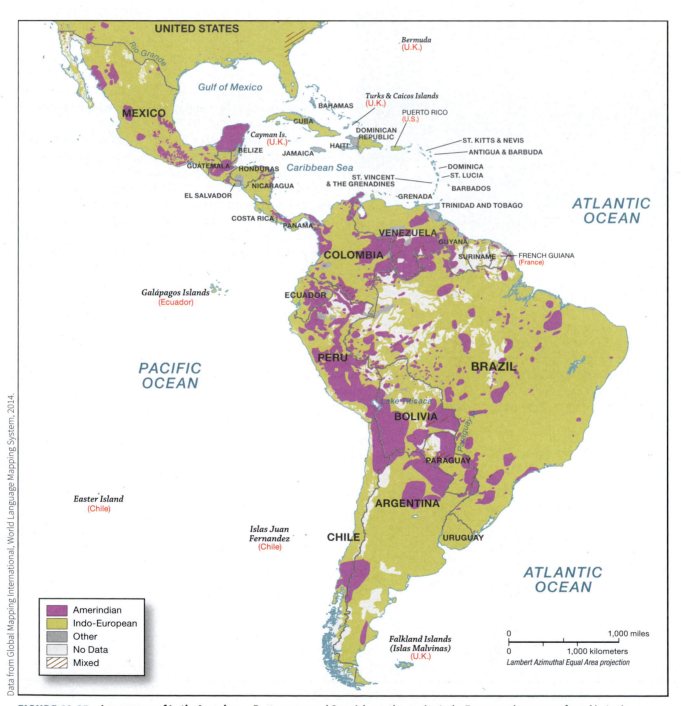

Data from Global Mapping International, World Language Mapping System, 2014.

Legend:
- Amerindian
- Indo-European
- Other
- No Data
- Mixed

FIGURE 12.25   Languages of Latin America.   Portuguese and Spanish are the major Indo-European languages found in Latin America. Amerindian languages are still found throughout the region, especially in western South America and southern Mexico.

FIGURE 12.26   Religions in Latin America.   Catholicism is the predominant religion in Latin America, as both Spanish and Portuguese colonizers brought their religion, Catholicism, to the region. Since the 1960s the region has experienced a surge in the number of Pentecostal Protestants with the highest growth in Brazil, Chile, Argentina, and Central America.

Stereotypes of Latin America suggest that gender roles are particularly strict. The idea of **machismo**, or strong masculine roles that make one manly, are perceived to be extreme in the region. Gender roles are not static. Heads of state in the region have traditionally been men, but since 1999, several women have been elected to office. Argentina elected Cristina Fernández de Kirchner as its first female president in 2007 (and second female president after the nonelected Eva Peron). Kirchner joins other female Latin American heads of state, including Mireya Moscoso, president of Panama (1999); Michelle Bachelet, president of Chile (2006); and Laura Chinchilla, president of Costa Rica (2010). As the economy and migration patterns have changed in Latin America, so too have gender roles.

Women traditionally employed in the informal economy increasingly found employment in the formal economy in the late twentieth century. Women largely staff Latin America's export-oriented industries, including maquiladoras (foreign-owned assembly plants), in Mexico (Wright 1997) and the cut flower industry in Ecuador (Coulson 2004). While jobs in the export sector are being made available to women, they are often accompanied by unfair labor conditions and unequal compensation, which reflects both local and global sexism. Female workers also experience sexual harassment and assault (Wright 1997).

Maquiladoras are tied directly to the American and Canadian economies. Maquiladoras produce finished products from components and raw materials that are imported into Mexico duty free. Upwards of 80 percent of the finished products made in maquiladoras are exported to the United States for consumption.

The maquiladora program began in 1965 when the Mexican government created the Border Industrialization Program (BIP). Initially, the program applied to regions within 12 miles (20 kilometers) of the Mexican border and allowed materials to be imported duty free. During the first two decades of the program, maquiladoras were concentrated in border cities, including Matamoros, Juárez, Nuevo Laredo, and Tijuana. Since 1985, the Mexican government has allowed maquiladoras to be built anywhere in the country except the three major urban areas that are already too congested.

The peak of maquiladora production followed the passage of the North American Free Trade Agreement (NAFTA) in 1994. Before passage of the agreement, about 1,800 maquiladoras operated in Mexico. Within five years, nearly 4,000 were in operation. As of 2013, approximately 3,000 maquiladoras operated (Figure 12.27).

Maquiladoras, most of which are owned by American companies, employ 1.2 million Mexican workers and account for one-third of all manufacturing jobs in Mexico.

The vast majority of workers in maquiladora factories are female both because they are typically paid less and because employers preferred female workers. This preference for female workers is largely based on gender stereotypes held by Mexican and American factory managers that women are more dexterous and therefore better at assembly work (Figure 12.28). Because of the fast pace of assembly lines, many workers suffer from repetitive use injuries. Maquiladora goods are exported to the United States, generating profits for American companies; American consumers then purchase these goods with little to no understanding of the occupational hazards involved with their production.

Change often comes slowly to gendered roles and opportunities. In a study in El Hatillo, Nicaragua, American geographer Julie Cupples (2004) describes how women responded in a strategic and calculated way to the constraints of sexism and poverty. Women were primarily engaged in unpaid work in the community or in agriculture, whereas men controlled the more profitable work. An outside aid agency supported a milk cow project in the community (primarily for home consumption). The project enabled women to fill their maternal roles of providing

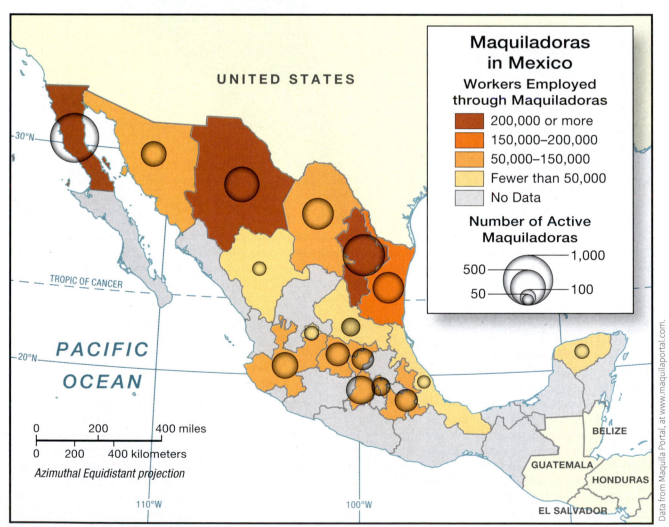

Data from Maquila Portal, at www.maquilaportal.com.

FIGURE 12.27 **Maquiladoras in Mexico.** Northern states in Mexico have more active maquiladoras and more workers employed through maquiladoras than central states.

FIGURE 12.28 Jerez, Mexico. Young Mexican women work as seamstresses in a maquiladora that produces clean room garments for high tech companies.

food and nutrition for the family without challenging the gendered division of labor in the family. One could argue that the project perpetuated traditional roles. But Cupples suggests that the project also empowered women, since it helped them address a critical need and to see that their problems were not insurmountable. She concluded that women in the community were "strategically working towards gender equality in a way which was *seen* as less disruptive to gender relations."

CONCEPT CHECK

1. **What** are the major racial and ethnic groups that make up Latin America?

2. **Why** is colonial heritage used as one of the major factors defining Latin America as a region?

3. **How** are women's roles in Latin America changing?

HOW HAVE EXTERNAL AND INTERNAL FORCES AFFECTED LATIN AMERICA'S DEVELOPMENT?

LEARNING OBJECTIVES

1. **Critically evaluate** the influence of outside actors on development in Latin America.

2. **Explain** how the international debt crisis and subsequent policies influenced development in Latin America.

3. **Discuss** how indigenous movements have influenced politics in Latin America.

As the first region of the world to be fully colonized by Europeans, Latin America has been a laboratory for oppression and extraction. Yet, such oppression has bred resistance and tremendous experimentation with new forms of development, politics, and governance. In this section we explore the complicated tango between internal and external actors that have made Latin America what it is today.

OUTSIDE ACTORS AND DEVELOPMENT IN LATIN AMERICA

Even after independence, which the majority of Latin American countries achieved in the nineteenth century (Figure 12.29), outside powers continued to be a force in the region. The United States long considered Latin America to be its own backyard. The Monroe Doctrine, issued in 1823 by United States President James Monroe at a time when many Latin American countries were on the verge of becoming independent, warned European powers to not interfere in the Western Hemisphere. The United States issued the Monroe Doctrine at a time when it was a very weak international power, and as a result, the doctrine can be seen, in a sense, as an articulation of American moral opposition to colonialism. Subsequent American leaders reinterpreted and applied the doctrine in a variety of ways. Some presidents, such as Theodore Roosevelt, used the doctrine to assert the right of the United States to intervene in order to stabilize the affairs of small countries in the Caribbean and Central America.

American companies have also seen Latin America as their backyard, whether to produce consumer goods in maquiladoras or to grow commodities for export to the United States. In the late nineteenth century, American business tycoons realized they could grow bananas very cheaply in the Caribbean and Central America, then sell

FIGURE 12.29 **Political geography of Latin America and the Caribbean.** The region of Latin America includes Mexico, seven countries in Central America, 13 island nations in the Caribbean, and 12 countries in South America.

them for huge profit margins in the United States. The American-owned United Fruit Company was infamously called the octopus (or "El Pulpo") because it freely interfered with Honduran national politics in order to ensure that local policies were favorable to its business model. United Fruit owned several large plantations in Honduras and Guatemala, as well as key elements of the national infrastructures, including the railroads and ports (Figure 12.30). In 1904, the American writer O. Henry coined the term **Banana Republic** to refer to a small country that is overly dependent on one export commodity (like bananas)

and is ruled by a corrupt elite. Honduras and Guatemala were the classic cases of this distorted form of development during much of the twentieth century, though United Fruit reached beyond these countries.

In the 1930s, the global depression led to a decline in foreign sales. Several Latin American countries exported primary products and imported almost all of their manufactured goods at the time. Latin American countries realized the lack of domestic production made their economies vulnerable, and they began to encourage domestic production of goods they needed.

FIGURE 12.30   **Puerto Cortes, Honduras.**   In this 1954 photo, United Fruit Company workers transfer bananas for export by ship.

Through **import substitution**, Latin American countries encouraged manufacturing of goods at home rather than importing them in order to break free from dependency. Argentine economist Raoul Prebisch theorized a model of industrialization through import substitution and witnessed the model diffuse throughout Latin America. Import substitution led to the nationalization of foreign-owned operations and the proliferation of state-run companies or **parastatals**.

Mexico has an enormous parastatal, or state-owned company, in the Pemex (Petroleos Mexicanos) oil corporation. In 1938, the government established Pemex and claimed monopoly rights over oil and natural gas. Pemex is the largest corporation in Latin America, with annual revenues of $86 billion (Figure 12.31). The government of Mexico taxes Pemex revenues at a rate of 62 percent, generating 40 percent of the Mexican government's revenues.

The future of Pemex's oil production is questionable because production at the Cantarell oil field, which accounts for 75 percent of the company's oil output, is declining. Mexico's oil is found on the coast of the Gulf of Mexico, formed at similar times and through the same process as the oil found on the northern Gulf of Mexico in the United States. Oil fields farther off the coast of Mexico could possibly be tapped, but Pemex pays so much revenue to the Mexican government that it lacks the funds necessary for high-tech exploratory research into new production fields.

From the 1950s through the 1980s, the Cold War influenced politics in the region. Two global powers, the United States and the Soviet Union, maneuvered to secure client states (countries that followed one of the superpowers) throughout the Global South, with Latin America being a key terrain. U.S. concern about the spread of communism was greatly heightened when the Cuban Revolution ousted its staunch ally Fulgencio Batista in 1959. Cuba's new president, Fidel Castro, aligned with the Soviet Union.

Fearful of a growing communist presence in Latin America, President John F. Kennedy launched a major foreign assistance initiative in the early 1960s known as the Alliance for Progress. While the initiative was ostensibly about development (e.g., building wells and schools), it was really about developing allies so that communism would not spread further in the region. In some cases, the United States resorted to covert means to support a leader of its liking or to overturn one whose policies it did not like. For example, Chile democratically elected Salvador Allende, an openly Marxist politician, president in 1970. In response, a CIA-backed military coup ousted Allende in 1973, which paved the way for the anticommunist Augusto Pinochet to take control. The United States achieved the short-term goal of putting an anticommunist

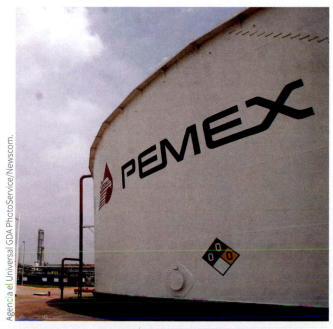

FIGURE 12.31   **Mexico.**   The Pemex logo adorns a storage facility at an oil refinery. The state-owned company has a monopoly on oil production and processing in Mexico and is investing billions to expand its refineries.

leader in power, but its success had long-term adverse consequences. Pinochet established a military dictatorship and governed with little regard for human rights, even using torture as a method of control until 1990.

THE INTERNATIONAL DEBT CRISIS, POLICY REFORM, AND DEVELOPMENT IN LATIN AMERICA

Commercial banks in Europe and North America lent massive funds to many developing countries in the 1960s and 1970s, including several in Latin America. Latin America's economy experienced high levels of growth after World War II, which made the region attractive for investment from the West. Western banks had **petrodollars**, revenue deposited by Middle Eastern countries from oil sales, to reinvest in order to turn a profit. Mexico, Argentina, and Brazil were some of the largest recipients of petrodollar loans (see Chapter 2).

Rising oil prices in the 1970s, coupled with inefficient state-run enterprises, led many Latin American governments, and the global financial system, to the brink of the so-called **Third World Debt Crisis** in 1982 (see Chapter 2). In early 1982, the Mexican government declared a moratorium on all debt repayments. Many commercial banks were alarmed because they had unwisely overinvested in a handful of Latin American countries. If Mexico continued to default, and other countries followed suit, then several banks could literally be wiped out. The U.S. government, World Bank, and International Monetary Fund (IMF) stepped in to help Mexico reschedule its loan payments and avoid defaulting, thereby averting a much more serious crisis.

While major creditors emerged from the Third World Debt Crisis relatively unscathed, the crisis had long-term implications for a number of Latin American countries. The key objective of the world's bankers was to ensure debtor countries kept paying down their loans. In order to do so, the World Bank and the IMF more or less forced indebted Latin American governments to implement a set of neoliberal policy reforms that called for increasing exports, slashing government expenditures, privatizing state-run enterprises, and devaluing national currencies (see discussion of structural adjustment policy in Chapter 2).

For better or for worse, Latin America was the original laboratory for the neoliberal economic policies subsequently applied across the Global South in the 1980s and 1990s. The first experiment with shock policy or shock therapy, a radical neoliberal reform, was undertaken in Chile in 1975 after the military coup by Augusto Pinochet. Pinochet reformed the economy based on the ideas of the economist Milton Friedman from the University of Chicago. Chilean economists who designed the policies had studied with Friedman and were collectively known as the "Chicago Boys." Their experiment in Chile, which emphasized the near-complete and quick withdrawal of government from the economy, was

an early form of the structural adjustment package that was broadly applied during the 1980s and 1990s.

Bolivia was the principal in the second experiment with shock therapy in 1985 as an attempt to stop hyperinflation. Although economists consider Bolivia's structural adjustments a success because it avoided hyperinflation, critics claim that the policies deepened poverty by leading to the closing of many state-run enterprises and increasing unemployment.

INTERNAL ACTORS AND DEVELOPMENT IN LATIN AMERICA

At the beginning of the twenty-first century, Latin America was "one of the regions of the world with the greatest inequality" according to the World Bank (2003). At that time, the wealthiest 10 percent of the people in the region earned "48 percent of the total income, while the poorest tenth" earned only 1.6 percent (World Bank 2003). The report cited unequal access to education and opportunities for Afro-Latin Americans and indigenous people as a major factor in creating unequal development.

Two main factors—government cash transfer programs aimed at redistributing wealth to the poorest in Latin America, such as Bolsa Familia, and improved levels of educational achievement—have helped to decrease the Gini coefficient in the region since 2000. In 2010, the Gini coefficient for Latin America was 0.5, which was down from 0.54 in 2000. A score of 0 means perfect equality in distribution of wealth, while a score of 1 means perfect inequality in wealth distribution. Although the Gini coefficient is moving toward more equality, generating economic development in Latin America is complicated by inequality in land ownership and prevalence of diseases (Figure 12.32).

BOLIVIA

More so than in any other world region, indigenous movements in Latin America are becoming a potent political force; these movements are particularly vigorous and successful in Bolivia, Peru, Ecuador, Guatemala, and Mexico. Bolivia offers an interesting case study of the rise of indigenous power. It is the poorest country in South America, with 37 percent of its population living in extreme poverty. The majority (65 percent) of its population is indigenous. The accumulation of rights and expression of power by indigenous peoples has gradually built momentum.

In 1952, a popular revolution led by Victor Paz Estenssoro in Bolivia toppled a conservative and oligarchic government and brought about several important changes. The new government nationalized the mines (Bolivia was and is a major producer of tin). It also abolished the hacienda system, redistributing much of this land to smaller producers. Most importantly for indigenous people, the government granted universal suffrage for the first time. Then, in 1964, a military junta overthrew President Paz Estenssoro at the outset of his third term. The coup ended the National Revolution and marked the beginning of nearly 20 years of military rule in Bolivia.

GUEST *Field Note*   Ethnographic Research in Mexico

DR. ERIC CARTER
Macalaster College

Before becoming a geographer, I traveled extensively around Latin America and fell in love with the distinctive cultures and diverse landscapes of the region. In formal field research in Argentina, Ecuador, and Mexico, I have employed a mix of methods based on archival research, ethnography, and landscape observation. The evidence for my book, *Enemy in the Blood: Malaria, Environment, and Development in Argentina*, comes mainly from archival documents, such as public health records, government reports, newspaper accounts, and topographic maps. Immersion in the timeworn documents of archives and libraries was vital to this project, since malaria is, for the most part, a thing of the past in Argentina. Yet ethnographic work and landscape observation helped me to imagine malarial environments and public health strategies from an earlier era.

For a couple of months in 2002–2003, I accompanied malaria control brigades and researchers studying malaria-carrying mosquitoes near the Bolivian border. This experience let me develop a much better sense of social and ecological characteristics of the region that have made it a hotbed for vector-borne diseases. Other local experts, or key informants, helped me visualize environmental change that had occurred in the region over the previous century, mostly from agriculture, forestry, and urban expansion.

All over Latin America, I have found that people in rural areas enjoy talking about their livelihood practices, that is, how they interact with the environment to produce food and

ERIC CARTER.

FIGURE 12.32   The Rio Grande in Jujuy, Argentina.   Earlier in the 20th century, this river bed was a notorious breeding ground for malaria-carrying mosquitoes.

make a living. Doing fieldwork, I've learned how to irrigate fields of corn in Mexico, how to plant manioc in northeast Argentina, and how to harvest shrimp in the mangrove estuaries of Ecuador. More recently, I have turned to ethnographic research of Latino immigrants in the United States, to understand how livelihood practices in their home countries shape values that may help them adapt to U.S. society. Whether it takes place abroad or in your own backyard, geographic field research requires, above all else, curiosity, patience, keen observation, and a place about which you feel passionate.

By 1985, Bolivia was experiencing hyperinflation, and the World Bank recommended structural adjustments, including devaluation of the currency, reprivatization of nationalized industries, and the layoff of some 30,000 tin workers. These policies led to considerable social strife and the rise of indigenous movements.

Between January 1999 and April 2000, indigenous and local cocoa growers led a series of protests in Cochabamba (Bolivia's third largest city). The World Bank extended a loan to the government under the condition that the city privatize the municipal water supply. When a private company took control of the water supply, it sought to raise rates an average of 35 percent to about $20 a month in a country where the minimum wage was $70 per month, and began turning off water to those who did not pay. The ensuing "Cochabamba Water Wars," marked a pivotal moment for mobilizing local

people against the government and the World Bank. Indigenous groups, labor unions, and coca growers all joined in public demonstrations against rate hikes (Figure 12.33). After weeks of blockades and violent confrontations between protesters and security forces, the investors left and the Bolivia ended its water privatization plan.

In 2003, many of the same protestors, coca growers, indigenous people, and labor unions, upset about the government's decision to export natural gas through Chile (an unpopular neighbor for many Bolivians), as well as changes in coca laws, mounted strikes and roadblocks, bringing the country to a standstill. The press deemed the protests the Bolivian "Gas War." Protestors demanded the resignation of the president, Gonzalo Sánchez de Lozada. Bolivian armed forces violently suppressed the protestors, leaving some 60 people in the city of El Alto

ASSOCIATED PRESS.

FIGURE 12.33 **Cochabamba, Bolivia.** The "Water War" was a series of protests by coca growers and indigenous farmers held between 1999 and 2000 over rising water costs resulting from privatization.

AIZAR RALDES/AFP/Getty Images.

FIGURE 12.34 **El Alto, Bolivia.** During the "Gas War" protests, indigenous Bolivians demanded the resignation of the president of Bolivia and the nationalization of natural gas reserves. The government violently cracked down on the protests, killing 60 protestors. In 2006, the first indigenous head of state in Latin America was elected by Bolivians.

dead in October 2003 (Figure 12.34). The protestors, who included future president Evo Morales, demanded full nationalization of hydrocarbon resources, and increased participation of Bolivia's indigenous majority, mainly composed of Aymaras and Quechuas, in the political life of the country. The president eventually resigned, and the people elected Evo Morales president in 2006. Morales is the first person of indigenous descent in Latin America to hold his country's highest office. He survived a recall vote in 2008 and won a second term at the end of 2009.

CONCEPT CHECK

1. **How** have outside actors attempted to influence Latin American politics in the post-independence era?

2. **What** types of policy reforms started to be implemented in Latin America after the international debt crisis in the early 1980s?

3. **In what ways** did local groups, including indigenous movements, in Bolivia react to structural adjustment reforms?

WHAT ARE THE CARIBBEAN'S UNIQUE DEVELOPMENT CHALLENGES AND OPPORTUNITIES?

LO

LEARNING OBJECTIVES

1. **Describe** the colonial history of the Caribbean subregion.

2. **Critically review** innovative development approaches in the Caribbean.

3. **Explain** the evolving economic geography of the Caribbean.

While the Caribbean shares much in common with the rest of Latin America, its slightly different colonial history and unique challenges and innovations make it a distinct subregion.

THE COLONIAL HISTORY OF THE CARIBBEAN SUBREGION

In 1492, Christopher Columbus landed on the Island of Hispaniola (now split between Haiti and the Dominican Republic), and thought he had sailed around the world and reached Southeast Asia. Over time, Europeans, realizing that the Caribbean was not in Asia, began to refer to the region as the **West Indies** (Figure 12.35). The Caribbean countries are typically divided into two groups of islands, the Greater Antilles and the Lesser Antilles. The **Greater Antilles** refers to the islands in the north and west, including the larger islands of Cuba, Jamaica, and Puerto Rico. The **Lesser Antilles** refers to the islands in the southeast, stretching in a crescent shape from the Virgin Islands southeast through St. Lucia and Grenada and west to Aruba.

While some countries in the Caribbean share Latin America's Iberian colonial heritage, most notably the islands of Cuba, Hispaniola, and Puerto Rico, several of the other islands (including the Haitian half of Hispaniola) fell under French, British, or Dutch colonial control. European powers scrambled for the Caribbean in order to establish sugar plantations, as the demand for sugar grew each year. Power battles, including piracy and sabotage, among the colonizers extended throughout the sixteenth, seventeenth, and eighteenth centuries. European diseases almost completely wiped out the indigenous population on the islands. Colonizers brought millions of enslaved Africans to the Caribbean to work on plantations. The few indigenous survivors of European diseases were enslaved and generally intermixed with Africans.

Haiti was the first colony in Latin America to gain independence (in 1804). In 1833, Great Britain abolished the slave trade. The Dominican Republic became independent in 1844 and Cuba in 1898. Colonized during the first wave of colonialism, many Caribbean countries did not achieve independence until the twentieth century, much later than the rest of Latin America and more in step with the second wave of decolonization. The importance of sugar production made Britain extremely reluctant to relinquish control of its Caribbean colonies. After World War II, during the second wave of decolonization, several British colonies gained independence, including Jamaica (1962), Trinidad and Tobago (1962), Barbados (1966), the Bahamas (1973), Grenada (1974), Dominica (1978), St. Lucia (1979), St. Vincent (1979), Antigua and Barbuda (1981), and St. Kitts and Nevis (1983). In the post-independence era, many of these countries have struggled to diversify their economies beyond sugar.

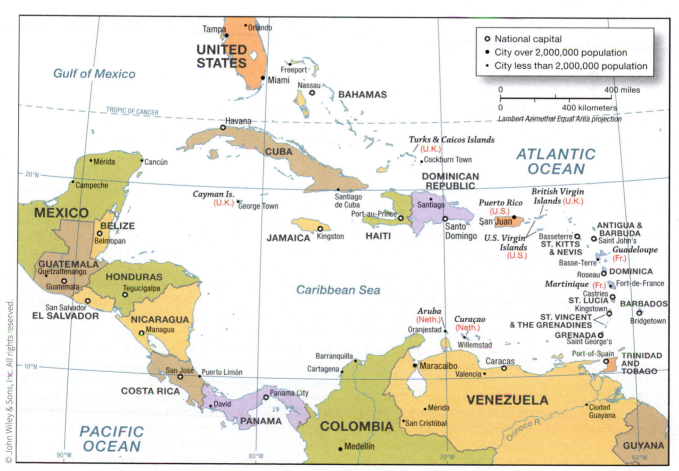

FIGURE 12.35 **Political geography of the Caribbean.** Several islands in the Caribbean are still territories or departments of former colonizers.

DEVELOPMENT APPROACHES IN CUBA

After more than 60 years of independence, a new government declared Cuba socialist in 1959, following a rebellion led by Fidel Castro. In its first year of power, Castro's government expropriated private property, nationalized public utilities, and tightened controls on the private sector. These changes provoked considerable tension with the United States, which saw Cuba as a client of its Cold War rival, the Soviet Union. Castro's Cuba received loans from the Soviet Union during the Cold War, avoiding World Bank loans and policies.

Post-revolution Cuba initially experienced a downturn in health measures in the 1960s when half of its 6,000 doctors left the country. However, the socialist government made universal health care a priority (along with education) in its planning and had trained sufficient medical personnel by the 1980s. Like the rest of the economy, Cuban medical care suffered from severe material shortages following the end of Soviet subsidies in 1991 and a tightening of the U.S. embargo in 1992. The country's significant investments in primary health have paid off over the long run, however (Farmer and Castro 2004).

Cuba now has the highest doctor-to-population ratio in the world and has sent thousands of doctors to work in other countries (Blue 2010). Primary care is available throughout the island, and infant and maternal mortality rates compare favorably with those in developed nations. The life expectancy in Cuba is 78.3 years (76.2 for males and 80.4 for females). These figures rank Cuba thirty seventh in the world and 3rd in the Americas, behind only Canada and Chile, and just ahead of the United States. Infant mortality in Cuba declined from 32 infant deaths per 1,000 live births in 1957 to 6.1 in 2000–2005 (compared to 6.8 in the United States). These numbers are particularly impressive given Cuba's relatively low per capita gross national income.

Cuba is also known for its innovative approaches to food production. Throughout the Cold War, Cuba's agriculture sector focused on producing sugar and tobacco, which were largely exported to the Soviet Union in exchange for food, oil, and medicines. By 1994, three years after the collapse of the Soviet Union, food production had declined by 54 percent from 1989 levels. Furthermore, average per capita caloric intake had dropped from 2,908 calories per day to 1,863 over the same period (below the USDA recommended intake of 2,100–2,300 calories per day). In response to this food crisis, which was further exacerbated by tightening U.S. trade restrictions, the government made increasing food production its number one priority (Torres et al. 2010).

The government allowed land that had previously been devoted to sugar production to be converted to food crops. The government also distributed underused state-held farmland to smallhold farmers. Since the country had also lost its access to externally supplied chemical fertilizers and pesticides from the Soviet Union, the Department of Agriculture also promoted an alternative model of agriculture employing organic fertilizers, biological pest control, and animal-drawn ploughs (Vandermeer et al. 1993). The government also allowed the opening of farmers' markets. These markets came to supply a third of the Cuban population's food, as well as up to 60 percent of the food in Havana.

Finally, the Cuban government launched a major organic farming and gardening initiative in the capital city of Havana. Academics sometimes define urban economies as those without primary activities. The rise of urban agriculture in Cuba and other cities (including Detroit, Michigan) undermines this conception of urban. While cities are often more densely populated than rural areas, they also often contain brownfields and dead spaces. These areas, which are typically former industrial, shopping, or residential areas, may be ideal for agriculture because they offer minimal transit costs to get a product to market and there is often plentiful labor in the area, particularly in neighborhoods with high unemployment. Increasingly, agricultural approaches that employ intercropping (the mixing of complementary crops to enhance insect control and soil fertility), use of organic inputs (such as manure or compost), and intensive cultivation methods (in terms of human labor) have been shown to be highly productive and energy efficient. Urban agriculture uses small plots, and with these practices, urban agriculturalists can produce a reliable, consistent crop.

THE EVOLVING ECONOMIC GEOGRAPHY OF THE CARIBBEAN

Following European contact, most Caribbean islands became plantation economies specializing in the production of commodities including sugar, tobacco, or cotton. While these crops were once only grown in conditions that favored the Tropics, the twentieth century brought innovations in agricultural production, including new seeds, technologies, irrigation systems, and pesticides that stabilized production in other regions.

Caribbean governments sought to diversify their economies after independence. Some countries turned to specialty crops, such as bananas and coffee, which were particularly profitable if a country had a preferential trade agreement with a European ally. However, the European Union refined its agricultural policies, which forced the phase-out of exclusive agreements with Caribbean suppliers. Expanding beyond agriculture, some Caribbean countries have sought to process or refine raw materials they mine in order to garner further revenues, such as bauxite in Jamaica (which is made into aluminum) or oil in Trinidad and Tobago, or to engage in light manufacturing or assembly. More than any other industry, however, tourism has become one of the economic mainstays in the Caribbean, accounting for 80 percent of GNP in some countries.

Tourism is an inherently problematic business and can be interpreted as both good and bad for the region.

Richard Cummins/Lonely Planet Images/Getty Images.

FIGURE 12.36   St. John's, Antigua, and Barbuda. A cruise ship docked for an afternoon at St. John's, the capital of Antigua and Barbuda.

While tourism is a major source of foreign revenue, it can only be established after considerable investment in hotels, roads, and other infrastructure. The problem with infrastructure investments is that money is concentrated in a few areas, especially ports, which leaves the rest of the country out and leads to regional disparities. Investments in tourism infrastructure divert funds from education, health, and agriculture where greater benefits for locals could be realized. In response to these criticisms, some argue that the infrastructure developed to support tourism benefits other sectors of the economy and provides jobs for locals.

One benefit of using tourism as a form of economic development is that the industry is employment intensive and creates jobs. Critics argue that jobs in tourism are low paying and seasonal, providing few benefits. Another argument in favor of tourism is that through tourism visitors from wealthier parts of the world interact with and become more aware of accomplishments and problems in the Caribbean. The counterpoint to that argument is that tourism gives tourists a partial and distorted view of the Caribbean region by only showing the glamorized resorts and not the daily hardship of many residents. Finally, while some contend that tourism helps preserve local culture, others hold that tourism promotes a caricature of Caribbean culture.

Adding to the discussion of whether tourism in the Caribbean is beneficial or detrimental for the region, two other developments must be considered. First, much Caribbean tourism is cruise oriented, meaning that large numbers of tourists offload onto towns for a day and then return to their ship (Figure 12.36). The problem with cruises is that the majority of tourists have only a superficial experience of a country and often spend the bulk of their money

with the cruise company and on the ship (Klak 2007). To add insult to injury, large cruise companies dispose of waste on the open seas, creating environmental hazards (Johnson 2002).

The second development that must be considered is the expansion of ecotourism in the region, which some scholars argue are less problematic and more sustainable (Pulsipher and Holderfield 2006) than large-scale tourism. Researchers contend that smaller scale eco-tourism can benefit a place financially without undermining local cultures. However, eco-tourism can undermine local cultures by commodifying them (see Chapter 7) for tourists to 'consume.' The scale of eco-tourism also impacts its sustainability. If too many tourists tromp through a beautifully preserved landscape, they damage the ecosystem and change the behaviors of animals. For example, tourists who leave behind food trash or directly feed animals make animals dependent at least in part on human food. Also, large companies can own eco-tourist sites and set up the business so most of the tourist dollars go to the company and not to the local people.

CONCEPT CHECK

1. **In what ways** is the Caribbean's colonial history different from that of other parts of Latin America's colonial history?

2. **What** are the strengths and weaknesses of Cuba's development innovations?

3. **What** are the pros and cons of the changing mix of economic activities in the Caribbean?

● SUMMARY

What Is Latin America and the Caribbean?

1. Latin America and the Caribbean is a region that stretches from the Rio Grande River on the northern border of Mexico to Cape Horn at the southern tip of South America and also from the Baja peninsula in western Mexico east to the island of the Caribbean. Two massive mountain chains, the Sierra Madre in Mexico and the Andes Mountains in South America extend from north to south through the region. Climates vary widely in the region, as it stretches over so many degrees of latitude and varying elevations in the mountains have different climate types.

2. Latin America was a hearth of the agricultural revolution, with agriculture beginning with squash cultivation in Mexico as early as 10,000 years ago. Latin American farmers also domesticated corn (maize), potatoes, sweet potatoes, tomatoes, peanuts, and beans. After 1500 CE, agriculture in Latin America intensified as demand for tropical cropland to produce sugar and coffee grew. The Columbian Exchange (Triangular Trade Network) centered on the Atlantic with the trade of cash crops from Latin America and North America, finished products from Europe, and gold and enslaved people from Africa. Along with goods and people, cultural traits diffused along the Triangular Trade Network. Enslaved Africans brought agricultural practices from Africa to Latin America that helped make the cultivation of rice and coffee successful in the region. European colonization fundamentally changed land ownership in the region. Colonizers granted large agricultural estates, now called latifundia, to conquistadors and those close to the crown. Indigenous Latin Americans who did not succumb to European diseases were left with small subsistence plots, minifundia.

3. The Amazon Rainforest is the largest continuous rainforest in the world and with that distinction comes a great degree of biodiversity. Farmers in Latin America use a number of practices to preserve biodiversity in the region. Ecotourism, or nature-based tourism, is one method of preservation that encourages humans to hike and photograph tropical landscapes instead of turning forests to agricultural land. One major benefit of ecotourism is the preservation of endemic species, those found only in one particular area. Another approach is conservation, which calls for land and species to be used purposefully and within their biological limits. Conservation encourages small scale agricultural production in biologically diverse areas so entire swaths of land are not converted to cropland.

Who Are Latin Americans?

1. Latin America has a diverse composition of racial and ethnic groups because hundreds of thousands of indigenous people died from European disease after 1500 at the same time that migrants from abroad arrived in the region. Europeans, Africans, and Asians all migrated to Latin America, supplanting indigenous peoples especially in the Caribbean and eastern sides of Central and South America. The largest concentrations of indigenous people in Latin America today are in and along the Andes in South America and in southern Mexico.

2. South and Central America are called Latin America because they were largely colonized by Spain and Portugal following the voyage of Columbus in 1492. Spain and Portugal are both located on the Iberian peninsula in Europe; both countries are primarily Catholic; and, both the Spanish and Portuguese languages are derived from Latin. Today, Catholicism is the primary religion in Latin America, and Spanish and Portuguese are the dominant languages used.

3. Gender roles, society's assumptions about the roles and duties of women and men, are in flux in Latin America. While the region is traditionally known for machismo, or masculine dominance in society, the region is now home to several female heads of state and women are gaining an increasing role in the formal economy.

How Have External and Internal Forces Affected Latin America's Development?

1. Most Latin American countries gained independence in the nineteenth century. During the same century, the United States became more heavily involved in the politics and economics of Latin America, beginning with the Monroe Doctrine in 1823, which warned European countries to stay out of the region. American companies, including the United Fruit Company, became directly involved in the economies of the region. During the Cold War, the Soviet Union also engaged in the region, especially in Cuba.

2. Following World War II, western banks and international financial institutions issued loans to Latin American countries primarily to build infrastructure, including roads and dams. Rising oil prices in the 1970s along with inefficient state-run industries and loans that needed to be repaid brought Latin America into financial crisis in the 1980s. Countries in the region addressed the Third World Debt Crisis in different ways. Mexico stopped repaying debts and Chile under Pinochet withdrew the government from the economy through radical neoliberal reforms. International financial institutions, including the World Bank and International Monetary Fund (IMF) implemented as series of neoliberal reforms, called structural adjustments, in response to the Third World Debt Crisis.

3. Indigenous groups have responded to structural adjustments to become contending political players in the region. At the turn of the century when the government of Bolivia privatized and raised the cost of water resources and then altered export policies for natural gas, indigenous people protested. Bolivia's president eventually resigned and Bolivians elected Evo Morales, who is the first person of indigenous descent to hold a Latin American country's highest office.

What Are the Caribbean's Unique Development Challenges and Opportunities?

1. The Caribbean subregion has a slightly different colonial history and set of development challenges. Several islands in the Caribbean are well suited for producing sugar, and between the sixteenth and eighteenth centuries, European countries competed to colonize the region to gain a share of the burgeoning global trade in sugar. Although most of South and Central America became independent at the end of the first wave of colonialism in the 1800s, much of the Caribbean remained colonies of Europe until the second half of the twentieth century, at the end of the second wave of colonialism.

2. While most governments in the Caribbean are influenced by European and American economic and political policies, Cuba aligned itself with the Soviet Union during the Cold War and continues to take an alternative path toward development since the fall of the Soviet Union. Cuba can no longer rely on one consumer for its goods, and in response the government has diversified by converting fields used for sugar production to food crop production and initiating organic farming and gardening programs in the city of Havana in order to provide a stable food supply for Cubans.

3. During the era of European colonialism, Caribbean economies were based on plantation agriculture. While agricultural production of tropical crops continues to be important in the region, tourism has grown in prominence. Tourism brings jobs and revenues into the region but not without cost. Most of the profits made through tourism leave the region and jobs are seasonal and relatively low pay. Tourism also diverts resources, including fresh water, away from locals for consumption by tourists. Some Caribbean countries are now building ecotourism industries as an alternative form of tourism.

Geography in the Field   YOUR**TURN**

Havana, Cuba. An urban garden makes use of open space in the city.

© John Birdsall/AGE fotostock.

The Cuban agricultural experiment, which has dramatically increased local food production through organic means, has been heralded as a development success story to be emulated by others in the region and around the world (e.g., Miller 2007). This garden in the municipal Plaza de la Revolución in Havana, Cuba, demonstrates that success. Here farmers grow organic vegetables to serve the people who live in nearby neighborhoods. Fifteen people work to cultivate a wide variety of vegetables and herbs, and another five are employed to care for the facility. Tape-recorded sounds are used to frighten away birds, and only organic products are used for pest control.

Thinking Geographically

- Much like the benefit of eco-tourism depends on the scale, so too does the benefit of organic farming. If Cuba had large, plantation-like organic farms instead of urban gardens like this one, how would the impact change?

- As U.S. relations with Cuba change, tourism may increase in Cuba. How might an increase in tourism change Cuba's farming methods?

Read:
http://www.cityfarmer.info/2008/01/22/organic-cuba-without-fossil-fuels-the-urban-agricultural-miracle.

KEY TERMS

biomes	latifundias	endemic species	gender
intertropical convergence Zone	minifundias	conservation	machismo
	conquistadors	triple heritage	banana republic
Amazonia	haciendas	Iberian	import substitution
ultisols	agricultural dualism	race	parastatals
oxisols	slash and burn agricultural	Mestizo	petrodollars
cold water desert	black earth	colonialism	Third World debt crises
domesticated	monoculture	mercantilism	West Indies
Columbia exchange	mutualism	encomienda	Greater Antilles
Mesoamerica	ecotourism	syncretic religion	Lesser Antilles
historical cultural ecology	preservation		

CREATIVE AND CRITICAL THINKING QUESTIONS

1. How have colonialism, indigenous practices, and physical geography interacted to produce unique agricultural systems in the Latin American **context**?

2. What is Latin America's **identity** known as triple heritage, and why have some components of this identity been emphasized over others?

3. The Amazon is one of the most biologically diverse ecological zones on the planet. Does economic **development** have to take place at the expense of the environment and biodiversity? If you were asked to devise a plan to address the loss of biodiversity in this region, would you pursue a conservationist or preservationist approach and why?

4. One of the most significant developments in Latin American politics in the last decade is the rise of indigenous political movements. Such movements exist around the world to varying degrees. Can you hypothesize why we observe active indigenous political movements in some world **regions** and not in others?

5. As the sole overtly socialist country in the Western Hemisphere, Cuba is unique in many ways. Will Cuba's approaches to development, including providing health care, turning agricultural fields into food crop production, and engaging in urban farming better balance **unequal exchange** than **tourism**?

6. What are the strengths and weaknesses of **tourism** as an economic development strategy in the Caribbean? If you were president of a small island country, how would you structure tourism to make it most beneficial for your people?

7. If you were to visit Cuba today, how could you see Cuba's alternative approaches to food production in the **cultural landscape**?

8. How have **migration** flows to Latin America since 1500 created a complex set of racial **identities** in the region?

SELF-TEST

1. The Greater Antilles are found in which subregion of Latin America:
   a. Amazon
   b. Andes
   c. Caribbean
   d. Southern Cone

2. The Atacama Desert (in Chile) is the driest place on the planet because:
   a. it is in the rainshadow of the Andes Mountains
   b. of the impact of the cold Humboldt current
   c. of the impact of the cold Falkland current
   d. All of the above
   e. a and b

3. As the tropical rainforest is the most productive biome in the world, its soils have huge agricultural potential.
   a. True
   b. False

4. All of the following crops were originally domesticated in Latin America except:
   a. potatoes
   b. maize
   c. peanuts
   d. coffee
   e. manioc

5. All of the following except _____ are past and present agricultural field arrangements found in Latin America.
   a. taungya
   b. latifundias
   c. minfundias
   d. chinampas
   e. haciendas

6. The national parks in Costa Rica have sometimes been referred to as "diamonds in a sea of devastation" because:

   a. Costa Rica's national park system is so much stronger than that found in other Central American nations
   b. diamonds were discovered in Costa Rica's Monte Verde cloud forest
   c. many of Costa Rica's beautiful national parks are surrounded by intensive land use for agriculture, forestry and other uses
   d. Costa Rica's national parks have allowed endemic species to survive in otherwise very challenging circumstances

7. The main factor that allowed the Iberians (Spanish and Portuguese) to conquer Latin America in about 40 years' time was:
   a. the element of surprise
   b. superior military tactics
   c. superior military fire power
   d. Old World diseases and subsequent population decline

8. Race is more of a social construction than a category backed up by real biological differences. From a genetic standpoint, there is often as much difference between individuals within a particular racial category (say, black or white) as there is between individuals in different categories.
   a. True
   b. False

9. The Spanish and Portuguese essentially invented a new form of colonialism in Latin America. While there were colonial empires in antiquity (such as the Egyptians, Greeks, and Romans), the nascent European colonialism developed in Latin America was different because:
   a. these lands were much farther away from the metropole (or power center) than previous colonial ventures
   b. there was a cultural and religious component to Iberian colonialism that was new
   c. the highly extractive nature of Iberian colonialism, driven by mercantilist ideologies, marked a distinct departure from previous colonial endeavors, which had tended to

be city states spawning new city states that traded with one another

d. All of the above

10. All of the following are examples of how gender roles are changing in Latin America except:

a. While Latin American immigrants to the United States have historically been male, the majority are now female—suggesting that women are now increasingly called upon to work in the cash economy and be breadwinners

b. Argentina recently (in 2010) passed a law making gay marriage legal

c. Female heads of state have been elected in Argentina, Chile, and Panama

d. Many of Latin America's export-oriented industries, such as maquiladora assembly plants in northern Mexico along the U.S. border, or the cut flower industry in Ecuador, are largely staffed by women. Not only are these jobs in the export sector being made available to women, but the labor conditions and compensation are some of the best in the world

11. Geographer Judith Carney has been critical of the dominant conceptions of the Columbian Exchange (CE). All of the following are elements of her critique except:

a. The evidence suggesting that Old World seeds and crops flourished in the New World is weak

b. The CE downplays the role of Africans (in favor of Europeans) in the exchange

c. The CE narrowly conceptualizes the process as the transfer of plants and seeds

d. The CE does not give adequate attention to the African knowledge systems that allowed Old World food crops to survive in the New World

12. All of the following are critiques of tourism as a means of development in the Caribbean except:

a. Tourism creates menial and poor-paying jobs. This perpetuates certain stereotypes about Caribbean people

b. Most tourists get a partial and distorted view of the Caribbean region

c. Tourism often promotes a caricature of Caribbean culture

d. Tourism creates an employment-intensive industry that creates lots of jobs

e. Many of the goods used by tourists are imports. The majority of profits leave the country

13. The rise of the indigenous peoples' movement in Bolivia was marked by a number of important milestones. These include all of the following except:

a. Evo Morales was elected president of Bolivia in 2006, the first person of indigenous descent in Latin America to hold his country's highest office

b. The Water Wars of 1999–2000

c. The Gas War of 2003

d. Indigenous people were granted universal suffrage for the first time in Bolivia during the administration of Evo Morales in 2008

14. Cuba is known for which of the following development achievements:

a. free and universal health care

b. a higher life expectancy than the United States

c. a dramatic increase in food production for local consumption using largely organic methods after the end of its subsidized trade with the Soviet Union in 1991 and a tightening the U.S. trade embargo in 1992

d. All of the above

e. a and c

15. Latin America remains one of the least urban of the major world regions.

a. True

b. False

ANSWERS FOR SELF-TEST QUESTIONS

1. c, **2.** e, **3.** b, **4.** d, **5.** a, **6.** c, **7.** d, **8.** a, **9.** d, **10.** d, **11.** a, **12.** d, **13.** d, **14.** d, **15.** b

13

THE PACIFIC

In 2011, Samoa moved from the east side of the International Date Line to the west. The more than 180,000 people in the country skipped December 30. Samoans who went to sleep before midnight on December 29 woke up the next morning on December 31.

The International Date Line, which designates where each new calendar day begins, cuts right through the Pacific Ocean, at approximately 180 degrees longitude, halfway around the world from the Prime Meridian, which runs through Greenwich, England. When you cross the Pacific from east to west, your calendar moves forward one day, and when you cross from west to east, you "lose" a day, as your calendar moves back. Countries that lie near the line can choose whether to be west or east of the line.

In 2009, the people of Samoa changed the side of the road on which they drive. Samoans had driven, American style, on the right side of the road, but on one September morning, police officers stopped traffic and moved cars to the left side.

Each of these moves, to the west side of the International Date Line and to the left side of the road, has moved Samoa figuratively closer to Australia and New Zealand and away from the United States. Australia and New Zealand are the country's largest trading partners and being in synch with those countries benefits the Samoan people, both in business transactions, which now occur on the same day, and in the used car market, since used cars from Australia and New Zealand can now be sold in the Samoan market.

The Pacific is more interconnected through communication, transportation, and trade than at any point previously, and connections with major countries around the rim of the Pacific are bringing the whole area from Asia, Australia, New Zealand, Oceania, and the Americas into a vast region of its own.

Apia, Samoa. Cars drive against traffic signs painted on the road a day before the country abandoned a century of driving on the right side of the road and started driving on the left side. The United States colonized Samoa, which is why they were once driving on the right side of the road and located east of the International Date Line. By moving to the left side of the road and to the west of the International Date Line, Samoans are choosing to move closer to Australia and New Zealand, their largest trading partners and the countries to which Samoans often migrate.

ASSOCIATED PRESS.

CHAPTER OUTLINE

THRESHOLD CONCEPT in this Chapter

Time-space compression

WHAT IS THE PACIFIC REGION?

LEARNING OBJECTIVES

1. **Describe** how the islands of the Pacific were formed.
2. **Explain** the distribution of climates in the region.
3. **Determine** why plastic pollution is accumulating in the Pacific.

The Pacific Ocean covers one-third of Earth. It is easiest to appreciate the massive size of the Pacific Ocean when you look at it on a globe. Scientists estimate that the Pacific Ocean (65 million square miles or 168 million square km) is larger than all of the landmasses on earth combined (57 million square miles or 147 square km). From east to west, the Pacific extends a width of 12,300 miles (19,800 km) from Indonesia to South America, and the full circumference of earth is 24,900 miles (40,072 km) at the equator. The country of Kiribati alone covers an expanse of the Pacific Ocean about as large as India.

THE RING OF FIRE

The Pacific region is at the heart of the **Pacific Rim**, also known as the **Ring of Fire** after the active volcanoes and earthquakes that outline the Pacific Ocean along major subduction zones where denser oceanic plates, whether it be the Pacific plate or a smaller oceanic plate, are subducting underneath less dense continental plates (Figure 13.1).

The leading edge of the oceanic plate melts under heat and pressure, recycling the crust into magma, which eventually erupts into volcanoes that form islands. Most islands of the Pacific run parallel to these subduction zones. Almost all of the region's islands around the Pacific Ocean, from the Aleutians in the far northeast of Alaska to New Zealand in the far southwest, are volcanic islands formed along subduction zones. A **volcanic arc** is a line of volcanoes that is located on the continental plate side of the subduction zone (Figure 13.2), above the magma source created from the leading edge of the oceanic plate.

The volcanoes of the Pacific are not only found in islands that extend above sea level. Amazingly detailed maps of the ocean floor reveal thousands of volcanic islands under the surface, as well (Figure 13.3).

The islands of the Pacific can be divided into three different kinds, and in each case volcanoes are an important part of their formation: continental (volcanic islands found on continental plates), high (formed by volcanoes or limestone cliffs), and low islands (made of coral reef) (Figure 13.4).

The formation of the islands in the Pacific, whether by volcano, limestone, or coral reef, is part of the oral tradition of the indigenous in the region. From a hearth region extending from Tonga to Samoa and then east to the Society Islands, geographer Patrick Nunn found a common myth that the god Tangaloa "fished up" the islands from beneath the surface to bring them above sea level and make them livable for humans. (Figure 13.5). Nunn describes Tangaloa as the "father of the gods in the pantheon of most Pacific Islanders" (2003, 352). He found a pattern in which indigenous in the hearth region of the myth used the "fishing up" myth to explain the creation of limestone islands that were lifted through tectonic activity. At the same time, myths about a god "throwing down" an island from the sky were most common on volcanic islands in the region.

Continental Islands

Moving north over millions of years, the Australian plate converged with the Eurasian and Pacific plates approximately 25 million years ago, creating continental

Reading the **PHYSICAL** Landscape

The Pacific Ring of Fire

The Pacific Rim is one of the most tectonically active places on Earth, with numerous earthquakes and volcanoes that occur along the plate boundaries around the Pacific plate (Figure 13.1a).

The active volcanoes and frequent earthquakes of the Pacific have impacted humans in a variety of ways, and given the region the name *Pacific Ring of Fire*. Volcanoes such as Mount Ruapehu in New Zealand (Figure 13.1b), Mount Pinatubo in the Philippines (Figure 13.1c), and Mount St. Helens in the State of Washington in the United States (Figure 13.1d) have erupted explosively, sometimes causing death and destruction. For example, 800 people died with the 1991 Mount Pinatubo eruption, and 57 died with the eruption of Mount St. Helens. Some of the worst earthquakes on record have also occurred along the Pacific Ring of Fire. The 1906 San Francisco earthquake (Figure 13.1e), for example, destroyed much of the city and killed hundreds of people. The 1960 Valdivia earthquake in Chile,

(Continued)

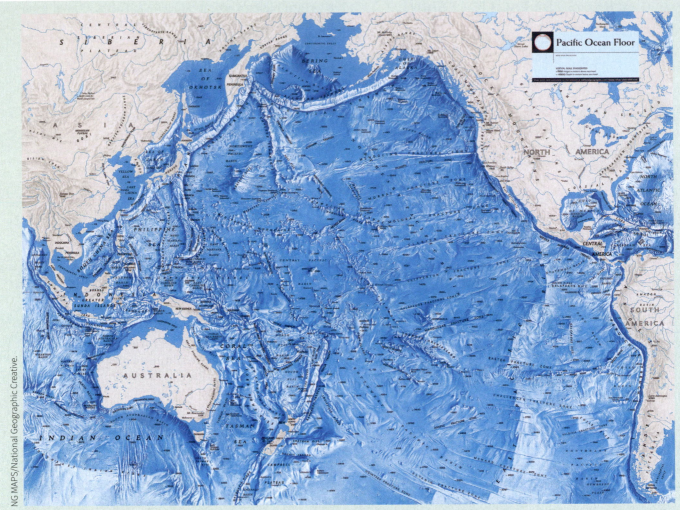

NG MAPS/National Geographic Creative.

FIGURE 13.1a **Topography of the Pacific Ocean.** The mountains on the ocean floor in the Pacific were formed by volcanoes. The lines of trenches on the map, such as the long trench stretching from the Kuril peninsula in Russia, past Japan and then toward Australia, mark subduction zones where the oceanic plate is moving under a continental plate. Earthquakes occur when plates move at subduction zones, and arcs of volcanoes form on the continental side of subduction zones. Islands in the middle of the Pacific plate were generally formed on hot spots.

© Philartphace/iStockphoto.

FIGURE 13.1b **Mount Ruapehu, New Zealand.** Part of a national park, Mount Ruapehu is a ski and hiking destination.

© Alberto Garcia/Corbis.

FIGURE 13.1c **Mount Pinatubo, the Philippines.** Pinatubo is a stratovolcano that erupted in 1991. The mountain rose about 5,725 feet above sea level before the eruption, and almost 500 feet of the volcano was blasted away by the eruption.

John Barr/Liaison/Hulton Archive/Getty Images.

FIGURE 13.1d   Mount St. Helens, Washington, United States. The volcano erupted in May 1980, blowing a mushroom cloud of ash thousands of miles into the air that destroyed nearly one hundred fifty square miles of forest.

Library of Congress Prints and Photographs Division Washington, D.C. 20,540 USA.

FIGURE 13.1e   San Francisco, California, United States. The city was transformed into a landscape of ruins after a 1906 earthquake and fire.

which is the strongest earthquake ever recorded with a magnitude 9.5 on the Richter scale, killed up to 6,000 people and generated a tsunami that spread rapidly across the Pacific Ocean. A massive earthquake occurred off the coast of Alaska in 1964 (Figure 13.1f) and generated a tsunami that killed people as far away as California. Smaller, more numerous earthquakes continue to occur along the Pacific plate boundaries. Sudden releases of tectonic stress keep the Pacific Ring of Fire a restless region and continue to create and shape new landscapes.

© Bettmann/Corbis.

FIGURE 13.1f   Anchorage, Alaska, United States. The region experienced a tremendous earthquake in 1964, which rocked the city and dropped 4th Avenue and a row of cars some 20 feet below street level.

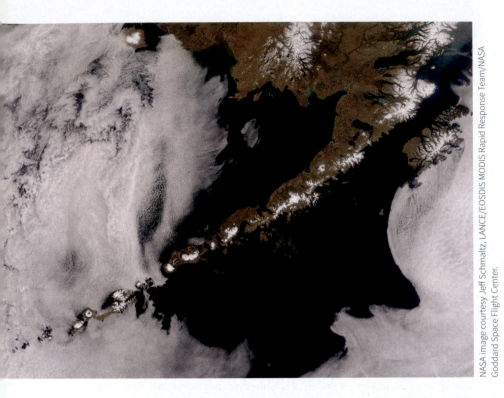

NASA image courtesy Jeff Schmaltz, LANCE/EOSDIS MODIS Rapid Response Team/NASA Goddard Space Flight Center.

FIGURE 13.2   Aleutian Islands, Alaska. The Aleutian Islands are a volcanic arc of islands formed on the continental side of a subduction zone, where the Pacific plate is descending under the North American plate at a rate of between 2 and 3 inches a year.

⊕ USING **GEOGRAPHIC** TOOLS

Mapping the Ocean Floor

The National Oceanic and Atmospheric Association (NOAA) began mapping the ocean floor in 1968. Originally, NOAA produced nautical charts with details about depth at only certain locations. The advent of satellite imagery and sonar has enabled the development of intricately detailed maps of the ocean floor. According to *National Geographic*, "most of the waters and ocean floor remain unexplored." To map the "unexplored" ocean floor, NOAA has used ships equipped with sonar since 2004 to ping the ocean floor continuously in a fan shape, traveling back and forth across the ocean. The sonar not only measures depth, but also the type of surface in order to map the habitat on the ocean floor. The ocean is vast, covering 70 percent of the Earth. The Pacific Ocean reaches a depth of 36,070 feet (10,994 m) in the Mariana Trench, as revealed on this detailed map of the ocean floor.

Thinking Geographically

1. This map is created using sonar. Some features on the map look more detailed and others appear smoother. How would the number of times a particular feature has been mapped by sonar affect the level of detail with which it appears on this map?

2. Look online for coverage of the team of researchers from the University of Hawaii who created this map. In December 2014 they released video image of fish species living in the trench. How have these fish evolved differently than species that live closer to the surface?

Nathan C. Becker and Patricia Fryer, University of Hawaii.

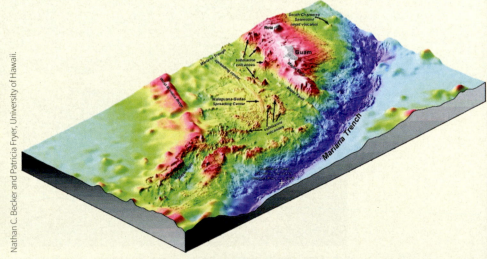

FIGURE 13.3   Oblique view of the Southern Mariana Arc.

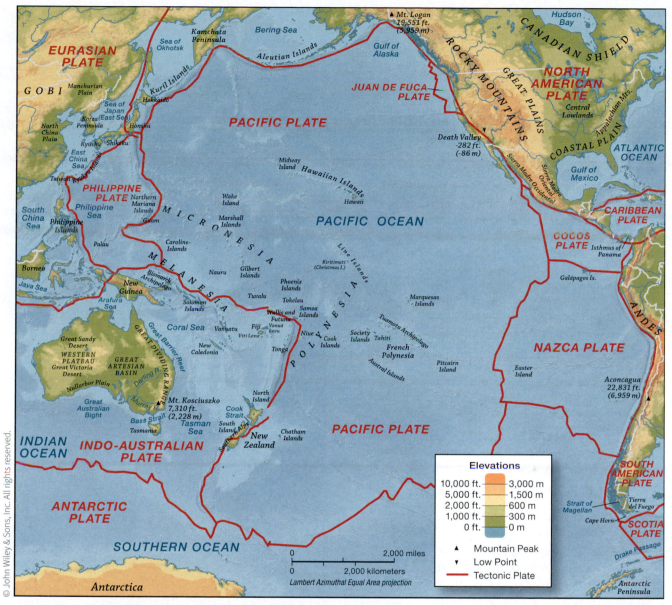

FIGURE 13.4 **Physical geography of the Pacific.** Plate boundaries encircle the Pacific region. Continental islands like New Zealand and New Guinea form along plate boundaries. Also along plate boundaries, uplift of seafloor created by plate movement forms limestone islands, such as Tonga and Samoa. In the middle of the Pacific Plate, the volcanic islands of Midway and Hawaii form where a hot spot in Earth's mantle generates magma. Low islands such as Tuvalu and the Marshall Islands are atolls, formed from coral reef around an extinct, eroded volcano.

islands in the Pacific including New Zealand, New Guinea, New Caledonia, and the Chatam Islands. New Zealand and New Guinea both have active volcanoes along the plate boundaries. New Zealand is located on a transform plate boundary where the Australian plate is moving north and the Pacific plate is moving west. New Guinea is located on the plate boundary between the Australian plate and a much smaller subplate, the Bismarck plate. New Caledonia and the Chatam Islands are both part of the Zealandia continental island, which also includes New Zealand.

The highest point on the North Island of New Zealand is Ruapehu, a volcanic mountain that stands over 9,000 feet (2,743 m). Geologists estimate that Ruapehu, which is still active, began erupting 250,000 years ago. Three volcanic peaks including Ruapehu make up the Tongariro National Park, also known as "fire and ice" for its combination of active volcanoes and glaciers (Figure 13.6).

Australia has small mountain ranges, including the Great Dividing Range, which were formed through volcanic activity around the time the Australian plate converged with the Pacific plate. Australia is located in the middle of the Australian plate; so, scientists have typically thought of the continent as being tectonically inactive since it is not near plate boundaries. However, geologists have found evidence of historic tectonic activity in Australia in the dormant volcanoes of the east and in earthquakes that periodically take place still today.

The earthquakes that happen in Australia today occur because immense pressure is being put on the upper layer of the continent's crust as the plate is squeezed eastward from the Indian Ocean plate on its west and westward from the Pacific plate on its east (Figure 13.7). The southern boundary of the Australian plate is a midoceanic ridge in the Antarctic Ocean. Along the ridge, the seafloor is diverging, or spreading apart, much like the midoceanic ridge in the Atlantic Ocean. This spreading continually pushes the Australian plate northward while it is also squeezed from the east and west, thus generating earthquakes in the upper layer of the continent's crust.

High Islands

High islands are marked with a peak surrounded by lower elevations that extend to sea level. When you combine the elevation from the seafloor to sea level with the elevation from sea level to mountain peak, some high islands of the Pacific are taller than Mount Everest. For example, Mauna Kea on the island of Hawaii is 32,000 feet (9,750 m) from ocean floor to sea level, and another 13,796 feet (4,205 m) from sea level to peak. In comparison, Mount Everest stands at 29,029 feet from sea level to peak (8,840 m).

High islands are formed either as volcanoes or as **limestone** cliffs. Volcanic high islands originate on the ocean floor and build up over thousands of years to elevations above sea level. The rocks of these high islands are basalt, which is made from cooled magma following volcanic eruptions. Limestone high islands and cliffs are formed when tectonic activity lifts limestone above sea level. Limestone is a sedimentary rock that is made of the shells of sea life that have accumulated over time and compacted into rock under pressure of the overlying layers.

JASON EDWARDS/National Geographic Creative/National Geographic Stock.

FIGURE 13.5   Tonga.   Limestone arches eroded by waves and water dot the coast of Tonga. A common creation myth in the Pacific tells the story of the god Tangaloa fishing up the limestone islands.

Low Islands and Atolls

Low islands are made of coral, a living sea creature that secretes calcium carbonate to build its skeleton. The calcium carbonate material amasses over time on the tiny creatures, and they extend and grow, creating a coral reef.

Steve Clancy Photography/Getty Images.

FIGURE 13.6   Tongariro National Park, New Zealand.   An UNESCO World Heritage site, the volcanic peak Ruapehu is named for a mythical Maori woman who betrayed her husband, Taranaki, with Tongariro. Taranaki fled and Ruapehu "sighs" because she still loves her husband. The most active volcano and the park itself are named for Tongariro, who still "smokes and smoulders with anger" over the fact that he will never get Raupehu back.

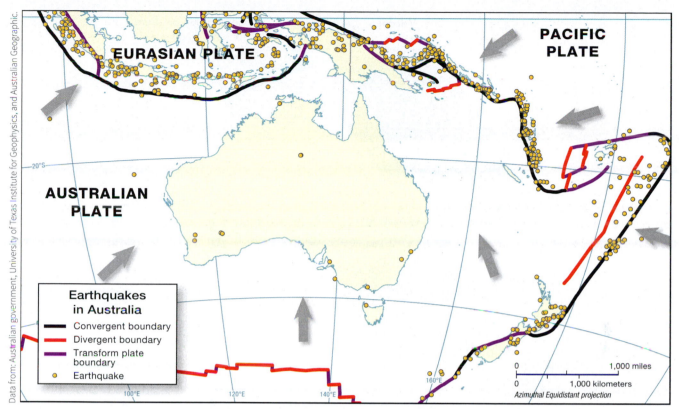

Data from: Australian government, University of Texas Institute for Geophysics, and Australian Geographic.

FIGURE 13.7 **Tectonic pressure in Australia.** The Indian and Pacific plates are moving toward the Australian plate, creating pressure on the plate from both sides. Additionally, the divergent plate boundary between the Australian plate and the Antarctic plate creates pressure on the Australian plate from the south. Earthquakes in Australia occur along the upper crust of the continent where pressure accumulates.

Coral thrives in clean, warm oceans, typically between 30°N and 30°S latitude.

In the Pacific, once the volcano of a high island located at a warm latitude becomes dormant, coral starts to grow around the island forming a fringe (Figure 13.8). Over time, the volcano in the center subsides into earth's crust, and erosion begins above sea level. As the high island sinks, a lagoon starts to form between it and the coral, which is now considered a barrier island. The developing lagoon is a perfect place for sea life to accumulate. When the high island sinks below sea level, the lagoon is complete, and the low islands of coral surrounding it become an **atoll**. Kiribati, the Marshall Islands, Tokelau, and Tuvalu are all example of atolls.

PACIFIC CLIMATES

Climates of the Pacific vary by latitude and between low and high islands. The average annual temperature of Pacific Islands varies by latitude with warmer temperatures near the equator and cooler average temperatures near the poles. The maritime location of the islands means that temperature ranges (differences between high and low temperatures) are relatively small.

Annual precipitation amounts also vary between high and low islands, with high islands receiving more total precipitation than low (Figure 13.9). High elevations of volcanoes or limestone cliffs create an orographic effect, which occurs when air lifts over a mountain and cools. The cooling air reaches its dew point and precipitation falls on the windward side of the mountain, creating densely vegetated, even rainforest landcover (Figure 13.10). On the leeward side, which is protected from the wind, dry, warming, air descends and creates an arid to semiarid climate. Low islands do not create a lifting effect; so, they do not experience orographic precipitation. On low islands, precipitation falls when storms come off the ocean.

Climate of Continental Islands

New Zealand and New Guinea both have relatively large mountain ranges. The latitude of New Guinea, extending between the equator and 10°S, is tropical, which helps create a tropical rainforest climate on the island. A warm ocean current runs along the northern side of the island, which along with consistent incoming solar radiation over the course of the year creates relatively warm temperatures and consistent precipitation (Figure 13.11).

The latitude of New Zealand is much closer to the poles, extending between 35°S and 46°S. At these southern latitudes in the Pacific, New Zealand is affected by a cold ocean current from the South Pacific. The mountains of

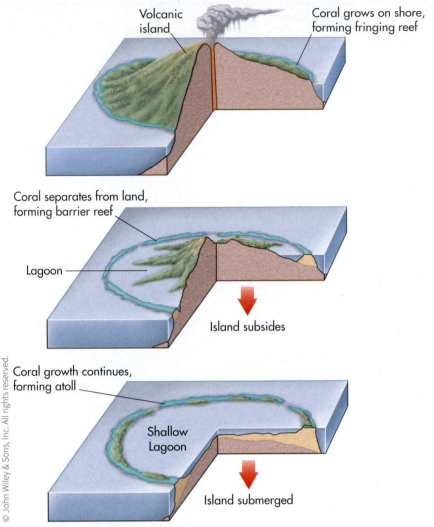

Volcanic island

Coral grows on shore, forming fringing reef

Coral separates from land, forming barrier reef

Lagoon

Island subsides

Coral growth continues, forming atoll

Shallow Lagoon

Island submerged

FIGURE 13.8   Atoll formation.   This diagram breaks the formation of an atoll into three steps. First, coral forms in a fringe around a volcanic island. Next, the volcano subsides and a lagoon forms between the coral and volcano. Finally, the volcano disappears beneath sea level after eroding and subsiding. A ring of coral remains above sea level with a shallow lagoon in the middle.

New Zealand's north and south islands are both inundated by cool, moist winds. As a result of the combination of latitude, elevation, and ocean currents, the climate in New Zealand is much colder than New Guinea's. The presence of mountain ranges in both areas helps create orographic precipitation, and the climate follows suit with moist areas on the windward side and drier climates on the leeward sides of the islands.

Australia's climate is diverse because the continent extends over a wide span of latitude, from 10°S to 40°S. Australia's best known climate is the Outback, which is a semiarid to arid climate region in the country's interior. Because Australia straddles the Tropic of Capricorn, a subtropical high-pressure cell associated with the Tropic of Capricorn affects the region. In a high-pressure system, winds from aloft (at the higher elevations in the **tropo-sphere**) descend toward Earth's surface and then air flows

outward in a counterclockwise direction in the Southern Hemisphere.

Australia's Outback is so dry because warm air needs to rise for precipitation to occur (Figure 13.12). Earth certainly absorbs incoming solar radiation and heats up in the Outback, but that warm air cannot easily rise because of the dominance of the high-pressure system, which is pushing the air down and outward. The world's most massive deserts, including the Sahara in Africa, are found where high-pressure systems dominate.

Both the northern side of Australia and the southern side of New Guinea are susceptible to tropical cyclones. New Zealand is too far from the Tropics to be affected by cyclones. Because Australia and New Guinea are south of the equator, the path of tropical cyclones flows across the Pacific from east to west and then deflects toward the left, and moves counter-clockwise around the low-pressure system

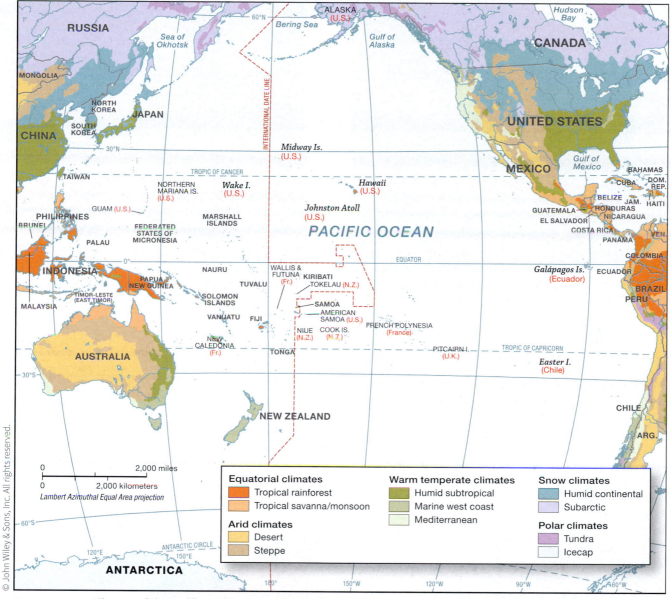

Equatorial climates
- ■ Tropical rainforest
- ■ Tropical savanna/monsoon

Arid climates
- ■ Desert
- ■ Steppe

Warm temperate climates
- ■ Humid subtropical
- ■ Marine west coast
- ■ Mediterranean

Snow climates
- ■ Humid continental
- ■ Subarctic

Polar climates
- ■ Tundra
- □ Icecap

FIGURE 13.9 **Climates of the Pacific.** The vast majority of islands in the Pacific are located in the tropics, between the Tropic of Capricorn and the Tropic of Cancer. Closest to the equator, the islands have rainforest climates, and closer to the Tropic of Cancer and the Tropic of Capricorn, the islands experience tropical savanna/monsoon climates with distinct wet and dry seasons.

of the cyclone. As a result, the eastern side of Australia can be affected by cyclones traveling poleward, just like the east coast of the United States is affected by hurricanes even though they originate at more tropical latitudes.

Climate of High Islands

Situated in the middle of the Pacific Ocean, with abundant moist air, high islands receive a reliable supply of

FIGURE 13.10 **Big Island of Hawaii, United States.** Waterfalls cascade into the Pacific Ocean, surrounded by rainforest vegetation, on the Hamakua Coast, the windward side of the island.

FIGURE 13.11   Kwato Island, Papua New Guinea. Lush vegetation fills this small island off the southeast coast of the island of New Guinea.

FIGURE 13.12   Outback, Australia. Central Australia is influenced by a subtropical high pressure cell, which creates a dry climate in the Outback.

precipitation as moist air rises over interior mountains, cools, condenses, and precipitation follows. On islands where a dominant wind blows predictably from one direction, the windward side of the island will receive precipitation and the leeward side of the mountain will be much drier.

Hawaii is composed of eight main islands that extend from the northwest to southeast between 23°N to 19°N, just south of the Tropic of Cancer. The trade winds, which run from a high-pressure cell near the Tropic of Cancer to the low pressure of the Intertropical Convergence Zone (ITCZ), shape the precipitation pattern in Hawaii. Because of the trade winds, northeasterly winds influence the Hawaiian Islands and shape the climate, dependably bringing moisture-laden air from the Pacific onto the islands of Hawaii. Moisture-laden air lifts over the peaks formed by volcanoes, and precipitation falls on the windward side. As air descends on the leeward side, precipitation totals fall (Figure 13.13). Comparing Hawaii's precipitation map with a map of elevation, vegetation, and beaches confirms a general trend: the northeastern sides of the islands are densely vegetated and the southwestern sides of the islands are home to many of Hawaii's most famous beaches.

Hawaii's state license plate includes a rainbow, and if you walk the beaches on the leeward side of one of Hawaii's beautiful islands, you are likely to see a rainbow in the late afternoon (Figure 13.14). With the sun in the western sky and moisture or precipitation in the air after the heat of the day, rainbows appear frequently when the sun is at your back.

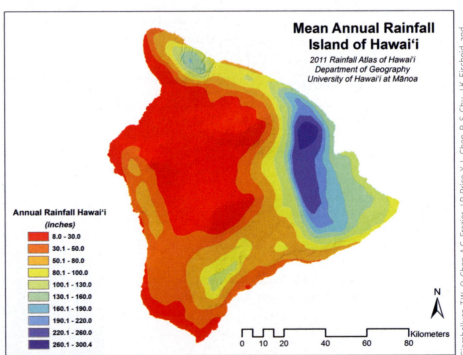

FIGURE 13.13   Annual rainfall in Hawaii. Geographers at the University of Hawaii at Manoa used GIS to create this map of annual rainfall and an online, interactive atlas of rainfall in Hawaii.

© nagelestock.com/Alamy.

FIGURE 13.14 **Oahu, United States.** A late afternoon rainbow forms over Waikiki Beach on Oahu, Hawaii.

Climates of Low Islands

Without elevation to create orographic precipitation, low islands typically lack fresh water and therefore "sustain very little in the way of useful plants or animals" (McNeill 1994, 4). Low islands are particularly susceptible to tsunamis and cyclones because little slows or stops seismic sea waves or storm surges from inundating the low topography.

For example, the average elevation of Tuvalu is 6.5 feet (2 m). Located south of the equator, the tropical island does receive rain during its summer, when the ITCZ is located in the Southern Hemisphere. The nine islands and lagoons of Tuvalu definitely suffer from what nineteenth-century poet Samuel Taylor Coleridge described as "water, water, everywhere, nor any drop to drink." The islands have no streams, rivers, or lakes. People capture rainwater as the main source of potable, drinking water, and some of the islands have fresh ground water tables from accumulated rainwater.

Despite the dryness of low islands, the presence of coral reefs can be helpful for the production of certain crops. A good example is found off the coast of Japan on the island of Kojima. Although Japan is culturally part of East Asia, Japan's physical geography is more similar to that of the Pacific region than to the rest of East Asia. Kojima is a tiny, low island that Japan protects from erosion by building seawalls and jetties. Several Japanese

families have small homes on the island where they live in the summer months to harvest and dry kombu, a kelp used in Japanese cooking. The residents lay the kelp out on the flat, dry island to be dried by the sun.

FLORA, FAUNA, AND FARMING

Tourist brochures advertise pristine, virtually untouched physical landscapes in the Pacific. However, over centuries, settlers in the Pacific Islands and Australia altered their environment, especially to develop agriculture. The peoples of the Pacific used fire on both low and high islands as a means of preparing the soil for shifting cultivation, or slash and burn agriculture, and Pacific islanders transported species across islands for centuries.

The first human migrants in the Pacific created what historians call "transported landscapes" on the islands, bringing flora, fauna, and farming practices (including slash and burn) to new places. Historian J.R. McNeill describes the Polynesian settlement of the Pacific as having two steps. First, the Polynesians "exploited and depleted the resources that appeared easiest to use on an island," and then they brought new resources and agricultural products to the island (1994, 4).

Polynesians introduced "four animals (the rat, dog, chicken, and pig) and several edible plants (for example coconut, taro, and breadfruit) now widespread throughout the Pacific" (McNeill 1994, 5). In New Zealand, which Polynesians settled approximately 1,000 years ago, humans changed the landscape from what environmental historians estimate was 85 percent forested to a landscape that is more than 50 percent grassland for livestock raising.

The level of diffusion and interaction of species throughout Melanesia likely reflects the spatial interactions between people in the region. Through studies of plant life, biologists have found that the islands close together and close to Australia and New Guinea share common species but that "as the distance between islands becomes greater, the total number of animals and plant species found on each island decreases" while the number of endemic species increases (Loope 1998, 747).

Despite the diffusion of species across islands over centuries, native species in the region are often **endemic**, which means they are only found on the island where they originated. With no natural predators, species of birds thrived, and with few fires or invasive species to compete for water and sun, species of fauna also thrived. Even today, the island of Palau is populated by 3,334 species

FIGURE 13.15 **Peleliu, Palau.** The giant white-eye (Mega-zosterops palauensis) is an endangered endemic plant found only on the Palau islands of Babelthuap, Urukthapel, and Peleliu.

of plants (about half the plants found on the islands) that are endemic (Parfit 2003) (Figure 13.15).

PLASTIC POLLUTION

As people came to populate the Pacific, they generally tacked their sails to follow prevailing winds. The winds drive the ocean currents, which carry seeds and new species across the ocean. Prevailing winds and ocean currents now shape the path of plastic trash as it moves through the Pacific Ocean.

Plastic came into widespread use in wealthier regions of the world in the 1920s for kitchen, furniture, and office goods. Plastic diffused broadly after World War II when less expensive, purely synthetic forms of plastic "democratized" the consumption of goods from hair combs to kitchen appliances (Figure 13.16). Plastics break up in the ocean but do not decompose. They are pulled together in the North Pacific around the North Pacific Subtropical High and move over the course of the year with prevailing winds. This mass of plastics is called the "Great Pacific Garbage Patch." It is located between California and Hawaii but is not permanently located in one place.

The National Oceanic and Atmospheric Administration (NOAA) explains in "De-Mystifying the Great Pacific Garbage Patch," the high-pressure system in the Pacific is

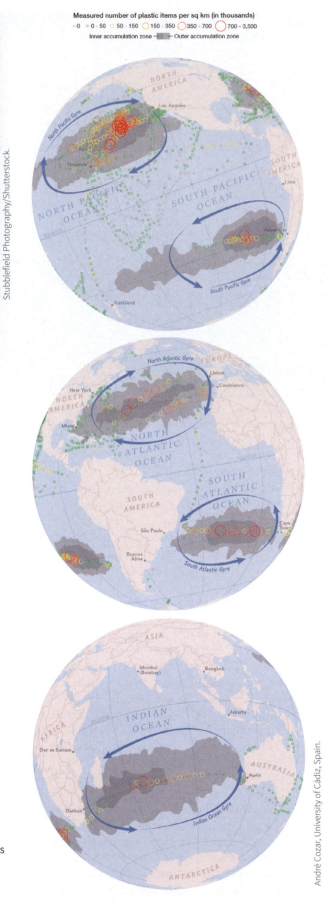

Measured number of plastic items per sq km (in thousands)
0   0 - 50   50 - 150   150 - 350   350 - 700   700 - 3,500
Inner accumulation zone ▬▬ Outer accumulation zone

FIGURE 13.16 **Great Pacific Garbage Patch.** Also known as the Pacific Ocean garbage gyre, a high pressure cell drives the ocean currents in the North Pacific, moving plastics tossed into the ocean. The top map shows a large concentration of plastic in the northern Pacific, and the other two maps show smaller concentrations of plastic in the Atlantic and Indian Oceans.

Courtesy of Kim R. De Wolff.

FIGURE 13.17   Kamilo Beach, Hawaii, United States.   Known locally as "Plastic Beach," Kamilo Beach on the Big Island of Hawaii is littered with plastic because ocean currents drive plastic toward the beach, especially in the summer months.

strongest during the Northern Hemisphere's summer and migrates south during the Northern Hemisphere's winter. The gyre of plastic trash is moving, covering a massive area and disrupting the ecosystem of the Pacific.

Widespread plastic debris is having a large impact on the Hawaiian Islands because when the Pacific High is at its strongest in the summer months, the high pressure cell is located relatively far north in the Pacific, as is Hawaii. As a result, plastic is drawn to the beaches of the Big Island of Hawaii where one such beach is now known in the local vernacular as "Plastic Beach" (Figure 13.17).

The impacts of plastic debris on the food chain are not fully understood, but scientists have reported that 9 percent of fish sampled in the Pacific have plastic in their stomachs and one species of marine insect is now using floating small pieces of plastic as a surface to lay its eggs. Ridding the Pacific of plastic debris is difficult because most pieces are smaller than the size of fingernails and the mass of plastic covers an area about the size of Texas. The only way to pick up the plastic would be to use fine mesh nets, which would also skim the surface marine life, including the base of the ocean's food chain—phytoplankton.

CONCEPT CHECK

1. **What** are the differences in the formation process between high and low islands?

2. **How** do prevailing winds affect the climates in the region?

3. **Why** is the Pacific Ocean Garbage Gyre not visible on satellite images?

WHO ARE THE PEOPLE OF THE PACIFIC?

LEARNING OBJECTIVES

1. **Explain** how the order of populating the Pacific region is reflected in links among indigenous cultures.

2. **Describe** the main flows of migration for Pacific Islanders since World War II.

3. **Determine** how Australia's migration policy has changed.

4. **Compare** residential segregation patterns in New Zealand and Australia.

Approximately 3,000 years ago, low sea levels permitted people to easily cross a land bridge that connected Southeast Asia to New Guinea, and New Guinea to Australia. Migrating out of New Guinea, people moved onto nearby islands. By 650 CE, humans had migrated as far northeast as Hawaii. Curiously, although New Zealand is located only 1,400 miles (2,253 km) from Australia, it took humans another 550 years to get there, making it the last place to be populated in the region, around 1200 CE (Figure 13.18).

The World Bank estimates that the entire population of the islands of the Pacific at 3.4 million people, which is roughly equivalent to the population of Panama in Central America (Figure 13.19). Australia's population is 22.6 million, and New Zealand is home to 4.4 million people.

INDIGENOUS CULTURES

The region of the Pacific is traditionally divided into three smaller regions: Melanesia, Micronesia, and Polynesia. We use these regions to discuss the indigenous peoples of the Pacific and cultural attributes, including languages. Polynesia, which was the most recent area populated, shares clear cultural linkages within the region. Melanesia has historical connections but is incredibly linguistically diverse, and Micronesia was settled in four time periods from different directions. Peoples and cultures of Micronesia are diverse, as the small island chains are historically self-sufficient and

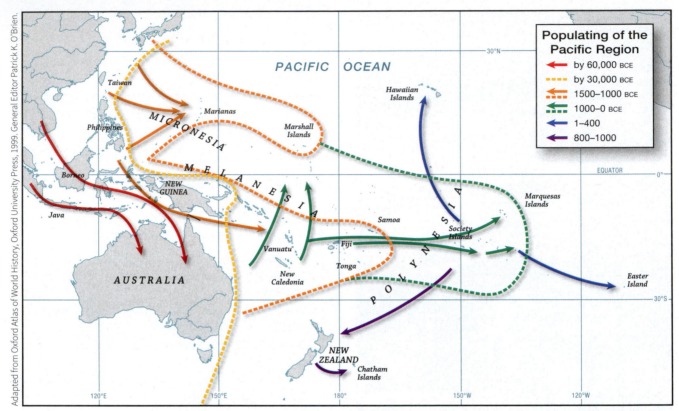

FIGURE 13.18 **Populating of the Pacific.** People migrated from Eurasia to populate Australia as early as 60,000 BCE and the Pacific islands as early as 30,000 BCE. In a further push between 1500 BCE and 1000 CE, humans migrated to populate the entire region.

have experienced relatively little spatial interaction over time. Shared experiences of Japanese colonial control starting in the 1850s and strong influence by the United States after World War II have brought some islands of Micronesia closer together.

Melanesia

Linkages among indigenous peoples of the Pacific follow the paths of migration their ancestors took. Humans first entered Australia around 60,000 BCE. Indigenous people who migrated and settled Australia, New Guinea, and the islands of Melanesia are closely related, but their languages are distinct because thousands of years have passed since they peopled the region.

Islands in Melanesia are primarily volcanic (high islands) and are in tropical latitudes. The plush vegetation and variety afforded by different elevations, consistent heating, and dependable precipitation enabled communities to live relatively independently of each other across islands. Distinct communities with little spatial interaction formed their own languages in their localities over time. The people of Melanesia speak more than 1,000 distinct languages, and 700 of these languages cross language families (Figure 13.20). In a report on cultural landscapes in the Pacific, UNESCO researcher Anita Smith described

Melanesia as "the most linguistically diverse region of the world where people living in adjacent valleys speak completely different languages" (2009, 21).

Micronesia

Micronesia has more than 2,300 relatively tiny islands that extend over an area in the Pacific that is larger than the United States. The vast majority of the area is water, as the combined land area of all of Micronesia is less than 1,600 square miles (3,000 sq km), which is smaller than the tiny European country of Kosovo. Among the countries of Micronesia are Palau, Guam, the Marshall Islands, and the Gilbert Islands. Finding the relatively small islands of Micronesia in order to populate them happened by chance and in four phases over time, which has created differences in cultures within Micronesia.

Approximately 3,500 years ago, people entered the Mariana Islands in Micronesia from the west, most likely the Philippines. The archaeological record reveals that the Caroline Islands and Palau were populated approximately 3,000 years ago, also from the west. Two thousand years ago, the Marshall Islands and the eastern atolls of the Caroline Islands were populated from eastern Melanesia. Finally, Micronesia is influenced by Polynesia, as the two southernmost atolls, Kapingamarangi and

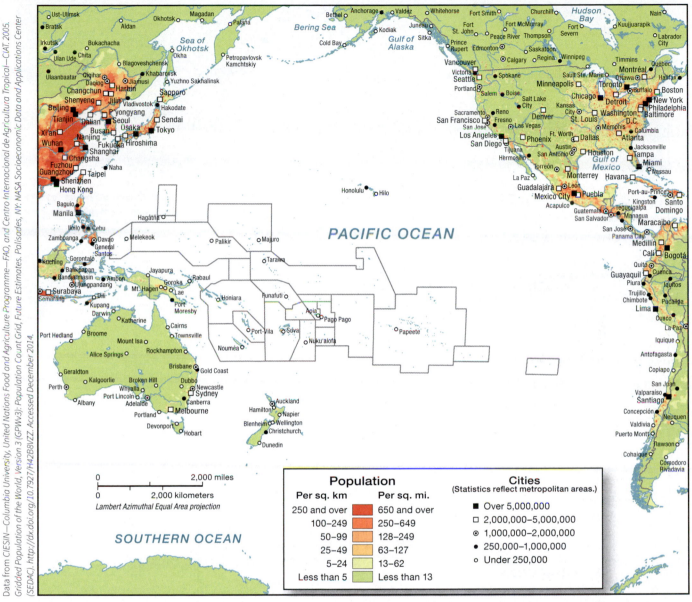

Data from CIESIN—Columbia University, United Nations Food and Agriculture Programme—FAO, and Centro Internacional de Agricultura Tropical—CIAT. 2005. Gridded Population of the World, Version 3 (GPWv3): Population Count Grid, Future Estimates. Palisades, NY: NASA Socioeconomic Data and Applications Center (SEDAC). http://dx.doi.org/10.7927/H42B8VZZ. Accessed December 2014.

Population

Per sq. km	Per sq. mi.
250 and over	650 and over
100–249	250–649
50–99	128–249
25–49	63–127
5–24	13–62
Less than 5	Less than 13

Cities
(Statistics reflect metropolitan areas.)

■ Over 5,000,000
□ 2,000,000–5,000,000
◉ 1,000,000–2,000,000
● 250,000–1,000,000
○ Under 250,000

FIGURE 13.19 **Population density of the Pacific.** The islands of the Pacific region are lightly populated, as are Australia and New Zealand. In the entire region of the Pacific, major cities in East Asia, Southeast Asia, and the Americas are connected through trade.

Nukuoro, were populated by Polynesians starting 1,000 years ago (Smith 2009, 24). The language map of Micronesia demonstrates the diversity of influences and the geography of the flow of people into the region across four phases.

Polynesia

Polynesia extends over a vast triangular region of the Pacific from New Zealand to Hawaii to Easter Island. Despite the distance among islands in this vast area of the Pacific, the Polynesian people are closely linked linguistically and culturally.

Researchers study indigenous Pacific languages to reinforce or revise theories on the peopling of the islands. A prevailing theory among anthropologists is that Polynesians from Samoa migrated to Hawaii and populated those islands and then moved southward, eventually to New Zealand. William Wilson, a linguistics professor at the University of Hawaii Hilo, believes that Polynesians migrated from the high island of Samoa northwest to a grouping of atolls off the coast of the Solomon Islands and lived there, separated from Samoa, before migrating to Hawaii. Wilson posits that this **step migration** accounts for the distinct differences between the Samoan and Hawaiian languages as well as for the fishing innovations found among the first Polynesians in Hawaii, of which there is no evidence in Samoa. Indigenous Samoans depended more on agriculture than on fishing; thus, Wilson

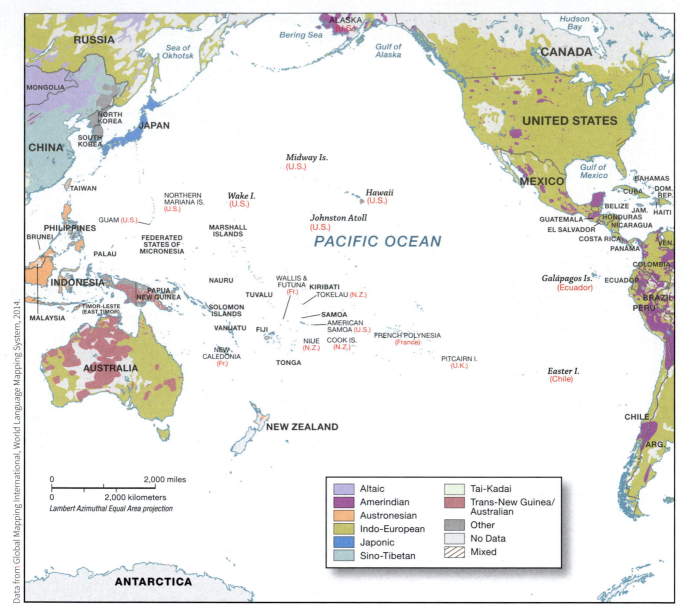

Data from Global Mapping International, World Language Mapping System, 2014.

FIGURE 13.20 **Languages of the Pacific.** Both the historical migration of people into the Pacific and the relative lack of spatial interaction among islands over time have led to a diversity of languages.

contends that the fishing innovations in Hawaii came through **relocation diffusion** from the atolls off the coast of the Solomon Islands, where indigenous people depended more on fishing.

From Hawaii, Polynesians populated Tahiti Marques, Rapanui, and New Zealand. Maori are the indigenous peoples of New Zealand, and the cultural practice of the **haka** has diffused throughout Polynesia and continues to this day in such venues as New Zealand All Blacks Rugby games and University of Hawaii football games (Figure 13.21). The haka dance, traditionally used to prepare for battle, is choreographed with the men standing in front of the women. The men take different postures in unison with sharp movements transitioning from one posture to the next. Dancers also speak the haka, stating the words spoken by Te Rauparaha, a Maori chief, who escaped death (the stories of how he escaped and when he wrote the haka vary by source) and celebrated by performing the haka.

Near and Remote Oceania

To better understand the cultural linkages among the people of the Pacific, the regions of Near Oceania and Remote Oceania can be more useful than Melanesia, Micronesia, and Polynesia. Near Oceania includes areas west of the eastern Solomon Islands, and Remote Oceania is east of there. People migrated into Near Oceania approximately 30,000 years ago (some as early as 60,000 years ago), primarily crossing

ASSOCIATED PRESS.

FIGURE 13.21a   **Manchester, England.**   New Zealand's Kevin Locke, formerly of the All Blacks, center right, performed the haka with teammates before the World Cup Final International Rugby League match against Australia at Old Trafford Stadium in 2013.

Kent Nishimura/Getty Images Sport/Getty Images.

FIGURE 13.21b   **Honolulu, Hawaii.**   Lance Williams of the Hawaii Warriors and teammates performed the haka before the start of the first quarter of a NCAA game against Nevada at Aloha Stadium in 2012.

land bridges (Figure 13.22). People migrated into Remote Oceania only within the last 3,500 years.

The rise of the Lapita civilization was the impetus for the movement of people beyond Near Oceania into Remote Oceania around 1300 BCE. The Lapita language originated in Taiwan. Their pottery traces back to northern Philippines, and their cultural influence is evident throughout the region of Lapita settlement. Archaeologists map the Lapita region by charting where Lapita pottery has been unearthed (Figure 13.23a and b). The archaeological record of the Lapita includes pottery, agricultural practices, obsidian use, crops and animals,

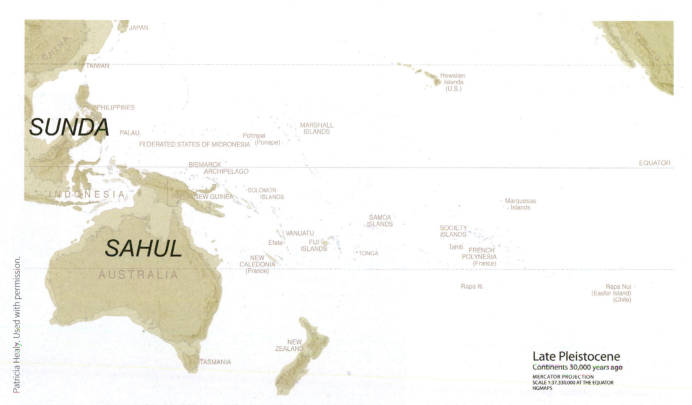

Patricia Healy. Used with permission.

Late Pleistocene
Continents 30,000 years ago
MERCATOR PROJECTION
SCALE 1:37,330,000 AT THE EQUATOR
NGMAPS

FIGURE 13.22   **Former land bridges in Near Oceania.**   About 30,000 years ago, sea levels were much lower than they are now, and Southeast Asia and Indonesia were connected with Australia and New Guinea by land bridges. The darker tan areas show the outlines of the continents as they are currently, and the lighter tan show the portions of the continents that were above sea level and formed land bridges 30,000 years ago.

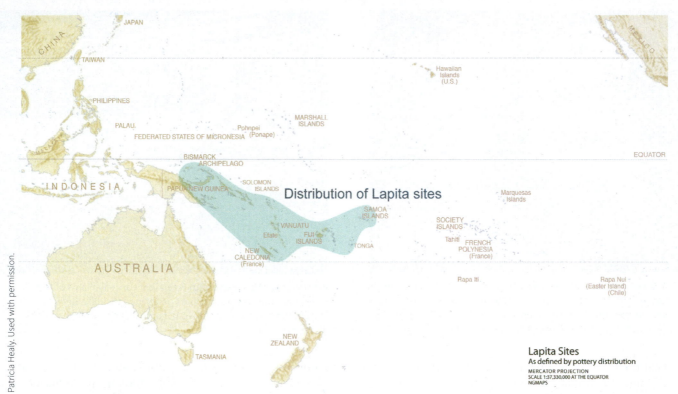

Patricia Healy. Used with permission.

Distribution of Lapita sites

Lapita Sites
As defined by pottery distribution
MERCATOR PROJECTION
SCALE 1:37,330,000 AT THE EQUATOR
NGMAPS

FIGURE 13.23a **Lapita region.** The Lapita region includes places where Lapita pottery has been found in the archaeological record. Lapita sites date back to 1500 BCE in New Guinea. Archaeologists trace the migration of Lapita cultural traits as far east as Samoa by dating the arrival of Lapita pottery. The movement of Lapita pottery beyond the eastern Solomon Islands marks the movement of people from Near Oceania to Remote Oceania.

and burial practices. Evidence suggests that the Lapita purposefully migrated eastward and into Remote Oceania, taking with them staple crops and animals, just as the pioneers in the American West (Smith, 2008).

The divide between Near and Remote Oceania is the eastern extent of the Solomon Islands. When the Lapita left the Solomons and set out for the Santa Cruz Islands and Vanuatu, they traveled 230 miles (370 km) and "at least 150 of those miles the Lapita sailors would have been out of sight of land, with empty horizons on every side" (Smith 2008, 4). From there to Fiji, they had to travel 500 miles (804 km) of ocean. Researchers are confounded by the direction the Lapita traveled. The prevailing trade winds are from the southeast in the region of Lapita expansion; so, the Lapita would have needed advanced seafaring abilities to tack into the wind, or they had to take advantage of a change in winds——either seasonal from the monsoon or climatically from an El Niño event—to sail east.

Lapita influence is not limited to the region of Lapita settlement. The Lapita were the ancestors of the Polynesians, who 1,500 years ago purposely migrated throughout the rest of Remote Oceania, a thousand years after Lapita migration ended. The expansion of Polynesian explorers over the last 1,500 years is reflected

STEPHEN ALVAREZ/National Geographic Creative/National Geographic Stock.

FIGURE 13.23b **Republic of Vanuatu.** A newly reconstructed piece of Lapita pottery is displayed in a museum.

in the connectedness among languages in Polynesia. When English explorer James Cook reached Hawaii in 1778, he was shocked to hear how the language was so similar to that spoken in Tahiti, especially when he realized Hawaiians and Tahitians knew little to nothing of each other at the time.

COLONIZATION OF THE PACIFIC

The first Europeans to reach the Pacific arrived in 1520 on three ships led by the explorer Ferdinand Magellan, who named the ocean because it looked calm after he had been weaving through the straits at the southern tip of South America, which were also later named for him.

During the first wave of colonialism, between 1500 and 1850, the Dutch were the primary colonial influence in the western Pacific, and they extended their reach into the region from a base in Southeast Asia, Amboina (present-day Papua New Guinea). During the 1700s, European explorers, including Spanish, French, and British, traveled the Pacific by ship (Figure 13.24), and the Spanish claimed and colonized islands in Micronesia.

During the second wave of colonialism, from approximately 1850 until after World War II, Germany claimed several islands around Guam and also the Samoa Islands; the United States claimed Hawaii, the Midway Islands, Wake Island, Johnston Islands, and Palmyra; the British claimed most of the southern Pacific, near their colonies in Australia and New Zealand; and, France held New Caledonia, Tahiti, Society Islands, Marquesas, and several other islands in the southern Pacific. In the same time period, Japan controlled the islands closest to them and then expanded their colonial claims in the region after 1900 (see Chapter 8).

The history of colonialism is reflected in the distribution of languages and religions in the Pacific (Figure 13.25). European colonizers, especially Spanish and British, brought their religions to the region and established Catholic and Protestant churches and missions, respectively.

Since the 1880s, Mormon missionaries have actively sought converts in the Pacific, with a particular focus on the Maori of New Zealand (Figure 13.26). The Maori tie their creation story to people coming from a long distance away. Through conversion to European Christianity, Maori identified with the Israelites and identified themselves as coming "from the land of Canaan" (Underwood 2000, 136). The sacred text of the Church of Jesus Christ of Latter-Day Saints (LDS), the Book of Mormon, teaches that several groups migrated from Israel to the Americas and one of those groups later left for the Pacific. Mormon theology teaches that this group was the "ancestors of the Polynesians" (Underwood 2000, 137). Since the 1880s, many Maori have fused the Book of Mormon and the Maori creation story to fully integrate the Mormon religion into their Maori identity, thereby creating a **syncretic religion**.

The theological connection between Mormonism and Polynesians has led to the widespread diffusion of the LDS church across island countries in the Pacific, especially Polynesian countries. According to Brigham Young University, approximately 40 percent of Tongans are Mormon, giving it the largest proportion of Mormons in the world with its 44,819 followers. Over 418,000 Mormons lived in the Pacific as of 2000, including 99,000 in Australia, 89,000 in New Zealand, 71,000 in Samoa, and 55,000 in Hawaii.

DECOLONIZATION OF THE PACIFIC

Between World Wars I and II, Japan colonized most of the islands of Micronesia as part of its imperialist program in East Asia and the Pacific. After World War II, the United Nations placed the islands of Micronesia under the protection of its Trusteeship Council, which was designed to help transition colonies to a political status of their choosing. The United Nations Trusteeship Council typically

© Christie's Images/Corbis.

FIGURE 13.24 Voyage of Captain Cook. Scottish artist Isaac Cruikshank used watercolor to paint "Captain Cook Landing in Owyhee," an old English spelling of Hawaii.

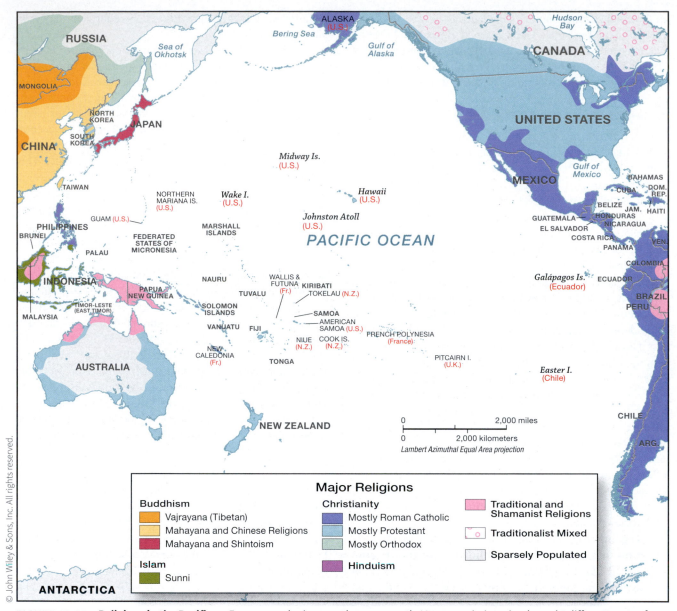

Major Religions

Buddhism
- Vajrayana (Tibetan)
- Mahayana and Chinese Religions
- Mahayana and Shintoism

Islam
- Sunni

Christianity
- Mostly Roman Catholic
- Mostly Protestant
- Mostly Orthodox

Hinduism

- Traditional and Shamanist Religions
- Traditionalist Mixed
- Sparsely Populated

FIGURE 13.25 Religions in the Pacific. European colonizers, and more recently Mormon missionaries, brought different sects of Christianity to the Pacific.

FIGURE 13.26 Hamilton, New Zealand. This Latter Day Saint (LDS or Mormon) Temple built in the 1950s was the first LDS temple built in the Southern Hemisphere. Although Maori religious beliefs have combined with teachings of the Mormon Church to create a syncretic religion, the architecture of this temple does not reflect Maori aesthetics. The temple design, with the single spire, is quite common for LDS temples throughout the world; in fact, this temple is an architectural duplicate of the LDS temple in Bern, Switzerland.

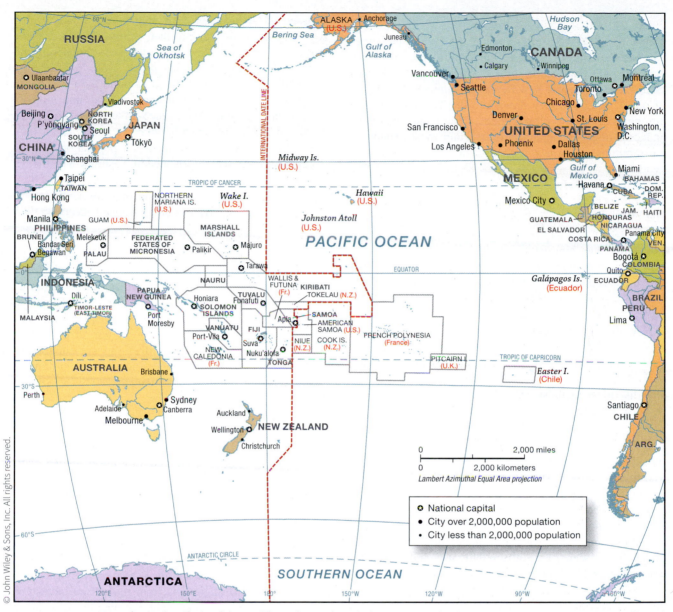

FIGURE 13.27 **Countries and capitals of the Pacific.** The Pacific includes a mixture of independent countries and territories affiliated with other countries.

assigned a "protectorate" to the colony to help mentor them to choose a political status. The choices offered to the colonies were: (1) independence, (2) integration with another country, or (3) a special relationship with another country (Figure 13.27).

Today the political status of the islands varies. Independent countries in the Pacific include Fiji, Palau, Vanatu, Tonga, Samoa, Solomon Islands, and Papua New Guinea. These comprise 6 of the 16 independent countries and self-governing states that are part of the Pacific Islands Forum, which is an organization committed to cooperation in order to establish sustainable economies and security in the region.

Two Pacific territories are fully integrated with other countries. Easter Island is a province of Chile, and Hawaii is a State of the United States. Hawaii chose to integrate with the United States and became the country's

fiftieth state in 1959. The question of integration is not fully settled for all Hawaiians. The Hawai'ian Sovereignty Movement began in the 1970s and still operates. Some members advocate for full independence of Hawaii, and others push for land reform so that native lands are returned to native ownership.

Some parts of the Pacific have special statuses with independent countries and are not fully independent or fully integrated into another country. For example, Guam and the Marshall Islands are protectorates of the United States. French Polynesia, New Caledonia, and Wallis and Futuna hold special status with France and are designated as overseas possessions. The United Kingdom holds onto the Pitcairn Islands as an overseas territory, and Australia continues to claim Norfolk Island as a territory (Smith 2009, 17).

MIGRATION IN THE PACIFIC

Since World War II, migration in the Pacific has escalated to the point that some ethnic groups, like Tongans, have a larger population abroad, primarily in the United States, Australia, and New Zealand, than they do at home (Connell 2012). Money flows into the Tongan economy through selling agricultural goods, foreign aid, and remittances. Tonga's economy is the second most dependent on remittances (money sent from migrants back home) worldwide.

Migration diffuses cultural practices, and it also fuels economies in the Pacific through remittances and trade in the informal economy. Pacific economies rely on remittances as a major source of revenue to pay for children's education and household goods. Migration also helps churn the informal economy when migrants bring or send home goods purchased in their destination countries. Trade in these goods has led "to thriving informal markets" (Skeldon 2000, 377) in the Pacific.

Within Tonga, rural to urban migration is building, as Tongans migrate from outlying, rural islands to the main island, which includes the capital city of Tuku'alofa. According to Connell, 70,000 of Tonga's 110,000 people now live in the main island, and 50 percent of them live "in town" (2012, 75).

MIGRATION IN AUSTRALIA AND NEW ZEALAND

Aborigines lived in Australia for thousands of years before Europeans arrived. The Australian Museum reports that in 1788, Aborigines, with an estimated population of 750,000, were the only people living on the continent. Before Europeans reached Australia, the number of different indigenous languages hovered around 200, whereas today "fewer than 20 are still in daily use, and even these are endangered" (Australian Geographic, 2).

Similar to Canada and the United States (see Chapter 11), Australia also attempted to assimilate indigenous people. The Australian government is working to reverse the effects of assimilation on the indigenous population. The government financially supports the National Indigenous Television channel, including programming aimed to teach children indigenous languages.

British Migration to Australia

In 1788, the First Fleet of British convicts, typically urban thieves, arrived at present-day Sydney and established the penal colony of New South Wales. The ship included more than 700 convicts (about 20 percent were women) and their children and more than 200 soldiers and their family members. The convicts, however, were not "free" upon arrival in Australia. The British regularly used transportation to a penal colony as a sentence. The government of Australia reports that 70 percent of the convicts were British or Welsh and another 24 percent were Irish.

Once in Australia, the sentenced convicts were compelled to work. The captain of the First Fleet, Arthur Philip, served as the governor of the British colony and "founded a system of labour in which people, whatever their crime, were employed according to their skills—as brick makers, carpenters, nurses, servants, cattlemen, shepherds and farmers" (Australian Government).

Male convicts who labored to build the new colony and female convicts who married and became mothers earned their freedom. New fleets of convicts arrived in 1790 and 1791. Convicts continued to arrive, and as late as the 1830s, about 6 percent of the population of New South Wales was still imprisoned in Australia. The British established additional penal colonies in present-day Tasmania in 1825, Western Australia in 1849, and Victoria in 1851. New South Wales colony stopped accepting convicts in 1850.

Asian Migration

Around the time New South Wales stopped welcoming transports of convicts, the number of Chinese migrants in Australia began to rapidly increase under a policy the Australian government describes as a "system of indentured labour." Chinese migrants were drawn to work in agriculture and market gardening, and some aspired to strike it rich in Australia's gold rush. By the late 1800s, Sydney and Melbourne both had Chinatowns.

The British government officially ended slavery in 1808 but still needed labor in their colonies, especially for labor-intensive industries of agriculture and mining. During the 1800s, the British encouraged migration among their colonies through a system of indentured labor. Indians and Malaysians, who hailed from heavily populated colonies, migrated to Kenya, South Africa, and Australia, British colonies in need of more laborers. Through relocation diffusion, Hindu temples and crematoriums can be found in Kenya, South Africa, and Australia. The 21st century communities of Indians and Malaysians in these now independent countries date back to migrations from the 1800s. Indentured laborers were typically men "who left their homes with the hope of returning," though most became permanent settlers or died before returning (Skeldon 2000, 369).

MODERN AUSTRALIA

Australia became a recognized, independent country in 1901, when six colonies were assembled in a federation as a constitutional commonwealth. As in Canada, the monarch of England is the official head of state in Australia, but the government was established and operates under the constitution passed in 1901.

In the same year that Australia passed its constitution, the government passed the **White Australia Policy**, which sought to keep out Chinese and other non-European migrants. Canada and the United States also gave preference to European migrants at this time. Canada ended its preferential system in 1962, and the United States in 1965. The Australian government ended the White Australia Policy in 1973. Australia further opened its doors to more Asian immigrants in 1988, at a

living in the country were migrants. Of the 5.3 million, 20.8 percent were from the United Kingdom; 9.12 percent from New Zealand; 6.1 percent from China; and 5.57 percent from India. Between 2007 and 2011, the greatest proportion of migrants from Asia to Australia came from India. In the same time frame, the economic links between Asia and Australia have deepened so that Australia's top two trading partners are in Asia, with 22.6 percent of Australia's exports destined for China and 16 percent for Japan (Figure 13.29).

RESIDENTIAL SEGREGATION

Seventy percent of the 4.4 million people in New Zealand live on the North Island. The 2011 New Zealand Census reported that about 5 percent (682,200) of the country's 4.4 million people self-identified as Maori; 7 percent as Pacific Islanders, 10 percent as Asian, and the majority (77 percent) as European or New Zealander.

Residential segregation of Maori and Pacific Islander minority populations is relatively common in New Zealand. Through a statistical analysis at a fine resolution of block and neighborhood scales, geographers found that as the proportion of Maori in a city's population increased, the

FIGURE 13.28   Melbourne, Australia.   A woman wearing a gray hoodie over traditional South Asian clothing, salwar kameez, crosses a pedestrian bridge.

time when Australia was approximately 90 percent people of European descent.

Since 1988, the number of migrants from Asia to Australia has increased (Figure 13.28). In the 2011 census, the Australian government reported that 5.3 million people

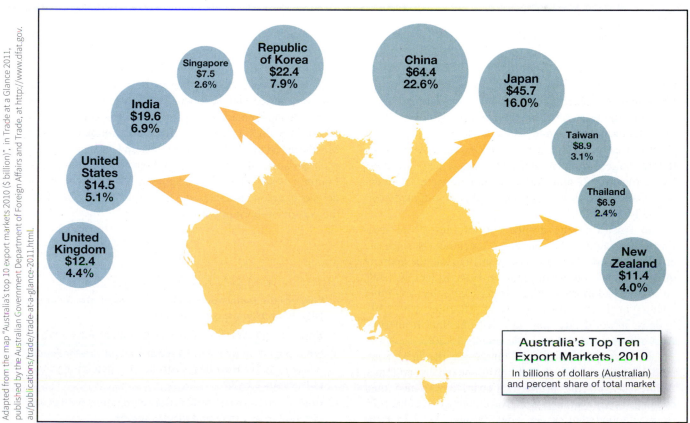

Adapted from the map "Australia's top 10 export markets 2010 ($ billion)", in Trade at a Glance 2011, published by the Australian Government Department of Foreign Affairs and Trade, at http://www.dfat.gov.au/publications/trade/trade-at-a-glance-2011.html.

Singapore
$7.5
2.6%

Republic of Korea
$22.4
7.9%

China
$64.4
22.6%

Japan
$45.7
16.0%

India
$19.6
6.9%

Taiwan
$8.9
3.1%

United States
$14.5
5.1%

Thailand
$6.9
2.4%

United Kingdom
$12.4
4.4%

New Zealand
$11.4
4.0%

Australia's Top Ten Export Markets, 2010
In billions of dollars (Australian) and percent share of total market

FIGURE 13.29   Australia's exports by value.   The largest proportion of Australia's exports are destined for China, followed by Japan, Korea, India, the United States, and the United Kingdom.

Reading the **CULTURAL** Landscape

Ethnic Neighborhoods in Australia

Like other world cities that attract large numbers of migrants, including New York and London, Sydney's growth as a world city has coincided with a rise in immigrants to the city from throughout Asia and the Pacific. Lebanese migration to Sydney dates to the country's 1976 civil war, and Lebanese restaurants like the one in this photograph are found in primarily Lebanese neighborhoods.

Large numbers of Asian migrants in Sydney are a more recent phenomenon. Asian migrants, especially Indian and Chinese, are clustered in medium- to high-income neighborhoods in Sydney, and the cultural landscape of those ethnic neighborhoods reflects the more recent arrival and higher income status of these ethnic groups.

Peter Charlesworth/LightRocket via Getty Images.

FIGURE 13.30　Sydney, Australia.　A restaurant specializing in Australian and Lebanese food advertises fish and chips, kebabs, hamburgers, tabouli, and Coke.

residential segregation of Maori increased. Where the Maori make up a larger proportion of the population, "Maori are more likely to be disadvantaged in the lower strata of the labor market, and hence in the housing market too" (Johnston et al. 2005, 128). In Auckland and Wellington, the two largest cities on the North Island, the researchers found that Maori and Pacific Islanders share "residential space" and as a group are highly segregated residentially.

In Australia, the number of migrants from non-European countries has increased precipitously since the end of the White Australia Policy in 1973 and the opening of the country to more migrants in 1988. In fact, in 1991, Australia "had more immigrants per capita than any other country in the world except Israel" (Burnley 1998, 49). Sydney, a world city, has attracted migrants from rural Australia, Asia, and the Pacific (Figure 13.30).

In Sydney, Arabic and Chinese are the most common languages other than English spoken by migrants. Arabic speakers are primarily Lebanese who fled their country's civil war in 1976. Australian geographer I.H. Burnley found that, as in New Zealand, the larger the population of an ethnic group in Australia, the higher its

residential concentration in urban areas. Unlike New Zealand, however, residential segregation in Australia does not always come from a lower socioeconomic status (Burnley 1999). In Sydney in 1991, the most concentrated areas of Mandarin, Hindi, Indonesian-Malay, and Japanese speakers were areas of medium- to high-socioeconomic status. Immigrants speaking these languages had higher levels of university graduation rates and higher family incomes than the Australian-born population.

CONCEPT **CHECK**

1. **Why** are the indigenous peoples of Hawaii, the Solomon Islands, and Samoa culturally related?

2. **How** did migration from Great Britain shape Australia?

3. **What** migration policy did Australia adopt after independence in 1901? **How** has Australia changed immigration policies since 1970?

4. **How** are patterns of residential segregation similar and different in Australia and New Zealand?

HOW IS THE REGION INTERCONNECTED THROUGH TRADE?

LEARNING OBJECTIVES

1. **Describe** how ship traffic changed in the Pacific from the time of Magellan to today.

2. **Explain** what time–space compression is and how it is reflected in the economies of the Pacific.

3. **Determine** whether the Pacific region is similar to the historic Mediterranean and Atlantic regions.

In 1956, the first container ship set sail, a revolution that met the economies of scale required to make shipping across the vast Pacific Ocean profitable. Seeking access to the spices of Southeast Asia (see Chapter 7), the Spanish explorer Magellan headed west from Europe, rounded the tip of South America, and around 1500 became the first European explorer to enter the Pacific from the east. Nevertheless, the Pacific route for trade from Europe was largely set aside in favor of the more efficient Triangular Trade Network in the Atlantic for more than four centuries. The routes recorded in the shipping logs of the British, Dutch, and Spanish from 1750 to 1800 demonstrate that the Atlantic Ocean functioned as the center of global trade during this period (Figure 13.31). A twenty-first-century map of global shipping lanes (Figure 13.32) demonstrates how the new technology of container ships has interconnected so much more of the world through trade. The map also reflects the vital role of Pacific trade in globalization and capitalism today.

TECHNOLOGY, TRADE, AND COLONIALISM

Between 1500 and 1780, approximately 450 European ships traveled through the Pacific Ocean for trade. Sailors charted latitude easily by sun angles. To track longitude, navigators needed to know the time on the ship and the time at the home base simultaneously. To measure time, navigators used measurements based off the location of the moon, a method that was not always accurate. By the 1730s, inventors in Great Britain had developed reliable naval chronometers designed to keep time at the home location on a ship. But it was not until after 1780 when chronometers became less expensive and increasingly common that ships shifted from the lunar method of charting longitude to the chronometer (Figure 13.33).

Between 1780 and 1880, the Pacific region entered a trade period shaped by the travels of British Captain James Cook (who reached the Pacific in 1769). As historian McNeil explains, once "Cook and his contemporaries" were

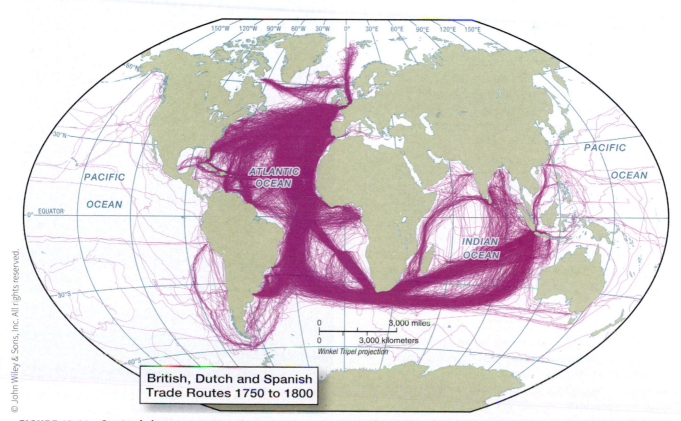

British, Dutch and Spanish Trade Routes 1750 to 1800

FIGURE 13.31 **Sea trade between 1750 and 1800.** James Cheshire of Spatial Analysis mapped the routes recorded in the shipping logs of the British, Dutch, and Spanish from 1750 to 1800. This map compiles the routes of all three countries to demonstrate how the Atlantic Ocean functioned as the center of global trade during this period.

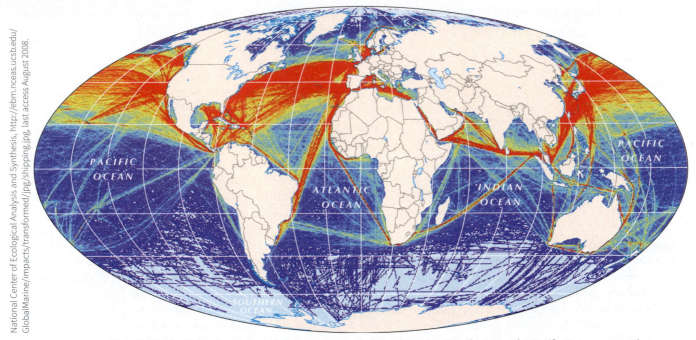

National Center of Ecological Analysis and Synthesis, http://ebm.nceas.ucsb.edu/GlobalMarine/impacts/transformed/jpg/shipping.jpg, last access August 2008.

FIGURE 13.32 **Global shipping lanes.** Modern global shipping lanes interconnect primarily across the Pacific Ocean, as massive container ships encircle the world.

"armed with chronometers," they "could describe any location with precision and return to it directly if desired" (1994, 12). Europeans mapped the Pacific and increased trade and the diffusion of crops and invasive species in the region.

The impact of growing European trade in the Pacific was similar to the impact of Europeans in the Americas (see Chapter 11). Crops and invasive species were not the only things to diffuse. Europeans who migrated to or traded with the Pacific brought European diseases to which the indigenous in the region had not been exposed. From Hawaii to New Zealand, the Pacific Islands experienced large-scale depopulation in the 1800s.

Island cultures and environments changed in response to depopulation and forced labor migration of indigenous across islands. Between the 17th and 19th centuries, islanders in Micronesia were exposed to European diseases through first contact and continuing trade. A smallpox outbreak on the Marianas Islands in 1854 ravaged the population, killing about 40 percent of the people. In addition to smallpox, influenza, measles, and typhoid fever spread relatively easily among the indigenous who had no prior exposure to the diseases.

TIME–SPACE COMPRESSION

In 1913, the United States Postal Service published a map of rates charged for parcel post mail from Washington, DC (Figure 13.34). The map demonstrates that the U.S. Postal Service assumed directional symmetry, looking only at distance between locations and not at ease of travel, such as nodes in transportation routes or the presence or absence of roads or mountains, in establishing postal rates. The parcel post rate map demonstrates the traditional geographic con-

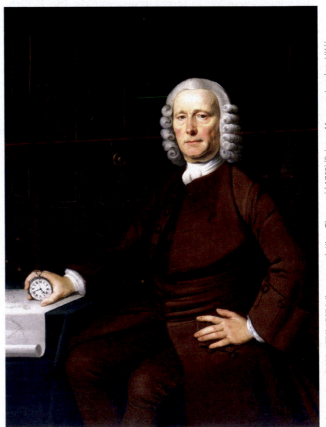

John Harrison (1693–1776) 1767 (oil on canvas), King, Thomas (d.1769)/Science Museum, London, UK//Bridgeman Art Library.

FIGURE 13.33 **The chronometer.** The artist Thomas King painted English horologist and inventor John Harrison who invented the gridiron pendulum, which solved the problem of how to determine longitude at sea. One version of Harrison's chronometer is shown in the background on the right side of this painting.

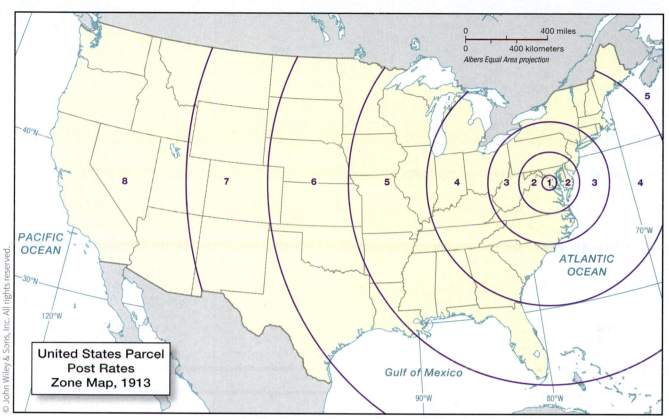

FIGURE 13.34   **United States Postal Service rate zones for parcel post mail.**   The United States Postal Service determined rates based on directional symmetry in 1913, assuming distance as the primary factor and not considering transportation nodes or connectedness.

cept of **distance decay**, which hypothesizes that places equidistant from a hearth will have the same level of connection and likelihood of diffusion.

Distance decay assumes the actual distance between two places describes the level of connectedness between the places. If we apply distance decay to Washington, DC, using the postal map as guidance, the level of connection between Washington, DC and New Orleans, Louisiana will be the same as between Washington, DC and Duluth, Minnesota.

In 1962, geographer William Bunge published the ground breaking book *Theoretical Geography*, analyzing movement, central place, and networks. In his concluding chapter, Bunge examined a map of parcel post mail rates and determined that the concentric circles on the map did not accurately reflect the network of connectedness among the cities. Bunge described the "shortest completely connected network" as connecting the "shortest route from any point to all others" and demonstrated this network as shown in D on Figure 13.35. Unlike Figure 13.34, this diagram does not assume directional symmetry or that distance alone accounts for connections. Comparing Figure 13.35D to Figure 13.32 reveals that although Bunge was studying highways, today's global shipping routes—more than highways or railroads—create the "least cost highway system to the user" (Bunge 1962, 185).

In the "shortest completely connected network" of global shipping, the points of connection rise in importance while interior places and coastal locations that are not connection nodes fade. The nodes or points of connection are major shipping ports around the world's oceans. Connections between these world cities by shipping transport, passenger airline flights, Internet connections, telephone conversations, and social media links have amplified what geographers call **time–space compression**. Deepening connections between world cities have compressed the distance between them to the degree that physical distance alone no longer accounts for the level of connection between two places. As a result of improvements in transportation and communication technologies, some places on Earth are more connected, while other places are left farther behind from lack of connection. The most connected places are linked across space that is "folded and refolded, like origami," distorting time and space through "wormholes" that "re-create the world's geographies" (Warf 2011, 438).

PACIFIC TRADE

Container ships transport more goods farther and faster and compress relative distance in the Pacific, and trading ports in the rest of the world are also increasingly compressed. Geographer Barney Warf theorizes that time–space compression is perhaps misnamed because the term

---

 time–space compression   increasing connectedness between world cities from improved communication and transportation networks.

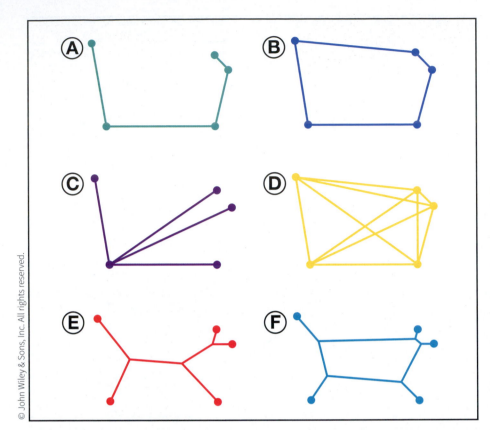

FIGURE 13.35   Minimum distance networks among five nodes. Geographers and other spatial thinkers have theorized how to create efficient networks, and this graphic shows six alternatives for connecting five nodes. A connects all nodes by starting at one point. B shows the classic traveling salesman problem by connecting the five nodes and returning to the starting point. C connects one node with each of the other four. D is the shortest completely connected network drawn by geographer William Bunge in 1962. E and F show two different ways to connect the five nodes using the minimum distance.

implies a "death of distance" that is not accurate. Rather, he suggests that the rapid improvements in transportation and communication technologies "continually" generate "new spatialities" or organizations of space that have "expanded, not compressed" trade over long distances (Warf 2011, 435). Competition in trade and economies of scale makes trade the easiest and least expensive among the most connected places. Globalization, then, has amplified the relative distance of Pacific Islands that are not among the most linked places.

Places in the Pacific that are not linked to the world's most powerful ports through technological "wormholes" or "origami folds" may be more closely linked economically than before, but their remote location still impacts economic development. A lack of connectedness to the network and the relative remoteness of small island developing states in the Pacific still hamper economic development.

Australian researchers Donovan Storey and Shawn Hunter studied the economic development of small island developing states (SIDS) in the first decade of the 2000s. Their description of the atoll country of Kiribati pinpoints the development issues of Pacific Islands, including "limited arable soils," relative isolation, and "environmental vulnerabilities" (2010, 168). Kiribati is made up of

33 atolls, coral reef islands largely organized around a central lagoon (Figure 13.36). The per capita gross national income (GNI) of Kiribati is $1,300.

FIGURE 13.36   Marakei, Kiribati. The Markei atoll in Kiribati displays the classic circular form of an atoll with a chain of coral reef islands surrounding a central, enclosed lagoon. The Marakei atoll is one of few in Kiribati that displays this classic form, as most of Kiribati is comprised of long, narrow islands.

FIGURE 13.37   **New York, United States.**   Fiji water, bottled on the Pacific island of Fiji, is more expensive than many other bottled waters but is known for its crisp, clean taste.

In many ways, the increasing connectedness of the world economy makes the economies of SIDS more vulnerable to shifts in supply and demand. Francis Hezel's study of the vulnerability of Pacific Island economies explains that development in the world economy depends on exports, but many island countries, including Kiribati, have "relatively few products and limited quantities" of products to export to trading partners. Hezel contends that "Everyone is vulnerable in today's global economy, but small is especially vulnerable" (2010, 10).

The value of Kiribati's exports is equal to 6 percent of the country's gross domestic product (GDP), and the value of Fiji's exports is equal to 36 percent of the country's GDP. Fiji's export economy is diversified, including manufactured clothing, gold, fish, and bottled water (Hezel 2012) (Figure 13.37). The combination of Fiji's exports and their tourism economy has created a per capita GNI of over $3,000 and helps Hezel, in his study, conclude that "Fiji seems to be the only Pacific nation to have prospered because of its exports" (2010, 12).

FOREIGN INFLUENCE ON PACIFIC ECONOMIES

Although formal colonialism has ended, to the degree that political entities in the Pacific have decided on independence, integration, or a special territorial relationship

with a state, many economies of Pacific countries still rely heavily on foreign aid, military presence, and investment. Guam's economy is driven by a large U.S. military presence (Figure 13.38), which owns and occupies 30 percent of the island and accounts for about 30 percent of the economy (Bureau of Economic Analysis). The island of Guam is sometimes referred to as the United States' "unsinkable aircraft carrier," and its role in the U.S. military has only increased as the American government has relocated Marines from Japan and South Korea to Guam.

Pacific islands have tried to develop niche economies in order to produce or provide a good that is in high demand. For example, the country of Vanuatu focused on exporting organic beef to East Asia. However, it is difficult to capitalize on agricultural pursuits in niche markets for two main reasons. First, if one country finds success in a niche market, such as organic beef, another country can tap into the same niche and flood the demand with too much supply or shift production to a place with lower transportation costs. Second, in places where a new crop has been introduced, agriculture has not had a significant positive impact on the economy because "agricultural products are generally high volume, low value, and expensive to transport to foreign markets" (Hezel 2010, 14).

When a Pacific economy is built around producing a good that needs to be exported, the economy is dependent on trading partners, other countries that purchase the

FIGURE 13.38   **Guam.**   US Navy service members from the USS Albuquerque lay cables offshore in Guam. Guam is home to US Navy and Air Force bases, which cover about 30 percent of the island.

goods. Pacific countries are increasingly relying on the economies of other countries through remittances. Pacific islanders migrate to other parts of the Pacific, Australia, the United States, or Europe to find work and then send wages back to their home country to family and extended family.

THE PACIFIC AS A NEW GEOPOLITICAL CENTER

In 1995, just six years after the fall of the Berlin Wall, political geographer Alexander Murphy suggested that the next geopolitical order would have three centers: western Europe, North America, and Pacific Asia. Although his discussion largely overlooked the economic rise of China, his pronouncement of **Asia Pacific** as a new center of gravity had staying power.

Both Murphy in 1995 and geographer Paul Blank in 1999 analyzed the levels of historic and current cultural, economic, and political integration in the Pacific to decipher whether the Pacific was becoming a new economic and political region, much like the Mediterranean during the Roman Empire era or the Triangular Trade Network of the Atlantic during European colonial era. Blank observed that 1980 marked the first time the "value of transpacific commerce outweighed that of transatlantic commerce" (1999, 265). He reasoned that the end of the Cold War and the "economic reopening of China" created conditions "ripe for the resumption of Pacific integration" (1999, 271). He predicted a future Pacific with influential port cities connecting islands and coasts across the Pacific culturally and economically as the Mediterranean did during the Roman Empire.

Counter to Blank's analysis, Murphy found a lack of historical integration in the Pacific and predicted political integration would be difficult because of the history of Japanese hegemony and colonialism in the region. Murphy acknowledged the widespread use of English in trade around Asia Pacific and the economically integrating linkages among overseas Chinese in the region. However, he suggested that Asia Pacific lacked an external threat that "helped spur integration in other cases, most notably Europe" (Murphy 1995, 129).

Australia's defense minister ignored the issue of integration in Asia Pacific and suggested that in the twenty-first century Asia Pacific will become a center of gravity in the world (Smith 2011), as major economies including China, India, the United States, and the ASEAN countries increasingly influence the world economy. Additionally, Smith suggested that Asia Pacific extends to Russia, the United States, and India, arguing that Asia Pacific is a global focal point militarily. The region, he said, is home to "four of the world's major powers and five of the world's largest militaries—the United States, Russia, China, India, and North Korea."

Applying Barney Warf's theorization of time–space compression reveals that Asia Pacific has risen as a center of gravity or integrated region not because distance has been made meaningless but because connections across the region have created "wormholes" and "folded and refolded space like origami." The pace at which new wormholes are formed, old ones are abandoned, and origami folds are created and then refolded is unprecedented. The clip of change is unlike anything experienced in the Mediterranean or the Atlantic previously. And just as collapsing the space of the Mediterranean or Atlantic before felt unprecedented and at times overwhelming, so too can the collapsing of space in the Pacific today.

CONCEPT CHECKS

1. **How** has ship traffic changed in the Pacific from the time of Magellan to today?

2. **What** is time–space compression, and is the Pacific both compressed and distanced through globalization?

3. **Compare** and **contrast** the Pacific region with either the historic Mediterranean or Atlantic regions.

HOW IS GLOBAL CLIMATE CHANGE AFFECTING THE PACIFIC REGION?

LEARNING OBJECTIVES

1. **Describe** the greenhouse effect.

2. **Explain** how additional energy from incoming solar radiation creates global climate change.

3. **Determine** why Australia is feeling the effects of global climate change.

Climate scientists refer to the Pacific and Australia as a "canary in the coal mine" or as a "petri dish" because they are already feeling the effects of global climate change on a grand magnitude. Atolls in the Pacific, like Tuvalu, have little to no high ground so rising sea levels can inundate these islands and also bring salt water into the water tables, making it impossible to use the water for drinking water or to irrigate crops. Since 2000, Australia has experienced torrential rains followed by nine years of unprecedented flooding, massive cyclones, and raging wildfires. In response to these catastrophes, Australia's global climate change advisor warned "you ain't seen nothing yet" (Farr 2011).

Climate change occurs when more energy is added to and trapped in the lowest layer of Earth's atmosphere, the troposphere. This section of the chapter explains the science of climate change, the different ways climate change is already impacting parts of the region, and how people and governments are navigating through the current and coming changes.

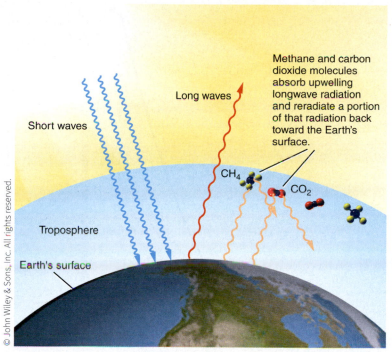

FIGURE 13.39 **Greenhouse effect.** Incoming solar radiation is relatively short-wave. Earth absorbs the energy and emits energy at a longer wavelength. The long waves move up through the troposphere and hit greenhouse gases at the top of the troposphere (tropopause). Some are absorbed, some make it through to the stratosphere and some are reflected back to earth. The greenhouse gases at the top of the troposphere act like a greenhouse, reflecting energy back to earth and keeping Earth about 59°F (15°C) warmer.

THE GREENHOUSE EFFECT

Greenhouse gases in the upper troposphere, including methane and carbon dioxide, work to keep earth warmer by acting as a glass "lid" to the lowest layer of the atmosphere, the troposphere (Figure 13.39). Ultraviolet radiation, which is short wave, travels through the "glass" and reaches earth. The sun's energy is absorbed, and Earth reemits the energy as longwave radiation. The greenhouse gases at the top of the troposphere absorb and reradiate energy back toward Earth. Scientists contend that as a result of the greenhouse effect, a naturally occurring process, Earth is 59°F (15°C) warmer, which makes the planet livable for humans.

The basic premise of climate change science is that human-made greenhouse gases, emitted primarily through fossil fuel burning, are amplifying the greenhouse effect to unprecedented heights. Too many greenhouse gases are trapping too much energy in the troposphere. Since 1850, the energy in Earth's atmosphere has increased by "1.6 to 2.4 W/m², which is about 1% of the total solar energy flow absorbed by the Earth and atmosphere" (Foresman and Strahler 2011, 86) (Figure 13.40).

The extra energy is causing "global warming," which means the average temperature of the world is increasing. Since 1980, the average temperature on Earth has increased by 1.1°F (0.6°C). Media reports about climate change used to only focus on this number, the rise in average temperatures. Actually, 1.1°F does not seem like a large increase; if one is comparing a 87°F summer day to a 88.1°F summer

day, the difference is hardly noticeable. As a result, many people interpreted "global warming" as putting a 1.1°F blanket around the world, warming it up just "a little bit." This thought process, however, is not accurate, and that is why climate scientists now use the term *global climate change* instead of *global warming*. The change in temperature is not the end of the story; the real problem is the extra energy retained in the atmosphere.

TRAPPED IN THE TROPOSPHERE

Energy from the sun, incoming solar radiation, fuels the atmosphere, wind, storms, and weather. The energy we feel, sense, or can measure with a thermometer is called **sensible heat**. The energy we cannot measure with a thermometer that is either added or released to change the state of matter from solid to liquid to gas (or the opposite) is called **latent heat**. When water changes from liquid to gas (water vapor), it requires the addition of latent heat. In the process of changing from liquid to gas, the molecules holding the water together need to be broken apart so that the molecules can move more freely, as they do in a gaseous state compared to a liquid state. Latent heat is the energy the molecules absorb that change water from liquid to gas. If you keep a thermometer in water and watch it boil, the temperature of the water will stay constant for a period when the water is boiling. The water will be adding energy from the heat source, but it will be using the energy to change the state of water from liquid to gas.

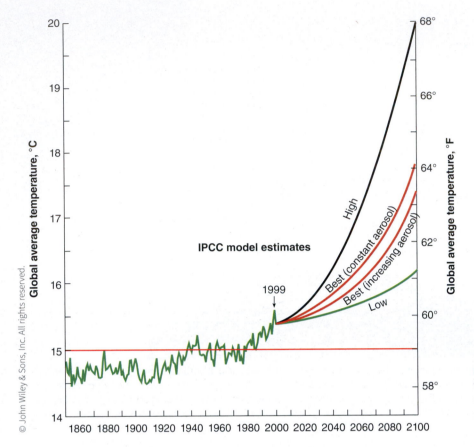

FIGURE 13.40 **Computer projections for future warming.** Using computer models, climate change scientists have predicted several scenarios for how much Earth will warm in coming years.

When climate change scientists predict the future of Earth's temperatures, they use climate models. Figure 13.40 shows four different scenarios for the period after 1999 when this particular model was run. Each line shows a different estimate from the Intergovernmental Panel on Climate Change. Climate change scientists use massively complicated computer models to predict the future of Earth's atmosphere. The computer models process millions of data points over Earth at different time periods. In the case of this diagram, scientists knew what Earth's temperature change had been up to the point when the model was run. Scientists work backwards to check which of their computer models fits best with what actually happened to Earth's temperature. Then, using the accumulation of historical data to date, the scientists run future scenarios, each based on a different combination of possible outcomes.

When the amount of energy in Earth's atmosphere increases, the potential for change in weather and climate is great, and the ripple effects are profound. Energy fuels the global system of winds. So, a change in the amount of energy in Earth's atmosphere can change the circulation of winds. Energy fuels low-pressure systems, which can generate severe storms, including blizzards, tornadoes, and hurricanes. Energy fuels weather systems and the movement of fronts. Places that are normally wet can become dry, and places that are normally dry can become wet.

THE IMPACT OF CLIMATE CHANGE ON TUVALU AND AUSTRALIA

The regional and local impacts of climate change vary based on how the increase in energy in the troposphere and the circulation of air and ocean currents impacts that particular place. Countries from Tuvalu with a population of 12,000 to Australia with a population of 23 million are already dealing with changing availability of fresh water, a rise in wild fires in areas riddled with drought, increasing intensity of storms, and the inundation of salt water through surface flooding, which mixes with fresh water estuaries and seeps into water tables.

The atoll islands of Tuvalu have graced the covers of textbooks, popular magazines, and brochures for environmental groups. Tuvalu is widely discussed because its elevation is only 6.5 feet (2 m) above sea level and with current predictions in sea level rise, the country would be covered by the Pacific as early as 2050. Climate change is already impacting Tuvalu, as coasts are eroding, salt water floods are inundating coastal estuaries, the availability of potable water through rainwater is shifting, new pests and diseases are affecting crops, changes in sea surface temperatures are impacting fish stocks, and increasing climate hazards from tropical cyclones and fires are already being felt (Tuvalu Ministry of Natural Resources 2007).

In response to changes in fresh water availability and rising sea levels, Tuvaluans are choosing to migrate. The popular media calls these migrants "climate change refugees," but using this term paints the people of Tuvalu as passive and reactionary, when in fact the residents and government have been preparing for climate change that was created at the global scale by the largest CO_2 producers in the world and its impacts on their small, remote country since the earliest warnings of climate change (Figure 13.41). Tuvalu signed the United Nations Convention on Climate Change in 1992 and has worked since then to develop environmental strategies and adaptation programs to sustain the population, economy, and environment and has received funding from the United Nations and other countries toward its efforts.

Australia is located between 9°S and 30°S latitude, which means the entire continent is affected by the Hadley cell migration, which moves with the ITCZ. The ITCZ (see Chapter 6) is a belt of a low-pressure system that forms along the equator and migrates north and south of the equator with the direct rays of the sun. In a low-pressure system, surface air flows into the low and then air lifts at the low, flowing aloft toward the tropopause. Warm air rising at the ITCZ fuels the Hadley cells, or belts of wind that move air from the surface aloft at the ITCZ and then poleward where air descends toward the surface at about 30°N and S of the equator (creating the Subtropical Highs), and then flows along the surface from the Subtropical Highs toward the ITCZ (Figure 13.42).

GUEST *Field Note*   Climate Change in Tuvalu

DR. CAROL FARBOTKO

Adaptive Social and Economic Systems, Commonwealth Scientific and Industrial Research Organisation, Australia

Tuvalu is an independent country of 12,000 people in the central Pacific. Comprised of low-lying islands, its land territory in its entirety is at risk from rising sea levels associated with climate change. My work in Tuvalu involved interviewing Tuvaluan people about climate change and the concept of climate refugees. I also observed the activities of journalists and environmentalists who came to the islands from around the world during seasonal king tide flooding to witness climate change impacts.

While Tuvaluan people have serious concerns about climate, they reject the label 'climate refugee' because it positions them as passive victims and does not address the responsibility of industrialised countries to reduce greenhouse gas emissions. Tuvaluans see international migration as a solution of last resort to climate change. However, international media and environmental organisations are prematurely depicting Tuvalu as a place in environmental crisis, often suggesting migration as refugees is the only option for Tuvaluans. These representations marginalise authentic Tuvaluan voices and yet ventriloquise them for enviro-political purposes: to present international climate change migration as a problem and to convince climate sceptics in business and politics that climate change is occurring.

Courtesy of Carol Farbotko, Adaptive Social and Economic Systems, Commonwealth Scientific and Industrial Research Organisation, Australia.

FIGURE 13.41   Funafuti, Tuvalu.   Journalists film high tide flooding in the capital of Tuvalu.

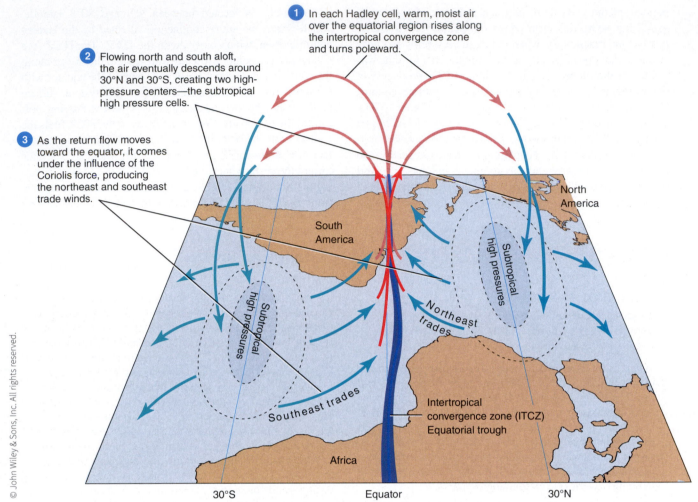

1 In each Hadley cell, warm, moist air over the equatorial region rises along the intertropical convergence zone and turns poleward.

2 Flowing north and south aloft, the air eventually descends around 30°N and 30°S, creating two high-pressure centers—the subtropical high pressure cells.

3 As the return flow moves toward the equator, it comes under the influence of the Coriolis force, producing the northeast and southeast trade winds.

North America

South America

Subtropical high pressures

Subtropical high pressures

Subtropical high pressures

Northeast trades

Southeast trades

Intertropical convergence zone (ITCZ) Equatorial trough

Africa

30°S   Equator   30°N

FIGURE 13.42   Hadley cells.   The global system of winds includes Hadley cells, which start with warm air rising at the low pressure of the Intertropical Convergence Zone (ITCZ) along the equator. Along the top of the troposphere, air flows toward 30°N and 30°S. There, air descends toward the surface, creating high pressure belts. Air flows from the surface High back to the ITCZ, completing the Hadley Cell.

The Outback has a semiarid to arid climate because the Subtropical High dominates that area of Australia over most of the year. Moist areas in southern Australia are coastal locations that receive precipitation from storms coming in off the Indian or Pacific Ocean. Northern Australia's climate is linked to the climate of Southeast Asia, as the north coast of Australia receives seasonal moisture with the summer monsoon and is dry during the winter monsoon (see Figure 13.9).

A change in the amount of energy coming into the atmosphere changes the energy going into the Hadley cells both north and south of the equator (Figure 13.43). More warm air rising along the ITCZ means more potential for precipitation (whether more days

Map by NOAA Climate.gov, based on AVISO data provided by Mark Merrifield. Adapted from Figure 3.28(a) in State of the Climate in 2013.

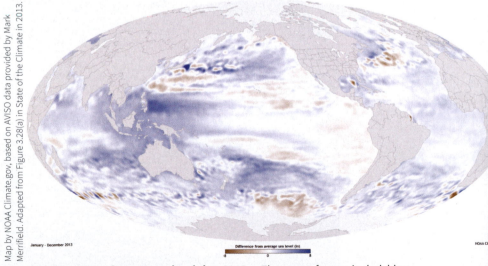

January - December 2013

Difference from average sea level (in)

NOAA Clim

FIGURE 13.43   Average sea level rise, 2013.   The areas of ocean in dark blue are places where sea level in 2013 rose up to 8 inches above the average sea level from 1993 to 2010. The areas of ocean in darkest red are places where sea level fell up to 8 inches in 2013 relative to the 1993–2010 average.

FIGURE 13.44 Murray River, Australia. Water from the Murray River has been depleted as a result of a drought attributed to climate change in Australia. The flow of the river (brighter blue) now stops before reaching the ocean (bottom right).

AMY TOENSING/National Geographic Creative/National Geographic Stock.

or more intense). In the 1980s, climate change scientists concluded that increased energy in the atmosphere would intensify tropical cyclones.

Australia is seen as the first large, relatively wealthy country to be deleteriously impacted by global climate change across the planet. Australia also demonstrates how the added energy and warmer temperatures do not simply warm the continent 1.1°F. The impact of additional energy on the continent varies by region and over time. For example, one of the major rivers that flows through the eastern side of the continent lost so much water from the combination of a drought and evaporation that in 2002 the river stopped flowing before it reached the ocean (Figure 13.44). The year 2002 marked only the second year of a massive drought known in Australia as the Millennium Drought. After nine years of drought, in 2009, wildfires "torched more than a million acres and killed 173 people" (Goodell 2011, 2). The drought ended after ten years with "torrential rains and flooding" (Gleick and Heberger 2012, 2).

Australia's Response

An article in *Scientific American* explains that Australians responded to the drought by decreasing their per capita water use in cities by 37 percent between 2002 and 2008, and by the five largest cities in the country "spending $13.2 billion to double the capacity of desalination, enough to meet 30 percent of current urban water needs" (2012, 2). Additionally, the government has "continued with plans to restore rivers and wetlands by cutting withdrawals from the Murray–Darling river basin, which is the main agricultural area, by 22 to 29 percent" and by buying back irrigation rights from users in order to keep more water flowing in Australia's rivers (2012, 22). The Australian government

predicts the growing frequency and length of droughts will impact food production, cutting it by an estimated 15 percent.

The Australian government is concerned that one of the reasons it is feeling the impact of climate change so much more severely and rapidly than other countries located at similar latitudes is that the country is emitting too many greenhouse gases. The government recognizes Australia has the "highest rate of carbon emissions per person in the developed world" (Australian Government, climate change). Like China, Australia primarily uses coal, which is found in abundance in the country, to fuel electricity plants that power its cities. Coal "provides 80 percent" of the country's electricity and "serves as its leading export" (Goodell 2011, 3).

In addition to the loss of food production and the increased frequency and intensity of natural disasters, the government of Australia recognizes that it will likely lose some of its natural treasures and greatest tourist attractions to climate change. One of the most immediate concerns is the Great Barrier Reef, which is located off the northeast coast (Figure 13.45). The oceans are warming, which impacts the algae growing in the reefs. Current estimates are that the reef, which is composed of living organisms, will die by 2050.

CONCEPT CHECK

1. **What** is the greenhouse effect and **how** has the increase in greenhouse gas emissions created global climate change?

2. **How** does additional energy in the troposphere impact weather systems?

3. **Why** is the Australian government concerned about global climate change, and **what** is the government doing to try to cope with and lessen the effects of climate change?

JC Photo/Shutterstock.

FIGURE 13.45 Great Barrier Reef, Australia. The world's largest coral reef stretches over 1,800 miles (3,000 km) along the northeast coast of Australia.

SUMMARY

What Is the Pacific Region?

1. The Pacific region is larger in area than all of the landmasses on Earth combined. Islands in the Pacific are continental, volcanic, limestone, or coral reef. High islands are those with mountains in the center, typically volcanic or limestone. Low islands are coral reefs that grew around a volcano. The volcano erodes and leaves behind a lagoon in the middle of the coral reef atoll.

2. The prevailing winds between 30°N and the equator flow from northeast to southwest. Between 30°N and the equator, prevailing winds flow from southwest to northeast. Winds flowing off of the ocean bring moist air onto islands, the air lifts over mountains in the center of high islands, condensation occurs, and precipitation follows. The air descends on the leeward side of the mountains and creates drier climates with comparatively less vegetation than the windward side.

3. The Pacific Ocean garbage gyre migrates from north to south as the winds and ocean currents shift seasonally. Plastics do not decompose fully; they simply break up into smaller and smaller pieces. The Pacific Ocean garbage gyre is made up of small pieces of plastic covering an area the size of Texas, which makes it impossible to see from satellites.

Who Are the People of the Pacific?

1. The indigenous peoples of Hawaii, the Solomon Islands, and Samoa are culturally related as a result of the migration patterns people followed through the Pacific. The most recent area of the Pacific to be populated was Polynesia. Although the islands of Polynesia cover a large expanse of area, the languages are closely related.

2. Beginning in the 1700s, Great Britain sent convicts to Australia to establish penal colonies and free up jail space in Great Britain. The convicts had to work for their freedom in Australia.

3. Australia had a "White Australia" policy until 1975, allowing immigration from Europe but not from other regions of the world. Starting in the 1980s, Australia opened the country to migrants from Asia. In recent years, migrants from India and China have comprised the largest proportion of migrants from Asia.

4. Australia and New Zealand have growing populations of Pacific Islander migrants, who along with indigenous peoples, have moved primarily to cities. Geographers have found that the larger the population of minorities in an Australian or New Zealand city, the greater the residential segregation found between minorities and the European descendants in the city.

How Is the Region Interconnected through Trade?

1. The chronometer, invented in the 1700s and broadly used by the 1800s, enabled European explorers to chart and map islands in the Pacific and then return to the precise locations later on. European exploration brought European diseases to the Pacific, similar to North America, in the 1800s. Some islands in the Pacific experienced depopulation as a result of disease diffusion.

2. The Pacific region has the largest area in the world, and as a result of improvements in air travel and shipping, parts of the region are tightly interconnected in global trade. Other areas of the region are more distanced, in a sense, because they lack access to transportation and direct connection to the intertwined global economy. People in developing economies in the Pacific are migrating to Australia and New Zealand to gain employment, often sending remittances to their families.

How Is Global Climate Change Affecting the Pacific Region?

1. The Greenhouse effect is a naturally occurring phenomenon in earth's troposphere. Greenhouse gases accumulate at the top of the troposphere. The sun emits shortwave energy, Earth absorbs the shortwave energy and reemits it as longwave radiation. Greenhouse gases absorb and reemit the longwave radiation back toward earth (like the glass in a greenhouse). This process keeps earth at a livable temperature, 86°F warmer than it would be without the greenhouse effect.

2. Energy fuels the global system of winds and weather systems, including severe storms such as hurricanes, blizzards, and tornadoes. The rise in the amount of energy in Earth's atmosphere, caused by humans emitting greenhouse gases, is a source of concern for climate change scientists.

3. The Pacific, including Australia and New Zealand, is already being impacted by global climate change, as warming temperatures are generating higher rates of evaporation and more frequent and severe storms in parts of the region. The physical geography of the region, including thousands of low islands made up of coral around extinct volcanoes, makes the region particularly susceptible to rising sea levels.

YOUR TURN ▶ Geography in the Field

A tourist examines a stone, standing 8 feet (2.4 m) tall, made of cut limestone. Stones like this large one and the others next to it have been the currency on the Micronesian island of Yap since the eighteenth century when the stones were imported by boat from Palau, a distance of about 300 miles (484 km). All of the 11,000 people of Yap own land, and foreigners are not allowed to own land. But a massive Chinese tourist agency, the Exhibition Travel Group (ETG), that specializes in providing travel opportunities for wealthy Chinese is working to secure a number of 99-year leases from Yapese land owners in order to build an enormous hotel and shopping complex (up to 20,000 rooms have been proposed). The people of Yap are divided about whether to allow the ETG proposal to move forward, even in a scaled-back form without a casino. They recognize that opening its doors to large-scale tourism will fundamentally change the geography of Yap.

Thinking Geographically

- Yap and other islands in Micronesia currently receive aid from the United States at an annual rate of $1,000 per capita. The American aid is slated to be cut back. How will a shift from U.S. aid to Chinese corporate influence change Yap's economy and its cultural landscape?

© John Elk III/Alamy.

Yap, Micronesia. Cut limestone imported from Palau has been used as currency in Yap since the eighteenth century.

- Yap is a tropical island where taro, sweet potatoes, and bananas grow relatively abundantly and are used primarily for local consumption. A ship of imported produce and other goods arrives in Yap once a month. How will the food supplies and diets of Yapese change if the island becomes a major tourist destination?

Read:

Frangos, Alex. "Is Yap Ready for the World? A Tiny Pacific Island Wonders What It Will Lose by Welcoming a Giant Tourist Resort. *Wall Street Journal*, March 9, 2013. http://online.wsj.com/article/SB10001424127887324712504578131864269855132.html.

Thompson, Adam. "Tourism in Yap and Micronesia: Will China Run the Show?" *Asia Pacific Bulletin*, February 7, 2013. http://www.eastwestcenter.org/sites/default/files/private/apb199.pdf.

KEY TERMS

Pacific rim	troposphere	syncretic religion	sensible heat
ring of fire	endemic	white Australia policy	latent heat
volcanic arc	step migration	distance decay	
limestone	relocation diffusion	time-space compression	
atoll	haka	Asia Pacific	

CREATIVE AND CRITICAL THINKING QUESTIONS

1. The Pacific Ocean covers one-third of earth. Pollutants, including trash and radioactive waste, have been disposed of in the ocean by humans. Explain how the Pacific, from pollutants to climate change, demonstrates the concept of an **anthropocene**.

2. Papua New Guinea exports goods at a value of 80 percent of the country's gross domestic product (GDP). At the **scale** of Papua New Guinea, describe the physical geography of the country and the kinds of products produced on the island. Then explain how the **situation** of Papua New Guinea in the Pacific enables the island to have more exports than other Pacific island countries.

3. The Outback in Australia is heated by solar radiation, but the dominating high-pressure system prevents warm air from rising, so precipitation rarely occurs. Predict how climate change may affect precipitation patterns in Australia and the Pacific **region**.

4. Find the University of Hawaii Manoa's interactive atlas of rainfall online. Examine the map and change the layers. Notice the direction of the prevailing northeasterly winds. If Hawaii were located at the same latitude in the Southern Hemisphere, the prevailing winds would come from the southeast. How would the climate of Hawaii change if the **site** of the island chain were in the Southern Hemisphere?

5. Researchers use archaeology and linguistics to help them uncover routes of **migration**. The newest theory of Polynesian migration holds that Hawaii was populated by Polynesians who followed step migration from Samoa to the Solomon Islands before reaching Hawaii. What archaeological and linguistic evidence could you look for to support the theory that the pre-Hawaiians lived in the Solomon Islands for some time before migrating to Hawaii?

6. Ancient sacred **sites** of the Maori culture include the volcanic peak Ruaphu. Modern sacred sites of the Maori include Mormon temples and Christian churches. What role might pilgrimage and **migration** play in maintaining both of these types of sacred sites—natural and human made?

7. The end of European colonization began in the post–World War II era and continued to 1994 when Palau gained independence. Although the Pacific is no longer formally colonized, how do former colonizers and foreign countries continue to influence the Pacific **region** today?

8. **Migrants** to Australia since 1988 have increasingly come from Asia. How would you expect to see evidence of the relocation **diffusion** of Chinese and Indian migrants in the cultural landscape of Australia today?

9. The invention and widespread **diffusion** of the chronometer changed shipping in the 1800s, and the invention of the container ship revolutionized shipping after 1956. Recognizing that some goods are heavy, bulky, or in enough quantity that they need to be carried by sea to achieve economies of scale, predict what may be the next innovation to revolutionize shipping and how it will affect the geography of trade.

10. How would the loss of the Great Barrier Reef, the bulldozing of traditional cultures, and the pollution dumped into the Pacific Ocean affect **tourism** and **development** in the Pacific region?

SELF-TEST

1. The International Date Line marks:
   a. the new day, to the west of the line from the old day to its east
   b. the new day, to the east of the line from the old day to its west
   c. sunrise, to the west of the line from sunset to its east
   d. sunrise, to the east of the line from sunset to its west

2. Australia has a large number of earthquakes today primarily because:
   a. Australia is a volcanic arc located on a subduction zone of the Pacific plate
   b. the continent of Australia is being squeezed from the west and the east through plate movements
   c. the Australian continent was glaciated 14,000 years ago, and the continent is going through isostatic rebound
   d. active volcanoes on the Australian continent generate large-scale earthquakes

3. Atolls are comprised of _____ and are formed around _____ volcanoes.
   a. marble/active   b. marble/dormant
   c. limestone/active   d. limestone/dormant

4. Orographic precipitation on high islands:
   a. generates rain on the windward side
   b. creates dry conditions on the leeward side
   c. depends on the prevailing winds
   d. All of the above

5. The last country in the Pacific Ocean to be occupied by humans was:
   a. Hawaii   c. New Zealand
   b. Fiji   d. Australia

6. More than 1,000 distinct languages are spoken in Melanesia because:
   a. people migrated to the region around 1,500 years ago during Polynesian migration
   b. people migrated to the region around 30,000 years ago and had little spatial interaction with each other
   c. Melanesia is composed of thousands of tiny islands with little spatial interaction between them
   d. migrants from all over the world were attracted to the gold and other resources in Melanesia, bringing their languages with them through relocation diffusion

7. The Lapita civilization is important to understanding the cultural geography of the Pacific because the Lapita:
   a. acted like pioneers and purposefully settled beyond Near Oceania and into Remote Oceania
   b. Lapita pottery is found on nearly every island in Micronesia
   c. Lapita language is widely spoken in Melanesia
   d. Captain James Cook defeated the Lapita at the Sandwich Islands, ushering in the European colonial era in the Pacific

8. After World War II, colonies in the Pacific region were allowed to choose their political status. Each of the following demonstrates one of the three choices offered by the United Nations Trusteeship Council except:
   a. The Solomon Islands chose to become an independent country
   b. The Hawaiian Islands chose to become integrated into the United States
   c. New Zealand chose to become integrated into Australia
   d. Guam chose to become a protectorate of the United States

9. The Australian government passed the White Australia Policy in 1901 with the primary goal of keeping migrants from _____ out of Australia.
   a. Italy   c. China
   b. New Zealand   d. Kenya

10. True or False:
    Residential segregation of Maori and Pacific Islander minority populations is relatively common in New Zealand.

11. Between 1500 and 1950, the center of global trade was the _____, and since the 1950s, the center of global trade has shifted to the _____.
    a. Pacific/Atlantic
    b. Atlantic/Pacific
    c. Mediterranean/Indian Ocean
    d. Indian Ocean/Mediterranean

12. Distance decay assumes directional symmetry and that the actual distance between two places describes the level of connections between the places. Time–space compression recognizes that:
    a. some places or nodes are more connected by transportation and communication than others
    b. places that are not connected are distanced in the modern world economy
    c. the distances between global cities is compressed because of deepening connections between cities
    d. All of the above

13. True or False:
    Small countries like Kiribati are especially vulnerable in the rapidly changing world economy because they have few products and limited quantities of the products.

14. Greenhouse gases are a concern for climate change scientists because the gases:
    a. absorb more energy from the sun
    b. trap energy in the troposphere
    c. are created almost exclusively by anthropogenic sources
    d. destroy the ozone layer in the stratosphere

15. The country with the highest rate of carbon emissions per person in the developed world is:
    a. the United States   b. Canada
    c. Australia   d. the United Kingdom

ANSWERS FOR SELF-TEST QUESTIONS

1. a, **2.** b, **3.** d, **4.** d, **5.** c, **6.** b, **7.** a, **8.** c, **9.** c, **10.** True, **11.** b, **12.** d, **13.** True, **14.** b, **15.** a

14

WORLD REGIONS AND WORLD CITIES

Looking at a neoclassical clubhouse on a horse racetrack, one might suspect the scene is from France or the United States, both of which are known for horse racing. The buildings and signs in the background of this photo are clues that the clubhouse is actually located in Shanghai, China. In the 1850s, British entrepreneurs built the Shanghai Race Course for the pleasure of British colonizers and European businessmen, adding the clubhouse in the 1930s. Shanghai housed British, French, and American sectors in the nineteenth century, which were established for the purposes of global trade.

After China's communist revolution in 1949, Mao Zedong banned gambling, closed the Shanghai Race Course, and turned it into a People's Square for political rallies. In the 1990s, city government filled the center of the oval space with a government building and an exhibition center. The former clubhouse, which now houses not jockeys but Chinese art, stands out as European architecture in an East Asian world city.

If you could hover in a helicopter over the sprawling city of 23 million people and see the neighborhood, you would notice that green space surrounds the former Shanghai Race Club and that the sidewalk has an oval curve following the lines of the old track.

The city of Shanghai grew at a rate of 600,000 people per year between 2008 and 2013. World cities, like Shanghai, are centers of gravity in globalization. People, jobs, trade, ideas, and money are drawn to world cities. Layers of history, from trading partners to waves of migration, are revealed in the cultural landscapes of world cities. People and businesses are drawn, economically, socially, culturally, and politically to world cities as they become more entwined. The cities most linked to a world city become a network, a new kind of region of connectedness, that creates opportunities for diffusion among the most linked peoples and places.

Shanghai, China. The Shanghai Art Gallery is located in the former clubhouse of the Shanghai Race Club. Once China became communist in 1949, the government banned gambling, and the race track was transformed into People's Park. Layers of history, from British colonial presence in the nineteenth century to the rapid growth of Shanghai into a world city in the twenty-first century, are shown in this photograph.

Walter Bibikow/Photolibrary/Getty Images.

WHY DO WORLD REGIONS CHANGE?

World regions are dynamic and are molded by shared experiences and similar forces that people and states experience over time. For example, European colonialism from 1500 to 1950 and Japanese colonialism from the late 1800s leading up to World War II affected people and places throughout East Asia. People in eastern China, the Korean Peninsula, and the islands of the Pacific did not share the exact same experience with either European or Japanese colonialism, but they were affected at the same time. The shared experience of colonialism is part of what shapes the region of East Asia.

While much of the world region of East Asia experienced European and Japanese colonialism, the outcome of that experience was not the same for everyone. People and places negotiate the same process or circumstance in different ways. Identity, place, culture, and economics all shape how a group of people navigate an experience. Keeping with the example of colonialism, its impact varied not just by place but by the people within each place.

In the past, geographers saw world regions as natural entities that stood alone and apart from each other. Traditionally, world regions were thought of as large groups of smaller spatial units (like cities and countries) that shared distinct characteristics and were easy to distinguish from one another. New regional geographers see world regions as constructed, imagined entities that are institutionalized by international actors (see Chapter1).

In *Understanding World Regional Geography*, we reconcile old and new regional geographies by seeing regions as a balance of real and constructed. Regions do have actual differences in languages, religions, political systems, colonial histories, and other spatially distributed characteristics. At the same time, world regions are constructed, imagined entities that seem permanent because governments, agencies, corporations, and individuals institutionalize them by reporting data, delivering programs, solving problems, and representing the world by region.

Geographers think in terms of three kinds of regions: formal, functional, and perceptual. Formal regions are distinguished by common cultural or physical criteria. We can draw formal regions around climate type, languages, religions, and other characteristics of people and places. Using ArcGIS Online to explore the dynamic webmaps in *Understanding World Regional Geography*, you can appreciate that dividing the world into formal regions is problematic in that the pattern of regions can change when you alter the scale.

By clicking on any map of East Asia and entering ArcGIS Online, you can see that few experiences or cultural traits are confined by the frame of the region on the map. East Asian languages extend into eastern Russia. Panning around the map, you can see that Buddhism extends across much of Asia, especially into Southeast Asia. Furthermore, a formal region at the world scale can look different when you change from the global to regional scales.

Functional regions are conglomerations of places that share a common purpose, often political, economic, or social. One place can be in hundreds (or thousands!) of overlapping functional regions, from city limits to newspaper delivery areas. Functional regions vary in size from relatively small areas (a school district) to larger areas (a court's jurisdiction) to areas larger than a group of countries (the European Union) to the world (the United Nations). By functioning in world regions, entities such as corporations, nongovernmental organizations, and governments help entrench or institutionalize world regions. For example, the World Food Program defines and institutionalizes functional regions because it uses world regions to distribute goods and services (see Chapter 1), and even FIFA, the International Federation of Association Football, which oversees soccer's World Cup, uses regions to determine who qualifies for the tournament (Figure 14.1).

Perceptual regions are the ideas we carry in our minds about similarities and differences across people and places. Each of us has perceptions about where a region, for example, South Asia, ends and where another region, Southeast Asia, begins. We also have ideas about what it means to be from a certain region or what formal characteristics the region has. How a people or place experienced history and how they make sense of that experience today has helped mold perceptual regions.

How people in a region share a perception of their history, present, and future is reflected in cultural landscapes—both by what is visible and what is missing. Going back to the example of Shanghai, the People's Square has no reference to the Japanese colonialism the city endured from 1939 to 1949, a bitter memory for Shanghai and neighboring Nanjing. Likewise, the South Korean government reconstructed the Gwanghwamun Gate in Seoul to remove the imprint of a painful and even longer period of Japanese colonialism (see Chapter 8). In these cases, what is absent from the immediate cultural landscape

Adapted from map of FIFA World Cup regions, at http://www.fifa.com/worldcup/groups/preliminaries/index.html.

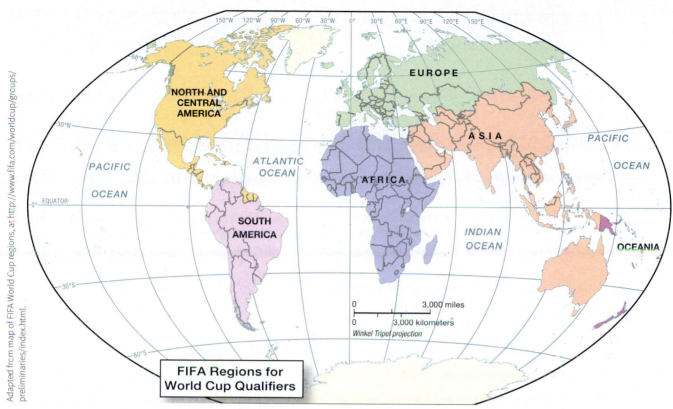

FIGURE 14.1   FIFA regions for World Cup Qualifiers.   The International Federation of Association Football (FIFA), the governing body of international soccer, uses world regions to determine who plays in the men's and women's World Cup.

reflects the desire to negate part of history, often while boosting a sense of nationalism.

How people experience, negotiate, and remember helps shape world regions, as do the religions, languages, economies, political systems, and other formal characteristics of a region. So, are the experiences of globalization rendering a new map of world regions? And is that new map based on world cities? Looking at networks is one way to discern how a new map may unfold.

WORLD CITIES AS WORLD REGIONS

In 1997, political geographer Peter Taylor embarked on a quest to quantify the growth of world cities and map the networks among them. Before Taylor, researchers had assumed that a hierarchy among world cities existed, but they had not worked to quantify them. Taylor recognized a central problem was the lack of data tabulated by city, as most data are reported by country. It was easy to find import and export data for China, for instance, but it was difficult to find it for Shanghai. Taylor defined **world cities**

world cities highly connected cities that function as nodes in global networks.

as the geographical entities "that appear as the organizing nodes of worldwide networks" (1997, 3).

Scholars associated with the Globalization and World Cities Research Network (GaWC), a United Kingdom-based research group, have published more than 400 articles reporting data and analyzing world cities and the networks among them. The GaWC gathers data on advanced producer services, including the very-high-value services in the banking, law, accounting, and advertising economic sectors. They looked at how services aimed at producers (corporations) are clustered in world cities and the flow of services across networks to distinguish which world cities have the greatest concentration of services and are the most connected through networks. Using all of these data, the GaWC Research Network classified world cities into alpha (with four subcategories), beta, and gamma world cities (Figure 14.2).

GaWC categorized world cities in 2012 by alpha++, alpha+, alpha and alpha−, beta, and gamma (Figure 14.3). Alpha world cities are the most integrated and connected in the world economy. Alpha++ world cities include London and New York, which "stand out as clearly more integrated than all other cities and constitute their own high level of integration" (GaWC). Alpha+ world cities "complement London and New York, largely filling in advanced service

Data from: "The World According to GaWC 2012," posted 13 January 2014. Globalization and World Cities Research Network, http://www.lboro.ac.uk/gawc/world2012t.html.

FIGURE 14.2   Map of Alpha, Beta, and Gamma World Cities.   The cities with the greatest concentration of international corporations and connectedness with other cities are classified as alpha. From there, each classification of world cities has a lower concentration of corporations and level of connectedness.

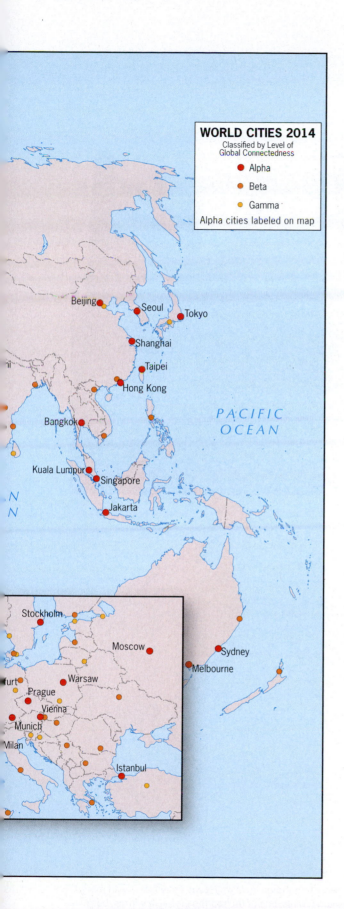

WORLD CITIES 2014
Classified by Level of
Global Connectedness

● Alpha

● Beta

● Gamma

Alpha cities labeled on map

needs for the Pacific Asia," and alpha and alpha−world cities are "very important world cities that link major economic regions and states into the world economy" (GaWC). Beta world cities link a state or region into the world economy, and gamma world cities link smaller states or regions into the world economy or function as important world cities whose economies are not built from advanced producer services.

Recognizing that globalization "has redrawn the limits on spatial interaction" and that how important one world city is depends on its connectivity with other world cities (Felsenstein et al. 2002, 1), scholars have also worked to enumerate and map the connectedness among world cities. For example, Figure 14.4 shows the world cities most linked to Johannesburg while Figure 14.5 shows Shanghai's network of world cities.

Could the networks of a given world city reveal a new spatialization (spatial arrangement) and perhaps regionalization of the world? People shape new spatializations or new world regions. If they are constructing and institutionalizing a new world regionalization, it will be reflected in shifts in people's perception and sense of belonging. Rather than identify themselves as part of a country or a region, people may identify themselves with a world city. People reference world regions in our conversations, government leaders discuss places as parts of world regions in speeches, and journalists refer to world regions in stories. To discern whether belonging, solving, and explaining are happening according to a new regionalization, listen for times when the media uses stories about world cities as a lens through which to understand what is happening in our world.

A new regionalization along world cities will be evident not only in perceptual regions but also in formal and functional regions where world cities are central. Formally, people and places within a given world city will share formal cultural traits, including language and religion. Functionally, world cities will serve as units of analysis for data collection and sites by which solutions are administered; flows of migration will be primarily within world city regions; paths of diffusion will follow networks among world cities; and problems would be defined and solved based on world cities (Figure 14.6).

USING GEOGRAPHIC TOOLS

Statistical Data Analysis

In order to classify world cities as alpha, beta, and gamma, geographers in the Globalization and World Cities Network first needed to assemble data related to globalization, economics, and networks that are reported by city instead of by country. In the 1990s, geographers assembled data about the location of global corporation headquarters by city and analyzed the data to help determine ranks of world cities. In the early 2000s, geographers focused on the locations of multinational corporations that specifically provide services to businesses (legal, accounting, marketing, and the like), defining a category of the world-economy called advanced producer services (Figure 14.3a).

Geographers are also statistically analyzing the networks among global corporations and among world cities to mathematically gauge the role world cities play in globalization—whether interactions to and from a city are geared toward interactions with other world cities or toward connecting their country and region into the world city network. To measure world cities statistically, analysts take the total universe of all advanced producer services, and they measure the presence of a given global business firm across all world cities in order to rank firms (Figure 14.3b).

(a)

Courtesy of Erin Fouberg.

(b)

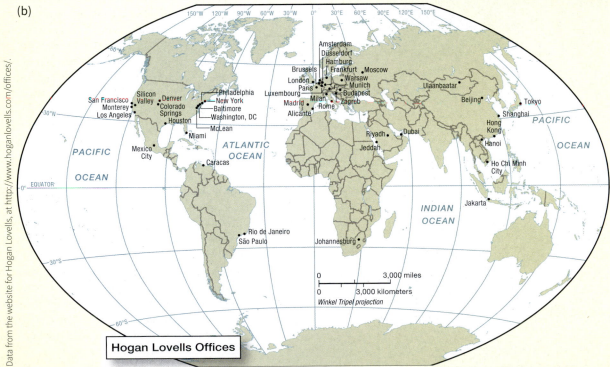

Data from the website for Hogan Lovells, at http://www.hoganlovells.com/offices/.

Hogan Lovells Offices

FIGURE 14.3a AND b   (a) **Denver, Colorado.**   The Denver headquarters of the international law firm Hogan Lovells is one of more than 40 offices the firm has globally. (b) **Location of Hogan Lovell offices.**   Hogan Lovells, one of the top ten global law firms, began as two separate law firms, one in Washington, DC, and one in London, which merged in 2010. The company's website is available in English, Chinese, French, German, Japanese, Russian, and Spanish.

They also measure the total amount of business producer services within world cities. Using these data, the Globalization and World Cities Network developed a set of statistical equations and paths of analysis that enabled him to categorize world cities as alpha, beta, and gamma every few years using updated statistics. Figure 14.2 represents the most recent results from their statistical analysis.

World regions reflect the reality of traits and flows, and they are also constructed and institutionalized by people. As world regions shift from traditional, contiguous, state-based regions to regions based on networked world cities, new geographies will incorporate and institutionalize traits and flows at the scale of world cities, thereby replacing the world regions in Figure 1.21 with world regions based on world cities.

Thinking Geographically

1. Compare and contrast the locations of Hogan Lovells offices with the map of world cities (Figure 14.2). Are the company's offices mainly found in alpha, beta, or gamma world cities?
2. Search the Hogan Lovells website. What kinds of law does the firm practice? How are these specializations reflected in the map of Hogan Lovells office locations?

CONCEPT **CHECK**

1. **Why** do geographers use regions?
2. **How** do functional regions affect formal regions and vice versa?
3. **What** role are world cities starting to play in world regionalization?

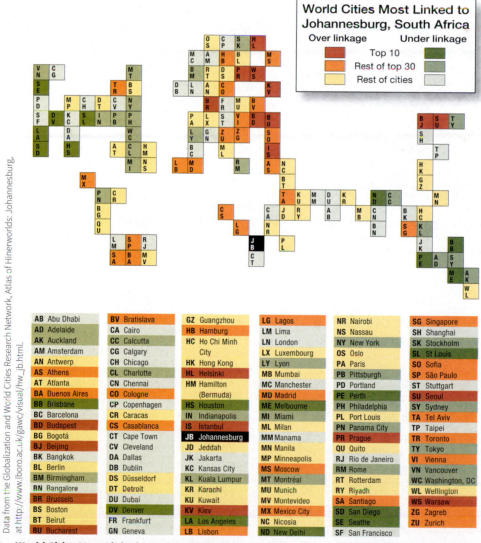

Data from the Globalization and World Cities Research Network, Atlas of Hinerworlds: Johannesburg, at http://www.lboro.ac.uk/gawc/visual/hw_jb.html.

AB Abu Dhabi	**BV** Bratislava
AD Adelaide	**CA** Cairo
AK Auckland	**CC** Calcutta
AM Amsterdam	**CG** Calgary
AN Antwerp	**CH** Chicago
AS Athens	**CL** Charlotte
AT Atlanta	**CN** Chennai
BA Buenos Aires	**CO** Cologne
BB Brisbane	**CP** Copenhagen
BC Barcelona	**CR** Caracas
BD Budapest	**CS** Casablanca
BG Bogotá	**CT** Cape Town
BJ Beijing	**CV** Cleveland
BK Bangkok	**DA** Dallas
BL Berlin	**DB** Dublin
BM Birmingham	**DS** Düsseldorf
BN Bangalore	**DT** Detroit
BR Brussels	**DU** Dubai
BS Boston	**DV** Denver
BT Beirut	**FR** Frankfurt
BU Bucharest	**GN** Geneva

GZ Guangzhou	**LG** Lagos
HB Hamburg	**LM** Lima
HC Ho Chi Minh City	**LN** London
HK Hong Kong	**LX** Luxembourg
HL Helsinki	**LY** Lyon
HM Hamilton (Bermuda)	**MB** Mumbai
HS Houston	**MC** Manchester
IN Indianapolis	**MD** Madrid
IS Istanbul	**ME** Melbourne
JB Johannesburg	**MI** Miami
JD Jeddah	**ML** Milan
JK Jakarta	**MM** Manama
KC Kansas City	**MN** Manila
KL Kuala Lumpur	**MP** Minneapolis
KR Karachi	**MS** Moscow
KU Kuwait	**MT** Montréal
KV Kiev	**MU** Munich
LA Los Angeles	**MV** Montevideo
LB Lisbon	**MX** Mexico City
	NC Nicosia
	ND New Delhi

NR Nairobi	**SG** Singapore
NS Nassau	**SH** Shanghai
NY New York	**SK** Stockholm
OS Oslo	**SL** St Louis
PA Paris	**SO** Sofia
PB Pittsburgh	**SP** São Paulo
PD Portland	**ST** Stuttgart
PE Perth	**SU** Seoul
PH Philadelphia	**SY** Sydney
PL Port Louis	**TA** Tel Aviv
PN Panama City	**TP** Taipei
PR Prague	**TR** Toronto
QU Quito	**TY** Tokyo
RJ Rio de Janeiro	**VI** Vienna
RM Rome	**VN** Vancouver
RT Rotterdam	**WC** Washington, DC
RY Riyadh	**WL** Wellington
SA Santiago	**WS** Warsaw
SD San Diego	**ZG** Zagreb
SE Seattle	**ZU** Zurich
SF San Francisco	

FIGURE 14.4 **World Cities Most Linked to Johannesburg, South Africa.** The world cities listed as over linkage, mostly in Europe, Africa, and Latin America, are most connected to Johannesburg, South Africa. The world cities listed as under linkage are least connected and are generally located in western North America, Australia and the Pacific.

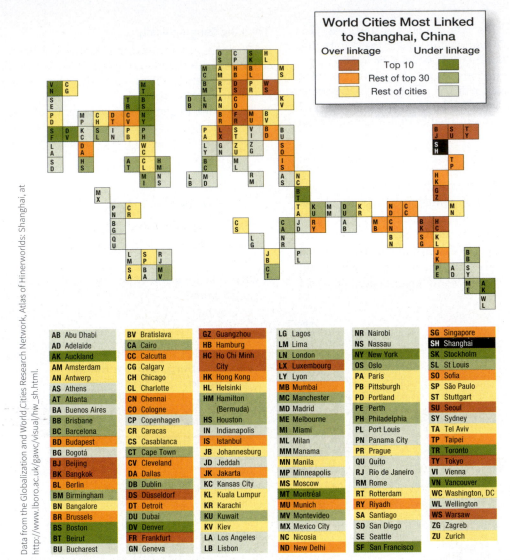

Data from the Globalization and World Cities Research Network, Atlas of Hinerworlds: Shanghai, at http://www.lboro.ac.uk/gawc/visual/hw_sh.html.

FIGURE 14.5   World Cities Most Linked to Shanghai, China.   The world cities listed as over linkage, mostly in Europe, Southeast Asia, South Asia, and North America, are most connected to Shanghai, China. The world cities listed as under linkage are least connected and are generally located in Africa, Australia and the Pacific, and Latin America.

HOW DO THRESHOLD CONCEPTS HELP YOU THINK?

LEARNING OBJECTIVES

1. **Explain** what a threshold concept is.

2. **Describe** how cultural landscape, site, and situation can be used to develop preliminary thoughts about a place that is new to you.

If you have ever watched the news or read a newspaper and thought: "I cannot possibly understand what is going on between Israelis and Palestinians"; "I cannot understand how China became so economically powerful in the last 30 years"; or "What is going on in Syria, and why are people being killed?" you know the feeling of thinking the world is simply too complex to understand.

We designed *Understanding World Regional Geography* and the online assignments to introduce you to the integrative way geographers gather and process information. To think geographically, you must ask geographic questions—*what is it? why is it there? so what?*—and apply geographic concepts to find answers.

Understanding World Regional Geography introduces you to dozens of geographic concepts that you can integrate and apply to real-world situations. We highlight 25 of the most integrative geographic concepts in the book and call them **threshold concepts** in geography. Each concept

GUEST *Field Note*   Food Security in World Cities

DR. LIAM RILEY

Balsillie School of International Affairs, Waterloo, Canada

Household food security is gaining increased attention in Africa's rapidly growing world cities. The African Food Security Urban Network (www.afsun.org) found about three quarters of households in low-income urban neighbourhoods were food insecure, meaning that they did not have consistent access to nutritious food sufficient to feed all household members. Harare is a world city that is exemplary of common food security problems in African urban communities. People have experienced a decline in living standards over the past three decades as a result of economic liberalization, political turmoil, and a sharp increase in urban unemployment. The result is that all except the wealthiest households rely on a mix of formal and informal sources of food. People buy food from supermarkets and convenience stores similar to those found in cities in more affluent countries, but they also purchase food at informal markets, roadside stands, and directly from farmers outside of the city. Many households also produce food for their own consumption, either in backyard gardens, on open land in the city, or by travelling to rural areas regularly to farm. For most households, accessing food in the city is a daily challenge made more difficult by related problems in housing, water, and sanitation.

FIGURE 14.6   Harare, Zimbabwe.   Informal markets help create a more stable food supply for people in the city.

challenges you to think about place, space, or landscape and to consider how people, ideas, values, economics, and cultures flow and how places and spaces are connected or networked.

Research on how students learn within a field of study has found that once students learn threshold concepts, they can start to think as practitioners or professionals would. Learning to apply threshold concepts across regions and scales helps you to think as a geographer would. Once you understand, apply, and personalize a threshold concept, you can begin to grasp its complexities. If you read through the list of threshold concepts in *Understanding World Regional Geography* (Table 14.1), you will not likely remember all of the concepts. You may be able to retain and apply between three and five of these concepts with confidence by the end of the course.

TABLE 14.1   Threshold concepts in *Understanding World Regional Geography*

CHAPTER	THRESHOLD CONCEPTS
Introduction to World Regions	Context Region Cultural Landscape Scale
Global Connections	Anthropocene Globalization Networks
Geography of Development	Development Unequal Exchange Mental Map
Subsaharan Africa	Site Situation
Southwest Asia and North Africa	Diffusion Hearth
South Asia	Green Revolution
Southeast Asia	Tourism Authenticity
East Asia	Commodity Chain
Europe	Population Pyramid
North and Central Eurasia	Identity
North America	Migration
Latin America and the Caribbean	Race Gender
Pacific Rim	Time–Space Compression
World Regions and World Cities	World Cities

To think geographically, a student needs to apply threshold concepts across classes and beyond the classroom. Which geographic concept takes a learner from novice over the threshold to thinking geographically depends on the student.

This list of threshold concepts is not the sum total of all geographic concepts. Geography has hundreds of other concepts and even draws from concepts in other disciplines to research and explain human, cultural, and physical phenomena. This list is based on research in geography education, but even so it is not definitive or exhaustive. Understanding one threshold concept in depth helps geographers better comprehend related geographic concepts.

For example, the threshold concept **anthropocene**, introduced in Chapter 2, recognizes the essential role of humans in shaping physical environment in the current era. In world cities, other geographic concepts, including urban heat island, are related to the larger threshold concept anthropocene (Figure 14.7). Humans shape and reshape world cities, replacing vegetation with pavement and buildings, which changes the albedo rate, or reflectivity rate of incoming solar radiation, of a city. By changing the land cover to surfaces that increase absorption of solar energy and runoff of precipitation in urban areas, the microclimates of world cities are literally heating up. The United Nations estimates that by 2025, 80 percent of the people in the world will live in urban areas, which means that diminishing the athropogenic impact of urban heat islands should be a priority in the short term.

Reading the PHYSICAL Landscape

The Urban Heat Island

Surface temperatures in urban areas are usually warmer than those in the surrounding countryside. Known as the *urban heat island,* this phenomenon occurs in part because built-up urban surfaces consist of metal, glass, asphalt, and concrete that absorb higher amounts of incoming solar radiation (14.7a). Rural areas, in contrast, are generally cooler because they are covered with soil, vegetation, and forest canopies that have a cooling effect because they provide shade and prevent runoff of precipitation (Figure 14.7b). Water absorbed by vegetation also has a cooling effect through transpiration. The combined effect of evaporation and transpiration (evapotranspiration) cools rural areas because energy is needed to turn liquid water into water vapor, and heat energy is pulled from surfaces to enable this change in state, thereby cooling the surface—just like sweating after a summer workout cools off your body. Cities tend to be warmer because water from precipitation runs off human-made surfaces in cities quickly, decreasing the likelihood of evapotranspiration (Figure 14.7c).

The built environment of Mumbai, India, changed drastically between 1992 and 2004. Mumbai has "more than doubled in size and population in the last 25 years as rural migrants have flooded in" (Guardian 2013). The government of India estimates that 18 million people living in Mumbai are going to have to cope with temperatures in May of 115°F (46°C), long before the hottest months of summer. The Energy and Resources Institute estimates that temperatures in Mumbai and Delhi have risen by 35 to 37.4°F (2 to 3°C) in 15 years (Guardian 2013). "The ongoing study, based on NASA satellite readings, also shows the cities to be 5 to 7°C warmer than in the surrounding rural areas on summer nights."

FIGURE 14.7a   Tokyo, Japan. Urban surfaces, such as this cityscape in the world city of Tokyo absorb more radiation.

J.W.F. Chu Photography/Getty Images.

Courtesy of Erin Fouberg.

FIGURE 14.7b   County Mayo, Ireland. Rural surfaces such as this Irish landscape are generally cooler than urban areas because less radiation strikes the ground due to tree cover, and there is more water in soil and on vegetation surfaces.

To reduce the impact of urban heat islands, cities are covering roofs with reflective materials instead of black tar (to reduce absorption rates and increase reflectivity); planting vegetation on roofs, in parking lots, and along buildings and roadsides to increase transpiration; and using "green" pavement to retain water instead of allowing it to run off.

FIGURE 14.7c   Delhi, India. Surface temperatures are represented by the color scale. Delhi exhibits an interesting pattern of having several small urban heat islands scattered throughout the city. The population of the city has grown by 40 percent in the last decade. The red areas of the map are the paved, densely built up parts of the city and are generally on the outskirts where pavement and apartment buildings have replaced farmland, and any planted trees have had little time to grow. The most intense heat island area, in the west, is a residential zone. On the north end of the river on the eastern side of the city, the urban heat island is clustered around the city's financial district. In the south central, the urban heat island is a dense residential area built after 1980.

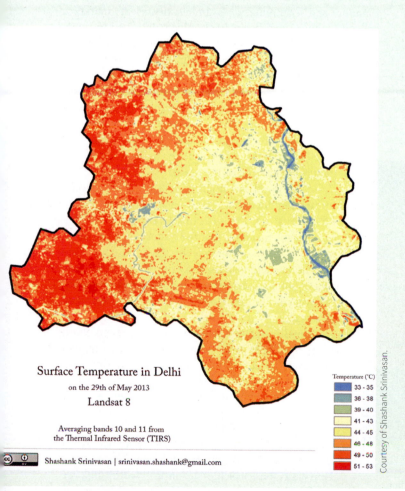

Surface Temperature in Delhi

on the 29th of May 2013

Landsat 8

Averaging bands 10 and 11 from
the Thermal Infrared Sensor (TIRS)

Shashank Srinivasan | srinivasan.shashank@gmail.com

Temperature (°C)

	33 - 35
	36 - 38
	39 - 40
	41 - 43
	44 - 45
	46 - 48
	49 - 50
	51 - 53

Courtesy of Shashank Srinivasan.

CULTURAL LANDSCAPE AS A THRESHOLD CONCEPT

Learning to think geographically requires practice. Applying threshold concepts in a book and in class is a good start. The next step is to go into the field, observe, and apply concepts. For example, to understand the threshold concept **cultural landscape**, one must practice reading cultural landscapes. Geographer Peirce Lewis described cultural landscape as "nearly everything that we can see when we go outdoors" and explained that "all human landscape has cultural meaning" (1979, 1). Cultural landscapes include a church on the top of a hill (Figure 14.8), a train platform (Figure 14.9), and a fast-food restaurant (Figure 14.10).

To practice using this concept, cultural landscape, start with the campus where you take classes. Think about the classroom building and how its **site** fits within the larger **situation** of the campus and the city or community. Walk around the neighborhood just off campus, look at the houses and apartments, and consider whether they were built before, during, or after the founding of the campus. Draw a **mental map** of the classroom building, campus, neighborhood, and streets. Label the map with as much detail as you can. Think about the parts of campus and the community that are "off limits" to anyone who does not "belong" there. How are those spaces distinguished in the cultural landscape to make visitors feel either welcomed or deterred?

What is interesting about this place and the cultural landscape? Take photographs of things that reflect the paths of migration or give you an idea of who built this cultural landscape. Look for what is common and mundane

Suzanne Plunkett/Bloomberg via Getty Images.

FIGURE 14.9　London, United Kingdom.　Commuters pass by Platform 9 3/4, a recreation of a scene from the Harry Potter series, in Kings Cross Station.

(what does not stand out) in the cultural landscape of your campus and community. Examine historical markers, plaques, statues, and permanent objects designed to mark significant people or events. Anything in the cultural landscape can be a clue to understanding a place, as landscape is an "unwitting autobiography, reflecting our tastes, our values, our aspirations, and even our fears, in tangible, visible form" (Lewis 1979). To understand any given part of a landscape, consider the whole, the context. How do these buildings, streets, green spaces, and markers fit together to create a **context** for understanding the cultural landscape of the place?

Such questions point to several of the axioms for reading the cultural landscape defined by Peirce Lewis. His first axiom is that landscape is a "clue to culture," meaning that the culture of a people is reflected in the ordinary, everyday landscape. One does not need to visit the most important memorials of a country to see culture in a landscape. Reading the everyday cultural landscape can tell us the story of a place.

Lewis's second axiom for reading the cultural landscape states that "nearly all items in a human landscape reflect culture in some way." In this vein, the paths of streets and locations of places relative to each other help us recognize meaning in nearly everything. The campus of Georgetown University is sited in northwest Washington, DC, on a hill overlooking the city and the Potomac River (Figure 14.11). To understand why Georgetown University has a prime site in this world city, we need to look at the timing of events. Located on the Potomac River, the town of Georgetown was a busy port on the Atlantic trade network, especially

© D A Barnes/Alamy.

FIGURE 14.8　Paris, France.　The Basilica of Sacre Coeur, built in the 19th century, stands atop Mount of Martyrs in the Montmarte area of Paris. From 1200 to the 19th century, the hill was home to another Catholic church and before that Romans had temples for Mars and Mercury on the hill. Evidence supports that before the Romans, Druids used Martyr's Hill as a place of worship.

FIGURE 14.10   **Seoul, South Korea.**   Seoul is home to about 10 million people and dozens of western chain donut shops, including Dunkin Donuts, Tim Horton's, Mister Donut, and Krispy Kreme.

for tobacco, when it was founded in 1751 (National Park Service). The hilltop where Georgetown is located was not prime real estate when the town was focused on the port on the Potomac River.

When Washington, DC, became the site for the U.S. capital in 1791, the campus of Georgetown was far enough from the new city's center that the plan for the city's roads and lay out stopped at Georgetown (Figure 14.12). Six years after the Civil War, in 1871, Congress enacted legislation that integrated Georgetown into Washington and established one city government over the entire area. In 1895, Congress legislated that streets in Georgetown be renamed to integrate with the avenue and street name conventions used in Washington, DC, requiring the neighborhood to change the toponyms of the streets.

Lewis's third axiom—that common things like streets and warehouses are difficult but important to study—and his fourth axiom—that history can be understood by studying contemporary landscapes—can be used together to grasp the meaning of mundane things such as transportation routes in a place. The cultural landscape of the Georgetown community has changed with transportation technologies and shifts in the economy. Streets in Georgetown were laid before automobiles, and the streets point to the centrality of the Potomac River and the Chesapeake and Ohio (C&O) Canal, which were important when the town was laid out.

Water Street (now partially K Street) ran along the Potomac River, and High Street (now Wisconsin Avenue) was the primary access line from the river into the neighborhood.

In the second half of the 1800s, streetcar technology diffused to Washington, DC. Evidence of the streetcars can be found just east of the university, on O and P Streets, where the city preserved the stone road and streetcar tracks in 2009. The streetcar tracks connected Georgetown with the Navy Yard southeast of the U.S. Capitol from the Civil

FIGURE 14.11   **Washington, DC, United States.**   Georgetown University sits on a hill above the Potomac River with the Key Bridge in the foreground.

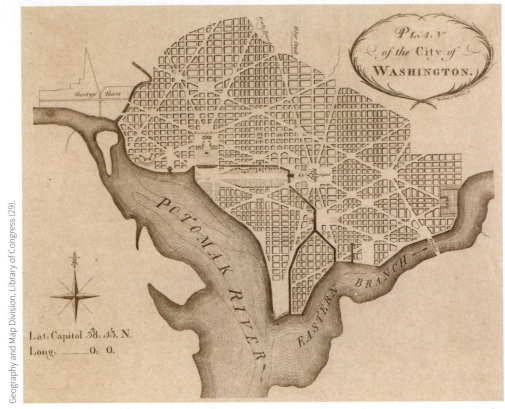

FIGURE 14.12   L'Enfant's Plan for Washington, DC.   President George Washington employed Frenchman Pierre L'Enfant to design the **urban morphology** (the layout of the streets and sites for primary buildings) of the city in 1791. The center of the capital was miles from Georgetown; so L'Enfant's plan for Washington, DC, stopped at Georgetown.

War until the rise of the automobile after World War II. In the automobile age, the cultural landscape of the waterfront area changed again. Before World War II, the city proposed building an elevated freeway over Water and K Streets along the Potomac in Georgetown. Washington, DC, eventually built the Whitehurst Freeway and opened it to traffic in 1949. The freeway bisected some mills and warehouses, forced the deconstruction of the Francis Scott Key home, and broke the site lines of the Potomac for many Georgetown residents (Figure 14.13). The Metro system opened in Washington, DC in 1976, but the commission did not plan a Metro stop for Georgetown because it would have required digging deep under the Potomac or building a bridge for the Metro across the river in order to connect the higher elevation of the Georgetown neighborhood with Rosslyn, Virginia.

Lewis's fifth axiom states that "elements of a cultural landscape make little cultural sense if they are studied outside of their geographic context." While Lewis was primarily concerned with the locational context of a landscape, reading cultural landscapes in world cities necessitates that understanding what we are seeing will increasingly need to take place in the global context. To make sense of one industrial building in the world city of Washington, DC, in this case Bomford's Mill, we can look at the site of a small town located at the fall line of the Potomac River and think of the building as reflecting the varying situation of the city (Figure 14.14).

FIGURE 14.13   Washington, DC, United States.   The Whitehurst Freeway is elevated above K Street right along the Potomac River.

• Reading the **CULTURAL** Landscape

Bomford's Mill

Along Water Street and K Street in Georgetown, a number of flour and cotton mills and warehouses were located between the C&O Canal and the Potomac River (Figure 14.14a). The mills were located between the C&O Canal, which sat at a slightly higher elevation and the Potomac River, which is at sea level. Lewis's sixth axiom of reading the cultural landscape explains that "most cultural landscapes are intimately related to physical environment"; thus, understanding the elevation differences between the canal and the river helps one to understand the cultural landscape. Gravity enabled the C&O Canal to sell water from the canal to the mills, and the mills used pipes to move water from the canals to their water wheels in order to power the flour mills and the cotton looms. Bomford's Mill still stands today along Potomac Street, which slopes south from the C&O Canal to Water Street in Georgetown. The building is named for George Bomford who built a flour mill on this site in 1832, which burned down in 1844. He built another building in 1845, but this time a cotton factory instead of a flour mill (Georgetown Metropolitan 2010). Using the Historic American Building Survey, the Georgetown Metropolitan reported that Bomford's cotton factory housed an "immense water wheel" that powered "3,000 spindles and 100 looms" and employed "over 100 men and women" (2010, 2). In 1866, Bomford's Mill became a flour mill again, in response to the Civil War, which had cut off the North's supply of cotton. Bomford's Mill produced flour from 1866 to 1969 under a few different owners. During this time frame, the Georgetown neighborhood was not the fashionable address it is today. When Senator John F. Kennedy lived in Georgetown in the 1960s, he helped revitalize the housing market and increased the desirability of the now high-end

© Rob Crandall/Alamy.

FIGURE 14.14a   Washington, DC, United States.   The C&O Canal runs through the Georgetown neighborhood. Old mills and warehouses line the canal, and many have been converted to condominiums and office buildings.

Library of Congress Prints and Photographs Division Washington, D.C.

FIGURE 14.14b   Washington, DC, United States.   Bomford's Mill operated as a flour mill under different company names during much of the nineteenth and twentieth centuries.

(Continued)

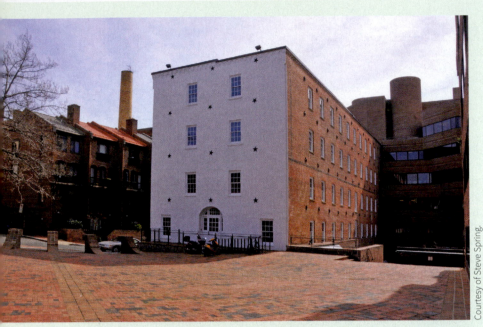

residential neighborhood. The last flour company to use the building as a mill was Wilkins-Rogers, who left in 1969 (Figure 14.14b). The company still sells flour under the label Washington Flour out of its production facilities in Maryland. Today, part of the mill has been renovated into an office building called The Flour Mill (Figure 14.14c), which houses a law office and public relations firm.

Courtesy of Steve Spring.

FIGURE 14.14c   Washington, DC, United States.   Bomford's Mill is now an office building called The Flour Mill.

The Bomford's Mill building first functioned as a cotton mill. The situation changed during the Civil War, when the flow of cotton into the city was cut off. The situation changed again when the city rebounded from the war and worked to grow a regionally interconnected economy. Instead of cotton, flour was milled in the building to feed people in the larger Washington area. As trade and communication flows increased from the 1960s on, the situation of this building changed yet again. Producer services were pulled to the centers of gravity of world cities, including Washington, DC. The building is now called The Flour Mill, and it houses a public relations firm and a law firm. A single industrial building in the cultural landscape reflects the changing situation of the Georgetown neighborhood in this world city.

THRESHOLD CONCEPTS AND WORLD CITIES

Whether studying Washington, DC, Shanghai, China, or Caracas, Venezuela, world cities have areas or regions that are "local" to them, from which they draw employees, customers, fans, belongingness, and revenue. World cities are nodes of commercial activity, reminiscent of the city-states of Athens and Sparta in ancient Greece or of Coba and Tikal, among pre-Columbian Maya. World cities, however, are distinctly different from historical city-states because they function as nodes in a **network** of global trade. World cities attract trade and concentrate areas of production of both industrial and service goods (especially producer services) among networks of global capital and labor.

CONCEPT **CHECK**

1. **Why** do threshold concepts matter when one is learning to think geographically?
2. **What** are cultural landscapes, and **why** do geographers read them?

HOW CAN WORLD REGIONAL GEOGRAPHY HELP YOU UNDERSTAND THE WORLD?

LEARNING OBJECTIVES

1. **How** does Generation Y receive information?
2. **Why** did the video *Kony 2012* diffuse quickly?
3. **What** perception of Uganda did *Kony 2012* seek to develop?

Five hundred years ago, the greatest library in Europe was at Queens College in Cambridge and had only 199 books in the entire collection. In 2013, publishers released more than 300,000 new titles or editions of books, and another 390,000 books were self-published in the United States alone. More than 1.1 million print on demand or Internet-access only books were published, with marketing done chiefly on the Internet. Five hundred years ago, a scholar could claim to have read every book in Europe's greatest library. Today, no one can claim to have read all of the more than 36 million books in the U.S. Library of Congress. At the turn of the millennium, in an article in the *Washington Post*, Librarian of Congress James Billington quipped that we are no longer in the Age of the Renaissance or the Information Age; rather, we live in the "Too Much Information Age."

Too much information. Generation Y, sometimes called the Millennials, includes young adults who were born between 1980 and 2000. This generation cannot remember a time before the Internet and probably never used a card catalog in a library. Generation Y is adept at quickly finding information on the Internet, the vast majority of which is less than 15 years old. All of this recent information can give society a short-term memory that lacks the depth of historical knowledge and geographic context. Does the flood of information you receive each day improve your understanding of the world? Not likely.

It is difficult to process information when you are continuously inundated with data that comes without context or explanation. It's enough to leave you wondering how a particular and complicated conflict or situation existed for so long unbeknownst to you! On March 5, 2012, a nongovernmental organization called Invisible Children, whose focus is to address the Lord's Resistance Army (LRA) conflict in Africa, released on YouTube a video called *Kony 2012* it had produced about the Ugandan warlord Joseph Kony (Figure 14.15). The video went viral, reaching a record-breaking 100 million views within six days. A number of factors contributed to the popularity of the video, including its high production value, the direct appeal of the video and call for action by Generation Y, the simplicity with which the LRA was presented, and the clear goal outlined by the video.

Kony 2012 diffused hierarchically. From the headquarters of Invisible Children in San Diego, California. Founder (and star of the *Kony 2012* video with his son) Jason Russell sent the video to teenage members of the organization, primarily American Christians. Gilad Lotan, vice president of Social Flow, a social media marketing firm, tracked the diffusion of *Kony 2012*. He studied the first 5,000 Twitter accounts to post the hashtag #Kony2012 and found the video trended first in Birmingham, Alabama, on March 1, 2012. Using an algorithm, Lotan charted Tweets and realized that the clusters of related Tweets represented "users from different physical locations" (2012, 4) (Figure 14.16). Invisible Children tapped into "networks of youth" they had "been cultivating across the US for years" (2012, 2). Using these networks, *Kony 2012* diffused not first to world cities but instead to small and medium-size cities in the United States, including Birmingham, Pittsburgh, Oklahoma City, and Noblesville (Indiana). From these initial "knowers," *Kony 2012* diffused contagiously among networks of friend and follower connections.

The second method of diffusion for *Kony 2012* was from "knowers" to celebrities and then hierarchically from celebrities to their followers. Invisible Children posted a list of selected celebrities on their website and made it simple (two clicks) for the early "knowers" to Tweet the video to the designated celebrities. Celebrities received a barrage of Tweets requesting them to reTweet *Kony 2012*. Word spread quickly through the hierarchy of celebrity followers, including Ryan Seacrest and Oprah Winfrey, and the hashtag #StopKony received "12,000 tweets per ten minutes at the height of events" (Lotan 2012, 2).

What, though, did these 100 million knowers actually understand about the Ugandan warlord Joseph Kony? The 29-minute video describes Kony as an enslaver of children, a rapist, and an indiscriminate murderer. In the video, Jason Russell describes the evolution of Kony's evil in a simplified form with little context. One of the stars of the video is a young man in Uganda named Jacob, and Russell draws parallels between Jacob and his own 6-year-old son living in California. He shows his son pictures of Jacob and pictures of Kony and describes Kony as a "bad guy" and says "what he does is he takes children from their parents and gives them a gun to shoot, and he makes them shoot and kill other people." The story

FIGURE 14.15 Nabanga, South Sudan. Joseph Kony, center front in this photo, the leader of Uganda's Lord's Resistance Army gave a news conference surrounded by his officers in 2006.

© STR/Reuters/Corbis.

Jacob Russell tells his son Gavin of Kony as a "bad guy" who "takes children" is complicated by several facts, including the one that Kony left Uganda in 2006 and that the country has been relatively peaceful since.

In the aftermath of the diffusion of the Kony 2012 video, journalists questioned whether the simplified story Russell and Invisible Children told about Uganda helped because it made millions aware of problems in Africa—or hurt because it provided no context for understanding and implied that killing one man will solve the problems in Central Africa. The 2012 video asserts that people need to rally behind capturing Kony in northern Uganda. The government of Uganda, however, countered (and foreign policy experts agreed) that Kony was pushed out of Uganda in 2006 when the war in the north ended and that he most likely now resides in the Democratic Republic of Congo. The rebel group Joseph Kony leads, the Lord's Resistance Army (LRA), is still "raiding and massacring and abducting in neighboring countries, but northern Uganda itself is peaceful" (NPR 2012).

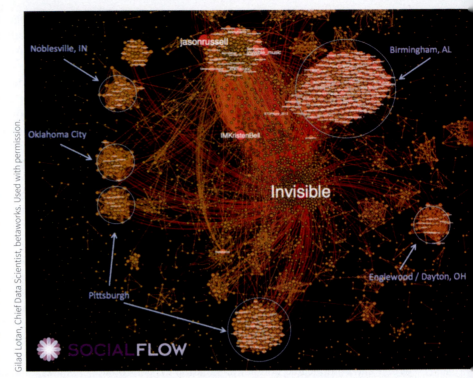

FIGURE 14.16   **Connections and Clusters of the First 5,000** *Kony 2012* **Tweets.**   Invisible Children tapped into networks of youth to diffuse the message of *Kony 2012*. The main nodes in their network, Birmingham, Pittsburgh, and Oklahoma City, are visible on the map.

Gilad Lotan, Chief Data Scientist, betaworks. Used with permission.

To understand how Joseph Kony, who is wanted for war crimes by the International Criminal Court, was able to grow and wield power, one must understand the regional context of the rise of Kony and the LRA. During British colonialism, between 1911 and 1962, the British introduced cash crops and built industries in southern Uganda, while the north became a "reservoir of cheap labour to be employed in the south" (Doom et al. 1999, 7). Also during the colonial period, the British recruited northern Ugandans to serve in the military. Members of the Acholi tribe were "transformed into a 'military ethnocracy'" (Doom et al. 1999, 7).

When Uganda became independent in 1962, the south had a stronger economy than the north, and people in the north tended to serve in the military, migrate to work in industries in the south, or raise livestock. A number of factors in Uganda and the broader context of East and Central Africa contributed to the rise of Joseph Kony.

First, the region has a number of young men who were in the army and were then demobilized. In Uganda, young men from the sparsely populated Gulu, Kitgum, and other northern districts found work in the military until current president Yoweri Museveni came to power with the help of rebel groups in 1986. The West saw Museveni as a progressive leader, but in 1998 he invaded the Democratic Republic of the Congo and in 2006 he ended presidential term limits and democracy. Rebel groups, including the Lord's Resistance Army, grew during the Museveni rule.

Through the LRA, Kony first sought to overthrow the Museveni government but then turned toward "cleansing" the north of the Alcholi tribe and imposing on Ugandans a religious rule based on a combination of Christianity and Kony's own brand of mysticism.

Second Kony's rise in power occurred in the context of Sudan's Civil War (1983–2005) and the politics of the Sudanese government (Figure 14.17). The government in northern Sudan waged a campaign of ethnic cleansing and genocide in southern Sudan and western Sudan (Darfur). Over 100,000 refugees from Sudan fled to northern Uganda between 1989 and 2005, helping to create even greater instability in the region where Joseph Kony operates. At the same time, Kony found a kindred spirit in the government of Sudan, which has financially supported, trained, and offered refuge for Kony and the LRA because the Sudanese government is "bent on destabilizing Uganda, which it says supports the Christian rebels in the southern Sudan" (Lorch 1995, 1).

Third, the rise of Joseph Kony and the LRA has also happened in the context of the Museveni government in Uganda, which "some diplomats and opposition members assert" have used Kony and the LRA to "continue to obtain further Western aid while delaying democratization" (Lorch 1995, 2). Human Rights Watch reports that the Museveni government is guilty of abusing the human rights of Internally Displaced peoples (IDPs) living in camps in northern Uganda.

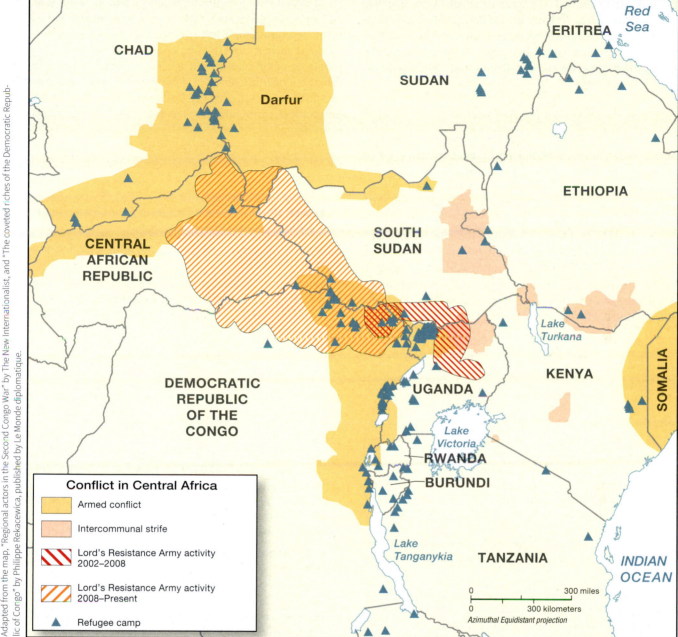

Conflict in Central Africa

- Armed conflict
- Intercommunal strife
- Lord's Resistance Army activity 2002–2008
- Lord's Resistance Army activity 2008–Present
- Refugee camp

FIGURE 14.17   Conflict in Central Africa.   The Lord's Resistance Army initially operated in northern Uganda but has shifted its operation to the Democratic Republic of the Congo, the Central African Republic, and South Sudan since 2008.

WORLD REGIONAL GEOGRAPHY AND KONY

How does world regional geography help you understand the context of *Kony 2012*, and does understanding matter when horrible acts committed by an evil person are involved? *Kony 2012* presents Kony as a "bad guy" in Uganda who needs to be captured in order to secure the region. World regional geography helps us think about and analyze the situation at different scales and in a regional context. Once we do so, we recognize that the region has layers of "bad guys" and interweaving conflicts. While capturing Kony, who is wanted for war crimes, is a worth-

while goal, it would not, as *Kony 2012* suggests, make life peaceful and content for all in the region.

Digesting *Understanding World Regional Geography*, exploring the dynamic webmaps using ArcGIS Online, and completing class assignments enable you to think geographically. The book equips you with 25 threshold concepts in geography for you to apply across multiple regions. You can understand how geomedia like *Kony 2012* creates a **perception of place**, in this case, Uganda. You can understand not only the site of northern Uganda but also how the interweaving political strife in East and

Central Africa created a context in which Kony and the LRA could operate. You can also understand how complex it is to "solve" the problems in the region and how capturing Kony, though a worthwhile goal, will not undo all the damage that made his rise possible or that continues to be done in the region.

UNDERSTANDING THE WORLD

You can use threshold concepts in geography to help you digest other, complicated problems that arise globally. For example, the context of the Iraq War and the Syrian Civil War can help explain how the brutal terror organization ISIS rose to power in a relatively short time. The solution to ISIS is not simple. In the context of globalization, once one portion of the ISIS problem is "'solved," the same issue could arise in another location or manifest itself in the same place at a later time. Globalization and networks make it difficult to find the roots of a crisis in Syria and Iraq, but thinking geographically helps us see the economic, political, and cultural contexts of complicated world problems.

Understanding is the first step to finding solutions, and thinking geographically helps you consider how a given response could have impacts at different scales—within families, locally, regionally, and globally. The problems of the world will not be "solved" in our lifetimes, but opening our eyes, applying threshold concepts, and thinking geographically lead to understanding, and in that context, the solutions we choose have a better chance of success.

CONCEPT CHECK

1. **Explain** whether the easy acquisition of information by Generation Y encourages thinking.

2. **Describe** the paths and types of diffusion the video *Kony 2012* followed.

3. **Explain** how thinking geographically can help you recognize that *Kony 2012* oversimplifies a complex problem.

SUMMARY

Why Do World Regions Change?

1. A world region is not a container separated from the rest of the world. A world region is a sum total of experience that shapes a perception shared by many within and outside the region. As a thoughtful global citizen, you see world regions as containing, connecting, shaping perceptions, and dynamic. Through reading *Understanding World Regional Geography* and taking world regional geography, your schematic of the world expands and you add detail and connections in your mind to paint a refined understanding of world regions.

2. Geographers use formal, functional, and perceptual regions to organize information about the world. Formal regions are those that share cultural or physical traits and generally do not have clearly defined borders, such as the Spanish speaking region of the Americas. Functional regions share a specific purpose and generally have a clearly defined border, such as a city. Perceptual regions are the ideas we carry in our minds of similarities among places and differences between regions.

3. Geographers have studied world cities, major nodes in globalization, since the 1990s. World cities are gaining in importance as centers of economic might, political power, and creativity, which may signal a shift from world regions based on countries to world regions based on world cities.

How Do Threshold Concepts Help You Think?

1. Threshold concepts are geographic concepts that once learned will help you think more like a professional geographer. Understanding World Regional Geography focuses on 25 threshold concepts in geography. The discipline has many more than 25 concepts, and no set list of threshold concepts for each discipline

exists. As you read through the list of threshold concepts, focus not only on defining the concepts but practice applying them.

2. Cultural landscape is the visible human imprint on the landscape. One trait of threshold concepts is that they are often related. Practice reading cultural landscapes, and you will also be able to apply other threshold concepts such as site, the precise location of a place, and situation, the location of a place relative to the surrounding context.

How Can World Regional Geography Help You Understand the World?

1. Generation Y receives massive amounts of information though social networks and internet sources. The connectedness of networks coupled with the speed of communications technologies help information diffuse quickly.

2. An organization called Invisible Children produced the video *Kony 2012* to garner support for a campaign against Joseph Kony and the violent actions of his Lord's Resistance Army in Uganda and surrounding countries in Africa. Invisible Children tapped into networks of youth and opinion leaders on Twitter and Facebook to rapidly diffuse *Kony 2012*.

3. *Kony 2012* created a perception of Uganda as a lawless place; however, by the time the video went viral, Kony was no longer operating in Uganda and the country was working to rebuild. *Understanding World Regional Geography* and your world regional geography class have helped you revise your schemata of the world, add detail and connections in your mind, and paint a refined understanding of the world. You will now recognize when media, both popular and social, construct perceptions and you will think and analyze the stories they tell using your enhanced schemata.

Geography in the Field YOUR TURN

Rio de Janeiro, Brazil, is a world city that geographers have classified as a beta world city. Beta cities are "important world cities that are instrumental in linking their region or state into the world economy" (Globalization and World Cities 2010). Brazil is the seventh largest economy in the world, and Rio plays a central role in the Brazilian economy and in linking the city to the world economy. This photograph of Rio de Janeiro shows the famous Copacabana Beach and high-rise residential buildings and tourist hotels facing the Atlantic Ocean. In the foreground of the photograph is a shopping district, and in the background is Ipanema Beach and the Lagoon of Rodrigo de Frietas. Nestled in the Morro de Sao Joao, Morro Dos Cabritos, and Morro Do Camergalo Mountains in this photograph are favelas, homes built upon the side of the mountains to house migrants and low-income workers. The structures were originally built as temporary residences but have become permanent. A 1937 ordinance prevented favelas from being mapped, but the government is now mapping and giving addresses to the favelas

in recognition of their permanent fixture on this world city's landscape. Rio de Janeiro hosted the World Cup in 2014 and will host the Olympics in 2016. To prepare, the government bulldozed parts of the city, including Porto Maravilha (which is behind the viewpoint of the person taking this photograph) to make way for Olympic stadiums and high-end office and residential complexes, one of which was built by Donald Trump.

Thinking Geographically

- Beyond the scope of this photo, in the distance, lies Vila Autodromo, which is a favela of approximately 900 homes and 4,000 residents. The Olympic Training Center was built on the site of the racetrack (Autodromo) abutting Vila Autodromo. Examine the map of favelas and Olympic sites at: http://urbanruminations.blogspot.com/2011/05/olympic-sized-displacement-in-rio.html. Vila Autodromo is located in the Barra region of the map. Community leaders in Vila Autodromo protested the plans to level their neighborhood. Search the Internet for information about the Vila Autodromo campaign. Did the long-time residents win their campaign to keep their neighborhood? How did the residents jump scale and make their local issue a concern to the entire city, country, and world?

- As part of the preparation for hosting the World Cup and the Olympics, Rio de Janeiro has been pushing drug gangs out of the favelas in order to lower the crime rate in the city. Being a world city enables Rio to attract major global events, changing the situation of Rio in a global context. How does this global context affect the local scale of the city of Rio and the site of its favelas?

Read:

Downie, Andrew. "Rio Gives Its Favelas a Pre-Olympics Makeover." *Time*. September 6, 2011. http://www.time.com/time/world/article/0,8599,2091817,00.html.

Romero, Simon. "Slum Dwellers Are Defying Brazil's Grand Design for Olympics." *New York Times*. March 4, 2012. http://www.nytimes.com/2012/03/05/world/americas/brazil-faces-obstacles-in-preparations-for-rio-olympics.html?pagewanted=all&_r=0.

Rio de Janeiro, Brazil. View of the city from Sugar Loaf Mountain.

© John Warburton-Lee Photography/Alamy.

KEY TERMS

world cities	cultural landscape	mental map	perception of place
threshold concepts	site	context	
Anthropocene	situation	network	

CREATIVE AND CRITICAL THINKING QUESTIONS

1. Does the **diffusion** of goods and services provide evidence to support a new spatialization of world regions that is based on world cities?

2. How people in a region share a perception of their history, present and future, is reflected in **cultural landscapes**—both by what is visible and by what is missing. Search the

Internet for a description of important tourist destinations in one of the beta or gamma world cities (see Figure 14.2). Choose a world city that was colonized by Europe or Japan. Look for evidence in the tourist sites of how the colonial experience is visible or missing from the cultural landscape.

3. Figure 14.17 depicts the hierarchical **diffusion** of the *Kony 2012* video. Using the map of world cities in Figure 14.2 to help you, list the world cities that appear on the Kony 2012 diffusion graphic. Examine your list and hypothesize why these **world cities** were nodes in the diffusion of *Kony 2012*.

4. Celebrities also played a key role in the diffusion of *Kony 2012*. Think through the kind of network created by Twitter. Hypothesize how the **context** of information flow is fundamentally different from that in the age of radio, television, or some other communication device.

5. Rio de Janeiro is a beta global city, according to the Globalization and World City Network categorization. How did being the **site** of the 2014 World Cup affect the status or **world city** rank of Rio?

SELF-TEST

1. British colonizers built the Shanghai Race Course in the 1850s. Today, the former horse race track is home to:
   a. a series of factories that largely produce textiles
   b. rice paddies
   c. a park and an art museum
   d. the United States embassy

2. New regional geographers see world regions as:
   a. static
   b. defined by language
   c. constructed
   d. easy to distinguish

3. A region that has a shared purpose, such as the city limits of Chicago, is best described as a _____ region.
   a. formal
   b. functional
   c. perceptual
   d. vernacular

4. Political geographer Peter J. Taylor defined world cities as the organizing nodes in:
   a. worldwide networks
   b. capital cities
   c. agricultural production
   d. universalizing religions

5. The two alpha+++ world cities are:
   a. Tokyo and Shanghai
   b. Paris and London
   c. London and New York
   d. New York and Shanghai

6. Threshold concepts in a discipline are the concepts and ways of thinking that a student needs to learn and apply in order to:
   a. pass an Advanced Placement exam
   b. think like a professional in the discipline

   c. get a Ph.D. in a discipline
   d. find the lost city of Atlantis

7. Geographer Peirce Lewis wrote axioms for reading the cultural landscape, which include all of the following except:
   a. The most important clues in a cultural landscape are government buildings and memorials
   b. Most cultural landscapes are intimately related to physical environments
   c. The culture of a people is reflected in the ordinary, everyday landscape
   d. History can be understood by studying current cultural landscapes

8. In L'Enfant's plan for Washington, DC, the most important building in the urban morphology of this world city is the:
   a. White House
   b. Supreme Court
   c. National Cathedral
   d. Capitol

9. The Librarian of Congress in the United States called the current era in which we live the:
   a. Nouveau Renaissance
   b. Third Urban Revolution
   c. Too Much Information Age
   d. Second Industrial Revolution

10. Joseph Kony was a warlord who lived in:
   a. India
   b. Mexico
   c. Colombia
   d. Uganda

ANSWERS FOR SELF-TEST QUESTIONS

1. c, **2.** c, **3.** b, **4.** a, **5.** c, **6.** b, **7.** a, **8.** d, **9.** c, **10.** d

GLOSSARY

Ablutions Ritual bathing or cleansing followed as a religious practice or instruction.

Absolute location The precise location of a place, usually defined by latitude and longitude.

Accessibility Ease of flow between two places.

Acqua alta frequent flooding or "high water" in Venice, Italy that occurs in response to rising sea levels in the Adriatic Sea.

Acquis Communitaire the European Union's set of laws and regulations. Countries that join the European Union must negotiate agreement to these laws and regulations.

Agency Ability of local actors and individuals to voice an opinion or affect change.

Agglomeration effect cost advantages created when similar businesses cluster in the same location.

Agricultural dualism The practice of working for a larger farm while maintaining a smaller farm.

Agropastoralists People who make their livelihood through a combination of raising livestock and farming.

Amazonia The largest rainforest in the world, found in the Amazon region of South America.

Animist a traditional or indigenous religion where animals or objects take on significance.

Anthropocene The current geologic era in which humans play a major role in shaping Earth's environment.

Arab Spring A movement against authoritarian leaders in the Middle East, with its hearth in Tunisia, secondary diffusion in Egypt, and further diffusion since the spring of 2011.

Asia Pacific Trade region including Southeast Asia, Oceania, Australia, New Zealand, and East Asia.

Assimilation A policy used to change the culture of indigenous peoples by encouraging or forcing them to end their cultural practices and adopt the cultural practices of the assimilator.

Atlantic World the lands and sea lanes around the Atlantic Ocean that were connected through trade during the Age of Exploration from 1400s to the 1900s.

Atoll A coral island chain, typically in a ring formation with a central lagoon, formed around a high island.

Authenticity the idea that one place or experience is the true, actual one.

Avatar An incarnation, or earthly manifestation, of a god.

Backward linkage relationship between a company and its suppliers.

Backwash effect Negative economic impact in one region that stems from positive economic impact in another region.

Balkanization The splitting of a country into smaller countries along national lines, named for the Balkan Peninsula.

Banana Republic A small country whose economy is almost completely dependent on one commodity.

Basin and range topography A series of parallel valleys (basins) and mountains (ranges) created where crust thins as it is pulled apart. Along normal faults in the crust, the mountains rise and the valleys drop.

Biomes A region with similar vegetation.

Black earth Patches of fertile black soil made by pre-Colombian farmers by adding charcoal, bone, and manure to otherwise infertile soils.

Bodhisattva a follower of Buddhism who is intent upon becoming enlightened.

Boreal forest A vegetative region marked by coniferous trees that can withstand the cold climate.

Bourgeoisie The wealthy class; those who owned the means of production.

Brownfield an abandoned industrial site.

Buddha an enlightened being in Buddhism, including the Buddha, Siddhartha Gautama.

Bunkering port: designed to fuel ships, especially large container ships.

Cartography The art and science of making maps.

Caste system Social structure of South Asian society that dates back to the Indus civilization whereby people are born into their place in society.

Central business district (CBD) The zone of a city where businesses cluster and around which a city and its infrastructure are typically built.

Centrally planned economy Government ownership, planning, oversight, and distribution of production within a country's economy.

Chain migration A migration flow in which migrants communicate with family and friends at their place of origin about the virtues of the destination, encouraging others to follow their migration path.

Channelization reconstruction of a river or stream to follow a straighter path.

City states Ancient cities that had state-like properties, including a military, a leadership class, and a territory over which they claimed control.

Climate change The generation of new climate patterns as a result of human activities.

Cold War the period from World War II until 1991 when two superpowers, the United States and the Soviet Union, politically postured against each other without fighting a direct, active war.

Cold water desert A desert formed where the prevailing wind coming onto the land is dry and cold, making precipitation unlikely.

Collectivization Taking land, resources, and industries out of the hands of private individuals and into the hands of the government, owning and operating it for the whole.

Colonialism Taking over a territory and physically controlling its government and economy.

Columbia Exchange Centered on the Atlantic Ocean, the triangular trade network of goods and enslaved people among Europe, Africa, and the Americas, dating back to 1500.

Commodification the buying, selling, and trading of goods or services that were not previously bought, sold, or traded.

Commodity chain steps in the production of a good from its design and raw materials to its production, marketing, and distribution.

Confucianism A political philosophy that values benevolence, loyalty, and diligence within the structure of the family and the government.

Conquistadors Spaniards who worked for the crown to conquer the Americas.

Conservation An approach to maintaining biodiversity by using the land within its biological limits.

Context The physical and human geographies creating the place, environment, and space in which events occur and people act.

Contiguous places that are physically touching.

Continent an extensive, contiguous, discrete landmass.

Continental shelf the area under the sea where the continent continues up to the continental slope.

Continental slope the graded area under the sea that connects the continental shelf with the deep sea.

Continent an extensive, contiguous landmass made of the lithospheric layer of the earth that is surrounded by water.

Core processes production methods that add wealth to an economy through high technology, education, and wages.

Creole Blend of two more ethnic or nationalist groups.

Crony capitalism an economic growth model in which the state plays a large role in creating contracts and giving tax breaks to local industries, which are typically owned by people with strong connections to the government, thereby propping up the industries.

Cultural brokers in tourism, the resort companies, tour operators, and guides who present a local culture in a certain way and commodify opportunities for tourists to observe or experience a local culture.

Cultural landscape The visible human imprint on the landscape.

Dacha Plots of land in the countryside, usually with a home or cabin, where Russians spend their summers.

Dependency theory Andre Gunder Frank's theory that wealthy countries set up exploitative economic relations whereby poorer countries became dependent on the wealthy and the wealthy benefited economically.

Deposition When forces of erosion slow down or halt, materials, including rock, soil, and vegetation, are left behind.

Development Improvement in the economy and well-being of a place relative to another place. Improved wealth or progress in other socioeconomic variables, including education and health.

Development through resettlement A Soviet-era policy to move Russians into the far reaches of the Soviet Union.

Diffusion The spread of an idea, innovation, or technology from its hearth to other people and places.

Digital divide The growing gap in access to Internet and communication technologies between connected and remote places.

Distance decay The decreasing likelihood of diffusion with greater distance from the hearth.

Domesticated A plant purposefully grown by humans for consumption or feed. Also animals purposefully reared by humans for consumption or labor.

Double exposure The combined effect of two global processes on one local place.

Drawdown water that recedes from a beach when a tsunami wave forms. A depression occurs at the shore approximately 5 to 10 minutes before the crest of the wave hits the beach.

Dualism A relationship between two places or regions whereby one develops by exploiting the other.

Ecotourism Nature-based tourism that seeks to conserve the environment and improve the economic position of local populations.

Edge cities Concentrated business districts outside of a city. Edge cities are typically sited on major infrastructure hubs and attract shopping and housing districts.

Enclave an outpost or home base for an ethnic group living in a country other than their home country.

Encomienda A Spanish system whereby a conquistador was granted responsibility for a number of indigenous people with the expectation the conquistador would teach the indigenous the Spanish language and Catholic faith in return for labor.

Endemic A species, plant or animal, found only in the location (typically an island).

Endemic species Plants and animals native to and found only in a small area of a biome.

Entrepôt entry point or gateway port for trade in a region.

Environment Earth processes and human actions combined to create a physical context.

Environmental determinism A set of theories that use environmental differences to explain everything from intelligence to wealth.

Environmental security The theory that many ethnic and political conflicts originate from environmental concerns or problems.

Erosion The removal of rock, soil, and vegetation from a landscape, caused by water, wind, waves, glaciers, or gravity.

Ethnic religion A religion into which people are born and whose followers do not actively seek converts.

Ethnic tourism tours designed to show remote areas of a country to tourists who can observe ethnic groups or local cultures and their unique lifestyles.

European Union an organization of 28 European states that share common economic and agricultural policies and is expanding common policies in human rights, security, and politics.

Expansion diffusion The spread of an idea or innovation from its hearth across space without the aid of people moving.

Fair trade A trade network developed to encourage environmentally sound production practices and to give greater compensation to the producers of the product, often farmers.

Fall line A change in the bedrock under a river, typically designated by falls. Divides the navigable part of the river (below the fall line) from the interior.

Female infanticide killing of a female child under the age of 1.

Fertile Crescent Region in Mesopotamia and Anatolia where agriculture began.

Fieldwork Observations researchers make of physical and cultural landscapes with a focus on seeing similarities and differences.

First agricultural revolution The transformation of societies from hunting and gathering to purposeful raising of food, feed, and fiber.

First wave of colonialism From the late 1400s to 1850s, when Europeans colonized the Americas and coastal Africa.

Forced migration Movement of people across country borders where the migrants are coerced to relocate.

Fordist Manufacturing system in which raw materials are brought into a central location and component parts and the final product are produced at the same location and then shipped globally.

Formal economy Economic productivity in agriculture, mining, industry, and services that is counted or taxed by government. Trade is through formal channels, often using credit.

Formal region An area of land with common cultural or physical traits.

Forward linkage Relationship between a company and its purchasers or distributors.

Fracking System of drilling deep holes into shale layers of earth and then pumping water and chemical into the wells to extract the natural gas or oil deposited in the shale.

Functional region An area of land defined as sharing a common purpose in society.

Functional zonation The division of the city into regions by use or purpose.

Gender Socially constructed notions of roles, behaviors, and activities that are appropriate for men and women.

Gender imbalance a lack of balance between the number of girls and number of boys born, as a result of human intervention favoring one gender.

Gentrification Renewal or rebuilding of a lower income neighborhood into a middle to upper class neighborhood, which results in driving up property values and rents and the dispossession of lower income residents.

Geographic information system (GIS) A system of computer hardware and software designed to show, analyze, and represent geographic data (data that have locations).

Geography The study of earth, its people, cultures, and environments.

Ghost schools Schools that exist either on paper but not on the ground or on the ground but are empty of teachers, books, and students.

Glacier large mass of ice, formed from compacted snow, which expands over land when snow accumulates faster than it melts.

Globalization processes heightening interactions, increasing interdependence, and deepening relations across country boundaries.

Global language a language with widespread use across the world that is used for communication among a diversity of speakers.

Global outsourcing International companies that specialize in securing contracts and then completing the work by outsourcing the lowest skilled parts of the task.

Global sourcing expertise in helping a business determine where and how to produce a good as well as efficient coordination of production and delivery of the good.

Global South Countries that are less well off economically, typically located south of economically wealthy countries.

Glocalization Adapting a global phenomenon to suit a local need or context.

Greater Antilles The larger Caribbean islands found in the northwest part of the Caribbean Sea, including Cuba, Jamaica, and Puerto Rico.

Green Revolution Intensified agriculture that uses engineered seeds, fertilizers, and irrigation to increase intensive agricultural practices.

Gross national income (GNI) A measurement of a country's wealth that takes into account wealth generated for a country both inside and outside of the country's borders.

Guest workers migrants invited into a country to work temporarily and who are expected to return to their home country.

Gulag Soviet-era prison camp typically sited in remote areas.

Haciendas Large land-holdings granted to conquistadors during the Spanish colonial era.

Haka Traditional Polynesian dance used to prepare for battle and now commonly used in sporting events by predominantly Polynesian teams.

Harmattan winds Dry northeasterly winds.

Hearth An area or place where an idea, innovation, or technology originates.

Hegemon The state that is the world power and shapes the tone and rules of politics and trade in a particular era.

Hierarchical diffusion The spread of cultural traits and innovations along a structured network from most important person or place to the next most important person or place.

High-pressure system an area on earth's surface where atmospheric pressure is greater than surrounding areas, causing surface winds to flow out from the high-pressure area to surrounding lower-pressure areas.

Hill tribes Indigenous peoples in Southeast Asia who live primarily in the mountainous areas of the region.

Hinterland Territory outside of the city proper from which agricultural goods were produced to supply the city.

Historical Cultural Ecology A subfield of geography that examines historical farming systems and historical human-environment interactions.

Hominids All species closer to humans than to primates. *Homo sapiens* are the most recent species among the hominids.

Hukou China's registration system that tracks demographic data and limits where Chinese can live.

Human Development Index (HDI) A composite index derived from statistics covering life span and health, education, and standard of living.

Human–environment A subdiscipline of geography that studies the reciprocal relationship between humans and environment.

Human geography The geographic study of human phenomena.

Humid continental Climate region located away from oceans and at midlatitudes, which has a relatively large temperature range and seasonal precipitation.

Humid subtropical climate region located on east coasts of continents near the tropics with high precipitation in the summer.

Hutment factories Small-scale manufacturing conducted in slums, typically relying on intensive hand labor and low cost machines.

Iberian southwestern peninsula of Europe, where Spain and Portugal are now located.

Identity How we make sense of ourselves.

Import Substitution Government policies to protect and encourage domestic production of goods.

Industrial Revolution a series of innovations and mechanization in agriculture and manufacturing that increased production of goods. Occurred first in Great Britain in the 18th century.

Informal economy Portion of the economy that is not taxed or regulated by government. Goods and services are exchanged based on barter or cash systems and earnings are not reported to the government.

Initial advantage Economic boost experienced by a region that is the hearth of production for a good.

Insolation Incoming solar radiation. Solar radiation is energy from the sun that reaches earth. The amount of insolation varies by latitude and season.

Institutionalize The establishment of something as normal or a given through repeated recognition and use by governments and non-governmental organizations.

Intercropping Mixing complimentary crops, those that benefit from each other's attributes and byproducts, in the same field.

Internal migration Movement within the borders of a country.

International financial institutions (IFIs) Privately and publicly held corporations and nongovernmental and intergovernmental organizations set up to promote international trade.

International Monetary Fund International financial institution that monitors policies affecting exchange rates between currencies. Traditionally has a European president.

Intertropical Convergence Zone (ITCZ) A global low-pressure belt that migrates between the Tropics of Capricorn and Cancer over the year, with the direct rays of the sun. The ITCZ largely drives seasonal wind and moisture patterns in tropical areas of Africa.

Iron Curtain figurative division between Eastern and Western Europe during the Cold War.

Irredentist movement A desire to link a minority nation in one country with a majority nation in an adjoining country.

Islamic City The urban form of a city that reflects the importance of Islam to the city and the historical growth of Muslim cities over time.

islands of development cities in developing regions where foreign investment is concentrated and to which rural migrants are drawn.

Isostatic rebound when a plate, previously weighed down by glaciers, uplifts after the glaciers melt away. The process takes thousands of years and can generate small earthquakes and changes in elevation.

Jadidism A Muslim intellectual movement tied to Sufism in Turkey that sought to improve life for Muslims, including empowering women.

Just in time Production system in which parts are delivered as needed to the assembly line so that parts are not warehoused, stored, or overproduced.

Karst topography Limestone formations shaped when rainwater mixes with the calcium carbonate in the rock, creating a carbonic acid. Features include erosional landforms including caves and sinkholes and depositional landforms including stalactites and stacks.

Kleptocracy A government system in which the leaders steal wealth from the state.

Land cover Categories of vegetative and physical properties of earth, for example, forest or grassland.

Landlocked State that shares its borders only with other states, not with oceans or seas.

Latent heat Energy that is added or subtracted to change the state of matter.

Latifundias Large agricultural estates that produce crops for export, many of which were granted to colonizers during the first wave of colonialism.

Lesser Antilles The smaller Caribbean islands found in the southeast part of the Caribbean Sea.

Limestone Sedimentary rock made from accumulated layers of shells of sea life that are compressed into rock under the weight of the overlying layers.

Lingua franca language used for trade or cultural interaction among people who speak diverse languages.

Livelihood Activities pursued to secure food, shelter, and cover expenses.

Local culture a group of people, usually living in close proximity, who share common cultural practices, festivals, and beliefs that are passed down across generations.

Loess windblown sediment.

Low birth weight Defined by UNICEF as birth weight of less than 5.5 pounds.

Machismo A set of gender roles that includes strongly masculine roles.

Madrasas Islamic schools with curriculum based on the Quran.

Mahayana Buddhism The sect of Buddhism called the 'greater vehicle' that teaches all of society should work together to reach enlightenment. Mahayana Buddhism diffused from northern India across China, Korea and Japan and into part of Southeast Asia.

Marginality The process of a political elite defining a people as the "other," thereby pushing them outside of decision making and often physically onto lands not wanted by the political elite.

Market economy An economy where privately owned companies respond to supply, competition, and customer demand to plan and produce goods.

Megalopolis An urban agglomeration that stretches from Washington, DC, in the south to Boston, Massachusetts, in the north.

Mental maps (also known as cognitive maps) Maps of an area made from memory and experience by individuals or groups.

Mercantilism An early form of capitalism based on trading large quantities of goods, using gold and silver as currencies.

Mesa A mountain or hill with a flat top and steep sides that is formed by erosion of horizontally layered rock in arid to semi-arid regions.

Mesoamerica The subregion of Latin America that includes Mexico and Central America, typically used in reference to Mayan and Aztec civilizations.

Mestizo A Latin American person of mixed ancestry.

Microcredit lending The practice of giving small loans to individuals, who operate within a community of other borrowers, to start businesses and cottage enterprises.

Midlatitude desert Arid climate zone adjacent to semiarid climate zones in midlatitude locations.

Mid-oceanic ridge Parallel mountain ridges formed where two plates are diverging or pulling apart on the ocean floor.

Migration Purposeful movement from a home location to a new place with a degree of permanence or intent to stay in the new place.

Mikroyan Soviet-era prefabricated apartment buildings found throughout the former Soviet Union and in Soviet-influenced countries of Eastern Europe.

Minifundias Small-scale subsistence farms, many of which date to the first wave of colonialism.

Minying firms companies in China that operate without government oversight.

Modernization The theory proposed by Rostow and others that countries follow the same path along stages of development.

Modifiable areal unit problem Statistical change that occurs when changing the size of spatial unit of analysis. For example, the voting pattern in a presidential election looks different when you change the unit of analysis from states to counties or counties to voting districts.

Monoculture agricultural production relying on planting a single crop.

Monsoon A prevailing wind coming from one direction for a long period of time.

Mountaintop mining Method of surface coal mining where the ridge or top of a mountain is blasted off to reach an underlying coal seam.

Multinational corporations Privately and publicly held companies with global presences.

Mutualism Coevolution of species that rely on each other for survival.

Nation A group of people with a shared past and common future who relate to each other and share a political goal.

Neocolonialism Economic and financial policies that give wealthier countries and organizations control over poorer countries.

Network A system of connections among people and places.

Nongovernmental organizations (NGOs) Privately funded institutions that aid in development and relief work.

Nonrenewable resource A resource extracted from the earth that took hundreds of thousands of years to form and cannot be made by humans.

North Atlantic Treaty Organization (NATO) an agreement among countries in Western Europe, Canada, and the United States to defend each other against a threat from the Soviet Union and its allies.

oligarch a small group of people who rule a country. In Russia, refers to a group of very wealthy business leaders who benefitted financially from the privatization of Russia's economy after the Soviet Union dissolved in 1991.

One-child policy antinatalist policy instituted by China in 1979 to slow population growth.

Orographic precipitation precipitation associated with lifting over a mountain range, where the windward side of the mountain receives precipitation but the leeward is relatively dry.

Outsourcing Hiring employees outside the home country of a company in order to reduce the cost of labor inputs for the good or service.

Overseas Chinese Chinese who migrated to Southeast Asia, often following chain migration and setting up ethnic neighborhoods called Chinatowns, for economic opportunities.

Oxisols Nutrient-poor tropical soils that are highly weathered, form in moist climates, and easily drain water.

Pacific Rim Region surrounding the Pacific Ocean, including East Asia, Southeast Asia, Australia, New Zealand, and western North, Central, and South America.

Parastatals State-run companies such as the Mexican oil corporation Pemex.

Participatory development Development projects designed to incorporate local knowledge, input, and solutions.

Pastoralism Livestock raising in which herds are moved seasonally from pasture to pasture.

People The uniqueness of individuals and groups of individuals, including how they identify themselves.

Perception of place How a place is envisioned.

Perceptual region An area of land that an individual perceives as being similar.

Periphery processes production methods add little wealth to an economy through low technology, education, and wages.

Permanent cultivation Planting crops in the same field repeatedly with rare periods of leaving the land fallow.

Petrodollars The system in place since the early 1970s whereby oil-producing countries, including Saudi Arabia, value their oil in U.S. dollars and create demand for U.S. dollars.

Physical geography The geographic study of physical earth phenomena.

Physical landscape The visible appearance of physical geographic processes on the landscape.

Place The uniqueness of a location, the character given to the location by people.

Plateau An elevated, flat region created by uplift along a fault, by glacial erosion, or by pooling of magma in volcanic eruptions.

Polders land reclaimed from the sea by building dikes around an area and then pumping water off the land.

Political ecology An approach to studying human-environment interactions in the context of political, economic, and historical conditions operating at multiple scales.

Population pyramid A graphic representation of the age and sex composition of a population.

Post-Fordist Just-in-time production where component parts are manufactured around the world, shipped to a single location, assembled, and then shipped anywhere in the world.

Preservation An approach to maintaining biodiversity by setting aside land and resources so they are not used for consumption.

Primate city the lead city in a country in terms of size and influence.

Princely states Territories in South Asia that remained outside of direct British colonization.

Proletariat The working-class population, which Marx and Engels called to revolution.

Pronatalist policy a government policy to encourage citizens to have more children.

Pull factors Circumstances a migrant considers when deciding to leave the home country.

Push and pull factors the considerations a migrant weighs when choosing to leave home (push factors) and where to go (pull factors).

Push factors Circumstances a migrant considers when deciding where to migrate.

Qanat A system of using gravity to bring water into a town or agricultural area.

Race Social constructions of differences among humans based on skin color that have had profound consequences on rights and opportunities.

Rainshadow effect the dry, leeward side of a mountain.

Region An area of Earth with a degree of similarity that differentiates it from surrounding areas.

Relative humidity the amount of water vapor in the air compared to how much water vapor the air can hold at that temperature.

Relative location The location of a place or attribute in reference to another place or attribute.

Relocation diffusion The spread of an idea or innovation from its hearth by the act of people moving and taking the idea or innovation with them.

Remittances money earned by an immigrant and sent home.

Remote sensing Method of collecting data by using instruments that are physically at a distance from the area of study.

Return migration reversal in a migration flow.

Rig Veda A religious text, comprised of over 1,000 poems and hymns, followed by many Hindus.

Ring of fire Tectonically active region of earthquakes and volcanoes surrounding the Pacific plate.

Rotational bush fallow system A regular system of shorter fallows in which farmland never completely reverts to natural forest or savanna vegetation.

Russification The policy of spreading the Russian culture and language through education and political action.

Sacred space A place that is infused with religious or spiritual meaning.

Satellite Countries the countries in eastern Europe that were officially independent but were largely controlled by the Soviet Union during the Cold War.

Scale The geographical scope (local, national, or global) in which we analyze and understand a phenomenon.

Schemata Structures people have in their brains through which they process and understand the world.

Scramble for Africa During the 1800s when European countries claimed colonies in Africa in order to control resources in Africa and to bolster their own nationalism.

Second wave of colonization From the 1850s to 1960s, when Europeans colonized Africa and Asia in the context of the Industrial revolution.

Secularism indifference to or rejection of religion.

Seed culture The agricultural practices associated with preserving seeds for the next year's harvest that are passed down through generations in communities.

Selective adoption Negotiating globalization by choosing to adopt policies and changes that work for your country or locality.

Sense of place Infusing a place with meaning as a result of experiences in a place.

Sensible heat Energy that can be "felt" and is measurable by a thermometer.

Sepoy Rebellion Uprising in 1857 by Indian soldiers against the British East India Company that marked the transition to direct colonial rule by the British in South Asia.

Settlement The act of residing in and establishing a permanent community by a group of people.

Shi'a Sect of Islam found commonly in Iran and its neighbors.

Shifting cultivation agricultural practice based on clearing and farming land for a time before moving on to a new parcel and allowing the first to fill in with native vegetation.

Shuttle trading The practice of a trader purchasing goods in another country and then transporting and selling the goods in his or her home country.

Siberian High a high-pressure system associated with a extremely cold air mass that forms in the winter over eastern Siberia.

Site The physical attributes of the location of a human settlement—for example, at the head of navigation of a river or at a certain elevation.

Situation The position of a city or place relative to its surrounding environment or context.

Slash-and-burn agriculture see **shifting cultivation.**

Slavic People who speak a Slavic language, whose ancient roots are in the area northeast of present-day Slovakia. Slavic languages are found primarily in east and central Europe and in northern Eurasia.

Small-hold cash-crop farming Small family farms that rely on their own labor.

Social networks established connections among groups of people and individuals flowing across space.

Sovereignty The legal authority to have the last say over a territory.

Space A boundless set of connections that people make meaningful in order to define the norms of everyday movement.

Spatial interaction The degree of connectedness or contact among people or places.

Spread effect Positive economic impact in one region that spurs positive economic growth in another region.

State a sovereign territory, recognized as a country by other states under international law.

Step migration A migration flow in which the migrants stay for short periods of time at different locations or steps along the path to their final destination.

Steppe Semiarid climate zone.

Striations Cuts in bedrock carved by glaciers that run the same direction as a glacier advanced.

Strong globalization A view that globalization after 1970 is fundamentally different because global corporations have become more important than states in the world economy.

Structural adjustments A set of requirements to open markets, privatize industries, and allow foreign direct investment in developing countries in exchange for loans from international financial institutions.

Stupas Sacred sites in Buddhism, typically designed in a dome shape and containing a relic of the Buddha or one of his saints.

Subcontinent A portion of a continent that was once separate and merged through continental drift.

Subduction A plate boundary where a more dense oceanic plate converges with a less dense continental plate, causing the oceanic plate to descend underneath the continental plate.

Subduction zone a tectonically active region where subduction is occurring.

Subsolar point The place on earth where the sun's rays hit directly at a 90 degree angle. Migrates over the year between 23.5°N and 23.5°S latitude.

Sultanates politically-defined territories governed by Islamic leaders called sultans.

Sunni The largest sect of Islam with approximately 80 percent of the world's followers.

Sustainable development Economic development that does not undermine the long term health of an environment.

Swidden agriculture see **shifting cultivation.**

Syncretic diffusion A process of diffusion where two cultural traits blend to create a distinct trait.

Syncretic religion Merging or combining two or more religions to form a new religion or set of religious beliefs.

Theravada Buddhism The sect of Buddhism that teaches an individual achieves enlightenment through good acts, religious practice, and service as a monk or nun. Theravada Buddhism diffused from northern India to southern India and Sri Lanka and then into Southeast Asia.

Third World Countries that were not aligned with the United States or the USSR during the Cold War.

Third World debt crises Economic instability in the 1980s caused when a number of developing countries could no longer afford to pay interest on loans.

Threshold concepts Geographic concepts, that once understood, help the knower think geographically.

Time–distance decay The likelihood of a trait or innovation diffusing decreases the farther away in time or distance it moves from the origin (hearth).

Time–space compression Increasing connectedness between world cities from improved communication and transportation networks.

Toponym Place name.

Torii a gate, typically red, at the entrance of a Shinto shrine, which defines the sacredness of the site.

Total Fertility Rate (TFR) the average number of children born to women of child-bearing age.

Tourism short-term travel for the purpose of recreation and relaxation.

Townships underdeveloped outskirts of cities designated for nonwhites to live during the apartheid era in South Africa.

Trade center a port or city on a river or land route that draws its economy mainly from trade.

Transboundary parks Nature preserves designed to cross borders to encourage cooperation between neighboring countries.

Transhumance A migration pattern in which livestock are led to highlands during summer months and lowlands during winter months to graze.

Transhumant pastoralists People who make their livelihoods raising livestock and move the livestock seasonally to access fresh water or to keep livestock in a suitable climate.

Transnational bridging nations.

Treaties Agreements between two or more countries, which are binding under international law.

Tree line The transitional border beyond which (either at higher elevations or higher latitudes), trees no longer grow.

Triple heritage The mixture of indigenous, African, and Iberian identities in Latin America dating back to the Colombian Exchange.

Troposphere Lowest layer of the atmosphere starting at earth's surface and extending approximately 11–12 miles (18–20 km) at the equator and 4 miles (6 km) at the poles).

Tsunami seismic sea wave caused by an earthquake, a volcano, or a landslide.

Tundra Climate zone located at high latitudes or high elevations, which is treeless and swampy and typically marked with permafrost.

Ultisols Nutrient-poor tropical soils that are reddish in color, strongly leached, and typically found in forests.

Unequal exchange Uneven relationship between low labor costs and high-value products.

United Nations An organization of more than 190 states with the mission of establishing international peace and security. Headquartered in New York City.

Universalizing religion A religion believed by its followers to have universal application and to which followers actively seek converts.

Urban morphology The layout of a city, including the sizes and shapes of buildings and the pathways of infrastructure.

Urban revolution The establishment and growth of cities. During the first urban revolution, people started developing and living in cities and during the second urban revolution, people migrated to the cities in large numbers to work in the factories established through the industrial revolution.

Urban sprawl The expansion of low-density urban areas around a city.

Vajrayana Buddhism form of Buddhism (sometimes seen as part of Mahayana and not a separate sect) based in Tibet (also called Tibetan Buddhism, Tantric, Mantrayana, or Lamaism) that teaches followers to discover the real and indestructible part of their humanity and to recognize the unity in dualities to reach enlightenment.

Vernacular a language used in everyday interaction among a group of people in a local area.

Vertical integration the merging of businesses that serve different steps in one commodity chain.

Virgin Lands Campaign A Soviet program that encouraged the farming of pasture and marginal lands.

Virtual water Water that is used for the production of goods that are consumed in one region but produced in another.

Volcanic arc A chain or line of volcanoes located on the continental side of a subduction zone.

Voluntary migration Movement of people across country borders where the migrant chooses to cross the border after weighing push and pull factors.

Vulnerability The susceptibility of a people and place to negative phenomena.

Warsaw Pact an agreement among the Soviet Union and several countries in Eastern Europe to defend each other against a threat from the North Atlantic Treaty Organization (NATO).

Washington Consensus Agreement among international financial institutions that policies opening markets in the developing world would lead ot economic development.

Weak globalization A view that traces modern globalization to circa 1500 and contends that globalization is a long-standing process.

West Indies The vernacular name used for the region of the Caribbean because Columbus believed he reached the Indies in Southeast Asia when he landed in Hispaniola in 1492.

White Australia Policy Australian government's official immigration policy from 1901 to 1973 that allowed only European, and preferably British, migrants legally into the country.

World Bank International financial institution that primarily lends money for development and infrastructure projections. Traditionally has an American president.

World cities Highly connected cities that function as nodes in global networks.

World-systems theory Wallerstein's theory that greater wealth is generated for countries that use core processes, including relatively more advanced technology and higher levels of education in producing goods.

World Trade Organization International organization whose main purpose is to promote free trade.

INDEX